High Brightness Light Emitting Diodes

SEMICONDUCTORS
AND SEMIMETALS
Volume 48

Semiconductors and Semimetals

A Treatise

Edited by R. K. Willardson
CONSULTING PHYSICIST
SPOKANE, WASHINGTON

Eicke R. Weber
DEPARTMENT OF MATERIALS SCIENCE
AND MINERAL ENGINEERING
UNIVERSITY OF CALIFORNIA
AT BERKELEY

In memory of Dr. Albert C. Beer, Founding Co-Editor in 1966 and Editor Emeritus of Semiconductors and Semimetals. Died January 19, 1997, Columbus, OH.

High Brightness Light Emitting Diodes

SEMICONDUCTORS AND SEMIMETALS

Volume 48

Volume Editors

G. B. STRINGFELLOW

DEPARTMENT OF MATERIALS SCIENCE AND ENGINEERING
THE UNIVERSITY OF UTAH
SALT LAKE CITY, UTAH

M. GEORGE CRAFORD

OPTOELECTRONICS DIVISION
HEWLETT-PACKARD COMPANY
SAN JOSE, CALIFORNIA

ACADEMIC PRESS
San Diego   London   Boston
New York   Sydney   Tokyo   Toronto

This book is printed on acid-free paper.

COPYRIGHT © 1997 BY ACADEMIC PRESS

All rights reserved.
NO PART OF THIS PUBLICATION MAY BE REPRODUCED OR TRANSMITTED IN ANY FORM OR BY ANY MEANS, ELECTRONIC OR MECHANICAL, INCLUDING PHOTOCOPY, RECORDING, OR ANY INFORMATION STORAGE AND RETRIEVAL SYSTEM, WITHOUT PERMISSION IN WRITING FROM THE PUBLISHER.

The appearance of the code at the bottom of the first page of a chapter in this book indicates the Publisher's consent that copies of the chapter may be made for personal or internal use of specific clients. This consent is given on the condition, however, that the copier pay the stated per-copy fee through the Copyright Clearance Center, Inc. (222 Rosewood Drive, Danvers, Massachusetts 01923), for copying beyond that permitted by Sections 107 or 108 of the U.S. Copyright Law. This consent does not extend to other kinds of copying, such as copying for general distribution, for advertising or promotional purposes, for creating new collective works, or for resale. Copy fees for pre-1997 chapters are as shown on the title pages; if no fee code appears on the title page, the copy fee is the same as for current chapters. 0080-8784/97 $25.00

ACADEMIC PRESS
525 B Street, Suite 1900, San Diego, CA 92101-4495, USA
1300 Boylston Street, Chestnut Hill, Massachusetts 02167, USA
http://www.apnet.com

ACADEMIC PRESS LIMITED
24–28 Oval Road, London NW1 7DX, UK
http://www.hbuk.co.uk/ap/

International Standard Serial Number: 0080-8784
International Standard Book Number: 0-12-752156-9

PRINTED IN THE UNITED STATES OF AMERICA
97 98 99 00 01 02 BB 9 8 7 6 5 4 3 2 1

Contents

LIST OF CONTRIBUTORS xi
PREFACE xiii

Chapter 1 Materials Issues in High-Brightness Light-Emitting Diodes
G. B. Stringfellow

I. Introduction	1
II. Techniques for Production of III–V Semiconductors for Light-Emitting Diodes	5
1. Liquid-Phase Epitaxy	6
2. Molecular Beam Epitaxy	7
3. Organometallic Vapor-Phase Epitaxy	8
III. Selection of Materials for High-Brightness Light-Emitting Diodes	12
1. Optimization of Alloys for Light-Emitting Diode Performance	14
IV. Fundamental Thermodynamic and Kinetic Considerations	17
1. Binary Compounds	17
2. Ternary and Quaternary Alloys	19
3. Doping	24
4. Kinetics of Pyrolysis and Growth Reactions	28
5. Lattice Mismatch	29
6. Surface Recombination	31
7. Low-Dimensional Structures	32
V. Specific Materials Systems	33
1. AlGaAs	33
2. AlGaInP	34
3. AlGaInN	34
VI. Conclusions	40
References	41

Chapter 2 Overview of Device Issues in High-Brightness Light-Emitting Diodes
M. George Craford

I. Introduction	47
II. Light-Emitting Diode Device Issues	50
1. Overview	50
2. Luminous Performance	53
3. Internal Quantum Efficiency	54
4. Light Extraction	56
III. Technology Status and Future Potential	58
1. Performance	58
2. Comparison of Light-Emitting Diodes to Other Light Sources	61
References	63

Chapter 3 AlGaAs Red Light-Emitting Diodes
Frank M. Steranka

I. Introduction	65
II. Device Design	68
1. Materials Parameters	68
2. Heterostructure Design	69
3. Optical Extraction	71
4. Active Layer Composition	72
III. Crystal Growth	74
1. Molecular-Beam Epitaxy and Metalorganic Chemical Vapor Deposition	74
2. Slow-Cooling Liquid-Phase Epitaxy	75
3. Temperature-Difference Liquid-Phase Epitaxy	77
IV. Device Fabrication	79
1. Contact Formation	79
2. Dicing	80
3. Selective Etching	81
V. Comparison with Other Types of Red Light-Emitting Diodes	81
VI. Temperature-Dependent Properties	83
1. I–V Curves and Forward Voltage	83
2. Electroluminescent Spectra	84
3. Efficiency	85
4. Minority-Carrier Lifetime and Analysis	87
VII. Reliability Characteristics	89
1. Low-Temperature Operation	89
2. High-Temperature, High-Humidity Operation	94
References	95

Chapter 4 OMVPE Growth of AlGaInP for High-Efficiency Visible Light-Emitting Diodes
C. H. Chen, S. A. Stockman, M. J. Peanasky, and C. P. Kuo

I. Introduction	97
II. General Overview of Organometallic Vapor-Phase Epitaxy of AlGaInP	99
1. Source Materials	99

	2.	Growth Conditions	102
	3.	Growth of Optoelectronic Devices	107
III.	Organometallic Vapor-Phase Epitaxy of AlGaInP—Critical Issues	112
	1.	Dopant Incorporation Behavior	112
	2.	Direct–Indirect Crossover	117
	3.	Ordering	119
	4.	Hydrogen Passivation of Acceptors	122
	5.	Oxygen Incorporation	127
IV.	High-Volume Manufacturing Issues	136
	1.	Reactor Issues: Delivery and Exhaust Treatment	136
	2.	Importance of Uniformity	137
	3.	Source Quality Issues	140
	4.	Color Control Issues	141
	5.	Yield Loss Categories	143
V.	Summary	144
	References	144

Chapter 5 AlGaInP Light-Emitting Diodes

F. A. Kish and R. M. Fletcher

I.	Introduction	149
II.	Active Layer Design	153
	1. Band Structure	154
	2. Double Heterostructure Devices	159
	3. Quantum Well Devices	169
III.	Current-Spreading in Device Structures	170
	1. Ohmic Contact Modifications	171
	2. Growth of P-Type Substrates	172
	3. Current-Spreading Window Layers	172
	4. Indium Tin Oxide and Other Approaches	176
IV.	Current-Blocking Structures	178
V.	Light Extraction	180
	1. Upper Window Design	181
	2. Substrate Absorption	183
	3. Distributed Bragg Reflector Light-Emitting Diodes	187
	4. Wafer-Bonded Transparent-Substrate Light-Emitting Diodes	195
VI.	Wafer Fabrication Techniques	206
VII.	Device Performance Characteristics	208
	1. Quantum Efficiency	208
	2. Luminous Efficiency	211
	3. Current–Voltage Characteristics	213
	4. Electroluminescence Spectra	215
	5. Reliability	217
VIII.	Conclusions and Outlook	219
	References	220

Chapter 6 Applications for High-Brightness Light-Emitting Diodes

Mark W. Hodapp

I.	Introduction	228
II.	Photometry and Color Measurement Principles	230

1. Photometry and Radiometry 230
2. Converting from Luminous Intensity to Luminous Flux 233
3. Color Measurement 247
III. Automotive Signal Lighting 251
1. Characteristics of the Automotive and Truck Market 251
2. Automotive Luminous Intensity and Color Requirements 252
3. Benefits of Light-Emitting Diode Technology 252
4. Luminous Flux Requirements 259
5. Optics Design Considerations 263
6. Electrical Design Considerations 268
7. Thermal Design Considerations 273
IV. Automotive Interior Lighting 277
1. Characteristics of the Automotive and Truck Market 277
2. Benefits of Light-Emitting Diode Technology 279
3. Lighting Considerations for an Instrument Cluster Warning Light 284
4. Cavity Design for an Instrument Cluster Warning Light 289
5. Legend Optimization of an Instrument Cluster Warning Light 294
V. Traffic Signal Lights 296
1. Characteristics of the Traffic Signal Market 296
2. Construction of a Traffic Signal Head 298
3. Comparison Between Incandescent and Light-Emitting Diode Traffic
 Signals 298
4. Luminous Flux Requirements 302
5. Electrical Design Considerations 307
VI. Large-Area Displays 319
1. Characteristics of the Large-Area Display Market 319
2. Alternative Large-Area Display Technologies 320
3. Benefits of Light-Emitting Diode Technology 324
4. Principles of Color Mixing 326
5. Character Height Considerations 329
6. Drive Circuit Considerations 330
VII. Liquid Crystal Display Backlighting 337
1. Characteristics of the Liquid Crystal Display Backlighting Market 337
2. Liquid Crystal Display Operation 338
3. Alternative Liquid Crystal Display Backlighting Technologies 340
4. Benefits of Light-Emitting Diode Technology 342
5. Direct-View Liquid Crystal Display Backlighting Techniques 345
6. Edgelighting Liquid Crystal Display Backlighting Techniques 347
7. Fiber-Optic Light-Emitting Diode Backlighting Techniques 351
References 355

Chapter 7 Organometallic Vapor-Phase Epitaxy of Gallium Nitride for High-Brightness Blue Light-Emitting Diodes
Isamu Akasaki and Hiroshi Amano

I. Introduction 357
II. Historical Overview 359
1. Hydride Vapor-Phase Epitaxy 360
2. Molecular-Beam Epitaxy 361
3. Organometallic Vapor-Phase Epitaxy 362

III. Design and Structure of Nitride-Based Superbright Light-Emitting Diodes 379
    1. Layered Structure 379
    2. Electrode 383
IV. Efficiency, Wavelength, and Lifetime 384
V. Laser Diodes 385
VI. Summary 387
    References 388

Chapter 8 Group III–V Nitride-Based Ultraviolet Blue-Green-Yellow Light-Emitting Diodes and Laser Diodes

Shuji Nakamura

  I. Introduction 391
 II. Gallium Nitride Growth 394
    1. Undoped Gallium Nitride 394
    2. n-Type Gallium Nitride 398
    3. p-Type Gallium Nitride 399
    4. Gallium Nitride p–n Junction Light-Emitting Diodes 405
III. InGaN Growth 409
    1. Undoped InGaN 409
    2. Impurity Doping of InGaN 413
IV. Light-Emitting Diodes 416
    1. InGaN/Gallium Nitride Double Heterostructure Light-Emitting Diodes . . 416
    2. InGaN/AlGaN Double Heterostructure Light-Emitting Diodes 419
    3. InGaN Quantum Well Structures 425
    4. InGaN Light-Emitting Diodes with Quantum Well Structures 426
    5. High-Brightness Green and Blue Light-Emitting Diodes 430
 V. InGaN Multiple Quantum Well Structure Laser Diodes 436
VI. Summary 440
    References 441

INDEX   445

CONTENTS OF VOLUMES IN THIS SERIES   455

List of Contributors

Numbers in parenthesis indicate the pages on which the authors' contribution begins.

ISAMU AKASAKI (357), *Department of Electrical and Electronic Engineering, Meijo University, Nagoya 468, Japan*

HIROSHI AMANO (357), *Department of Electrical and Electronic Engineering, Meijo University, Nagoya 468, Japan*

C. H. CHEN (97), *Optoelectronics Division, Hewlett-Packard Company, San Jose, California 95131*

M. GEORGE CRAFORD (47), *Optoelectronics Division, Hewlett-Packard Company, San Jose, California 95131*

R. M. FLETCHER (149), *Optoelectronics Division, Hewlett-Packard Company, San Jose, California 95131*

MARK W. HODAPP (227), *Optoelectronics Division, Hewlett-Packard Company, San Jose, California 95131*

F. A. KISH (149), *Optoelectronics Division, Hewlett-Packard Company, San Jose, California 95131*

C. P. KUO (97), *Optoelectronics Division, Hewlett-Packard Company, San Jose, California 95131*

SHUJI NAKAMURA (391), *Research and Development Department, Nichia Chemical Industries Ltd., Tokushima 774, Japan*

M. J. PEANASKY (97), *Optoelectronics Division, Hewlett-Packard Company, San Jose, California 95131*

FRANK M. STERANKA (65), *Optoelectronics Division, Hewlett-Packard Company, San Jose, California 95131*

S. A. STOCKMAN (97), *Optoelectronics Division, Hewlett-Packard Company, San Jose, California 95131*

G. B. STRINGFELLOW (1), *Department of Materials Science and Engineering, University of Utah, Salt Lake City, Utah 84112*

Preface

The progress in the performance of electronic and photonic devices in the 1970s and 1980s has been no less than spectacular. It has changed the function and appearance of nearly every piece of equipment used in the home, office, and laboratory. In the early 1970s, the red light-emitting diodes (LEDs) commonly used for displays in calculators and watches had conversion efficiencies of only approximately 0.1%. The technology evolved steadily until the mid-1980s, and since that time a technological revolution has occurred. Today, LEDs with quantum efficiencies exceeding 8% are available over the entire spectral range, from the red to the blue, and quantum efficiencies as high as 23% have been observed in the red–orange region. Luminous efficacies exceeding 15 lm/W (typical for incandescent lamps) have been observed for much of the spectral range, with values as high as 50 lm/W in the yellow–orange and 35 lm/W in the green. This has enabled the production of full-color displays with brightnesses sufficient for viewing in full sunlight. A study of how these improvements were achieved represents a tour de force of the advances in both materials and device technology that have affected such other important devices as injection lasers, transistors, and solar cells. These improvements have not come as a result of simple incremental advances in the technology used for the red gallium arsenide phosphide LEDs that were in use 15 years ago. Rather, they constitute a vivid demonstration of so-called atomic-scale engineering, that is, engineering the materials and devices by choosing the atomic-scale arrangements of up to four types of atoms in a single structure. The authors of the various chapters were carefully selected as the leaders in the development of the novel materials and device structures used for these high-brightness LEDs. Chapters are presented on the AlGaAs materials and devices developed some years ago for red LEDs, and still being improved; the AlGaInP materials and devices that led to major advances in LEDs, covering the color range from red to amber; and the very recently developed AlGaInN materials and devices that have led to dramatic improvements in the performance of green and blue LEDs. The final chapter discusses in detail the applications of these devices. The two introductory chapters by the volume editors are meant to supply the basic background in the areas of both materials and devices that will allow this book to be used as a stand-alone text for graduate-level courses or for the professional engineer or scientist wishing to learn of the latest progress in this fascinating area of technology. The rapid advance in this area cannot continue at the rate demonstrated in the recent past. The efficiencies obtained are approaching their theoretical maximum values, although in the AlGaInN system substantial further performance improvement seems plausible. There are no new materials systems evident on the horizon. Thus, the material presented in this book is expected to be relevant for decades to come.

G. B. Stringfellow
M. George Craford

CHAPTER 1

Materials Issues in High-Brightness Light-Emitting Diodes

G. B. Stringfellow

DEPARTMENT OF MATERIALS SCIENCE AND ENGINEERING
UNIVERSITY OF UTAH
SALT LAKE CITY, UTAH

I.	INTRODUCTION	1
II.	TECHNIQUES FOR PRODUCTION OF III-V SEMICONDUCTORS FOR LIGHT-EMITTING DIODES	5
	1. Liquid-Phase Epitaxy	6
	2. Molecular Beam Epitaxy	7
	3. Organometallic Vapor-Phase Epitaxy	8
III.	SELECTION OF MATERIALS FOR HIGH-BRIGHTNESS LIGHT-EMITTING DIODES	12
	1. Optimization of Alloys for Light-Emitting Diode Performance	14
IV.	FUNDAMENTAL THERMODYNAMIC AND KINETIC CONSIDERATIONS	17
	1. Binary Compounds	17
	2. Ternary and Quaternary Alloys	19
	3. Doping	24
	4. Kinetics of Pyrolysis and Growth Reactions	28
	5. Lattice Mismatch	29
	6. Surface Recombination	31
	7. Low-Dimensional Structures	32
V.	SPECIFIC MATERIALS SYSTEMS	33
	1. AlGaAs	33
	2. AlGaInP	34
	3. AlGaInN	34
VI.	CONCLUSIONS	40
	References	41

I. Introduction

The search for materials and structures for the fabrication of efficient light emitting diodes (LEDs) covering the entire range of wavelengths to which the human eye is sensitive has proven to be a Herculean task. A review of the development of high-brightness LEDs is an instructive example of the application of atomic-scale engineering at its best. This odyssey, which has been underway for more than three decades, is now reaching its final stages of fruition with the development of red diodes emitting at 629 nm, having extremely high external quantum efficiencies exceeding 17%; orange diodes emitting at 604 nm, with efficiencies of 11.5%; green LEDs emitting at 6.3%; and blue LEDs, with external quantum efficiencies exceeding 9% (Kish *et al.*, 1994) and Chapters 2, 3, 5, and 8 of this volume). Today, the performance of the brightest LEDs compares favorably with that of incandescent tungsten and halogen light bulbs.

The search for materials emitting in the visible region of the spectrum began in the early 1960s with the realization that the well-developed group IV elemental semiconductors germanium (Ge) and silicon (Si) were not suitable for visible LEDs. This is due to both their small bandgap energies, which results in the production of infrared (IR) photons, and their low conversion efficiencies because of the indirect nature of the band-edge electronic transitions; i.e., the electrons at the bottom of the conduction band and the holes at the top of the valence band have different values of wave vector or crystal momentum. Thus, in pure materials, a phonon must be involved to allow conservation of momentum in the electron-hole recombination process. The III-V compound semiconductors also have problems for visible LEDs. Many early compound semiconductors produced by bulk growth techniques—indium antimonide (InSb), indium arsenide (InAs), gallium antimonide (GaSb), indium phosphide (InP), and gallium arsenide (GaAs)—have direct bandgaps; however, the bandgap energies are too low for the production of visible photons. Gallium phosphide (GaP) has a larger bandgap energy, allowing the generation of green photons. However, as for the elemental semiconductors, the bandgap is indirect. Thus, no pure III-V compounds available at that time could be used to make efficient visible LEDs.

Two approaches were taken at different laboratories for the development of high-brightness visible LEDs. In some laboratories, efforts were mounted to make the indirect bandgap GaP emit photons efficiently by doping with isoelectronic impurities that localize the electronic transition, thus ameliorating the need for phonon participation (Bergh and Dean, 1976). This resulted in high-efficiency (15%) red light emitters by doping GaP with zinc-oxygen (Zn-O) pairs (Soloman and DeFevere, 1972), but with emission wavelengths of approximately 700 nm where the eye sensitivity is extremely small. Isoelectronic doping of GaP with low concentrations of nitrogen produces less efficient (0.1% to 0.3%) yellow and green (570 and 590 nm, respectively) emitters. This approach also can be used for enhancing the efficiency of the indirect bandgap of the GaAsP alloys, discussed later, grown on GaP substrates (Craford, 1977). The second approach is the use of alloy semiconductors to increase the bandgap energies of the binary alloys into the visible region of the spectrum while keeping the bandgap direct. The bandgap energy is plotted versus the lattice constant for the III-V compounds and ternary alloys in Fig. 1. The first example was the addition of phosphorus to GaAs, producing red-emitting GaAsP (Craford, 1977), with an external efficiency of approximately 0.1%. AlGaAs, which is always lattice-matched to GaAs substrates, was also developed for high-efficiency red LEDs (Cook *et al.*, 1987). More recently, the AlGaInP system has been developed to produce much higher efficiencies over the entire spectral range from red to green, as discussed in Chapters 2, 4 and 5. Each of these approaches is described in more detail below and in the chapters to follow.

Notably missing from the early successes was a material producing blue photons. This was not due to a lack of effort. The necessity of being able to produce all of the primary colors for the fabrication of full-color displays was recognized early. One approach for realizing the high bandgap ma-

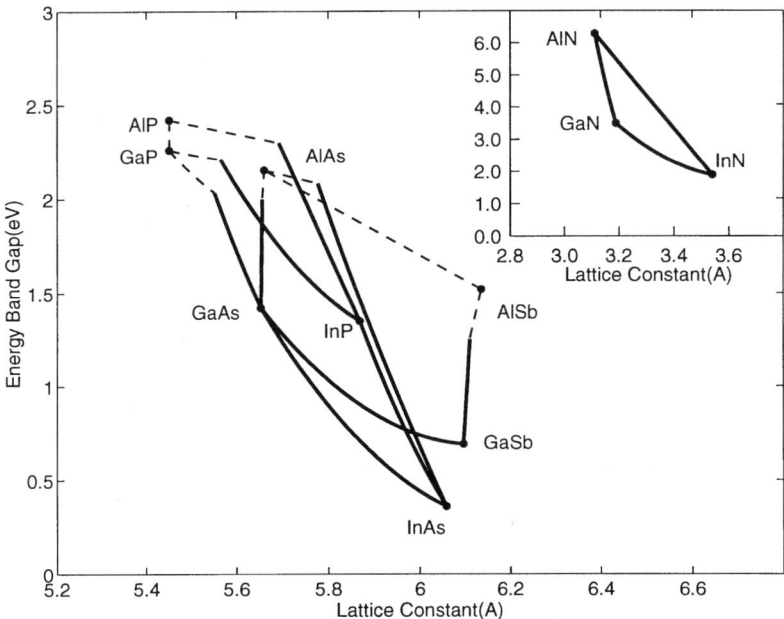

FIG. 1. Bandgap energy versus lattice constant for the high-bandgap III-V compounds and ternary alloys. AlP, aluminum phosphide; GaP, gallium phosphide, GaAs, gallium arsenide; AlAs, aluminum arsenide; AlSb, aluminum antimonide; GaSb, gallium antimonide; InP, indium phosphide; InAs, indium arsenide; AlN, aluminum nitride; GaN, gallium nitride, InN, indium nitride.

terials necessary for blue emitters was to use the II-VI semiconductors that were known to be highly efficient light emitters when pumped by high-energy photons or electrons. These materials were well-developed for use as the phosphors in cathode-ray tubes. However, the production of p-n junctions proved exceedingly difficult because of the propensity of these materials to self-compensate, whereby native defects are generated that hinder efforts to produce high levels of n-type (for some materials) or p-type (for other materials) doping (Bergh and Dean, 1976; Faschinger et al., 1995). The result was that high-quality p-n junction diodes were not produced until relatively recently (Haase et al., 1991). The development of p-n junctions in II-VI semiconductors has generated a great deal of excitement and activity. However, even now, the ease with which point defects are produced in these materials has placed a severe limit on the operating lifetimes of light-emitting devices. As a result, they are not commercially available at present. These materials are not dealt with in this volume.

Another material successfully used for blue emitters is silicon carbide (SiC). This IV-IV compound has many polytypes. The common 6H form has a bandgap energy of 3.03 eV; however, the bandgap is indirect. The first commercial blue LEDs were made in 6H-SiC. However, the efficiencies are considerably below those obtained for the nitrides (Morkoc et al., 1994). This has virtually eliminated SiC from the competition except as a substrate for gallium nitride (GaN), as discussed below.

The only III-V semiconductors capable of producing blue photons are the nitrides. The bandgap energies for indium nitride (InN), GaN, and aluminum nitride (AlN) are 1.89, 3.50, and 6.28 eV, respectively (Wright and Nelson, 1995), as seen in Fig. 1. These alloys are often assumed to be completely miscible. However, one might expect a region of solid immiscibility in the GaInN system, due to the large difference in the lattice constant (Stringfellow (1974)). In fact, recent calculations suggest that a large region of solid immiscibility exists at typical growth temperatures (Ho and Stringfellow, 1997a,b), as discussed in more detail in Part IV, Section 2, below. Nevertheless, both Osamura et al. (1972) and Nakamura et al., as described in Chapter 8, were able to grow alloys throughout the entire composition range with a smooth variation in bandgap energy with composition. However, the growth of alloys with significant In concentrations is found to be extremely difficult. Nagatomo et al. (1989) reported that the lattice constant obeys Vegard's law in GaInN alloys with InN concentrations of up to 42%. The bowing parameters are 0.53 eV and 1.02 eV for AlGaN and GaInN, respectively. These alloys also can be used for green and yellow LEDs, as discussed in Chapter 8. The values of bandgap energy for the nitrides are included in Fig. 1.

Early work on the epitaxial growth of GaN demonstrated that the volatility of N resulted in the production of Ga-rich layers, containing high concentrations of nitrogen vacancies, that were always highly n-type (Pankove, 1992). The lack of a lattice-matched substrate also caused problems. Growth on sapphire produces material with dislocation densities exceeding 10^{10} cm^{-2} (Lester *et al.*, 1995; Qian *et al.*, 1995), a value that would kill the luminescence from the conventional III-V arsenides and phosphides. The production of p-type GaN is a relatively new development, first demonstrated in 1989 by Akasaki *et al.*, 1989, 1990), as discussed in Chapter 7. This led to the first p-n junction LEDs emitting at a wavelength of 370 nm. The progress since then truly has been remarkable. Today, blue LEDs with external efficiencies of 9.1% (Nakamura *et al.*, 1995) have been produced in the laboratory and extremely bright blue GaN LEDs are available commercially, as discussed in Chapter 8. Despite high dislocation densities, the LEDs are also long-lived. A recent development has been the use of more nearly lattice-matched SiC substrates, having a lattice mismatch of 3.5% for GaN and 1% for aluminum nitride (AlN). This results in more than an order of magnitude reduction in the dislocation density to a value of approximately 10^9 cm^{-2} (Weeks *et al.*, 1995; Ponce *et al.*, 1995).

This chapter gives an overview of the basic issues involving the materials used for high-brightness LEDs. It lays the fundamental foundation for understanding the phenomena common to the materials discussed in this volume: AlGaInP, for the spectral range of red through green, AlGaAs for high-efficiency red emitters, GaN for high-efficiency blue emitters, and GaInN for high-brightness blue and green emitters. This serves as introductory material for the more detailed chapters on these materials to follow.

II. Techniques for Production of III-V Semiconductors for Light-Emitting Diodes

Most III-V semiconductor binary compounds can be grown as large boules directly from the melt. Slices from these boules can be used for the production of several electronic devices, such as field effect and bipolar transistors, by employing well-developed diffusion and ion implantation techniques for the formation of p- and n-type materials. They can also be used for the fabrication of relatively inexpensive homojunction LEDs. However, these are mainly GaAs and InP devices used in the IR region of the spectrum. The devices emitting at visible wavelengths, including the very high performance heterostructure and quantum well light-emitting devices, described in Chapters 3 through 8 of this volume, invariably involve alloy layers and complex structures, which cannot be produced by bulk growth techniques. Thus, the materials used for high-brightness LEDs must be produced by epitaxial growth techniques. Two of the earliest epitaxial techniques used for the growth of III-V materials are vapor processes using either the volatile group V halides or hydrides. These group V sources are combined with the group III halides, formed *in situ* at high temperatures due to their limited volatility at room temperature, to form the III-V materials. The group III hydrides generally are too unstable to be useful. These vapor-phase epitaxial (VPE) techniques were used for the early production of GaAsP and other materials used for LEDs. However, they proved incapable of producing the entire range of III-V alloys required for the high-brightness LEDs that are the topic of this volume. They also are rather awkward for the production of the heterostructure, quantum well, and superlattice structures required for the highest performance devices. Thus, they will not be mentioned further here. See Beuchet (1985) for an up-to-date review of this topic.

The other early epitaxial technique for the production of III-V semiconductors is liquid-phase epitaxy (LPE). This technique is capable of the production of the very highest quality materials for LEDs and lasers. It is also capable of the production of heterostructures. Thus, it will be reviewed briefly in this chapter, with more detailed results for the production of high-brightness red AlGaAs LEDs in Chapter 3.

Two more modern techniques that have become the workhorses for production of the most advanced devices, including LEDs, are molecular beam epitaxy (MBE) and organometallic vapor-phase epitaxy (OMVPE), also known as metalorganic vapor-phase epitaxy (MOVPE), organometallic chemical vapor deposition (OMCVD), and metalorganic chemical vapor deposition (MOCVD). These techniques are described in more detail in separate sections later.

1. LIQUID-PHASE EPITAXY

LPE is by far, the simplest technique mechanically. In the modern configuration, it consists of a graphite boat contained in a quartz tube filled with 1 atm of high-purity gas, typically purified hydrogen. The boat contains several wells, each containing a group III–rich melt with the proper ratio of group III elements, group V elements, and dopants to produce the desired III-V semiconductor when cooled in contact with a suitable substrate. The substrate is placed in a slot in a graphite slider that is moved sequentially to positions in contact with the various melts. The driving force for epitaxial growth is a supersaturation of the solution caused by either cooling the melt to a temperature below its liquidus temperature or by impoing a temperature gradient between source and substrate crystals (Stringfellow and Greene, 1971). This is a near-equilibrium process. Thus, equilibrium phase diagrams (Panish and Ilegems, 1972) are used to determine the liquid composition required to produce the desired III-V semiconductor at a specific growth temperature (Stringfellow, 1979). Extremely high purity layers are produced by this technique due to a natural purification process that leaves harmful impurities, such as oxygen, either in the liquid phase or in an oxide "scum" produced on the surface away from the substrate (Stringfellow, 1982). The two main advantages of LPE are the simplicity of the technique and the purity of the layers produced. It is also an excellent technique for the production of the very thick layers used in some high-brightness LED structures, as discussed in Chapter 3. The layers of AlGaAs (Ahrenkiel, 1993) and GaP (Bergh and Dean, 1976; Soloman and DeFevere, 1972) having the longest minority carrier lifetimes and the highest radiative recombination efficiencies have been produced using LPE. Thus, the use of LPE for growth of AlGaAs for high-brightness red LEDs is described in Chapter 3. However, LPE is not suitable for the growth of other desirable alloys such as AlGaInP, as described below. Neither is LPE suitable for the III-V nitrides. It also is awkward for the production of quantum well and superlattice structures.

2. Molecular Beam Epitaxy

The most powerful technique for the production of superlattice and quantum-well structures is MBE, developed in the late 1960s and early 1970s by Arthur and LePore (1969) and Cho (1971). It is basically a highly refined, ultrahigh-vacuum (UVH) evaporation process. The flux of an elemental source from a heated knudsen cell is controlled by the temperature of the source and the position (open or closed) of the externally controlled shutter. The shutters allow the beams to be rapidly modulated, resulting in the exquisite control required for the most elaborate low-dimensional structures. Although MBE can be used for the growth of a wide range of materials, a notable shortcoming is the difficulty experienced with the growth of the phosphides and, especially, the As–P alloys (Hirayama and Asahi, 1994). These problems led to the development of UHV techniques using group V molecular precursors such as arsine (AsH_3) and phosphine (PH_3) (Panish, 1980) as well as more advanced precursors (Jones, 1994). This technique is frequently referred to as gas source MBE (GSMBE). When group III molecular precursors are added to the system, it is usually referred to as chemical beam epitaxy (CBE) (Tsang, 1990). The GSMBE technique has produced excellent quality GaN and other nitrides for blue LEDs using molecular nitrogen dissociated in a plasma (Paisley *et al.*, 1989).

3. Organometallic Vapor-Phase Epitaxy

Without a doubt, the most versatile technique for the production of III-V materials and structures for electronic and photonic devices is OMVPE (Stringfellow, 1989a). It is the most recent technique to be developed for the production of high-quality III-V semiconductors, mainly due to the complexity of the process. However, the complexity of the technique also leads directly to its amazing versatility. It can be used for the production of essentially all III-V compounds and alloys. It is also conveniently used for the production of quantum well and superlattice structures, with interface abruptness and quality equal to or exceeding the results produced by other epitaxial growth techniques. It is also generally the preferred technique for large-scale, economical production of a wide variety of devices. Because this technique is used for the production of the materials described in Chapters 4, 5, 7, and 8, it is described in some detail subsequently. For an in-depth description of the OMVPE process, see Stringfellow (1989a).

The precursors used to produce the III-V semiconductors by OMVPE are typically either organometallic or hydride group V molecules combined with organometallic group III molecules. At the heated substrate, the molecules pyrolyze, sometimes in very complex ways, to produce the group III and the group V elements needed for formation of the desired III-V semiconductor. Because these precursor molecules are so unstable at typical growth temperatures and the III-V solid is so stable, the thermodynamic driving force for OMVPE typically is enormous. The thermodynamic driving force is compared for the various growth processes in Fig. 2. OMVPE is sometimes described as a highly nonequilibrium growth process, as is MBE (Stringfellow, 1991). In contrast LPE has a low thermodynamic driving force, and therefore is often described as a near-equilibrium process. Nevertheless, thermodynamic factors control much of what occurs during OMVPE for typical growth conditions (Stringfellow, 1984, 1989a, 1991). In the so-called diffusion-limited growth regime typically used for OMVPE growth of LED materials, which is described in more detail subsequently, the vapor phase adjacent to the growing solid–vapor interface is nearly in equilibrium with the solid being produced. Thus, thermodynamic analysis can be used to accurately describe the solid composition and stoichiometry. In many cases thermodynamic analysis also can be used to understand the incorporation

1 MATERIALS ISSUES IN HIGH-BRIGHTNESS LIGHT-EMITTING DIODES

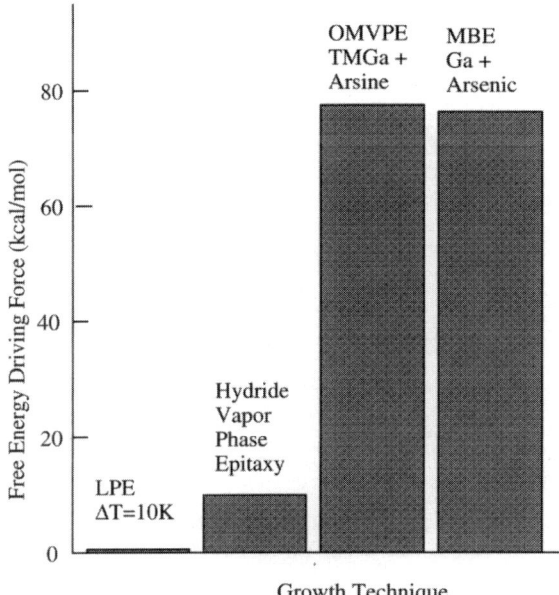

FIG. 2. Estimated thermodynamic driving force, the Gibbs free energy difference between reactants and products, for several epitaxial growth processes. All calculations are for the growth of gallium arsenide (GaAs) at 1000 K. LPE, liquid-phase epitaxy; OMVPE, organometallic vapor-phase epitaxy; MBE, molecular beam epitaxy; TMGa, trimethylgallium. (Reprinted from Stringfellow, 1991, J. Cryst. Growth 115; with the kind permission of Elsevier Science – NL.)

of dopants into the solid during OMVPE growth (Stringfellow, 1986, 1989a).

The high supersaturation can have beneficial effects. For example, under the proper growth conditions it allows the production of metastable III-V alloys that cannot be produced by near-equilibrium techniques such as LPE (Stringfellow, 1983). The high supersaturation combined with the volatility of the group V elements at typical growth temperatures results in an extremely useful property of OMVPE, namely that the group III distribution coefficient is frequently observed to be nearly unity (Stringfellow, 1984, 1985, 1989a). This is critical for the growth of AlGaInP alloys (Stringfellow, 1981, 1985), as described in Part IV, Section 2.

At very low temperatures, the OMVPE growth rate increases exponentially with increasing temperature, as seen in Fig. 3. The growth rate is limited by the pyrolysis rate of the group III precursor molecule(s). Thus, for example, the GaAs growth rate at low temperatures is more rapid for less stable preursors, such as triethylgallium (TEGa), than it is for more

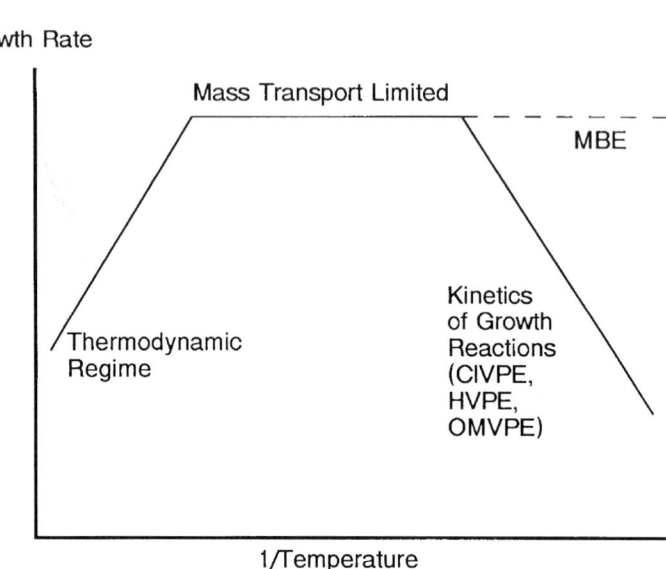

FIG. 3. Schematic diagram of growth rate versus reciprocal temperature for the organometallic vapor-phase epitaxy (OMVPE) growth of III-V semiconductors. MBE, molecular beam epitaxy; HVPE, hydride vapor-phase epitaxy; CIVPE, chloride vapor-phase epitaxy.

stable precursors that decompose more slowly, such as trimethylgallium (TMGa) (Stringfellow, 1989a). This extremely sharp temperature dependence of the growth rate (and solid composition for alloys with mixing on the group III sublattice) is undesirable because it results in layers that are nonuniform in both thickness and solid composition when a small temperature gradient exists across the substrate, as is almost invariably the case. It also is a cause of nonreproducibility if the temperature varies slightly from run to run.

Because of these problems, higher temperatures are normally used where the growth rate is nearly independent of temperature, as seen in Fig. 3. This requires growth at a temperature at which all of the group III precursors are completely pyrolyzed either in the vapor or on the substrate surface. In this case, the growth rate is controlled by the mass transport (normally diffusion) limited flux of the group III species to the solid–vapor interface.

At still higher temperatures, the growth rate is observed to decrease with increasing temperatures. This is frequently dominated by thermodynamic phenomena. For example, at high temperatures elemental In is able to desorb from the surface, thus decreasing the growth rate of In-V semiconductors (Turco and Massies, 1987). In other cases, the occurrence of parasitic reactions—including the nucleation and growth of III–V materials on the walls and homogeneous nucleation in the gas phase above the substrate—resuts in a reduction in growth rate (Stringfellow, 1989a).

Many characteristics of the OMVPE growth process and properties of the resulting materials are determined by the nature of the precursor molecules. From the previous discussion, the growth rate at low temperatures and the low temperature limit of the diffusion-limited growth regime are determined largely by the nature of the group III precursor molecules. Another significant effect of the precursor molecules is the incorporation of residual carbon. The methyl (CH_3) radicals produced on the surface by pyrolysis of R_3-III and R_3-V precursors, where $R = CH_3$, can result in the introduction of carbon into the epitaxial layer. This can be avoided to some degree by the production of atomic hydrogen on the surface typically from pyrolysis of the group V precursor molecules. The hydrides produce large quantities of atomic hydrogen on the surface during pyrolysis, leading to low carbon contamination levels. In fact, it is this beneficial process, combined with the ready availability of AsH_3 and PH_3, that led to the predominant positions of the group V hydrides in the early OMVPE literature. However, group V hydrides also have two less desirable characteristics. They are very stable, and therefore pyrolyze efficiently only at high temperatures. This results in nonuniformities in stoichiometry of the solid and dopant incorporation. However, the most important weakness for production operations is that group V hydrides are highly toxic. This combined with the fact that they are delivered from high-pressure cylinders from which large quantities could, in principle, escape quickly makes them highly dangerous. To rectify both problems, new precursor molecules have been developed in recent years. The most successful are tertiarybutylarsine (TBAs) (Chen *et al.*, 1987; Lum *et al.*, 1987; Komeno, 1994) and tertiarybutylphosphine (TBP) (Chen *et al.*, 1986; Komeno, 1994). Both have been used for the production of III-V semiconductors of quality equaling or exceeding that produced using the hydrides. These alternate precursors are already widely used, and only the low cost of the hydrides and the existing growth processes already in place for the production of commercial devices, which are difficult to change, are slowing the adoption of these sources for OMVPE. As is discussed in Part V, Section 3, the development of novel nitrogen precursors is one of the areas of ongoing research for the growth of the III-V nitrides. The development of new group III precursors is also an important topic for growth at low temperatures and the reduction of carbon contamination (Stringfellow, 1989a; Protzmann *et al.*, 1991; Jones, 1994).

III. Selection of Materials for High-Brightness Light-Emitting Diodes

Before describing the specific materials used for high-brightness LEDs, it is worthwhile to briefly review the main requirements for high-brightness LEDs, listed in Table I: high internal radiative recombination efficiency, efficient extraction of the photons from the material in which they are generated, ability to dope controllably to form p-n junctions, generation of photons at energies detectable by the human eye, and economical production of devices that can be profitably sold at reasonable prices. A more detailed description of the device aspects of high-brightness LEDs is provided in Chapter 5.

The efficient conversion of excited electron-hole pairs to photons requires that the radiative process be rapid and that all of the nonradiative recombination processes be relatively slow. The former is guaranteed by selecting direct bandgap materials having a large matrix element for the radiative transition. It is also important that electrons not populate the higher lying minima located at positions in k-space other than at the Γ point, at which the valence band maximum is located for the III-V semiconductors. This requires that the Γ point be many kT lower in energy than are the other minima. As discussed subsequently, this is a major factor

TABLE I

MATERIALS REQUIREMENTS FOR HIGH-BRIGHTNESS LIGHT-EMITTING DIODES

Requirement	Main Materials Factor
High radiative recombination efficiency	
Direct bandgap	Alloy composition
Large $E^\Gamma - E^X$	Alloy composition
Long minority carrier lifetime	Few NRCs
No surfaces for recombination	Heterostructures
Few line defects (except nitrides)	Lattice matched (all layers-substrate)
Efficient photon extraction	Transparent substrate
	Window heterostructure
Controlled doping	
Low residual doping	Source purity, reactor integrity
Wide doping range	No compensation by native defects or hydrogen
Both n- and p-type	No self-compensation
High eye response	Alloy composition
Inexpensive	Produced by OMVPE (or VPE, when possible)
	Thin layers

NRCs, nonradiative recombination centers; OMVPE, organometallic vapor-phase epitaxy; VPE, vapor-phase epitaxy.

determining the optimum compositions for alloy semiconductors.

Minimization of nonradiative recombination processes requires the minimization of recombination through-point, line, and planar defects. Common nonradiative point defects include deep states due to lattice vacancies, interstitials, and antisite or antistructural defects (Wolfe et al., 1989). Impurity atoms may also act as nonradiative recombination centers (NRCs). In many III-V semiconductors the presence of oxygen is notable for producing NRCs (Kisker et al., 1982; Kondo et al., 1994). Oxygen is particularly troublesome for Al-containing materials because of the high affinity of oxygen for Al. Elimination of oxygen has been a particular problem for vapor-phase growth techniques for which years have been required to eliminate oxygen from the source materials and to keep oxygen from leaking into the growth apparatus from both real and virtual leaks (Stringfellow, 1989a). Damage from the high-energy species produced in the plasma sources sometimes used for the growth of III-V semiconductors, particularly the nitrides, is another source of nonradiative recombination centers (Molnar et al., 1995).

Dislocations are common in III-V semiconductors, either propagating from the substrate into the epitaxial layers (Tu et al., 1992) or due to a difference in lattice constant between the substrate and the epitaxial layer (Dodson and Tsao, 1989; Jesser and Fox, 1990), as discussed in detail in Part IV, Section 5. Dislocations also are generated at heterostructure interfaces when the lattice mismatch and layer thickness exceed well-defined limits. Dislocations are known to act as nonradiative centers for many III-V semiconductors (Stringfellow et al., 1974). As a result, the minority carrier lifetime is found to decrease with increasing dislocation density (Ettenberg, 1974). The lifetime of GaAsP, containing approximately 10^5 cm^{-2} dislocations due to the lattice parameter mismatch to the GaAs substrate (Stringfellow, and Greene, 1969), is limited to approximately 10^{-9} sec. Much longer lifetimes and higher radiative recombination efficiencies are measured in materials with dislocation densities of less than 10^3 cm^{-2}.

With the high dislocation densities (approximately 10^{10} cm^{-2}) in GaN layers grown on sapphire substrates, where the lattice parameter mismatch is approximately 16%, the lifetimes of conventional III-V semiconductors would be extremely small. It is apparent that the dislocations in the nitrides do not act as efficient recombination centers for reasons that are not well understood.

Finally, for most III-V semiconductors, the dangling bonds at surfaces produce nonradiative recombination centers. Thus, the surface recombination velocity for GaAs is approximately 10^6 cm/sec (Kressel and Butler, 1977). In conventional homojunction LEDs, where the p-n junction is located within a few diffusion lengths of the surface, this results in low internal quantum efficiencies. In more advanced devices, a higher bandgap layer is placed between the p-n junction and the surface, thus preventing minority carriers from reaching and recombining at the surface. Similarly, a high bandgap layer between the active layer and substrate prevents loss of minority carriers due to recombination in the substrate. However, the interface between the layers in the heterostructure must have a low recombination velocity. This requires that the mismatch dislocation density be low, a result of having either very thin layers or lattice constants that are nearly matched (Kressel, 1980; Mullenborn et al., 1994).

Many photons generated near the p-n junction are not able to escape from the device. The absorption coefficient for band-edge photons is high in direct bandgap materials, and therefore many photons are absorbed before they can reach the top surface. The fraction absorbed can be reduced by using a window-type heterostructure, as mentioned previously, in which the photons pass through a higher bandgap material on their way to the surface. This reduces the absorption dramatically. Unfortunately, most of the photons reaching the top surface cannot escape because of the high refractive indices of the III-V semiconductors. Those intersecting the surface at an angle greater than the critical angle of approximately 16 to 18 degrees (for III-V semiconductors) are totally internally reflected. Most of these photons find their way into the substrate where they are absorbed. These are added to the photons exiting the p-n junction in the direction of the substrate. To avoid a major loss of phosons by absorption in the substrate, the substrate material must have a higher bandgap energy than does the active region. In the event that the constraint of lattice parameter matching prevents this, such as in the case of the AlGaInP–GaAs system, removal of the substrate is necessary. The fraction of photons escaping from the crystal can be increased to approximately 0.3—resulting in impressive increases in external quantum efficiency—when the substrate is removed, as discussed in Chapters 3, 4 and 5.

The term *alloy composition* is identified as the main materials factor for several categories in Table I. Thus, the next section is devoted entirely to this topic.

1. OPTIMIZATION OF ALLOYS FOR LIGHT-EMITTING DIODE PERFORMANCE

The basic approach used to determine the optimum alloy composition for LED applications is exemplified in a paper by Archer (1972), with a similar treatment by Bergh and Dean (1976). First, the rates of radiative and nonradiative transitions are compared. The radiative transition rate in p-type material is simply the concentration of electrons in the conduction band minimum at Γ divided by the radiative lifetime τ_R. The Γ electrons can also recombine nonradiatively with a lifetime τ_N. If all electrons in the indirect minimum (considered the X minimum in what follows, for notational simplicity) recombine nonradiatively, the resulting internal quantum efficiency η_i is simply

$$\eta_i = [1 + \tau_R/\tau_N(1 + n_X/n_\Gamma)]^{-1} \qquad (1)$$

where n_X and n_Γ are the excess electron concentrations (due to the excitation or injection process) in the X and Γ conduction band minima, respectively. This ratio is calculated by assuming a Boltzmann distribution:

$$n_X/n_\Gamma = N_X/N_\Gamma \exp[-(E_X - E_\Gamma)/kT] \qquad (2)$$

The quantities N_X and N_Γ are the densities of states in the indirect and direct minima, respectively. The energies of the X and Γ conduction band minima, E_X and E_Γ, are known quantities for most III-V alloys. The energies of the lowest conduction band minima relative to the valence band maximum are plotted versus lattice parameter, a linear function of alloy composition for III-V alloys, in Fig. 1. It should be mentioned that the indirect minimum need not be located at X. The value of the lowest energy minimum not located at Γ is plotted in Fig. 1 and used in the calculation.

The external quantum efficiency is simply η_i multiplied by the fraction of photons able to escape from the material f. At the time that Archer paper was written, the values of the two lifetimes appropriate for the GaAsP alloy system were approximately 1 ns and the value of f was approximately 10^{-2}, resulting in a maximum external quantum efficiency of 5×10^{-3}. The relatively short nonradiative lifetime is due to the dislocations in the GaAsP, present because of the large lattice mismatch with the GaAs substrate. The absorbing GaAs substrate is mainly responsible for the small fraction of photons escaping.

Archer performed such calculations using similar assumptions for several III-V alloy systems. The results indicated a much lower optimum performance for AlGaAs than for GaAsP LEDs. This is contrary to the subsequent experimental observations (Cook et al., 1987) due to two factors. (1) The value of E_X and E_Γ for AlGaAs used by Archer were later found to be in error. (2). The nonradiative minority carrier lifetime in the best AlGaAs is much longer than the value of 1 ns assumed by Archer. This is due to the high purity, particularly the freedom from oxygen contamination, of AlGaAs material grown by LPE (Ahrenkiel, 1993) and the lattice match with the GaAs substrate for all alloy compositions, which results in layers with very low dislocation densities. In fact, the best AlGaAs red LEDs have external quantum efficiencies of 18% at a wavelength of 650 nm (Cook *et al.*, 1987). For these extremely high external efficiencies, very thick layers and transparent substrates are used to improve the light extraction efficiency to value of approximately 30%. This topic is discussed in detail in Chapter 3.

The materials that emerged as the most promising from Archer's calculations were GaInP and AlInP. They were predicted to have peak brightness values more than an order of magnitude higher than those calculated for GaAsP. This is simply due to the larger value of bandgap energy at the direct–indirect crossover. However, the optimum GaInP and AlInP alloys are not lattice-matched to any binary III-V substrates, which prevents the desired performance levels from being obtained.

Later calculations by Stringfellow (1978) for the AlGaInP quaternary alloys lattice matched to GaAs substrates indicated a more than 50-fold improvement over the best GaAsP alloys. The calculations are conservative because they assume a ratio τ_R/τ_N of unity, following Archer. The lower dislocation density of the AlGaInP lattice matched to the GaAs substrate gives higher values of the nonradiative lifetime.

The achievement of the goal of high-performance AlGaInP LEDs was frustratingly slow in coming, mainly due to problems with epitaxial growth. As described below, the techniques used to produce the first LEDs, LPE, and hydride vapor-phse epitaxy (HVPE), are unsuitable for AlGaInP. The AlP is thermodynamically so much more stable than the InP that the alloy is difficult to grow by these "near-equilibrium" techniques. This problem was realized as early as 1976.[1] In fact, significant progress in the AlGaInP materials system had to await the development and refinement of the new epitaxial growth techniques, OMVPE and MBE. This is just one example of a fairly common phenomenon in the quest for advanced electronic and photonic devices: The development of a new epitaxial growth technique frequently enables the realization of entirely new materials and device structures.

Problems were also encountered with both residual oxygen an carbon contamination and intentional doping of AlGaInP to the levels required for high-performance LEDs and lasers. It was not until years later that reports

---

[1]The Advanced Research Projects Agency (ARPA) Materials Research Council Meeting on Epitaxy, July 12 and 13, 1976, LaJolla, California. At this meeting, G. B. Stringfellow and M. B. Panish made independent presentations of unpublished research related to the AlGaInP alloy system. Both agreed that the well-developed growth techniques of LPE and VPE would not work for thermodynamic reasons. Stringfellow proposed an OMVPE solution and Panish an MBE solution. The results were later published in part (Stringfellow, 1981, 1985; Casey and Panish, 1978).

of the first high-qulity material, produced by MBE, and laser devices appeared (Asahi *et al.*, 1982, 1983). Hino and Suzuki (1984) reported the first OMVPE growth of AlGaInP for laser devices. They resorted to the use of an air-lock to reduce oxygen contamination in their OMVPE system. Additional years were required for the development of OMVPE for the production of extremely high performance LEDs, the topic of Chapters 4 and 5 herein.

IV. Fundamental Thermodynamic and Kinetic Considerations

1. BINARY COMPOUNDS

The III-V compounds are stable because of the very strong, highly covalent bonds formed in the solid. In the idealized zinc-blende and wurtzite lattices, the cubic and hexagonal forms, respectively, observed for III-V compounds, each group III atom is in a tetrahedron surrounded by four group V atoms. Similarly, the group V atoms each have four group III neighbors in a tetrahedral arrangement. The bonds are mainly sp^3 hybridized covalent bonds. Due to the electronegativity difference between the group III and group V atoms some charge transfer occurs, resulting in a small fraction of ionic bonding (Philips, 1973).

The stability of the III-V compounds is much greater than are the stabilities of the individual elements in the vapor, as indicated in Fig. 2. This results in a large positive enthalpy of formation, which can be translated, via the law of mass action, to mean that the product of the group III and group V pressures in equilibrium with the solid is small (Stringfellow, 1984, 1989a, 1991). Due to the high supersaturation, nearly all of the element with the lowest concentration in the input vapor is depleted at the interface. Because equal numbers of group III and group V atoms are removed from the vapor to form the III-V solid, the pressure of the major component remains near the input pressure in most cases. For the arsenides and phosphides, the vapor pressure of the group V element is orders of magnitude greater than that of the group III element. It is for this reason that OMVPE growth typically occurs with an input V-III ratio (input group V pressure-to-input group III pressure) much geater than unity. Growth at high V-III ratios seldom results in the production of a second condensed phase, due to the high volatility of the group V elements at typical growth temperatures. At V-III ratios of less than unity, the less volatile group III element frequently forms a second condensed phase. Thus, a minimum group V pressure is required to produce a single-phase III-V solid. This pressure is ploted versus growth temperature in Fig. 4 for several III-V compounds of interest in this volume: GaP, InP, GaN, and indium nitride (InN). It should be noted that these are the *minimum* group V pressures. Growth with these group V pressures results in the solid on the most group III–rich side of the range of solid stoichiometry (Stringfellow, 1989a). For the arsenides and phosphides, growth typically occurs at group V input partial pressures that are orders of magnitude larger than the values plotted in Fig. 4. This results in a shift to a more group V–rich stoichiometry in the solid. These considerations are important for determination of the concentrations of native defects such as vacancies, interstitials, and antisite defects. It often also has a major effect on the incorporation of dopant atoms (Stringfellow, 1986; 1989a).

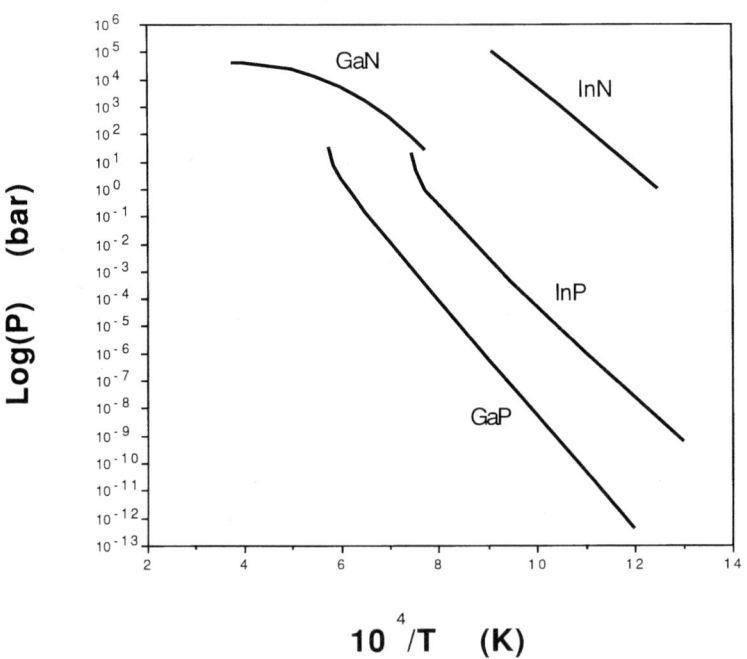

Fig. 4. Vapor pressure versus the inverse temperature. Pressure of the group V element in equilibrium with the III-V compound on the group III–rich side of the solid stoichiometric region. The data for gallium phosphide (GaP) and indium phosphide (InP) are from Stringfellow (1992), for gallium nitride (GaN) from Karpinski *et al.* (1984), and for indium nitride (InN) from MacChesney *et al.* (1970).

The pressures in Fig. 4 always increase with increasing temperture due to the highly exothermic nature of the growth reactions. The pressure at a given temperature also increases as the stability of the solid phase decreases. Thus, the group V pressure required to stabilize InP is higher than that for GaP. Similarly, the N pressure over InN is orders of magnitude higher than for GaN. The phosphides always require higher group V pressures than do the similar arsenides in order to stabilize the III-V solid, and the nitrides require even higher pressures. As discussed below, this represents a major problem for the growth of the nitrides, particularly for InN and GaInN alloys. Another important reason for the very high N_2 pressures over the group III–rich nitrides (30 atm at 1000°C for GaN) is the anomalous stability of the N_2 molecule. Growth would never be possible with N_2 as the nitrogen source. It is for this reason that NH_3 is commonly used as the nitrogen precursor. It pyrolyzes to a much larger extent at normal growth temperatures, giving many orders of magnitude higher atomic nitrogen concentrations in the vapor. Nevertheless, V-III ratios of over 1000 are required for GaN growth at 1000°C. Other less stable precursrs may produce even higher atomic N concentrations without the need for growth temperatures of 1000°C or greater, as discussed in Part V, Section 3. Alternatively, a plasma can be used to dissociate the N_2, producing large concentrations of atomic nitrogen. This approach is nearly always used for GSMBE growth of the nitrides (Morkoc et al., 1994; Paisley et al., 1989; Molnar et al., 1995; Davis, 1991) and also may be used for OMVPE growth, particularly for InN (Guo et al., 1994, 1995).

The phosphides are typically grown by OMVPE with V-III ratios of between 10 and 100. Values between 10 and 30 are common for the arsenides. For III-V antimonides, the less volatile antimony dictates the use of V-III ratios of near unity (Stringfellow, 1989a).

2. TERNARY AND QUATERNARY ALLOYS

Considerations of the relative stabilities of the III-V compounds have a profound influence on the growth of ternary III-V alloys. An example that is highly relevant for this volume is the system InP–AlP. Aluminum phosphide (AlP) is so much more stable than InP that the growth of InAlP alloys by LPE is virtually impossible (Stringfellow, 1981, 1985; Casey and Panish, 1978). The Al distribution coefficient, defined as the ratio of Al to (Al plus In) in the solid divided by the same ratio in the liquid, is plotted versus temperature in Fig. 5 for the solid–liquid equilibrium. The extremely large distribution coefficient is a direct reflection of the relative stabilities of the AlP and InP solids. As a consequence of the enormous Al distribution

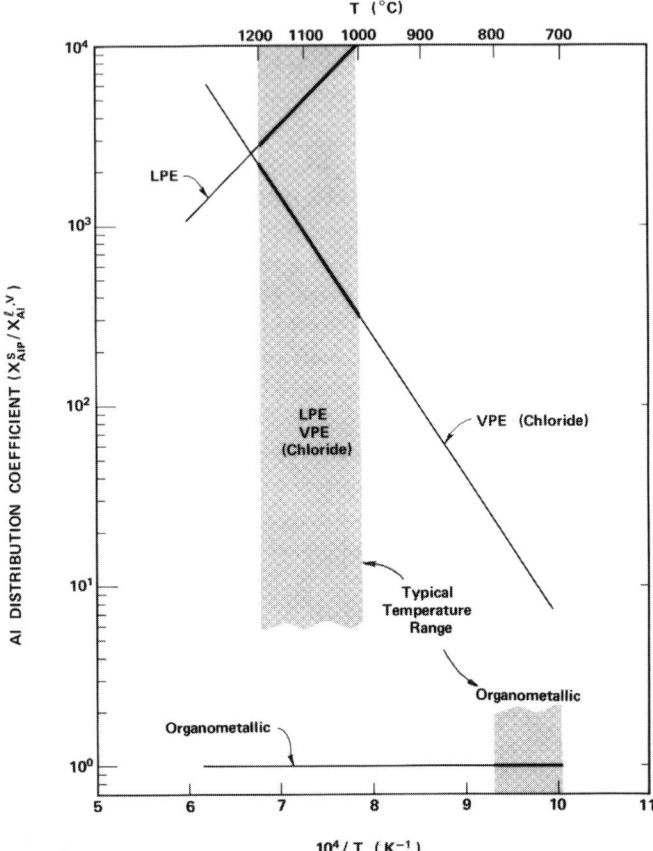

FIG. 5. Aluminum (Al) distribution coefficient versus temperature calculated for AlInP grown by several techniques. LPE, liquid-phase epitaxy; VPE, vapor-phase epitaxy; organometallic vapor phase epitaxy. AlP, aluminum phosphide. (Reprinted from Stringfellow, 1981 *J. Cryst. Growth* **115**, 1, with the kind permission of Elsevier Science – NL.)

coefficient, the liquid in equilibrium with the $Al_{0.2}Ga_{0.8}P$ solid, even at the impractically high growth temperature of 1000°C, would have an Al mole fraction of 10^{-4}. This makes LPE growth extremely difficult, because the tiny amount of Al in the liquid near the growing solid–liquid interface would be depleted immediately to produce the $Al_{0.2}Ga_{0.8}P$ solid. This results in two undesirable consequences: (1) the growth rate must be infinitesimally slow to produce a uniform solid, and (2) growth of an extremely thin epitaxial layer, even at this small growth rate, would totally deplete the Al from the liquid. Thus, the maximum epilayer thickness would be a few angstrons. Similar arguments apply to the hydride or chloride VPE (HVPE and ClVPE, respectively) growth of AlInP. The solid–vapor distribution coefficient for these techniques is included in Fig. 5. The large difference in stabilities of AlP and InP and the resulting large Al distribution coefficient again makes the VPE growth of layers of AlInP (and AlGaInP) virtually impossible. Similar arguments also apply to the VPE growth of AlGaAs (Stringfellow, 1978). This explains why neither ClVPE nor HVPE is used for either of these important alloys.

The thermodynamic problems described earlier can be overcome by using a growth technique operating at a very high level of supersaturation in the input phase and a large excess of the group V element (Stringfellow, 1978, 1989a). For example, the OMVPE growth of AlInP or AlGaInP is typicaly performed with an input V-III ratio of from 10 to 100. At the interface, the product of the group III and group V partial pressures must be orders of magnitude lower than the product of the input partial pressures in order to achieve the required near equilibrium condition, as described earlier. This combination dictates that all the group III elements will be severely depleted at the interface: Virtually all of the group III elements in the vapor at the interface are deposited onto the surface. For the OMVPE growth of AlInP, this gives an Al distribution coefficient of unity because both the Al and In are nearly totally depleted. The Al distribution coefficient is also independent of temperature, as seen in Fig. 5. For actual OMVPE growth conditions, other factors, such as differences in the diffusion coefficients in the vapor, evaporation of In from the surface at high temperatures, and parasitic reactions, can result in either lower or higher distribution coefficients. However, the experimental Al distribution coefficient is determined to be nearly unity (Yuan $et\ al.$, 1985). Similar arguments are valid for the MBE growth of AlInP when high V-III ratios are used. The flexibility of being able to produce materials such as AlGaInP represented one of the motivating factors for the erly development of OMVPE. The flexibility still represents a tremendous advantage of both OMVPE and MBE relative to the other growth techniques discussed.

Although both OMVPE and MBE can be used to grow a wide range of III-V alloys, problems with the stability of the solid phase can be troublesome. Consider the case of the OMVPE growth of GaInN alloys. The addition of In to GaN is desired to reduce the bandgap energy into the visible region of the spectrum and for the production of AlGaN–GaInN heterostructures, which result in the best LED performance. Both GaN and AlGaN can be grown readily by OMVPE at temperatures near 1000°C, with ammonia (NH_3) typically used to supply the nitrogen. However, even with high NH_3 partial pressures, the weak bonding in the InN prevents the deposition of a single condensed phase. As seen in Fig. 4, the nitrogen pressure required to stabilize InN is orders of magnitude larger than that required for GaN. It is partly for this reason that difficulties are experienced in producing GaInN layers with high concentrations of In at high growth temperatures.

Another aspect of ternary III-V phase diagrams not appreciated in the early literature is the tendency for phase separation at low temperatures. This is related to the enthalpy of mixing. For III-V alloys, both empirical evidence and phenomenological models indicate that the entropy of mixing increases with the difference in lattice parameter of the end components (Stringfellow, 1974). The mixing enthalpy is always greater than or equal to zero. A positive value means that like atoms prefer to cluster together at low temperatures where entropy effects are minimized. Thus, for temperatures below a certain critical temperature the equilibrium state of the system consists of two solid phases. This phenomenon of phases separation may even prevent the LPE growth of some alloys composed of compounds having widely different lattice constants, such as GaAsSb (Pessetto and Stringfellow, 1983). However, growth by OMVPE (Cherng et al., 1986a,b) or MBE (Waho et al., 1977) can be used to produce a nearly random, metastable solid. Even alloys with enormous regions of solid immiscibility, such as GaPSb and InPSb can be obtained over the entire range of solid composition (Jou et al., 1988a,b). Nevertheless, the tendency toward phase separation makes itself felt, leading to clustering by spinodal decomposition occurring at the surface during growth (Zunger and Mahajan, 1993). The process does not proceed far because of the relatively rapid growth rates, and spinodal decomposition does not occur in the bulk solid because of the extremely slow diffusion coefficients in III-V alloys.

Recent calculations (Ho and Stringfellow, 1997) indicate that the GaInN system also contains a sizable miscibility gap. At 1000°C the solubility of InN in GaN is less than 10%. However, these alloys also can be grown with compositions covering the entire alloy system by OMVPE (Osamura et al., 1972), as reported in Chapter 8, for the reasons discussed previously. Some evidence of phase separation has been observed (Singh et al., 1997).

Thus clustering resulting from the solid immiscibility in III-V alloys apparently leads to a distinct broadening of the band-edge photoluminescence (PL) emission peaks, which increases with increasing enthalpy of mixing (Cherng et al., 1986b). This is due to fluctuations in the local bandgap energy due to fluctuations in local composition. Clustering also produces adverse effects on the electron mobility in n-type alloys (Benchimol et al., 1983) due, again, to fluctuations in the bandgap energy and the local strain.

Nature may also relax the microscopic strain energy due to the difference in the sizes of atoms mixed on one sublattice by forming ordered structures. This is a phenomenon first discovered in metal systems. The large and small atoms are segregated into alternating atomic planes of a particular crystallographic orientation. The microscopic strain energy is reduced most in III-V alloys by formation of either the CuAu–I or the chalcopyrite structure where the crystallographic plane on which ordering is observed is $\{100\}$ or $\{210\}$, respectively (Stringfellow, 1989b; Zunger and Mahajan, 1993; Zunger, 1994). This is manifest as a doubling of the periodicity in a particular direction in the electron diffraction patterns. However, by far the most commonly observed ordered structure in III-V alloys grown epitaxially on (001) substrates is the CuPt structure, with ordering on the $\{111\}$ planes (Stringfellow, 1989b). It is now well established that the copper platinum (CuPt) structure is formed at the surface during growth. Only the (2×4) surface reconstruction, with rows of $[\bar{1}10]$ phosphorus dimers, stabilizes the CuPt ordered structure (Zhang et al., 1995). The degree of order can be controlled by varying the growth rate (Cao et al., 1989), temperature (Su et al., 1994b), V-III ratio (Kurtz et al., 1994; Murata et al., 1996a,b), and the exact substrate orientation (Su et al., 1994a). Ordering is virtually eliminated by growth at very high rates, high temperatures (or very low temperatures), and low V-III ratios. Perhaps the most effective method for eliminating ordering is to grow on substrate orientations other than (001). For example, growth on either (111)-oriented (Ikeda et al., 1988) or (511)B-oriented (Valster et al., 1991) substrates virtually eliminates ordering.

Control of ordering is vital for the AlGaInP and GaInP used for visible LEDs because formation of the CuPt ordered structure is known to result in a significant reduction of the bandgap energy (Su et al., 1993), thus moving the wavelength of red or orange LEDs to longer wavelengths where the eye response is lower. For $Ga_{0.5}In_{0.5}P$, the reduction in bandgap energy is predicted to be 320 meV (Wei and Zunger, 1990) for formation of a 100% ordered structure consisting of alternating $\{111\}$ planes of pure InP and GaP. In practice, the order parameter is not unity; however, the bandgap energy has been seen experimentally to decrease by as much as 160 meV (Su et al., 1993). This is a tremendously deleterious phenomenon for both visible LEDs and lasers. Thus, ordering is typically avoided by using either (111) or (511)B substrates. The reduction in ordering for growth on substrates with these orientations is believed to be related to the loss of the (2×4) surface reconstruction observed for growth of GaInP on the (001) surface (Murata et al., 1996c, 1997).

A possible useful application of the ordering phenomenon is the production of disorder–order–disorder heterostructures (Su et al., 1994c,d) and quantum wells (Schneider et al., 1994; Follstaedt et al., 1995), where the solid composition remains fixed and the bandgap energy is modulated by controlling the degree of order. These structures were first produced using modulation of the growth temperature to control the degree of order. Abrupt interfaces are produced in this way, resulting in quantum wells as thin as 1 nm, with a corresponding blue shift in the PL energy by more than 80 meV. Variation of the flow rate of the phosphorus precursr is also found to result in heterostructures, although the magnitude of the bandgap variation is smaller than for heterostructures produced by variation of the growth temperature (Chun et al., 1997).

3. Doping

a. Control of Background Doping

One of the difficulties in obtaining high-quality AlGaInP LED and laser devices by OMVPE was control of the background dopants carbon and oxygen as discussed briefly earlier. The carbon comes principally from the radicals produced during pyrolysis of the organometallic precursors in OMVPE. Methyl radicals are known to be efficient sources of carbon in the epitaxial layers (Stringfellow, 1989a), particularly for materials containing Al, due to the strong Al–C bond. This has resulted in efforts to produce new precursor molecules not having methyl radicals. Even more advanced precursors have no direct group III-carbon or group V-carbon bonds. Thus, trimethylaminealane (TMAA) has been successfully used as an Al precursor yielding material with low carbon contamination levels (Jones, 1993). The main problem with the use of this precursor is the low stability of TMAA, which may result in parasitic reactions that reduce the growth rate and produces less homogeneous material for OMVPE growth. The novel group V precursors TBAs and TBP are known to yield GaAs and InP, respectively, with reduced levels of carbon incorporation during OMVPE growth using trimethyl-group III precursors by providing an increased supply of atomic hydrogen on the surface (Haacke et al., 1991; Imori et al., 1991; Komeno, 1994). The atomic hydrogen produced during TBAs or TBP pyrolysis reacts with the methyl radicals from the group III sources to produce the stable and relatively innocuous methane. Other phosphorus precursors currently under investigation include trisdimethylaminophosphine (TDMAP) (Ryu et al., 1996), a molecule with no phosphorus–carbon bonds, and bisphosphinoethane (Chin et al., 1992; Kim et al., 1996). The dimethylamino radicals produced during pyrolysis of TDMAP are thought to serve the beneficial function of removing methyl radicals from the surface (Shin et al., 1994).

Oxygen is the other important residual dopant in AlGaInP grown by OMVPE. It is well known to be a nonradiative recombination center in AlGaInP (Kondo et al., 1994), and therefore must be avoided for the production of high-efficiency LEDs. Standard practice for many years has included the careful purification of the hydrogen ambient using palladium (Pd) diffusers. The system must also be extremely leak tight to avoid leaks of air into the vapor from which the solid is to be grown. Interlocks are sometimes helpful in preventing exposure of the inner parts of the reactor to oxygen during change of the substrate. Virtual leaks may exist due to adsorbed water on the silica (SiO_2) walls, and the susceptor, or both. Oxygen and water can also be "stored" in unpurged regions, for example the region between susceptor and the adjacent quartz wall for horizontal reactors. Another source of oxygen is the group III precursors themselves, which are known to frequently contain low concentrations of alkoxides (Terao and Sunagawa, 1984; Smith et al., 1993). These can be removed by careful purification after synthesis by the vendor, although this increases the cost of the precursor. In addition, oxygen is a contaminant in the group V precursors. Both the oxygen and water can be removed by filtering or gettering before the group V precursor enters the reactor. This can be accomplished using molecular sieves (Tsai et al., 1984), liquid metal bubblers (Shealy et al., 1983), and commercial purifiers using gettering agents that are extremely reactive with oxygen and water (Tom, 1986). Internal gettering, where the Al precursors present in the vapor are encouraged to react with residual oxygen and water, may also be a useful way of removing these contaminants (Stringfellow and Hom, 1979; Kuech et al., 1987).

Other sources of donors and acceptors are the native defects produced during epitaxial growth. This normally is not a significant problem for conventional III-V alloys. However, the presence of nitrogen vacancies, which act as donors, is a major problem for GaN and other nitrides due to the extreme volatility of nitrogen, as discussed earlier. This resulted in extremely high n-type doping levels for the epitaxial layers grown during the early GaN development (Maruska and Tietjen, 1969; Ilegems, 1972; Pankove, 1992), which prevented effective doping with acceptors to produce p-type material. The keys to reducing the residual donor concentration are seen from the thermodynamic factors discussed in Part IV, Section 3a: lower growth temperatures, to reduce the vapor pressure of nitrogen over GaN, and a high nitrogen pressure during growth. The nitrogen pressure can be increased by increasing the partial pressure of the nitrogen precursr, or by increasing the rate of pyrolysis to produce atomic nitrogen, or both. This can be accomplished by using ammonia, which is more effective than N_2. Even better is the use of N precursors such as hydrazine that pyrolyze more efficiently at low pressures (Gaskill et al., 1986). Alternatively, the N_2 can be dissociated using a plasma. This is an approach used for both MBE (Paisley et al., 1989; Morkoc et al., 1994; Molnar et al., 1995) and OMVPE (Guo et al., 1994, 1995) growth of the nitrides. Development of compact electron cyclotron resonance plasma units that can be conveniently attached to an epitaxial reactor has led to their widespread use in epitaxial growth systems. However, the plasma itself can represent a problem for the production of materials for high-efficiency LEDs, since highly energetic atoms from the plasma can damage the material, producing a significant concentration of nonradiative recombination centers. This must be avoided for the production of long minority-carrier lifetime layers (Molnar et al., 1995).

As mentioned above, the vapor pressure of nitrogen over InN is orders of magnitude larger than for GaN. Thus, the problem of residual n-type doping is more severe for GaInN alloys (Davis, 1991; Matsuoka et al., 1992; Morkoc et al., 1994). In fact, no attempt is typically made to make these materials p-type in LEDs (Nakamura et al., 1994, 1995), although Yamasaki et al. (1995) have reported p-type conduction in Mg-doped GaInN with an InN concentration of 9%.

b. Intentional Doping

The general features of both thermodynamic and kinetic aspects of dopant incorporation during OMVPE (Stringfellow, 1989a) and MBE (Heckingbottom, 1985; Tsao, 1993) growth are reasonably well understood. For simplicity, only the incorporation of neutral impurities are considered here. When the principal factor is the energy of replacing a host atom by a dopant, that is, when the pyrolysis rate for the dopant precursor is rapid, thermodynamic factors dominate. At low concentrations, the dopant concentration incorporated into the solid is typically directly proportional to the concentration in the vapor phase. The proportionality factor is the distribuion coefficient, which is frequently used as a measure of dopant incorporation. In the simplest model for dopant incorporation, two cases occur. For volatile dopants, increasing the temperature leads to a lower distribution coefficient resulting from the competition between incorporation of the dopant into the lattice and evaporation. When the dopant (e.g., Zn or Mg) incorporates onto the group III sublattice, an increase in the group V input partial pressure produces more group III vacancies and thus increases dopant incorporation. The reverse occurs for dopants incorporating on the group V sublattice. For the volatile dopants, the distribution coefficient is typically not dependent on the growth rate.

For nonvolatile impurities such as Si, neither the growth temperature nor the V-III ratio have any effect on incorporation. However, as the growth rate increases, the dopant becomes more dilute; that is, the dopant concentration decreases.

The previous discussion assumes rapid pyrolysis of the dopant precursor. For stable dopant molecules, dopant incorporation can be limited by the pyrolysis rate of the dopant precursor. For the case of the dopant Si supplied as silane (SiH_4), the strong Si–H bonds produce a stable precursor molecule that does not pyrolyze completely at normal growth temperatures. In this case, the distribution coefficient increases with increasing growth temperature. Use of the less stable Si dopant disilane results in a temperature-independent Si distribution coefficient (Kuech et al., 1984).

In addition to these considerations of the dopant distribution coefficient, which are dominant at low concentrations, the solid solubility limit becomes important for high dopant concentrations at which the dopant concentration in the III-V solid becomes independent of the partial pressure of the dopant in the vapor phase. This normally is limited by effects associated with the ionization of the dopant to produce free electrons and holes. The extreme increase in the energy of the system when the Fermi level rises into the conduction band for donors or falls to below the valence band-edge for acceptors typically gives a solubility limit for dopants that is low, well below 1%. Other factors affecting the maximum amount of dopant that can be incorporated into the solid include the size of the dopant atom, relative to that of the host atom, and the stability of the second condensed phase containing the dopant atom that precipitates when the solubility limit is exceeded.

Without question, the problems associated with obtaining high n- and p-type doping levels become worse as the bandgap energy increases. This is basically thermodynamic in origin. The Fermi level represents the average electrochemical potential of the free electrons in the semiconductor. Consequently, moving the Fermi level from the intrinsic level produces a significant increase in the energy of the system. Thus, the system can reduce its total energy by creating compensating defects, such as electrically active native defects and precipitates of a second phase, in order to moderate the movement of the Fermi level. A similar argument holds for both n- and p-type doping. Factors such as the energy required to create compensating defects determine the degree of self-compensation. For example, such factors determine whether a II-VI compound will prefer to be n- or p-type (Faschinger et al., 1995). Thus, these problems become much more severe as the color of the LED progresses from red to green to blue. The problems are virtually nonexistent for IR and red LEDs based on GaAs, AlGaAs, and GaAsP. However, for AlGaInP, problems surface that are associated with high levels of p-type doping (Kadoiwa et al., 1994; Ishibashi et al., 1994) and become much worse for the nitrides, as discussed earlier.

The most widely used acceptor dopant in III-V alloys is Zn. However, difficulties are experienced when the p-type doping level exceeds the mid-10^{17} cm^{-3} range in AlGaInP. Two compensation mechanisms have been associated with this probem. One is compensation by oxygen impurities acting as deep donors (Kadoiwa *et al.*, 1994) and the other is hydrogen passivation of the acceptor impurity (Kadoiwa *et al.*, 1994; Ishibashi *et al.*, 1994; Gorbylev *et al.*, 1994). Atomic hydrogen is ubiquitous for normal OMVPE growth conditions. It may come from dissociation of the H_2 carrier gas or, more frequently, from pyrolysis of the group V hydrides. The fact that neither mechanism involves native defects probably reflects the difficulty of forming vacancies, interstitials, and antisite defects in these highly covalent materials.

Compensation in GaN and the related high-bandgap alloys also occurs due to hydrogen passivation of Mg-doped material (Nakamura *et al.*, 1992). Thus, materials grown by OMVPE must either be annealed in nitrogen after growth (Nakamura *et al.*, 1991) to remove the hydrogen or irradiated by a low-energy electron beam during growth, as discussed in Chapter 7. This is not required for MBE growth using atomic nitrogen produced from a N_2 plasma where no atomic hydrogen is present on the surface during growth.

4. KINETICS OF PYROLYSIS AND GROWTH REACTIONS

The previous discussion emphasizes the need to reduce growth temperature to minimize the effects of native defects on the electrical properties. This is particularly true for the high bandgap materials such as the II-VI semiconductors and the III-V nitrides, especially InN. As a result, InN has been grown by OMVPE at temperatures as low as 400°C using a plasma N_2 source (Guo *et al.*, 1995; Matsuoka *et al.*, 1992). However, this approach has several associated problems. As mentioned earlier, pyrolysis of the precursors decreases at low growth temperatures. This results in reduced growth rates in Ga-containing and/or Al-containing semiconductors grown using TMGa and/or trimethylaluminum (TMAl) for temperatures below 500°C (Stringfellow, 1989a) as well as in the loss of control of solid composition. Small fluctuations or variations in the temperature result in nonuniformities of growth rate and solid composition. In addition, for systems without plasma cracking, reduction in temperature reduces the pyrolysis rate of the group V precursor, which reduces the effective V-III ratio at the interface. This, in turn, leads to loss of control of solid stoichiometry with the associated problems with both background and intentional dopant incorporation. These problems are common with the group V hydrides, AsH_3, PH_3, and NH_3, particularly the latter two, at normal OMVPE growth temperatures. These problems are alleviated somewhat by the use of more labile precursors such as TBAs (Stringfellow, 1989a; Chen *et al.*, 1987; Lum *et al.*, 1987; Komeno, 1994), TBP (Stringfellow, 1989a; Chen *et al.*, 1987; Komeno, 1994), and hydrazine (Gaskill *et al.*, 1986), all of which pyrolyze at much lower temperatures than required for the hydrides.

A more subtle problem associated with the use of low-growth temperatures is a degradation of the minority carrier properties of the material. It is well-known that at a fixed growth rate, a sequential reduction in the growth temperature results in transition from single-crystalline layers, produced at the highest growth temperatures, to polycrystalline material at lower temperatures, and, eventually, to amorphous materials at the lowest growth temperatures (Bloem, 1973; Stringfellows, 1980). In general, the crystalline quality—that is, the concentration of grain boundaries, dislocations, and other point and line defects introduced during growth—increases as the growth temperature decreases (in the low-temperature regime) and the growth rate increases. A simple model attributes this to a decrease in surface mobility as the temperature is reduced. If the atoms in the surface layer do not have sufficient time to find their lowest energy sites before being covered by the next layer, the result is a decrease in the structural perfection of the layer as the growth temperature is reduced at a fixed growth rate. Indeed, this general trend is observed for the low-temperature growth of GaInN layers. Matsuoka *et al.* (1992) found that increasing the growth temperature from 500 top 800°C resulted in a significant and measurable increase in the crystalline quality of the epitaxial layers.

One final kinetic problem, associated particularly with the use of the electron cyclotron resonance (ECR) source of atomic nitrogen for the growth of the III-V nitrides, is the limited flux available from the conventional sources. This limits the growth rates of GaN, for example, to approximately 0.2 μm/h for typical growth conditions (Moroc *et al.*, 1994; Molnar *et al.*, 1995). As a practical matter, this will limit the maximum growth temperature used for the plasma-generated nitrogen techniques for the growth of GaN and AlGaN to approximately 1000°C. This low growth rate may also be inconvenient for LED structures several microns thick that require tens of hours for growth. For conventional OMVPE grown at 1000°C using NH_3, four microns of GaN is grown in a half hour (Amano *et al.*, 1989, 1990).

5. LATTICE MISMATCH

Frequently, the lattice constant of a desired alloy differs from that of an underlying layer or the substrate. In this case, if the atoms matched perfectly at the interface, they could not be at their equilibrium spacings. Assuming that the substrate is very thick and that the epilayer is thin, the strain energy resulting from the lattice parameter mismatch increases linearly with the thickness of the lattice-mismatched epilayer. Eventually, the energy is so high that the coherence is disrupted and dislocations are generated at the interface between the two layers. Each so-called *mismatch dislocation* represents a row of unsatisfied bonds at the interface. Hence, formation of dislocation arrays increases the energy of the system; however, this energy does not depend on the epilayer thickness. Thus, beyond a critical thickness the energy of the system is dereased by formation of the dislocation array. This analysis is correct to first order, but oversimplifies a complex topic. For a more complete description of the generation of mismatch dislocations see Dodson and Tsao (1989) and Jesser and Fox (1990).

These mismatch dislocations interact in the interfacial plane, causing some of the dislocations to propagate in a direction out of the plane and into the epitaxial layer. This is the mechanism for production of the so-called *inclined* or *threading* dislocations found in epitaxial layers that are thicker than the critical thickness (Tamura *et al.*, 1992).

Both the mismatch and inclined dislocations can have deleterious effects on the properties of the epitaxial layers and heterostructures. For conventional III-V semiconductors, these dislocations are found to act as nonradiative recombination centers. Thus, the minority carrier lifetime and radiative recombination efficiency both decrease for high dislocation densities (Stringfellow *et al.*, 1974; Ettenberg, 1974).

During the 1980s the development of advanced photonic devices using conventional III-V semiconductors was based largely on the use of heterostructures where two single-crystalline layers, with different values of bandgap energy and refractive index, abut. As a result of the deleterious effects of dislocations, two of the most important guidelines for the fabrication of useful materials, particularly for the fabrication of photonic devices, are: (1) the epitaxial layer must have the same lattice constant as the substrate and (2) heterostructures must involve materials with identical lattice constants. Violation of these general rules results in many of the extreme difficulties experienced with the growth of III-V materials on lattice-mismatched substrates, for example, GaAs on Si. Even though enormous resources have been devoted to this problem, the fabrication of high-performance photonic devices in GaAs on Si has been elusive. This appears to be simply because the dislocations act as nonradiative recombination centers in conventional III-V semiconductors. In direct bandgap GaAs-doped p-type to $10^{19}\,cm^{-3}$, dislocation densities over $5 \times 10^6\,cm^{-2}$ result in a significant decrease in the minority carrier lifetime due to nonradiative recombination at the dislocations (Ettenberg, 1974). In materials with longer radiative recombination lifetimes the deleterious effect of dislocations begins at much lower dislocation densities. For example, for indirect-gap GaP, the radiative recombination efficiency begins to decrease measurably at a dislocation density of approximately 10^3 cm^{-2} (Stringfellow et al., 1974).

The recombination velocity at lattice-mismatched interfaces is also found to increase with increasing difference in lattice constant (Kressel and Butler, 1977; Kressel, 1980). The minority carrier lifetime at the GaAs–GaInP interface has been shown to be proportional to the square root of the density of misfit dislocations (Mullenborn et al., 1994). Amazingly low values of interface recombination velocity of less than 1.5 cm/sec have been reported for perfectly lattice-matched, high-quality GaAs–GaInP interfaces (Olson et al., 1989). This value of interfacial recombination velocity is significantly below those measured for other systems, such as AlGaAs–GaAs, perhaps because of the difficulties involved in completely avoiding oxygen contamination of the interface in systems containing Al, for reasons discussed earlier in Part IV, Section 3.a.

The problems associated with absorption of the emitted photons in the substrate, giving low external quantum efficiencies, could, potentially, be solved by growth of AlGaInP on transparent GaP substrates. However, this is untenable because of the lattice mismatch, which would generate dislocations that would degrade device performance. The solution is growth on lattice-matched GaAs substrates and subsequent removal of the GaAs. The GaP is wafer-bonded to the LED structure to give mechanical support. This results in an increased photon extraction efficiency of 0.3, as discussed in Chapter 5.

Dislocations are also troublesome in terms of device degradation (Thompson, 1980). The active regions of long-lived GaAs–AlGaAs lasers must be dislocation-free. The effects are less serious for GaInAsP devices.

Contrary to the previous discussion of conventional III-V semiconductors, for the nitrides dislocations are apparently not a problem. This accounts for the high LED efficiencies in materials with dislocation densities exceeding 10^{10} cm^{-2} (Weeks et al., 1995; Ponce et al., 1995). In this sense the nitrides are like the II-VI semiconductors in which neither surfaces, grain boundaries, nor dislocations are harmful. This is the reason that II-VI semiconductors are so efficient as phosphors, even though they consist of small-grain polycrystalline films. The reason for this behavior is not known. The degradation rate in the GaN LEDs is also small despite the extremely high dislocation densities.

6. SURFACE RECOMBINATION

For typical III-V semiconductors, minority carriers reaching the surface are lost to nonradiative recombination, as discussed in Part III. For example, the surface recombination velocity for GaAs is 10^6 cm/sec (Knessel and Butler, 1977). This parasitic recombination process lowers the internal recombination efficiency of the device. Two ways are available to minimize the deleterious effect of the surface. (1) The p-n junction can be placed many diffusion lenghs away from the surface by growing a very thick top layer. This is problematic because it involves the growth of thick layers, in itself an expensive process, and because these thick layers absorb photons emitted at the p-n junction before they can escape from the surface. This, of course, lowers the external efficiency. (2) The movement of the minority carriers to the surface can be prevented by the introduction of a higher bandgap layer between the p-n junction and the surface. This normally is the preferred approach. In fact, the highest performance light-emitting devices — both LEDs and laser diodes (LDs) — are double heterostructures. The heterostructures are used to confine the minority carriers (typically electrons) to regions near the p-n junction, as discussed in Chapter 3. The two high bandgap layers also act as window layers by reducing the number of emitted photons being absorbed before they can escape from the semiconductor. The three layers comprising the double heterostructure must be lattice-matched to avoid excess nonradiative recombination, as described above. This makes the $(Al_xGa_{1-x})_{0.5}In_{0.5}P$ system particularly useful. All compositions are lattice-matched to the GaAs substrate. The highest bandgap energy confining layer would be $Al_{0.5}In_{0.5}P$. The highest bandgap active layer typically would be the direct bandgap $(Al_xGa_{1-x})_{0.5}In_{0.5}P$, having a value of x slightly less than the value at which the material becomes indirect, 0.51, where the bandgap energy is 2.213 eV (Kish et al., 1994).

The high-performance blue LEDs are also double heterostructures consisting of an GaInN active layer sandwiched between high bandgap AlGaN layers. This structure gives high-performance blue LEDs, as discussed in Chapter 8.

7. LOW-DIMENSIONAL STRUCTURES

The radiative recombination lifetimes in III-V semiconductors can be improved by the formation of special low-dimensional structures such as quantum wells, quantum wires, and quantum dots. This is due to two factors: the forced increase in overlap of the wavefunctions for electrons and holes, and the changes in the shapes of the density of states distribution for the conduction and valence bands. The density of states at the bottom of the band increases in the following order: bulk (three dimensions of translational freedom), well (two dimensions), wire (one dimension), and dot (zero dimensions). This leads to higher recombination rates and a narrowing of the gain spectrum. These produce an increase in LED quantum efficiency and a decrease in threshold current density for lasers.

For the AlGaInP system, $Al_{0.35}Ga_{0.15}In_{0.5}P/GaInP$, quantum wells as thin as 9 Å have been grown by OMVPE. They give strong and narrow PL emission peaks (Jou *et al.*, 1993). Emission wavelengths as short as 545 nm were observed, the shortest ever reported for GaInP. Such structures allow the production of short-wavelength lasers and LEDs without the need for adding Al to the active region of the device, where recombination occurs. This avoids problems with carbon and oxygen contamination, as described previously, and produces highly efficient recombination.

Quantum well structures are also used to enhance the performance of LEDs fabricated in the nitrides. As reported in Chapter 8 herein, Nakamura *et al.*, reported the fabrication of high-performance yellow, green, and blue LEDs using AlGaN/GaInN/AlGaN quantum wells. Variation of the composition of the well layer produces the various colors.

V. Specific Materials Systems

1. AlGaAs

AlGaAs was the first material for which very high brightness LEDs were demonstrated, with efficiencies exceeding 1%. As seen in Fig. 1, the AlGaAs system is nearly lattice-matched to the GaAs substrates for all compositions. As the Al content increases, the bandgap becomes larger until the critical composition where the bandgap becomes indirect. With increasing composition, the wavelength of the photons emitted by the LED decreases and the eye response increases markedly. Thus, for visible LEDs, the composition must be pushed as close as possible to the crossover. This, in turn, requires extremely high material quality, that is, extremely long minority carrier lifetimes, as discussed in Part III. For high light extraction efficiency, the layers must be very thick. In fact, growing the layers thick enough to allow removal of the GaAs substrate increases the light extraction efficiency by a factor of two to three. The thickness of the structure grown is typically 125 μm for the highest performance devices. Together these requirements at present can only be supplied by material grown by LPE. The best LEDs, emitting at a wavelength of 650 nm, have external quantum efficiencies at room temperatue of 18% (Cook *et al.*, 1987). These devices and the techniques used for their fabrication are described in Chapter 3.

2. AlGaInP

The example used to illustrate many of the points discussed in the preceding sections has been AlGaInP. Thus, this section will be a brief summary of how all of the various factors converge to allow production of extremely high performance devices in this system. In addition, this material forms the main topic of Chapters 4 and 5.

The AlGaInP system was identified early as one of the most promising for high-performance LEDs (Stringfellow, 1978). The idea that this material would represent little difficulty because it would be a direct extension of the early work on GaAsP and AlGaAs was rapidly proven to be inaccurate. As discussed above, LPE, the technique used for the production of the high-performance red AlGaAs LEDs, was proven to be incapable of producing these materials. Similar thermodynamic problems prevented the use of the HVPE process, used for the commercial production of GaAsP LEDs, for the growth of AlGaInP. Thus, a new growth technique was required. Both OMVPE and MBE are capable of producing high-quality AlGaInP; however, OMVPE became the leading process for the commercial production of AlGaInP for high-brightness LEDs. The problems that had to be solved for the OMVPE growth of high-quality AlGaInP include growth lattice-matched to the GaAs substrate, with subsequent removal of the substrate to increase photon extraction efficiency; reduction of oxygen contamination; determination of methods for obtaining the high levels of p-type doping required; and minimization of CuPt-type ordering. Each of these problems is addressed in earlier sections, and therefore is not discussed further here.

The structure used for the high-performance LEDs consists of three layers. The top and bottom confining layers are AlInP lattice-matched to GaAs, heavily doped p- and n-type, respectively. Of course, the active layer is $(Al_xGa_{1-x})_{0.5}In_{0.5}P$, with x adjusted to determine the wavelength of the emitted light. The results are spectacular, with external quantum efficiencies of 20% at 630 nm (red-orange), 10% at 590 nm (amber), and 2% at 570 nm (green), as discussed in Chapter 2. These results are superior to the best AlGaAs results, in terms of the luminous efficiency, for red LEDs. They also are by far the best in the yellow, and are better than nitrogen-doped GaP in the green. Only recent reports of GaInN quantum well LEDs by Nakamura *et al.* exceed the performance of the AlGaInP LEDs in the green region of the spectrum, as discussed in Chapter 8.

3. AlGaInN

Although the nitrides of Al, Ga, and In are formally III-V semiconductors, they behave very differently from the conventional III-V semiconductors—

the phosphides, arsenides, and antimonides—and have been very late in developing. Most of the difficulties with these materials can be traced to the high volatility of nitrogen, discussed in Part IV, Section 1, combined with the requirement for high growth tempertures due to the large bond strengths. The combination of high growth temperatures and high nitrogen volatility leads to high concentrations of N vacancies in GaN. The nitrogen vacancies act as electron donor centers (Jenkins *et al.*, 1992); thus, the early GaN epitaxial layers contained background n-type carrier concentrations of greater than 10^{19} cm^{-3}. The thermodynamically motivated approaches to reducing the vacancy concentrations are the use of (1) lower growth temperatures and (2) higher atomic nitrogen concentrations in the nutrient vapor phase.

One factor influencing the choice of growth temperature for OMVPE is the temperature at which the N precursor, typically NH_3, pyrolyzes. A second factor is the minimum temperature required to provide sufficient surface mobility of atoms to allow the growth of highly perfect single crystalline layers at a reasonable growth rate, as discussed in the preceding Part IV, Section 4. Dryburgh (1988) predicts a lower temperature limit of 894°C for the epitaxial growth of single-crystalline GaN. However, we expect that somewhat higher temperatures will be required to produce high structural quality layers with excellent minority-carrier properties. This is borne out by the experimental evidence. For example, an apparent lower limit for the growth of high-quality GaN is approximately 1000°C. In addition, a clear improvement in the quality of GaInN layers is observed as the temperature is increased from 500 to 800°C (Matsuoka *et al.*, 1992). The minimum temperature required for the growth of this alloy was not calculated by Dryburgh; however, it is expected to be somewhat lower than for GaN, because of the lower average bond strength.

Efforts to produce high-quality epitaxial layers at the lowest possible temperature have included the use of a plasma to dissociate the thermally stable N_2, which gives an extremely high atomic nitrgen concentration, independent of the substrate temperature, and the use of alternate precursors that produce atomic nitrogen at temperatures much lower than 1000°C. Both have been discussed previously. The problem with the large n-type background concentration is compounded by the difficulty in finding suitabe p-type dopants and self-compensation mechanisms not involving N vacancies, such as H passivation.

Adding to these problems is the lack of a suitable lattice-matched substrate. Bulk growth of GaN is virtually impossible because of the extremely high nitrogen pressures required. Efforts to produce relatively large single substrate crystals by growth from a Ga-rich solvent have not met with a great deal of success, partly due to the low solubility of nitrogen in Ga without use of a bomb to produce extremely high nitrogen pressures.

For example, the mole fraction of nitrogen dissolved in Ga is 0.02 at a temperature of 1500°C and the pressure is 16 kbar (Elwell and Elwell, 1988)!

Most GaN layers are grown on sapphire substrates, even though the lattice parameter mismatch is approximately 16%. In addition, the thermal expansion coefficient of sapphire is significantly greater than that of GaN, as seen in Table II, which causes large stresses in the GaN epitaxial layer during cooling from the growth temperature. These problems lead to the extremely high dislocation densities ($>10^{10}\,\text{cm}^{-2}$) measured for GaN grown on sapphire substrates, as discussed above.

Despite these difficulties, tremendous progress has been made in the last few years, beginning with the success of Akasaki et al., as summarized in Chapter 7. They were able to nucleate the GaN on sapphire substrates by first growing an amorphous 50 Å layer of AlN before the OMVPE growth of the GaN layer at a temperature of 1020°C using the conventional precursors TMGa and NH_3. The authors were able to produce layers with background carrier densities of less than $10^{16}\,\text{cm}^{-3}$ in high-quality films. These layers are highly luminescent, despite the high dislocation densities. More recently, Nakamura et al. (1991) have demonstrated that GaN grown at 550°C can equally well be used as the buffer layer.

Akasaki et al. made an additional breakthrough when they demonstrated that with suitable postgrowth treatment they could produce p-type layers, using Mg doping with bis(cyclopentadienyl) magnesium (Cp_2Mg) (Amano et al., 1990), with hole concentrations as high as $3 \times 10^{18}\,\text{cm}^{-3}$. This led to the development of the first p-n junction, blue GaN LEDs. They emit at

TABLE II

Properties of High Bandgap Materials

Material	Lattice Parameters (Å)	Thermal Expansion Coefficients (K^{-1})	Bandgap Energy (eV)
Indium nitride	$a = 3.54$		1.95(D)
	$c = 5.70$		
Gallium nitride	$a = 3.189$	5.59×10^{-6}	3.45(D)
	$c = 5.185$	3.17×10^{-6}	
Aluminum nitride	$a = 3.112$	4.2×10^{-6}	6.28(D)
	$c = 4.982$	5.3×10^{-6}	
6H-silicon carbide	$a = 3.081$	4.2×10^{-6}	3.03(I)
	$c = 15.12$	4.68×10^{-6}	
Sapphire	$a = 4.758$	7.5×10^{-6}	
	$c = 12.99$	8.5×10^{-6}	

Values taken from Davis (1991) and Morkoc et al. (1994).
D, direct bandgap; I, indirect bandgap.

wavelenghs of 370 and 420 nm, somewhat below the bandgap energy.

In addition to the OMVPE approach to producing device quality GaN and related alloys, MBE has also been used with an ECR plasma source of nitrogen (Morkoc *et al.*, 1994; Molnar *et al.*, 1995). One advantage of this technique is tht atomic H is not present. Thus, no special annealing or electron irradiation treatments are necessary to activate the magnesium dopant. The GaN layers are typically grown on sapphire substrates at 800°C, following the low-temperature deposition of a nucleation layer, similar to the process developed for the OMVPE growth of GaN on sapphire. One disadvantage of the use of the microwave plasma is that the growth rate is limited to approximately 0.2 μm/h by the limited N flux from the standard plasma sources, as discussed above. It is in this way that p-n junction LEDs have been produced (Molnar *et al.*, 1995); however, to date, the performance of the LEDs fabricated in material grown by MBE has trailed that of the best devices made in material produced by OMVPE. This may be due to the limited nitrogen flux from the plasma, which limits the maximum temperature that can be used for the growth of single-phase GaN.

Unfortunately, the difficulties with high nitrogen pressures resulting in high n-type background concentrations are more of a problem for GaInN than for GaN. The weaker bonds in InN result in a lower dissociation temperature (MacChesney *et al.*, 1970). In other words, the nitrogen pressure at a given temperature is several orders of magnitude greater than for GaN, as seen in Fig. 4. Alleviating the problem somewhat is the expectation that the minimum temperature for growth of high-quality epitaxial layers of GaInN will be somewhat lower than for GaN, as discussed previously.

The same approaches used for GaN can be used for GaInN. Guo *et al.* (1994, 1995) and Wakahara and Yoshida (1989) have succeeded in growing InN on GaAs substrates at the low temperatures of 400 to 600°C by using OMVPE with a microwave-excited nitrogen plasma. However, the layers need a postgrowth annealing step to improve the material quality. Matsuoka *et al.* (1992) were able to grow higher quality GaInN alloys at higher temperatures by low-pressure OMVPE using ammonia, trimethylindium (TMIn) and TEGa on (0001) sapphire. The distribution coefficient of In was found to be much higher at low temperatures: The value of k_{In} is approximately unity at 500°C and 0.1 at 800°C. This is thought to be due to evaporation of In from the surface at high temperatures. The same phenomenon is observed for conventional III-V alloys containing In during growth in this temperature range by MBE (Turco and Massies, 1987; Allovon *et al.*, 1989) and CBE (H. H. Ryu, G. B. Stringfellow, and L. P. Sadwick, unpublished, 1996). Very high V-III ratios are required for growth of GaInN at the higher temperatures; however, the result was spectacular. The PL intensity was observed to increase by four orders of magnitude as the growth temperature was increased from 500 to 800°C. Of course, an increase in growth temperature also reduces problems with solid immiscibility. As discussed in Part IV, Section 2, the InN–GaN system has a large region of solid immiscibility, even at 1000°C (Ho and Stringfellow, 1997).

These breakthroughs have led to increased research on the III-V nitrides. Today, OMVPE has been used for the production of blue-emitting diodes with extremely high external quantum efficiencies of 5.4% (Nakamura, 1995), more than two orders of magnitude better than the best SiC blue LEDs, as discussed in Chapter 8. These high-performance devices are double heterostructures (DH) where GaInN is the active layer with AlGaN cladding layers. The nucleating layer is low-temperature GaN (510°C, 300 Å) and the p-type dopant is zinc, with a postgrowth anneal. These LEDs show little degradation after 10,000 hours at 20 mA. Increasing the In content of the GaInN active layer has also resulted in high efficiency (2.1%) green LEDs emitting at 525 nm. The quantum efficiencies of such LEDs have been improved even further by using single quantum well (SQW) p-AlGaN–GaInN–n-GaN structures. This has led to blue (450 nm) and green (520 nm) LEDs with the extraordinarily high external quantum efficiencies of 9.1 and 6.3%, respectively, as discussed in Chapter 8.

Gallium nitride with even higher structural quality can be grown on alternative substrates such as SiC. The lattice match between GaN and SiC is much better than for sapphire, as seen in Table II. The thermal expansion coefficient of GaN is also somewhat closer to that of SiC. The better lattice match apparently results in the order of magnitude reduction in dislocation density observed (Weeks *et al.*, 1995; Ponce *et al.*, 1995). Another important advantage is that the SiC substrates can be doped to make them conducting. This vastly simplifies the device structure. A major drawback to SiC substrates is that they are currently much more expensive than is sapphire. GaN–SiC LEDs are commercially available.

Another area offering improvement for the production of III-V nitrides is the use of alternate precursors. As discussed previously, a problem with the use of NH_3 is the high pyrolysis temperature required to break the relatively strong N–H bonds. This poses particular problems for the growth of the GaInN alloys. In addition, the presence of the high concentrations of atomic hydrogen in the films leads to problems with the activation of the p-type dopant, as discussed previously. Other more labile precursors that have been explored include hydrazine (H_2N_2) and related molecules. Hydrazine is quite unstable, pyrolyzing at 400°C. It has a convenient vapor pressure of 10 Torr at 18°C (Gaskill *et al.*, 1986; Mizuta *et al.*, 1986). However, because it is both explosive and highly toxic, it is extremely dangerous.

Dimethylhydrazine ((MeNH)$_2$) has been used with TMGa to deposit cubic GaN on GaAs substrates at low temperatures, from 550 to 650°C with V-III ratios of 160, significantly lower than those required for NH$_3$ at these temperatures (Miyoshi *et al.*, 1992). However, this compound is also extremely hazardous. A potentially less hazardous precursor, phenylhydrazine, has also been explored for the grwth of GaN with TMGa (Zhang *et al.*, 1994). Layers were successfully grown at the relatively low temperatures of 750 to 800°C with V-III ratios of only four. The problem with this precursor is the relatively low vapor pressure of (0.03) Torr at 25°C.

Jones *et al.* (1994a) have demonstrated the growth of AlN using nitrogen precursors similar to the singly alkyl–substituted AsH$_3$ and PH$_3$ precursors so successfully used for the OMVPE growth of conventional III-V semiconductors, as discussed in Part IV, Section 4. The precursors tertiarybutylamine (C$_4$H$_9$NH$_2$) and isopropylamine (C$_3$H$_7$NH$_2$) were combined with TMAl to form the AlN. The reaction occurs via *in situ* formation of an adduct between the Al and N precursors. Unfortunately, very high carbon contamination levels were observed. The carbon was presumed to come from the methyl radicals in TMAl; however, replacement of the TMAl by tertiarybutylaluminum led to no reduction in the level of carbon contamination (Jones *et al.*, 1994b). The use of similar precursors for the growth of GaN has been unsuccessful, producing only Ga droplets (Jones *et al.*, 1995). The growth of InN and GaInN with these precursors is expected to be even more problematic.

An Al precursor that is promising for the production of epitaxial AlGaN layers with lower carbon contamination levels is trimethylaminealane (TMAA), a recently developed Al precursor that has been successfully used for the growth of conventional III-V semiconductors. The use of TMAA with ammonia has been demonstrated to produce less carbon contamination than does the conventional TMAl (Khan *et al.*, 1994). It also has a higher vapor pressure than does the other frequently used Al precursor, triethylaluminum (TEAl). Trimethylaminealane combined with TEGa and NH$_3$ was used to produce layers of AlGaN over the entire composition range at 1000°C. The AlN layers had carbon contamination levels as low as 10^{17} cm^{-3}.

Precursors containing both the group III and the group V elements have also been explored (Jones *et al.*, 1995). However, as for the conventional III-V semiconductors, the resulting loss of control of stoichiometry is a problem. In addition, the volatility of these large and complex precursors is typically too low to be practical for OMVPE growth.

The potential for the use of alternate precursors for the growth of AlN, GaN, InN and their alloys is obvious. However, at this time the use of these alternate precursors has not resulted in significant improvements in the OMVPE process for the growth of these materials.

In summary, the nitrides are by far the most promising materials for blue light-emitting devices due to their already high efficiencies for p-n junction devices and their reliability. This is an inherent property of the nitrides due to the strong, covalent bonding. Defect production requires large energies. This is apparently the reason for the long device lifetimes even with the presence of high dislocation densities. The competing II-VI compounds appear to have fundamental problems. They are more ionic and the energy required to create native defects is considerably smaller. Thus, the problem of improving the device reliability may be a fight to overcome a fundamental problem, which does not bode well for the probability of ultimate success (Morkoc et al., 1994).

VI. Conclusions

The progress in the development of materials for high-brightness light emitting diodes during the last decade has been dramatic. The already well-developed AlGaAs was the first to demonstrate extremely high efficiency (18%) red light emission in thick layers grown by liquid-phase epitaxy with no GaAs substrate to absorb the emitted photons. The much more difficult materials system AlGaInP has progressed remarkably, overcoming problems with oxygen contamination, p-type doping, ordering, and photon absorption in the substrate to produce materials with external quantum efficiencies equaling the best AlGaAs values, but at shorter wavelengths, giving considerably higher brightness in the red region of the spectrum. The highest performance orange and yellow LEDs have also been produced in this material.

Perhaps even more spectacular has been the rapid pace of advancements in the III-V nitrides. The first p-n junction LEDs were produced as recently as 1990. Yet, the state of the art has now progressed to the point that blue LEDs have external efficiencies of 9.1% and the performance of green emitters fabricated using GaInN/AlGaN heterostructures is 6.3%, far exceeding even the performance of the AlGaInP materials.

These remarkable advances have come as a result of the patient work of researchers around the world on difficult materials issues such as control of background impurities and stoichiometric defects, the growth of layers on lattice mismatched substrates, the control of the formation of ordered structures, the controlled introduction of electrically active dopant impurities, and the use of complex structures such as heterostructures and quantum wells. The rate of advance even to the present day has been remarkable. However, based on the high-performance levels of today's LEDs, we suspect that the rate of progress will slow in the coming years.

ACKNOWLEDGMENTS

The author thanks the Department of Energy for financial support of part of the work described here.

REFERENCES

Ahrenkiel, R. K. (1993). *In* "Semiconductors and Semimetals" (R. K. Ahrenkiel and M. S. Lundström, eds.), Vol. 39, pp. 39–145. Academic Press, San Diego, CA.
Allovon, M., Primot, J., Gao, Y., and Quillec, M. (1989). *J. Electron. Mater.* **18**, 505.
Amano, H., Kitoh, M., Hiramatsu, K., and Akasaki, I. (1989). *Jpn. J. Appl. Phys.* **28**, L2112.
Amano, H., Kitoh, M., Hiramatsu, K., and Akasaki, I. (1990). *J. Electrochem. Soc.* **137**, 1639.
Archer, R. J. (1972). *J. Electron. Mater.* **1**, 1.
Arthur, J. R., and LePore, J. J. (1969). *J. Vac. Sci. Technol.* **6**, 545.
Asahi, H., Kawamura, Y., and Nagai, H. (1982). *J. Appl. Phys.* **53**, 4928.
Asahi, H., Kawamura, Y., and Nagai, H. (1983). *J. Appl. Phys.* **54**, 6958.
Benchimol, J. L., Quillec, M., and Slempkes, S. (1983). *J. Cryst. Growth* **64**, 96.
Bergh, A. A., and Dean, P. J. (1976). "Light-emitting Diodes." Clarendon Press, Oxford.
Beuchet, G. (1985). *In* "Semiconductors and Semimetals" (W. T. Tsang, ed.), Vol. 22A, pp. 261–298. Academic Press, Orlando, FL.
Bloem, J. (1973). *J. Cryst. Growth* **18**, 70.
Cao, D. S., Kimbal, A. W., Chen, G. S., Fry, K. L., and Stringfellow, G. B. (1989). *J. Appl. Phys.* **66**, 5384.
Casey, H. C., Jr., and Panish, M. B. (1978). "Heterostructure Lasers: Part B, Materials and Operating Characteristics," p. 44. Academic Press, New York.
Chen, C. H., Larsen, C. A., Stringfellow, G. B., Brown, D. A., and Robertson, A. J. (1986). *J. Cryst. Growth* **77**, 11.
Chen, C. H., Larsen, C. A., and Stringfellow, G. B. (1987). *Appl. Phys. Lett.* **50**, 218.
Cherng, M. J., Cherng, Y. T., Jen, H. R., Harper, P., Cohen, R. M., and Stringfellow, G. B. (1986a). *J. Electron. Mater.* **15**, 79.
Cherng, M. J., Jen, H. R., Larsen, C. A., Stringfellow, G. B., Lundt, H., and Taylor, P. C. (1986b). *J. Cryst. Growth* **77**, 408.
Chin, A., Martin, P., Das, U., Mazurowski, J., and Ballingall, J. (1992). *J. Appl. Phys. Lett.* **61**, 2099.
Cho, A. Y. (1971). *J. Appl. Phys.* **42**, 2074.
Chun, Y. S., Murata, H., Ho, I. H., Hsu, T. C., and Stringfellow, G. B. (1977). *J. Cryst. Growth* **170**, 263.
Cook, L. W., Camras, M. D., Rudaz, S. L., and Steranka, F. M. (1987). *Conf. Ser — Inst. Phys.* **91**, 777.
Craford, M. G. (1977). *IEEE Trans Electron Devices* **ED-24**, 935.
Davis, R. F. (1991). *IEEE Proc.* **79**, 702.
Dodson, B. W., and Tsao, J. Y. (1989). *Annu. Rev. Mater. Sci.* **19**, 419.
Dryburgh, P. (1988). *J. Cryst. Growth* **87**, 397.
Elwell, D., and Elwell, M. M. (1988). *Prog. Cryst. Growth and Charact.* **17**, 53.
Ettenberg, M. (1974). *J. Appl. Phys.* **45**, 901.
Faschinger, W., Ferreira, S., and Sitter, H. (1995). *J. Cryst. Growth* **151**, 267.
Follstaedt, D. M., Schneider, R. P., and Jones, E. D. (1995). *J. Appl. Phys.* **77**, 3077.

Gaskill, D. K., Bottka, N., and Lin, M. C. (1986). *J. Cryst. Growth* **77**, 418.
Gorbylev, V. A., Polyakov, A. Y., Pearton, S. J., Smirnov, N. B., Wilson, R. G., Milnes, A. G., Cnekaline, A. A., Govorkov, A. V., Leiferov, B. M., Borodina, O. M., and Balmashnov, A. A. (1994). *J. Appl. Phys.* **76**, 7390.
Guo, Q., Yamamura, T., and Yoshida, A. (1994). *J. Appl. Phys.* **75**, 4927.
Guo, Q., Ogawa, H., Yamano, H., and Yoshida, A. (1995). *Appl. Phys. Lett.* **66**, 715.
Haacke, G., Watkins, S. P., and Burkhard, H. (1991). *J. Cryst. Growth* **107**, 342.
Haase, M., Qui, J., DePuydt, J., and Cheng, H. (1991). *Appl. Phys. Lett.* **59**, 1272.
Heckingbottom, R. (1985). *J. Vac. Sci. Technol.*, *B*[2] **3**, 572.
Hino, I., and Suzuki, T. (1984). *J. Cryst. Growth* **68**, 483.
Hirayama, H., and Asahi, H. (1994). *In* "Handbook of Crystal Growth" (D. T. J. Hurle, ed.), Vol. 3, pp. 185–221. Elsevier, Amsterdam.
Ho, I. H. and Stringfellow, G. B. (1997a). *Appl. Phys. Lett.* **69**, 2701.
Ho, I. H., and Stringfellow, G. B. (1997b). *J. Cryst. Growth* (to be published).
Ikeda, M., Morita, E., Toda, A., Yammoto, T., and Kaneko, K. (1988). *Electron. Lett.* **24**, 1094.
Ilegems, M. (1972). *J. Cryst. Growth* **13/14**, 360.
Imori, T., Ninomiya, T., Ushikubo, K., Kondoh, K., and Nakamura, K. (1991). *Appl. Phys. Lett.* **59**, 2862.
Ishibashi, A., Mannoh, M., and Ohnaka, K. (1994). *J. Cryst. Growth* **145**, 414.
Jenkins, D. W., Dow, J. D., and Tsai, M. H. (1992). *J. Appl. Phys.* **72**, 4130.
Jesser, W. A., and Fox, B. A. (1990). *J. Electron. Mater.* **19**, 1289.
Jones, A. C. (1993). *J. Cryst. Growth* **129**, 728.
Jones, A. C. (1994). *J. Cryst. Growth* **145**, 505.
Jones, A. C., Auld, J., Rushworth, S. A., Williams, E. W., Haycock, P. W., Tang, C. C., and Critchlow, G. W. (1994a). *Adv. Mater.* **6**, 3.
Jones, A. C., Auld, J., Rushworth, S. A., Williams, E. W., Haycock, P. QW., Tang, C. C., and Critchlow, G. W. (1994b). *J. Cryst. Growthy* **135**, 285.
Jones, A. C., Whitehouse, C. R., and Roberts, J. S. (1995). *Adv. Mater.* **7**, 65.
Jou, M. J., Cherng, Y. T., Jen, H. R., and Stringfellow, G. B. (1988a). *Appl. Phys. Lett.* **52**, 549.
Jou, M. J., Cherng, Y. T., and Stringfellow, G. B. (1988b). *J. Appl. Phys.* **64**, 1472.
Jou, M. J., Lin, J. F., Chang, C. M., Lin, C. H., Wu, M. C., and Lee, B. J. (1993). *Jpn. J. Appl. Phys.* **32**, 4460.
Kadoiwa, K., Kato, M., Motoda, T., Ishida, T., Fujii, N., Hayafuji, N., Tsugami, M., Sonoda, T., Takamiya, S., and Mitsui, S. (1994). *J. Cryst. Growth* **145**, 147.
Karpinski, J., Jun, J., and Porowski, S. (1984). *J. Cryst. Growth* **66**, 1.
Khan, M. A., Olson, D. T., and Kuznia, J. N. (1994). *Appl. Phys. Lett.* **65**, 64.
Kim, C. W., Stringfellow, G. B., and Sadwick, L. P. (1996). *J. Cryst. Growth* **164**, 104.
Kish, F. A., Sternka, F. M., DeFevere, D. C., Vanderwater, D. A., Park, K. G., Kuo, C. P., Osenkowski, T. D., Peanasky, M. J., Yu, J. G., Fletcher, R. M., Steigerwald, D. A., and Craford, M. G. (1994). *Appl. Phys. Lett.* **64**, 2839.
Kisker, D. W., Miller, J. N., and Stringfellow, G. B. (1982). *Appl. Phys. Lett.* **40**, 614.
Komeno, J. (1994). *J. Cryst. Growth* **145**, 468.
Kondo, M., Okada, N., Domen, K., Sugiura, K., Anayama, C., and Tanahashi, T. J. (1994). *Electron. Mater.* **23**, 355.
Kressel, H. (1980). *In* "Semiconductor Devices for Optical Communications," (H. Kressel, ed.), p. 13. Springer-Verlag, Berlin.
Kressel, H., and Butler, J. K. (1977). "Semiconductor Lasers and Heterojunction LEDs." Academic Press, San Diego, CA.
Kuech, T. F., Venhoff, E., and Meyerson, B. S. (1984). *J. Cryst. Growth* **68**, 48.

Kuech, T. F., Wolford, D. J., Veuhoff, E., Deline, V., Mooney, P. M., Potemski, R., and Bradley, J. (1987). *J. Appl. Phys.* **62**, 632.
Kurtz, S. R., Arent, D. J., Bertness, K. A., and Olson, J. M. (1994). *Mater. Res. Soc. Symp. Proc.* **340**, 117.
Lester, S. D., Ponce, F. A., Craford, M. G., and Steigerwald, D. A. (1995). *Appl. Phys. Lett.* **66**, 1249.
Lum, R. K., Klingert, J. K., and Lamont, M. G. (1987). *Appl. Phys. Lett.* **50**, 284.
MacChesney, J. B., Bridenbaugh, P. M., and O'Connor, P. B. (1970). *Mater. Res. Bull.* **5**, 783.
Maruska, H. P., and Tietjen, J. (1969). *J. Appl. Phys. Lett.* **15**, 327.
Matsuoka, T., Yoshimoto, N., Sasaki, T., and Katsui, A. (1992). *J. Electron. Mater.* **21**, 157.
Miyoshi, S., Onabe, K., Ohkouchi, N., Yaguchi, H., Ito, R., Fukatsu, S., and Shirak, Y. (1992). *J. Cryst. Growth* **124**, 439.
Mizuta, M., Fujieda, S., Matsumoto, Y., and Kawamura, T. (1986). *Jpn. J. Appl. Phys.* **25**, L945.
Molnar, R. J., Singh, R., and Moustakas, T. D. (1995). *Appl. Phys. Lett.* **66**, 268.
Morkoc, H., Strite, S., Gao, G. B., Lin, M. E., Sverdlov, B., and Burns, M. (1994). *J. Appl. Phys.* **76**, 1363.
Mullenborn, M., Matney, K., Gorsky, M. S., Haegel, N. M., and Vernon, S. M. (1994). *J. Appl. Phys.* **75**, 2418.
Murata, H., Hsu, T. C., Ho, I. H., Su, L. C., and Stringfellow, G. B. (1996a). *Appl. Phys. Lett.* **68**, 1796.
Murata, H., Ho, I. H., Su, L. C., Hosokawa, Y., and Stringfellow, G. B. (1996b). *J. Appl. Phys.* **79**, 6895.
Murata, H., Ho, I. H., Hosokawa, Y., and Stringfellow, G. B. (1996c). *Appl. Phys. Lett.* **68**, 2237.
Murata, H., Ho, I. H., and Stringfellow, G. B. (1997). *J. Cryst. Growth* **170**, 219.
Nagatomo, T., Kuboyama, T., Minamino, H., and Omoto, O. (1989). *Jpn. J. Appl. Phys.* **28**, L1334.
Nakamura, S. (1995). *J. Vac. Sci. Technol.* **13**, 705.
Nakamura, S., Mukai, T., and Senoh, M. (1991). *Jpn. J. Appl. Phys.* **30**, L1998.
Nakamura, S., Iwasa, N., Senoh, M., and Mukai, T. (1992). *Jpn. J. Appl. Phys.* **31**, 1258.
Nakamura, S., Mukai, T., and Senoh, M. (1994). *Appl. Phys. Lett.* **64**, 1687.
Nakamura, S., Senoh, M., Iwasa, N., Nagahama, S., Yamada, T., and Mukai, T. (1995). *Jpn. J. Appl. Phys.* **34**, L797.
Olson, J. M., Ahrenkiel, R. K., Dunlavy, D. J., Keyes, B., and Kibbler, A. E. (1989). *Appl. Phys. Lett.* **55**, 1208.
Osamura, K., Nakajima, K., Murakami, Y., Shingu, P. H., and Otsuki, A. (1972). *Solid State Commun.* **11**, 617.
Paisley, M. J., Sitar, Z., Posthill, J. B., and Davis, R. F. (1989). *J. Vac. Sci. Technol.* A[2] **7**, 701.
Panish, M. B. (1980). *J. Electrochem. Soc.* **127**, 2729.
Panish, M. B., and Ilegems, M. (1972). *Prog. Solid State Chem.* **7**, 39.
Pankove, J. I. (1992). In "Non-Stoichiometry in Semiconductors" (K. J. Bachmann, H. L. Hwang, and C. Schwab, eds.). Elsevier, Amsterdam.
Pessetto, J. R., and Stringfellow, G. B. (1983). *J. Cryst. Growth* **62**, 1.
Philips, J. C. (1973). "Bands and Bonds in Semiconductors." Academic Press, New York.
Ponce, F. A., Krusor, B. S., Major, J. S., Plano, W. E., and Welch, D. F. (1995). *Appl. Phys. Lett.* **67**, 410.
Protzmann, H., Marschner, T., Zsebok, O., Stolz, W., Gobel, E. O., Dorn, R., and Lorberth, J. (1991). *J. Cryst. Growth* **115**, 248.
Qian, W., Skowronski, M., DeGraef, M., Doverspike, K., Rowland, L. B., and Gaskill, D. K. (1995). *Appl. Phys. Lett.* **66**, 1252.

Ryu, H. H., Stringfellow, G. B., Sadwick, L. P., Gedridge, R. W., and Jones, A. C. (1997). *J. Cryst. Growth* **172**, 1.
Schneider, R. P., Jones, E. D., and Follstaedt, D. M. (1994). *Appl. Phys. Lett.* **65**, 587.
Seki, H., and Koukitu, A. (1986). *J. Cryst. Growth* **78**, 342.
Shealy, J., Kreismanis, V. G., Wagner, D. K., and Woodall, J. M. (1983). *Appl. Phys. Lett.* **42**, 83.
Shin, J., Verma, A., Stringfellow, G. B., and Gedridge, R. W. (1994). *J. Cryst. Growth* **143**, 15.
Singh, R., Doppalapudi, D., Moustakas, T. D., and Romano, L. T. (1997). *Appl. Phys. Lett.* **70**, 1089.
Smith, L. M., Rushworth, S. A., Jones, A. C., Roberts, J. S., Chew, A., and Sykes, D. E. (1993). *J. Cryst. Growth* **134**, 140.
Soloman, R., and DeFevere, D. (1972). *Appl. Phys. Lett.* **21**, 257.
Stringfellow, G. B. (1974). *J. Cryst. Growth* **127**, 21.
Stringfellow, G. B. (1978). *Annu. Rev. Materi. Sci.* **8**, 73.
Stringfellow, G. B. (1979). In "Crystal Growth: A Tutorial Approach" (W. Bardsley, D. T. J. Hurle, and J. B. Mullin, eds.), pp. 217–239. North-Holland Pub., Amsterdam.
Stringfellow, G. B. (1980). In "Crystal Growth" (B. Pamplin, ed.), 2nd ed., pp. 181–220. Pergamon, New York.
Stringfellow, G. B. (1981). *J. Cryst. Growth* **55**, 42.
Stringfellow, G. B. (1982). *Rep. Prog. Phys.* **45**, 469.
Stringfellow, G. B. (1983). *J. Cryst. Growth* **65**, 454.
Stringfellow, G. B. (1984). *J. Cryst. Growth* **68**, 111.
Stringfellow, G. B. (1985). In "Semiconductors and Semimetals" (W. T. Tsang, ed.), Vol. 22, pp. 209–259. Academic Press, New York.
Stringfellow, G. B. (1986). *J. Cryst. Growth* **75**, 91.
Stringfellow, G. B. (1989a). "Organometallic Vapor Phase Epitaxy: Theory and Practice." Academic Press, San Diego, CA.
Stringfellow, G. B. (1989b). *J. Cryst. Growth* **98**, 108.
Stringfellow, G. B. (1991). *J. Cryst. Growth* **115**, 1.
Stringfellow, G. B. (1992). "Phase Equilibria Diagrams, Semiconductors and Chalcogenides." American Ceramic Society, Westerville. OH.
Stringfellow, G. B., and Greene, P. E. (1969). *J. Appl. Phys.* **40**, 502.
Stringfellow, G. B., and Greene, P. E. (1971). *J. Electochem. Soc.* **119**, 1780.
Stringfellow, G. B., and Hom, G. (1979). *Appl. Phys. Lett.* **34**, 794.
Stringfellow, G. B., Lindquist, P. F., Cass, T. R., and Burmeister, R. A. (1974). *J. Electron Mater.* **3**, 497.
Su, L. C., Pu, S. T., Stringfellow, G. B., Christen, J., Selber, H., and Bimberg, D. (1993). *Appl. Phys. Lett.* **62**, 3496.
Su, L. C., Ho, I. H., Stringfellow, G. B., Leng, Y., and Williams, C. C. (1994a). *Mater. Res. Soc. Symp. Proc.* **340**, 123.
Su, L. C., Ho, I. H., and Stringfellow, G. B. (1994b). *J. Appl. Phys.* **76**, 3520.
Su, L. C., Ho, I. H., and Stringfellow, G. B. (1994c). *Appl. Phys. Lett.* **65**, 749.
Su, L. C., Ho, I. H., Kobayashi, N., and Stringfellow, G. B. (1994d). *J. Cryst. Growth* **145**, 140.
Tamura, M., Hashimoto, A., and Nakatsugawa, Y. (1992). *J. Appl. Phys.* **72**, 3398.
Terao, H., and Sunagawa, H. (1984). *J. Cryst. Growth* **68**, 157.
Thompson, G. H. B. (1980). "Physics of Semiconductor Laser Devices," pp. 25–29. Wiley, New York.
Tom, G. M. (1986). *4th Microelectron. Tech. Symp.*, Semicon West.
Tsai, M. J., Tashima, M. M., and Moon, R. L. (1984). *J. Electron. Mater.* **13**, 437.
Tsang, W. T. (1990). *J. Cryst. Growth* **105**, 1.

Tsao, J. Y. (1993). "Materials Fundamentals of Molecular Beam Epitaxy." Academic Press, San Diego, CA.
Tu, K. N., Mayer, J. W., and Feldman, L. C. (1992). "Electronic Thin Film Science," Chapter 7. Macmillan, New York.
Turco, F., and Massies, J. (1987). *Appl. Phys. Lett.* **51**, 1989.
Valster, A., Liedenbaum, C. T. H. F., Finke, N. M., Severens, A. L. G., Boermans, M. J. B., vanden Houdt, D. W. W., and Bulle-Liewman, C. W. T. (1991). *J. Cryst. Growth* **107**, 403.
Waho, J., Ogawa, S., and Maruyama, S. (1977). *Jpn. J. Appl. Phys.* **16**, 1875.
Wakahara, A., and Yoshida, A. (1989). *Appl. Phys. Lett.* **54**, 709.
Weeks, T. W., Bremser, M. D., Ailey, M. D., Carlson, E., Perry, W. G., and Davis, R. F. (1995). *Appl. Phys. Lett.* **67**, 401.
Wei, S. H., and Zunger, A. (1990). *Appl. Phys. Lett.* **56**, 662.
Wolfe, C. M., Holonyak, N., and Stillman, G. (1989). "Physical Properties of Semiconductors," p. 88. Prentice Hall, Englewood Cliffs, NJ.
Wright, A. F., and Nelson, J. S. (1995). *Appl. Phys. Lett.* **66**, 3051.
Yamasaki, S., Asami, S., Shibata, N., Koike, M., Manabe, K., Tanaka, T., Amano, H., and Akasaki, I. (1995). *Appl. Phys. Lett.* **66**, 1112.
Yuan, J. S., Hsu, C. C., Cohen, R. M., and Stringfellow, G. B. (1985). *Appl. Phys. Lett.* **57**, 1380.
Zhang, G., Tong, Y., Xu, Z., Dang, X., and Liu, H. (1994). "Conference Digest, International Conference on Metal Organic Vapor Phase Epitaxy VII," p. 310.
Zhang, S. B., Froyen, S., and Zunger, A. (1995). *Appl. Phys. Lett.* **67**, 3141.
Zunger, A. (1994). *In* "Handbook of Crystal Growth" (D. T. J. Hurle, ed.), pp. 999–1047. Elsevier, Amsterdam.
Zunger, A., and Mahajan, S. (1993). *In* "Handbook of Semiconductors" (S. Mahajan, ed.), Vol. 3. Elsevier, Amsterdam.

CHAPTER 2

Overview of Device Issues in High-Brightness Light-Emitting Diodes

M. George Craford

OPTOELECTRONICS DIVISION
HEWLETT-PACKARD COMPANY
SAN JOSE, CALIFORNIA

    I. INTRODUCTION 47
   II. LIGHT-EMITTING DIODE DEVICE ISSUES 50
       1. *Overview* 50
       2. *Luminous Performance* 53
       3. *Internal Quantum Efficiency* 54
       4. *Light Extraction* 56
  III. TECHNOLOGY STATUS AND FUTURE POTENTIAL 58
       1. *Performance* 58
       2. *Comparison of Light-Emitting Diodes to Other Light Sources* 61
       References 63

I. Introduction

Red GaAsP light-emitting diodes (LEDs) were first introduced commercially by General Electric in 1962 following the work of Holonyak and Bevacqua (1962). Although both the volumes and performance levels of these devices were relatively low, the technology used led directly to the high-volume commercial introduction of LEDs by Monsanto and Hewlett-Packard in the late 1960s. The evolution of LED performance and device types is illustrated in Fig. 1. The vapor-phase epitaxial (VPE) grown GaAsP devices were soon joined by gallium phosphide (GaP) LEDs grown by liquid-phase epitaxy (LPE). A major increase in performance has occurred roughly every decade. Isoelectronic doping of GaP and GaAsP with nitrogen in the early 1970s resulted in increased performance and made green, orange, and yellow devices available in addition to red (Logan *et al.*, 1968; Groves *et al.*, 1971). High-performance AlGaAs devices also grown by LPE and introduced in the early 1980s provided another large performance

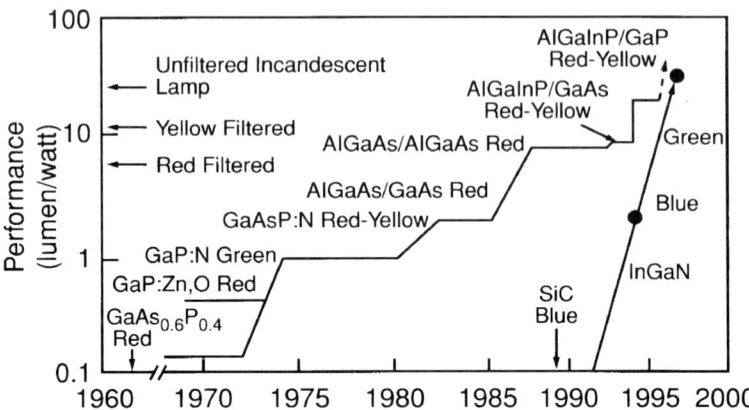

FIG. 1. Evolution of visible light-emitting diode (LED) performance with time. There has been about a tenfold improvement per decade in performance since high-volume commercial introduction.

increase for red devices (Alferov et al., 1975; Nishizawa and Suto, 1977). Devices of this type are reviewed in Chapter 3.

In the early 1990s AlInGaP devices grown by organo metallic vapor-phase epitaxy (OMVPE), were introduced that yielded a substantial performance improvement in the red-orange and amber spectral regions and potentially in the green (Kuo et al., 1990, Sugawara et al., 1991). Devices of this type are discussed in Chapters 4 and 5. A further twofold improvement has been obtained using a refined version of the AlInGaP devices that utilizes wafer bonding to attach a GaP substrate instead of the original lattice-matched gallium arsenide (GaAs) substrate on which the epitaxial layer was grown (Kish et al., 1994). The "transparent" GaP substrate enables more light to escape from the semiconductor chip. This issue of maximizing the "extraction efficiency" can be as important to final device performance as optimizing material growth, and internal quantum efficiency. These *transparent substrate* (TS) AlInGaP (TS-AlInGaP) devices are the highest performance LEDs yet developed.

High-performance blue-emitting LEDs have not been available until even more recently. In the late 1980s silicon carbide (SiC) devices were introduced that were useful for some applications but were much lower in performance than the high-performance devices of other colors. In late 1993, GaInN devices were introduced that gave much improved performance in the blue region (Nakamura et al., 1995). Green-emitting nitride devices also have been introduced, and therefore high-performance LEDs are now available that span the entire visible spectral region (Nakamura *et al.*, 1995). The history and technology of nitride devices are discussed in detail in Chapters 7, and 8. The persistent efforts by Akasaki and his group (Akasaki and Hashimoto, 1967; Amano *et al.*, 1986, 1989) over more than two decades set the foundation that enabled the commercial introduction of high-brightness devices by Nakamura and his team at Nichia Chemical.

Light-emitting diodes now compete effectively in performance with other types of light sources, particularly small incandescents, resulting in the emergence of a wide variety of new applications, as discussed in Chapter 6. This trend to higher performance and new applications can be expected to continue throughout the next decade and beyond.

The steadily improving performance of LEDs since the 1960s coupled with continuously decreasing costs has resulted in a large and growing commercial market. Figure 2 shows the market segmentation for compound semiconductor devices of all types. Visible LEDs are the largest compound semiconductor device type followed closely by infrared (IR) LEDs. Lasers are also a significant portion of the market. Thus, light-emitting devices completely dominate the device market. In 1994, the visible LED portion of

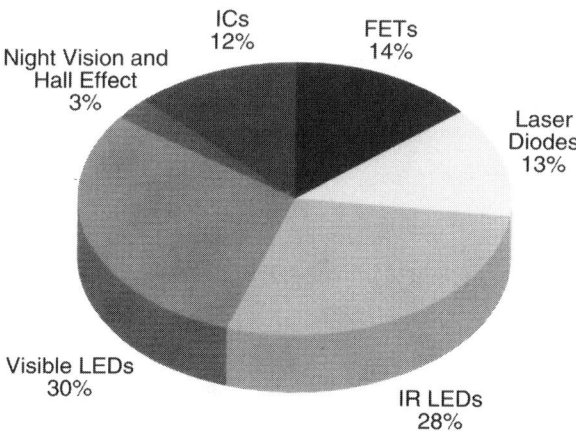

FIG. 2. The 1994 Compound Semiconductor Device Market ($4.4 billion total) segmented by device type. The market for visible light-emitting diodes (LEDs) was about $1.3 billion (after market reports published in 1995 by Strategies Unlimited). IC, integrated circuits; FETs, field-effect transmissions; IR, infrared.

the market was about $1.3 billion which corresponded to an estimated 30 billion visible LED chips being produced annually. Almost all these continue to be GaP and GaAsP chips. There are now probably about four billion AlGaAs chips produced, all using LPE. At this time there are less than one billion OMVPE grown AlInGaP and AlInGaN chips combined. The new high-performance devices remain in relatively low volume both because they are new and because the price is substantially higher than for the older technologies. Customers will not switch to a new technology unless that application requires the additional performance or the price is lower. Applications that require the higher performance are rapidly emerging as discussed in Chapter 6 and the device types discussed in this book can be expected to dominate the high-performance part of the LED market over the next 10 to 20 years.

It is interesting to note that in Fig. 1 the light output performance has been improving by nearly an order of magnitude every 10 years such that devices with a performance of more than 10 lm/W are now available compared with performance in the range of approximately 0.1 lm/W 25 years ago. A key question is whether another tenfold improvement can be realized. Performance at this level would establish LEDs as the most efficient light source available and could conceivably lead to penetration of the commercial lighting industry in the decades ahead. These issues will be briefly discussed in Part III, Section 2 of this chapter.

The progress in LED performance has been driven primarily by improvements in materials growth technology, summarized in Chapter 1, which have yielded higher purity and lower defect density materials with resulting higher quantum efficiencies. However, a variety of other issues such as cost, reliability, and extraction efficiency are also key drivers that determine commercial success and in some cases largely determine which growth technology is used. These issues are discussed in the following sections.

II. Light-Emitting Diode Device Issues

1. OVERVIEW

There are a number of recent reviews that can be studied for a more thorough discussion of LED device issues (Craford and Steranka, 1994; Kish, 1995; Haitz et al., 1995). Some of the key issues are discussed very briefly here in an attempt to make this volume reasonably self-contained for the convenience of general readers. The three main issues that must be addressed to make an LED technologically successful commercially are

Luminous performance
Cost-effectiveness
Reliability

The key components of the luminous performance are the internal quantum efficiency and the extraction efficiency. The definition of the luminous performance efficiency and approaches used to optimize it are discussed in Part II, Sections 2, 3, and 4.

Cost-effectiveness has been and continues to be a critical issue in determining whether LEDs can compete with other technologies for new markets. Typical high-volume LED chip costs are one cent or less, with an entire packaged LED lamp of the type shown in Fig. 3 selling for 10 cents or less. The newer high-performance AlInGaP and AlInGaN technologies are more expensive at this time; however, prices are expected to drop sharply as the volume increases. Cost-effectiveness is the reason that the older and much

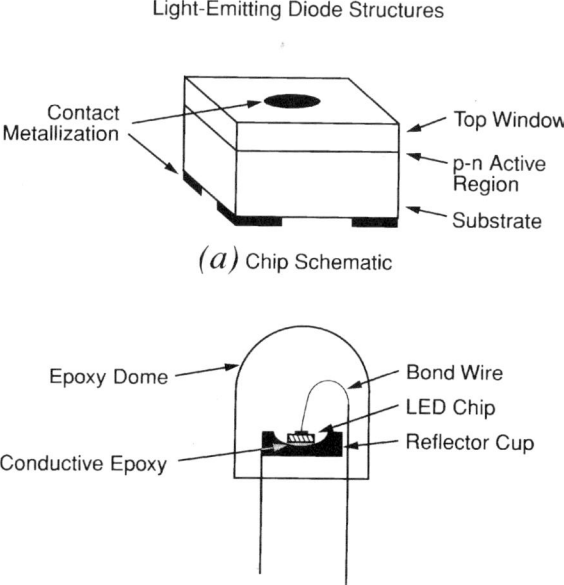

FIG. 3. Schematic illustration of a packaged light-emitting diode (LED) chip. (a) Chip schematic. (b) Package schematic.

less efficient technologies and growth processes continue to dominate the market in unit volume. It is also the reason that relatively primitive chip structures are used even though higher light output could be obtained with additional processing steps (and cost). Although OMVPE devices are higher in cost at this time, there appears to be nothing fundamental about OMVPE growth technology that precludes OMVPE devices from competing with LPE and VPE devices as long as the epitaxial structures do not require thick layers. This suggests that AlInGaN devices, which have thin epitaxial structures, could become quite cost-effective in the future, assuming larger scale reactors are developed and yields increase. Device prices in the range of 10 to 20 cents, rather than the present $1 to $2 will be required, and are probably realistic.

Reliability is another critical issue for LEDs. One of the key reasons customers are willing to go to the trouble to switch to LED sources, and sometimes to pay more for them, is that LEDs are reliable solid state sources that rarely, if ever, need to be replaced. This issue is discussed further in Chapter 6, but for many applications, such as traffic lights and signs, lifetime to half brightness of tens of thousands of hours are required and 100,000 hours (about 10 years) is desired. In addition, for outdoor operation, the devices must operate in very hot and very cold environments. This puts severe constraints on chip geometry, the choice of epoxy encapsulation that can be used (because it shrinks at low temperature and puts enormous stress on the chips) and on the shape and size of the lead frame and reflector that surrounds the chip. A schematic illustration of a packaged LED chip is shown in Fig. 3.

The nitride and phosphide materials are relatively robust and meet the reliability requirements without major difficulty. There were some problems with the first nitride devices released; however, these problems were related to packaging and processing and were not fundamental to the chip structure. Zinc selenide-(ZnSe-) based materials, however, which are a possible alternative technology for blue and green emitters, are inherently relatively soft, and it is not yet clear that a chip and package combination can be designed that will enable ZnSe devices to meet reliability targets. Consequently, devices of this type are not yet available commercially, and therefore are not discussed in detail in this book.

The AlGaAs devices fall between the nitrides and the ZnSe materials in terms of robustness, and a major technical effort over many years was required to achieve devices with acceptable performance. Even today, reliability problems occasionally occur with AlGaAs LEDs particularly in situations of high temperature and high humidity, or high current densities at low temperatures.

2. Luminous Performance

The luminous performance is the total flux output of an LED in lumens divided by the power input in watts. This is the same measure of performance commonly used to characterize other types of light sources. LEDs are generally characterized commercially in units of candela at a given current, typically 20 mA. A candela is an on-axis measure of lumens per steradian. The problem with candela measurements for comparing different technologies is that the candela value depends critically on the package design. The same LED chip can have more than an order of magnitude variation in candela reading depending on the viewing angle and other factors.

The luminous performance is obtained by multiplying the eye sensitivity curve (CIE curve), which is shown in Figure 4, by the radiometric power efficiency. Therefore, luminous performance (lm/W) $\cong 680\, V(\lambda) P_E$ where λ is the wavelength of the emitter and $680\, V(\lambda)$ is the CIE value. This is valid for a monochromatic source and approximately valid for an LED that emits in a relatively narrow emission band (typically near band-edge emission). More generally, however, it is necessary to integrate over the emission spectrum of the device. The power efficiency P_E is the power output divided by the power into the device, or

$$P_E = \text{(photons/sec)} \times \text{(emission energy/photon)}/$$
$$\text{(electrons/sec)} \times \text{(applied voltage)}$$
$$= \eta_{\text{ext}} \times E_r / V_A = \eta_{\text{int}} C_{\text{ex}} E_r / V_A$$

where η_{ext} is the external quantum efficiency; η_{int} is the internal quantum efficiency; C_{ex} is the extraction efficiency, which is the fraction of generated photons that escape from the chip; E_r is the energy of the emitted radiation (in electron volts); and V_A is the applied voltage.

$$V_A \cong E_g + \mathbb{R}$$

where E_g is the semiconductor energy gap and $\mathbb{R}$ is the resistive voltage drop in the device. since the resistive voltage drop is usually on the order of 0.2 V or less for low current, and since $E_r \cong E_g$ for band-edge or shallow impurity recombination, it is usually accurate within about 10% or so to write

$$\text{Luminous performance} \cong 680\, V(\lambda) \eta_{\text{int}} C_{\text{ex}}$$

Therefore the key issues in improving LED performance are generally the internal quantum efficiency and the extraction efficiency.

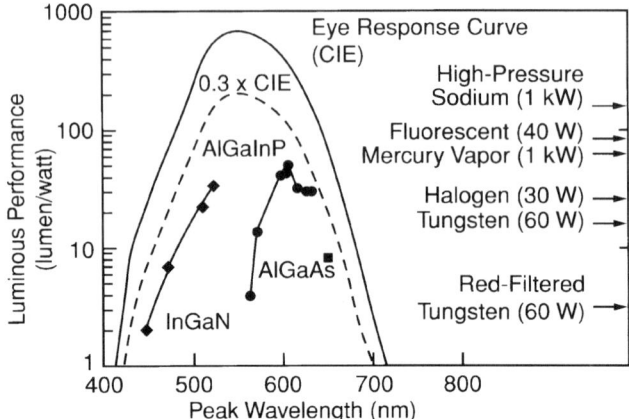

FIG. 4. Luminous performance of high performance light-emitting diodes (LEDs) versus wavelength. The eye sensitivity curve, (CIE), which is the maximum theoretic performance limit, is shown for reference. The dashed curve at 0.3 × CIE is a rough "practical limit" for conventional LED chip types.

3. INTERNAL QUANTUM EFFICIENCY

Three of the factors important for achieving high internal quantum efficiency are

- High purity and low defect density substrates and epitaxial structures
- A direct semiconductor energy gap covering the desired color region
- A lattice-matched materials system enabling the growth of heterostructure devices with low defect densities

All these materials-related issues are discussed in Chapter 1, as well as other review articles, and therefore they will be discussed only very briefly here.

The highest efficiency LEDs have internal quantum efficiencies approaching 100%, and achieving near 100% efficiency in the AlInGaP and AlInGaN systems is the primary goal of much of the ongoing LED research. One of the most critical issues in improving performance of OMVPE grown materials is the quality of raw materials. There is often a substantial variation from source to source and it is generally difficult to determine the cause of problem because the impurity concentrations are below the background detection limit.

The issue of direct versus indirect bandgap semiconductors has been an issue in LED development since the 1960s. Direct bandgap devices are clearly preferred because of their much higher radiative recombination rates and (potentially) efficiencies. However, the lack of a suitable growth technology delayed the introduction of LEDs containing aluminum (Al) and enabled GaAsP and GaP to achieve high-volume commercial success. The early GaAsP devices were direct bandgap devices; however, the efficiency was low primarily because of high defect densities resulting from the lattice-mismatched (GaAs) substrate. The shortest possible wavelength was about 650 μm (red) because of the onset of the direct–indirect transition. The addition of the isoelectronic impurity nitrogen enabled the fabrication of commercially viable GaAsP and GaP LEDs from the red through the green spectral regions even though the materials had indirect energy gaps that normally would not emit significant amounts of radiation. Surprisingly, devices of this type continue to dominate the market even though the efficiency is relatively low. The devices are simple homojunction emitters that are inexpensive and are good enough for many indoor applications. High-quality heterostructures are not possible in the GaAsP system because of the lattice mismatch between GaAs and GaP.

All of the new generation of high-performance AlGaAs, AlInGaP, and AlInGaN devices are direct-energy gap as well as lattice-matched devices, as discussed in Chapter 1. The AlGaAs system is limited to IR and red emitters because, just as in GaAsP, the alloy becomes indirect for shorter wavelengths. Similarly, the AlInGaP quaternary system is direct from red through amber but becomes indirect in the green, resulting in a sharp decrease in performance for green devices. The nitride system is direct for all compositions. The main problem with the nitride system is that there is no lattice-matched substrate, and therefore the defect density is high.

The simplest LED structure is a P-N homojunction, which can be formed either by diffusion, as is the case in GaAsP devices, or by growing separate P- and N-type epitaxial layers, as is generally done in other material systems. The homojunction is simple, inexpensive, and reproducible; however it is limited in efficiency. Generally, electron injection into the P-type region is desired for efficient radiative recombination. In homojunctions, however, hole injection into the N-type region also readily occurs, and this generally results in nonradiative recombination. In addition, self-absorption of the generated radiation is high in homojunctions because the entire structure has the same composition as does the active region.

A single heterostructure (SH) is an improvement over the homojunction, because by growing a narrower bandgap P-type region, hole injection into the N-type region can be minimized. A double heterostructure (DH) is a further improvement. Both electrons and holes can be injected from wide bandgap regions (confining layers) into an active region where they are trapped and cannot diffuse away. This gives rise to a high electron-hole density that results in fast recombination and tends to saturate any residual nonradiative traps that may be present. Further, a DH has reduced internal absorption because the relatively thin, typically about 1 μm, active region is the only part of the epitaxial structure that can reabsorb the emitted radiation.

A further refinement to the DH is the quantum well structure, which can be either a single quantum well (SQW) or a multiple quantum well (MQW) device. Quantum well structures serve to further reduce internal absorption, because the layers are still thinner (<0.02 μm), and to increase speed. Quantum well structures have not been widely used in visible LEDs because the difficulty and added expense involved in accurately controlling the growth of the very thin layers have made it more cost-effective to use DH structures. However, it appears that this situation will change because the best nitride device results have been obtained using a SQW structure. In this case an extremely thin GaInN active layer is a major advantage because, since it is not lattice-matched to the GaN or AlGaN confining layers, there would be a high density of interface defects if the layer were thick. Because the layer is thin, however, in this case believed to be approximately 20 nm, it is energetically favorable for the layer to "stretch" and conform to the confining layers. "Pseudomorphic" layers of this type have been widely used in IR injection lasers.

4. Light Extraction

The extraction efficiency is the fraction of generated light that escapes from the semiconductor chip into the surrounding air or encapsulating epoxy. This is a challenging problem because the chip has a much higher index of refraction, typically 3.4 compared with 1.0 for air and approximately 1.5 for epoxy. Because the critical angle for internal reflection is given by Snell's law as $\theta_c = \sin^{-1}(n_2/n_1)$, this results in a critical angle of 17% for air and 26 degrees for epoxy.

If we consider a single surface, the light escaping will be only the light that strikes the surface within an angle of 26 degrees (epoxy). The extraction efficiency for a single escape cone is

$$C_{ex} \cong \tfrac{1}{2}(1 - \cos\theta_c) \times (1 - [n_2 - n_1]^2/[n_2 + n_1])^2$$

where the first term is the fractional solid angle for the escape cone and the second term accounts for Fresnel reflection losses (assuming normal incidence). For extraction from a single surface into epoxy, C_{ex} is about 0.04.

Light-emitting diodes with thin epitaxial structures ($\leq 5\mu$) grown on absorbing substrates have only one escape cone, namely, the vertical top surface cone. Devices of this type typically have extraction efficiencies of even less than 0.04 because of "shadowing" from the top contact. If the epitaxial layer is grown thicker and is transparent then there is substantial edge emission and one can calculate that for 250 μm square chips with approximately 60 μm thick window layers, nearly a half cone of light can be extracted out of each edge. Because there are four edges, this gives a total of two "edge cones" to go with the top cone, for a total of three extraction cones or a total extraction efficiency of approximately 0.12. This approach has been used for AlInGaP devices, as discussed in Chapters 4 and 5. A still better approach is to eliminate the absorptive substrate altogether, and replace it with a transparent layer. This is the approach used in all of the highest performance devices. In the case of TS-AlGaAs, a 150 μm thick wide bandgap layer is grown on the substrate underneath the device and the original (absorptive) GaAs substrate removed. In the case of AlInGaP devices that are also grown on a GaAs substrate (for lattice matching reasons) a thick window is grown as described earlier, and then the GaAs is removed and a (transparent) GaP substrate is wafer-bonded to the structure in place of the GaAs.

In the case of AlInGaN devices, there is no lattice-matched substrate and the epitaxial layers are grown directly on either the Al_2O_3 or silicon carbide (SiC) substrates, both of which are transparent. This works quite well because the defects appear to have a negligible effect on the light output. However, because the epitaxial layers are thin and quite resistive, current spreading is an issue, and a conductive semitransparent metal film is deposited on top of the structure to solve this problem.

If a transparent substrate is used, C_{ex} can be 0.3 or even higher as is probably the case in the nitride devices on Al_2O_3 where there is very little absorption either in the epitaxial structure or in the substrate. If there were no absorption present, the light that is internally reflected would eventually be randomized by scattering from rough spots on the chip surfaces and would escape, giving a $C_{ex} \cong 1.0$. However, as a practical matter the ohmic contacts are absorptive, as is the active layer, and C_{ex} of about 0.3 is generally observed for DH structures with traditional (inexpensive) contacting schemes and relatively thick, 1 to 2 nm, active regions. Refinement of chip processing technology is expected to result in somewhat higher extraction efficiencies because applications are evolving that can justify the extra expense involved in developing the maximum possible efficiency. In particular, it is becoming viable to grow DH devices with thinner active regions (even 20-Å quantum well active layers in the case of the nitrides) that reduce internal absorption and can be expected to significantly increase extraction efficiency.

III. Technology Status and Future Potential

1. PERFORMANCE

Table I compares the performance of various types of LEDs in terms of external quantum efficiency and lm/W. The red AlGaAs devices have the highest typical external quantum efficiency of 15%, as discussed in Chapter 3. These DH devices grown by LPE represent a relatively mature technology that probably will not improve significantly. The extraction efficiency is estimated to be about 30%, which implies an internal quantum efficiency of about 50%. This approaches the maximum theoretic quantum efficiency allowed for red emission in the 640-μm range due to the proximity of the direct–indirect bandgap transition and the resulting population of the indirect minimum with electrons. Infrared AlGas emitters, which have similar device structures but are far from the direct–indirect bandgap transition, have external quantum efficiencies approaching 30%, implying internal quantum efficiencies of about 100%. Therefore, both the IR and red-emitting AlGaAs devices are believed to be approaching maximum performance using the existing chip shape and contact technology, which results in an approximately 30% extraction efficiency. This shows the great strength of LPE growth technology, which yields epitaxial structures with a high degree of perfection and reproducibility at a minimal cost. In addition, the TS-AlGaAs devices require an approximately 150 μm thick epitaxial layer so that the GaAs substrate can be removed, which would be difficult to do using other growth technologies.

The TS-AlInGaP devices are the highest performance LED device type in existence. The luminous performance is higher for the TS AlInGaP devices than for AlGaAs, even at similar quantum efficiencies, because the wavelength is shorter and the eye sensitivity is substantially higher. The AlInGaP devices are grown using OMVPE instead of LPE because of the difficulties discussed in Chapter 1. Although more expensive, OMVPE growth is necessary for this material system. However, the quantum efficiencies are still not approaching the theoretic limit, particularly for amber- and green-emitting alloys with high Al content. These issues are discussed in Chapters 4 and 5. The potential performance improvement that is possible changes with alloy composition; however a twofold or more improvement should be possible, with more improvement expected at shorter wavelengths.

The highest performance blue and green devices are nitride structures with SQW active regions. Devices of this type are discussed in detail in Chapter 8. These devices perform surprisingly well considering the extremely high

TABLE I

CHARACTERISTICS OF DIFFERENT TYPES OF HIGH-BRIGHTNESS LIGHT EMITTING DIODES (LEDs)

LED Type	Color (Peak Wavelength)	Structure	External Quantum Efficiency at 20 mA (%)	Luminous Performance at 20 mA (lm/W)	Reference
AlGaAs	Infrared (780–880)	DH	27	NA	Hewlett-Packard unpublished data.
	Red (650)	DH	16	8	Data are typical of the best commercial devices.
AlInGaP	Red-orange (636)	DH	24	35	Kish et al. (1996)
	Amber (590)	DH	10	40	Hewlett-Packard, unpublished data.
AlInGaN	Green (570)	DH	2	14	Kish et al. (1996)
	Green (525)	SQW	6.3	~18	Nakamura et al. Nitride workshop, Nagoya, Sept. 1995
	Blue (450)	SQW	9.1	~2	(and Chapter 8 herein).
AlInGaN	Green (517)	DH: co-doped with zinc and silicon	2.6	6.5	
	Blue (450)	DH: co-doped with zinc and silicon	5.5	5	Nichia Lamps
ZnTeSe	Green (512)	SQW	5.3 (at 10 mA)	17	Eason et al. (1995)
ZnCdSe	Blue (489)	MQW	1.3 (at 10 mA)	1.6	*Appl. Phys. Lett.* **66**, 115.

DH, double heterostructure; NA, not applicable; SQW, single quantum well; MQW, multiple quantum well.

($>10^{10}/\text{cm}^2$) defect densities present in the films due to their growth on sapphire (Al_2O_3) substrates which are very badly lattice-mismatched. Typically LEDs do not emit a significant amount of light at defect densities above approximately $10^7/\text{cm}^2$ (Lester et al., 1995). The densities typically must be below $10^4/\text{cm}^2$ to give good devices with quantum efficiencies in the several percent range. This is one of the many surprising and poorly understood issues related to the nitrides.

The performance improvement that can be expected in the nitrides is hard to estimate largely because of the unknown limitation imposed by the defects. They may have a negligible effect so that nearly 100% internal quantum efficiency may be possible; on the other hand, a limit may already have been reached but not recognized. Furthermore, the extraction efficiency and hence, the internal quantum efficiency have not been established. The extraction efficiency is probably quite high because Al_2O_3 is highly transparent and because the active layers are very thin, which should give low absorption. However, there is a thin metal film over the surface that is required to spread the current, and there are two large contact areas that are absorptive. If we make a very rough assumption that the extraction efficiency is about 50%, this would imply internal quantum efficiencies in the range of 10% to 20%. This would further imply that fivefold or more improvement is still plausible in these devices because they are direct semiconductors and about 100% internal quantum efficiency, in principle, should be achievable if the defects truly have no negligible impact on device performance. In any case the nitride SQW devices already have excellent performance characteristics and coupled with AlInGaP provide high-performance LED technologies for all colors.

Alternate technologies for high-performance blue and green emitters also are shown in Table I. Double heterostructures with silicon (Si) and zinc (Zn) co-doped active region also give excellent performance and, in fact, devices of this type were the first to become available commercially. A disadvantage of co-doped devices compared with SQW devices is that the emission spectrum is much broader and as a result the color purity is not as good. This can be an important consideration, particularly when mixing red, green, and blue emitters to get white light. These issues are discussed in more detail in Chapter 9. In addition, the co-doped recombination tends to saturate and shift in color at high current densities which is a limitation for some applications. Basically, the SQW approach appears to be better than co-doping in virtually every area, and therefore it should dominate unless some problem arises, such as latent reliability issues.

Another approach in the nitride area is to grow on silicon carbide (SiC) substrates instead of Al_2O_3 (Kong et al., 1996). Silicon carbide has the advantage of being conductive, enabling the fabrication of smaller chips

2 OVERVIEW OF DEVICE ISSUES IN HIGH-BRIGHTNESS LIGHT-EMITTING DIODES 61

with a single contact on top and the second contact on the bottom, which is the conventional LED approach. This would have some advantage because the chips could be mounted in conventional packages, and the smaller chips, in principle, could be less expensive if the market price for SiC, which is presently more than an order of magnitude higher than Al_2O_3, can be substantially reduced. The key issue, however, is performance. SiC has a better lattice match to GaN than Al_2O_3 does and may enable the growth of epitaxial films with lower defect densities. However, SiC is more absorptive than is Al_2O_3 due primarily to free-carrier absorption, and therefore the performance trade-off is not clear. The device results for co-doped structures, shown in Table I, are still not as good as the best results obtained with Al_2O_3 substrates. However, nitride technology is at an early stage of evolution and SiC substrates may yet become the substrate of choice, possibly for the fabrication of laser diodes if not LEDs.

The other technology type included in Table I is the II-VI technology based primarily on zinc selenide (ZnSe) (Eason et al., 1994). This technology has the advantage of having lattice-matched substrates, GaAs and ZnSe, available and of being able to grow lattice-matched epitaxial heterostructures with good device properties. The primary disadvantage of the material system is that the formation and motion of defects occurs easily and at low energies. This has caused severe reliability problems in lasers and so far no LEDs have been demonstrated that have reliability characteristics suitable for commercial production. This problem may be a fundamental limitation. It is certainly a formidable challenge in view of the reliability requirements placed on LEDs. Because the nitrides already appear to be providing a successful solution to the high-performance blue and green LED requirements, it does not appear that it will be necessary to deal with the II-VI reliability issues for commercial LED applications. Therefore, the II-VI approach to blue and green emitters is not covered in detail in this book.

2. COMPARISON OF LIGHT-EMITTING DIODES TO OTHER LIGHT SOURCES

Figure 4 shows the luminous performance of the key types of high-performance LEDs plotted as a function of emission wavelength. The eye sensitivity curve is also shown for reference as well as the performance of various types of other light sources. As discussed in Part II Section 1, high LED performance is now possible for all colors. AlInGaP technology should dominate in the amber, red-orange, and quite possibly the red region, depending on how effectively AlInGaP and TS-AlGaAs compete on a long-term cost-performance basis. At present, AlGaAs devices are widely used in applications requiring high-performance red LEDs; however, as

AlInGaP continues to improve, it is beginning to replace AlGaAs in a variety of applications.

In the green region the AlInGaP performance drops sharply. Some decrease is expected because of the approach of the direct–indirect transition; however, the decrease occurs at lower energies and is more severe than is predicted theoretically, as discussed in Chapters 4 and 5. This decrease may be related to oxygen incorporation, which is often a problem as the Al concentration increases. If this situation can be improved, green AlInGaP devices may find an important set of applications as long as they can be produced at low cost. However, AlInGaP devices will have difficulty competing with green nitride emitters if the prices are comparable because the nitrides should be brighter and can cover the color spectrum from yellow-green to deep blue-green.

The nitride SQW emitter with a performance of approximately 18 lm/W is an outstanding device and is within about a factor of two of the best TS AlInGaP amber and red-orange devices. The blue SQW device at 2 lm/W is lower in performance due to the reduced eye sensitivity at 450 μm, although the quantum efficiency is excellent at 9%.

The best red through green LEDs have somewhat higher luminous performance than do tungsten incandescent lamps, which are typically in the 15 lm/W range, and are substantially better than filtered tungsten lamps, which typically perform at 3 to 4 lm/W in the red region. By mixing red, green, and blue LEDs, one can now make a white source that also has a luminous performance approaching a white tungsten source. Although the blue is lower in performance, a relatively small fraction of blue light is required. If LED sources continue to improve, which appears plausible, it may become possible to fabricate LED solid state light sources that provide a substantial energy savings over conventional lamps. If 100% internal quantum efficiency can be obtained and if an extraction efficiency of 30% is assumed, then LED performance would fall on the dashed curve shown in Fig. 4, which is over 100 lm/W for colors in the orange to green spectral region. At that level LEDs would exceed even fluorescent lamps in performance.

If the extraction efficiency could be improved, LEDs could move even closer to the ultimate theoretic limit, which is the eye sensitivity curve (CIE) curve. The CIE curve corresponds to every injected electron resulting in an emitted photon, or 100% Q.E. and 100% extraction efficiency. Achieving the theoretic limit is not plausible; however, lasers with extraction efficiencies around 50% have been observed in long wavelength materials, and nitride LEDs grown on (highly transparent) Al_2O_3 substrates may also approach or even exceed 502 extraction efficiency. Therefore it is conceivable that very high performance solid-state light sources could be realized. At present, however, LEDs are cost-effective only as low power sources with outputs of about 0.1 W. Highly efficient cost-effective white light sources operating in the range of tens of watts (or even watts) appear to be far in the future. In the meantime LEDs will continue to move into a variety of markets previously occupied primarily by relatively low-power filtered incandescents, as discussed in Chapter 6.

REFERENCES

Alferov, Zh. I., Andreev, V. M., Garbuzov, D. Z., and Rumyantsev (1975). *Sov. Phys.—Semicond.* (*Engl. Transl.*) **9**, 305.
Akasaki, I., and Hashimoto, H. (1967). *Solid State Commun.* **5**, 851.
Amano, H., Sawaki, N., and Akasaki, I. (1986). *Appl. Phys. Lett.* **48**, 353.
Amano, H., Kitoh, M., Hiramatsu, K., and Akasaki, I. (1989). *Conf. Ser.—Inst. Phys.* **106**, 725.
Craford, M. G., and Steranka, F. M. (1994). *Encycl. Appl. Phys.* **8**, 485.
Eason, D. B., Yu., Z., Hughes, W. C., Roland, W. H., Boney, C., Cook, J. W., Jr., Schetzina, J. F., Cantwell, G., and Harsch, W. C. (1995). *Appl. Phys. Lett.* **66**, 115.
Groves, W. O., Herzog, A. H., and Craford, M. G. (1971). *Appl. Phys. Lett.* **19**, 184.
Haitz, R. H., Craford, M. G., and Weissman, R. H. (1995). In "Handbook of Optics" (M. Bass, ed.), Vol. I, p. 12.1.
Holonyak, N., Jr., and Bavacqua, S. F. (1962). *Appl. Phys. Lett.* **1**, 82.
Kish, F. (1995). *Encycl. Chem. Technol.* **15**, 217.
Kish, F. A., Steranka, F. M., DeFevere, D. C., Vanderwater, D. A., Park, K. G., Kuo, C. P., Osentowski, T. D., Peanasky, M. J., Yu, J. G., Fletcher, R. M., Steigerwald, D. A., Craford, M. G., and Robbins, V. M. (1994). *Appl. Phys. Lett.* **64**, 2839.
Kish, F. A., Vanderwater, D. A., DeFevere, D. C., Steigerwald, D. A., Hofler, G. E., Park, K. G., and Steranka, F. M. (1996). *Elec. Lett.* **32**, 132.
Kong, H. S., Leonard, M., Bulman, G., Negley, G., and Edmond, J. (1996). *Mater. Res. Soc. Symp. Proc.* **395**.
Kuo, C. P., Fletcher, R. M., Osentowski, T. D., Lardizabal, M. C., and Craford, M. G. (1990). *Appl. Phys. Lett.* **57**, 2937.
Lester, S. D., Ponce, F. A., Craford, M. G., and Steigerwald, D. A. (1995). *Appl. Phys. Lett.* **66**, 1249.
Logan, R. A., White, H. G., and Wiegmann, W. (1968). *Appl. Phys. Lett.* **13**, 139.
Nakamura, S., Mukai, T., and Senoh, M. (1970). *Appl. Phys. Lett.* **64**, 1687.
Nakamura, S., Senoh, M., Iwasa, N., Nagahama, S., Yamaka, T., and Mukai, T. (1995). *J. Appl. Phys.* **34**, L1332.
Nishizawa, J., and Suto, K. (1977). *J. Appl. Phys.* **48**, 3484.
Sugawara, H., Ishikawa, M., and Hatakoshi, G. (1991). *Appl. Phys. Lett.* **58**, 1010.

CHAPTER 3

AlGaAs Red Light-Emitting Diodes

Frank M. Steranka

OPTOELECTRONICS DIVISION
HEWLETT-PACKARD COMPANY
SAN JOSE, CALIFORNIA

I. INTRODUCTION	65
II. DEVICE DESIGN	68
1. *Materials Parameters*	68
2. *Heterostructure Design*	69
3. *Optical Extraction*	71
4. *Active Layer Composition*	72
III. CRYSTAL GROWTH	74
1. *Molecular-Beam Epitaxy and Metalorganic Chemical Vapor Deposition*	74
2. *Slow-Cooling Liquid-Phase Epitaxy*	75
3. *Temperature-Difference Liquid-Phase Epitaxy*	77
IV. DEVICE FABRICATION	79
1. *Contact Formation*	79
2. *Dicing*	80
3. *Selective Etching*	81
V. COMPARISON WITH OTHER TYPES OF RED LIGHT-EMITTING DIODES	81
VI. TEMPERATURE-DEPENDENT PROPERTIES	83
1. *I-V Curves and Forward Voltage*	83
2. *Electroluminescent Spectra*	84
3. *Efficiency*	85
4. *Minority-Carrier Lifetime and Analysis*	87
VII. RELIABILITY CHARACTERISTICS	89
1. *Low-Temperature Operation*	89
2. *High-Temperature, High-Humidity Operation*	94
References	95

I. Introduction

Gallium arsenide (GaAs) is a direct-energy-gap semiconductor with an energy gap (E_g) in the near-infrared (IR) portion of the electromagnetic spectrum ($E_g = 1.424$ eV), whereas aluminum arsenide (AlAs) is an indirect-energy-gap semiconductor with an energy gap in the yellow-green portion of the spectrum ($E_g = 2.168$ eV). As the AlAs mole fraction x of aluminum gallium arsenide ($Al_xGa_{1-x}As$) is raised from zero, the energy gap of the resulting compound increases from that of gallium arsenide (GaAs) to that of AlAs. A plot of the AlGaAs conduction band minima Γ, X, and L as a function of the AlAs mole fraction is shown in Fig. 1. The lowest-lying conduction band switches from Γ to X at an AlAs mole fraction of approximately $x = 0.45$, and the semiconductor switches from having a direct-energy gap to having an indirect-energy gap. This transition occurs at an energy of 1.98 eV, which is in the visible red portion of the spectrum.

The electroluminescent efficiency of a semiconductor typically falls by more than three orders of magnitude as it switches from being a direct-energy-gap semiconductor to an indirect-energy-gap semiconductor. This prevents the AlGaAs materials system from being used to make efficient orange or yellow LEDs. However, very high efficiency red and IR light-

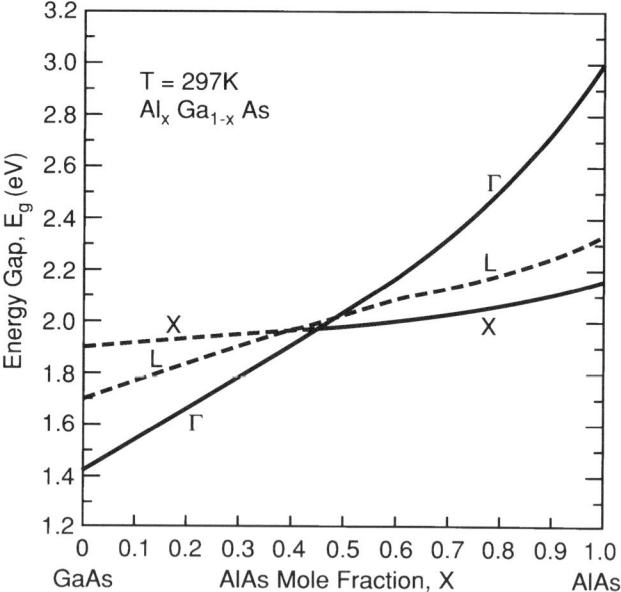

FIG. 1. Energy of the $Al_xGa_{1-x}As$ conduction band minima Γ, X, and L as a function of the aluminum arsenide (AlAs) mole fraction x. The ternary switches from having a direct-energy gap (Γ lowest) to having an indirect-energy gap (x lowest) at an AlAs mole fraction of $x = 0.45$ and an energy of 1.98 eV. AlAs, aluminum arsenide; GaAs, gallium arsenide; $Al_xGa_{1-x}As$, aluminum gallium arsenide. (Reprinted from Casey and Panish, 1978, *Heterostructure Lasers*, p. 17; with the permission of Academic Press.)

emitting diodes (LEDs) can be fabricated with this materials system (Alferov et al., 1975; Nishizawa et al., 1973; Cook et al., 1988; Ishimatsu and Okuno, 1989). In large part, this is due to the fact that GaAs and AlAs differ in lattice constant by less than 0.2% at 25°C (Fig. 2). This close match in lattice constant enables the growth of very high quality AlGaAs films on GaAs substrates and the deposition of heterostructure devices with very low interface defect densities (Casey and Panish, 1978; Kressel and Butler, 1977). Thus, complex device structures can be grown to optimize device performance.

Early work (Rupprecht et al., 1967; Alferov et al., 1973) indicated that high-performance visible LEDs could be made with the AlGaAs materials system; however, more than a decade of development was required before high-volume liquid-phase epitaxy (LPE) reactors were created that were capable of producing high-quality multilayer device structures (Nishizawa and Suto, 1977). Such LEDs did not become commercially available until the early 1980s.

AlGaAs red LEDs were the first LEDs to exceed the luminous efficiency level of filtered incandescent light bulbs (see Fig. 1 Chapter 2). This enabled LEDs to begin replacing light bulbs in many red exterior lighting applications such as in center high mount stop lights (CHMSLs) on motor

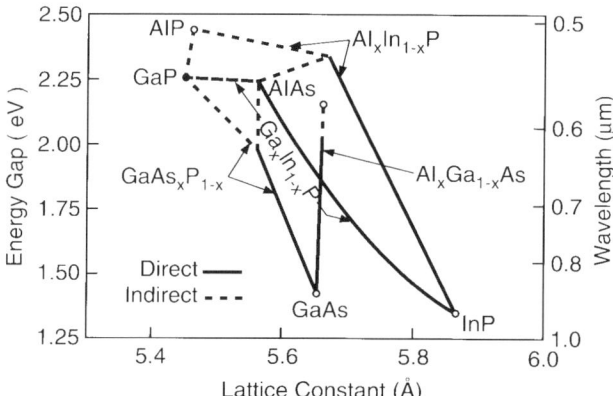

FIG. 2. Energy gap of several III-V semiconductors as a function of the lattice constant. The solid lines indicate direct-energy gap regions, and the dashed lines indicate indirect-energy gap regions. Gallium arsenide (GaAs) and aluminum arsenide (AlAs) have very similar lattice constants, differing by only 0.2% at room temperature. GaP, gallium phosphide; AlP, aluminum phosphide; InP, indium phosphide; GaAsP, gallium arsenide phosphide; GaInP, gallium indium phosphide; AlGaAs, aluminum gallium arsenide. AlInP, aluminum indium phosphide. (Reprinted from Craford and Steranka, 1994, *Encycl. Appl. Phys.* **8**, 498; with the permission of the American Institute of Physics.)

vehicles and in outdoor message signs. Today, over 2.5 billion AlGaAs red LEDs are produced each year, and the incandescent-replacement trend is accelerating with the recent development of high-brightness yellow, green, and blue LEDs, which are described in several of the other chapters in this volume.

A description of the different types of AlGaAs red LED structures and some of the basic parameters that are important in device design are presented in Part II. Part III contains a description of the crystal-growth techniques that have been used to produce AlGaAs red LEDs. A description of the AlGaAs wafer fabrication process is covered in Part IV and a performance comparison between AlGaAs red LEDs and other types of red LEDs in Part V. Part VI contains a description of the temperature-dependent properties of AlGaAs red LEDs as well as a discussion of the recombination kinetics in these devices. Finally, Part VII contains a description of the reliability characteristics of AlGaAs red LEDs and a brief discussion of on-going work to further improve these characteristics.

II. Device Design

1. MATERIALS PARAMETERS

Often, n-type and p-type materials of a given semiconductor can exhibit different luminescence efficiencies even though their doping levels are comparable. This is due to differences in the deep levels and crystal defects that can change with the type of dopant used. For AlGaAs, both photoluminescence (PL) (Alferov et al., 1973) and cathodoluminescence (CL) (Varon et al., 1981) studies show that the luminescent efficiency of n-type material falls more rapidly than does that of p-type material as the direct–indirect crossover is approached. The exact reason for this difference between n-type and p-type materials is uncertain; however, it may be due to the high concentration of D_x centers that are known to form in n-type $Al_xGa_{1-x}As$ at $x > 0.25$ (see, for example, Kumagai et al., 1984). Nonetheless, for high-efficiency red LEDs, the conclusion is that, if possible, recombination should be forced to occur in p-type material.

The electrical conductivity of AlGaAs with high Al content is typically quite poor. Table I shows the mobility, doping level, and resistivity values obtained from van der Pauw measurements on n-type and p-type AlGaAs single layers grown on semi-insulating substrates by LPE. The mobility values range from 31 to 389 $cm^2/V \cdot sec$. Because of the low mobilities, thick (greater than 15 μm) layers are typically required between the top contact

TABLE I
RESULTS OF VAN DER PAUW MEASUREMENTS ON LIQUID-PHASE EPITAXY–GROWN AlGaAs SINGLE LAYERS

Aluminum Mole Fraction	n- or p-type	Dopant	Free Carrier Density (cm^{-3})	Mobility ($cm^2/V \cdot sec$)	Resistivity (ohm-cm)
0.37	p	Zinc	4.0×10^{17}	99	0.16
0.85	n	Tin	8.3×10^{16}	112	0.67
0.85	n	Tellurium	1.5×10^{17}	389	0.1
0.85	p	Zinc	3.3×10^{17}	60.4	0.31
0.85	p	Zinc	5.0×10^{17}	40.2	0.31
0.85	p	Magnesium	8.7×10^{17}	31	0.23

layer and the active layer to enable current to spread out from under the top contact and reduce contact shadowing.

2. HETEROSTRUCTURE DESIGN

In a normal homojunction LED under forward bias, electrons are injected into the p-type material and holes are injected into the n-type material. Electron-hole recombination then takes place on both sides of the p-n junction, and light is generated in both the p-type and n-type regions. By forming a p-n junction between two materials where the n-type material has a wider energy gap than does the p-type layer, minority-carrier injection into the n-type material can be eliminated. Such a device is called a single heterostructure (SH) LED, and a schematic of the energy band diagram for this case along with those for homostructure and double heterostructure (DH) LEDs are shown in Fig. 3.

For AlGaAs heterostructures, the direct-energy-gap discontinuity is split between the conduction band and the valence band in roughly a 60:40 ratio (Watanabe *et al.*, 1985). However, at AlAs mole fractions greater than $x = 0.45$, the lowest conduction band switches from Γ to X (*see* Fig. 1), and the indirect–energy gap increases much more slowly than does the direct-energy gap with increasing mole fraction. The result is that the effective conduction-band discontinuity is very small or even negative (Type II alignment), as described by Dawson *et al.* (1986), and it plays very little role in the device transport characteristics. The discontinuity in the valance band then adds much more to the hole-potential barrier than the conduction-band discontinuity adds to the electron-potential barrier, and hole injection

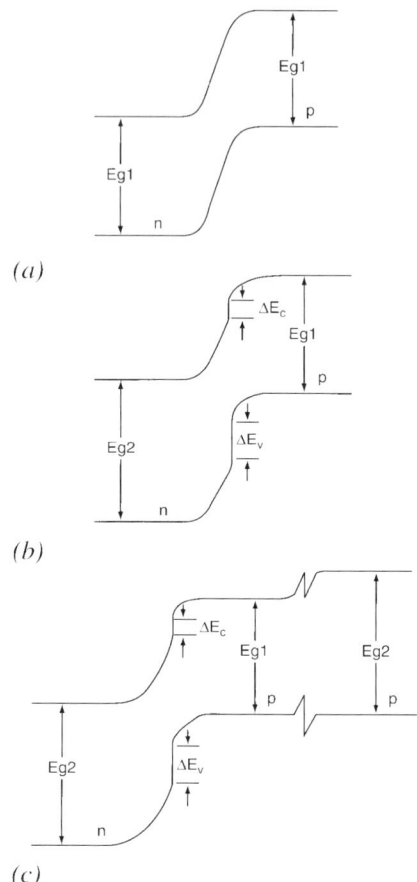

FIG. 3. Energy band diagrams for (a) homojunction, (b) single heterojunction (SH), and (c) double heterojunction (DH) AlGaAs red light-emitting diodes (LEDs). For these LEDs, the discontinuity in the valance band far exceeds that in the conduction band and gives rise to single-sided injection in both the SH and DH LEDs AlGaAs, aluminum gallium arsenide. (Reprinted from Steranka *et al.*, 1988 *Hewlett-Packard J.* **39**, 85; with the permission of the Hewlett-Packard Company.)

into the wider-energy-gap material is eliminated. For an $Al_{0.35}Ga_{0.65}As$ active layer, an $Al_{0.65}Ga_{0.35}As$ composition layer provides a sufficiently large hole potential barrier to result in single-sided injection, and thus layers with these compositions are typically used in SH AlGaAs red LEDs. Because the wider energy gap material is transparent to the light generated in the active layer, the SH device structure also dramatically reduces internal absorption inside the semiconductor and improves the light-extraction efficiency.

The defice efficiency can be further improved by putting a second wide-gap layer on the other side of the active layer and forming a DH device, as shown in Fig. 3(c). Because the Fermi levels in the two p-type layers must align, a barrier is formed in the conduction band that traps the minority-carrier electrons in the active layer. To provide effective confinement for electrons in an $Al_{0.35}Ga_{0.65}As$ active layer, a p-type confining layer with an Al mole fraction of $x > 0.75$ must be used. In addition, it is important for the doping level in the p-type confining layer to be at least as high as that in the active layer to avoid reducing the magnitude of the electron potential barrier. With such a p-type confining layer, the active layer can be made quite thin (typically less than 2.5 μm), improving device efficiency. It does so by further reducing the internal absorption of light in the active layer and by increasing the minority-carrier density in the active layer at a given drive current (this helps saturate some of the nonradiative recombination centers). DH AlGaAs red LEDs are approximately twice as efficient as the SH version; however, they are significantly more difficult to grow because they require the ability to accurately control the thickness of very thin layers.

3. Optical Extraction

For light to escape from the semiconductor into the epoxy encapsulant, it must strike the semiconductor–epoxy interface at an angle less than or equal to the critical angle for total internal reflection. Because the semiconductor has an index of refraction of approximately 3.5 and typical clear epoxy encapsulants have indices of refraction near 1.5, the critical angle is only $\theta_c = \sin^{-1}(n_1/n_2) = 25$ degrees. Thus, only light rays emitted in a narrow cone can escape from the chip, as shown in Fig. 4(a). The fraction of the total light generated that is in this cone is given by $(1 - \cos\theta_c)/2$ or approximately 5%. Fresnel losses reduce this by another 20%, and therefore only 4% if the light generated in an LED such as that shown in Fig. 4(a) can escape from the chip. If the wide-energy-gap window layers are thick, light from the "side cones" can also escape, as shown in Fig. 4(b). As has been mentioned in Section 1, thick epitaxial layers also help spread the current out from under the top contact, and therefore both SH and DH AlGaAs LEDs currently in production have wide-energy-gap window layers greater than 30-μm thick grown on top of the active layer.

The best device structure from the point of view of optical extraction is one in which the epitaxial layers are grown on a transparent substrate, such as that shown in Fig. 4(c). In this case, six full cones, or approximately 24% of the light generated inside of the chip, can escape on the first pass. AlGaAs red LEDs with structures like this are created by epitaxially growing very thick (greater than 150 μm) wide-energy-gap window layers beneath the

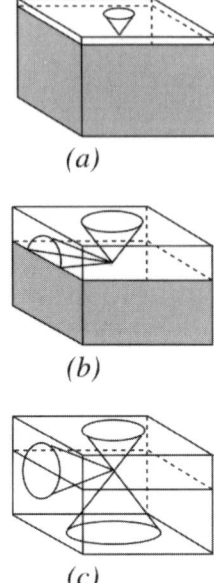

FIG. 4. Schematic showing light-escape cones for three light-emitting diode (LED) device geometries. An LED chip with thin light-emitting layers grown on an absorbing substrate is shown in (a). Only light emitted upward in a narrow cone can escape from the chip. If thick window layers are grown on top of the light-emitting layers, the "side-cones" also can contribute to the light-extraction efficiency, as shown in (b). (c) Light-escape cones for a transparent substrate LED. In this case, six full cones can escape from the LED on the first pass. (Reprinted from Kish, 1995, *Encycl. Chem. Technol.* **15**, 217; with the kind permission of Elsevier Science – NL.)

device layers and then chemically removing the original GaAs substrate (Ishiguro *et al.*, 1983). The resulting double heterostructure–transparent substrate (DH-TS) LEDs are twice as bright as double heterostructure–absorbing substrate (DH-AS) LEDs and approximately four times brighter than SH AlGaAs LEDs. Because of the complex epitaxial growth and device-processing steps, however, they are also more expensive to produce. All three types of AlGaAs red LEDs are currently on the market (Fig. 5).

4. ACTIVE LAYER COMPOSITION

As the direct-energy-gap to indirect-energy-gap transition is approached, the internal efficiency of AlGaAs LEDs falls rapidly. This occurs because more and more of the injected minority-carrier electrons reside in the

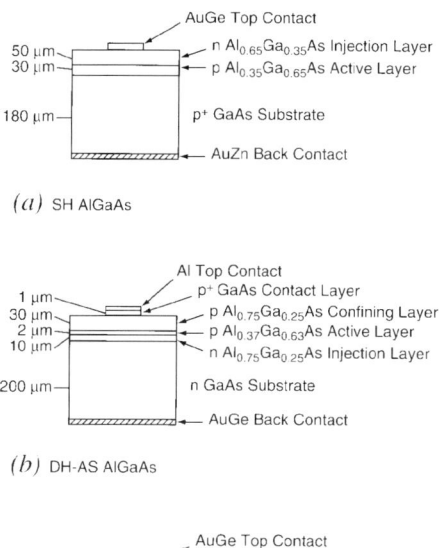
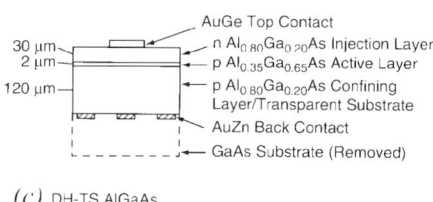

FIG. 5. Chip structures of the three types of AlGaAs light-emitting diodes (LEDs) currently on the market. (a) In single heterostructure (SH) AlGaAs LEDs, both the p-type active layer and the n-type injection layer are very thick (greater than 30 μm). In the double heterostructure–absorbing substrate (DH-AS) AlGaAs structure (b) and the DH transparent substrate (DH-TS) AlGaAs structure (c), a thin (<2.5 μm) active layer is sandwiched between wide-energy gap injection and confining layers. The DH-AS structure can be grown with either a *p*-type top layer (as shown) or with an *n*-type top layer. The original gallium arsenide (GaAs) substrate is removed from the TS-DH AlGaAs substrate. AuGe, gold germanium; AuZn, gold zinc; AlGaAs, aluminum gallium arsenide.

indirect minima and thus are unavailable for radiative recombination. However, the sensitivity of the human eye increases rapidly with increasing photon energy in the red portion of the spectrum. This results in a fairly broad wavelength region over which the photometric efficiency is fairly constant (640 to 660 nm), and different manufacturers have chosen slightly different optimum wavelengths. The human eye has very poor color discrimination capability in the red portion of the spectrum, and these differences in wavelength are almost indistinguishable.

III. Crystal Growth

Several crystal-growth techniques have been employed to grow AlGaAs red LEDs including LPE, metalorganic chemical vapor deposition (MOCVD), and molecular beam epitaxy (MBE). However, the advantages of thick epitaxial layers described in Sections 1 and 3 and the difficulty of growing high-quality AlGaAs layers with high Al content have resulted in LPE being the only technique currently employed for the high-volume production of AlGaAs red LEDs. Both the slow-cooling LPE technique first proposed by Nelson (1963) and a temperature-difference method similar to that described by Nishizawa *et al.* (1973, 1975) and Stringfellow and Greene (1971) are currently in use.

1. MOLECULAR-BEAM EPITAXY AND METALORGANIC CHEMICAL VAPOR DEPOSITION

Molecular beam epitaxy is a crystal-growth technique in which material is deposited by evaporating individually controlled source materials onto a heated substrate in an ultrahigh vacuum chamber (see Cho and Arthur, 1975). Molecular beam epitaxy is capable of producing high-quality AlGaAs heterostructure devices with very uniform thickness and doping characteristics, and it is often employed for the growth of AlGaAs laser diodes such as those used in compact disc players. The design considerations for laser diodes, however, are significantly different from those used for LEDs. In particular, current confinement beneath the contact metallization is a desirable property for a laser structure and one that severely limits the performance of an LED. The growth rates in MBE are very slow (0.1 to 2 μm/h), and therefore the growth of the very thick layers necessary for current spreading and optical extraction in an efficient AlGaAs red LED is impractical. In addition, the lowest reported interface-recombination velocities in MBE grown DH devices with high Al content (Ahrenkiel *et al.*, 1991) are approximately an order of magnitude larger than are those obtainable by LPE (Steranka *et al.*, 1995). It is therefore doubtful that MBE grown structures could attain the same efficiencies as do the LPE devices.

Crystal growth by MOCVD began in the late 1960s (Manasevit and Simpson, 1969), and a great deal of work in the 1970s focused on improving MOCVD source purity and growth technology (Stringellow, 1981). This work finally resulted in the successful room temperature operation of AlGaAs laser diodes in 1977 (Dupuis and Dapkus, 1977). Because MOCVD is a vapor-phase growth process, it is easier to scale it up to larger growth areas than it is to scale up LPE. In addition, the uniformity and surface morphology of MOCVD grown layers are superior to those attainable with LPE. For these reasons, MOCVD was seriously investigated as an alternative technique for the growth of AlGaAs red LEDs. The best efficiency results achieved, however are roughly a factor of two below those obtainable by LPE (Kellert and Moon, 1986). The efficiency of these LEDs varied significantly as a function of drive current which implied that there were a significant number of saturable nonradiative centers present in the active layer of the MOCVD material. Low levels of oxygen and water still present in the source gases were hypothesized to be the source of the nonradiative centers.

2. SLOW-COOLING LIQUID-PHASE EPITAXY

Detailed reviews of the slow-cooling LPE process have been published elsewhere [see Stringfellow (1979) and the books by Casey and Panish (1978) and Kressel and Butler (1977)], and therefore only a brief discussion is presented here. The slow-cooling LPE process involves the growth of single-crystal epitaxial layers from liquid Ga solutions. The first step is usually to bake out the Ga for the growth melts in a purified hydrogen atmosphere for several hours. This is done to eliminate residual oxygen or other impurities that may still be present in the as-purchased Ga (Dawson, 1974). Next, Al metal and GaAs are added to the melts to supply the Al and As necessary for the growth of ternary AlGaAs. Different amounts of Al and GaAs are added to the melts for different composition layers according to the Al–Ga–As phase diagram (Panish and Ilegems, 1972). Either zinc (Zn) or magnesium (Mg) are typically used as p-type dopants and tin (Sn) or tellurium (Te) are used as n-type dopants. The melts are then raised to just above the growth temperature (800 to 900°C). At this temperature, the Al and the dopant atoms completely dissolve in the liquid Ga, and the melts are left there for several hours to ensure that they reach equilibrium. GaAs substrates are then pushed under the melts, and the furnace temperature is slowly lowered (0.1 to 1.0°C/min). As the temperature is lowered, the melts become slightly supersaturated and AlGaAs layers grow on the substrates.

To grow devices with multiple layers, a horizontal graphite slider arrangement is often employed (Fig. 6). Melts for the different layers are in different cavities. The substrate is positioned under one cavity for the growth of one layer and then pushed into the next cavity for the growth of the next layer. The time that the growth surface is exposed during the transfer from one melt to the next must be kept to a minimum to prevent the surface from oxidizing and thus forming defects at the layer interfaces.

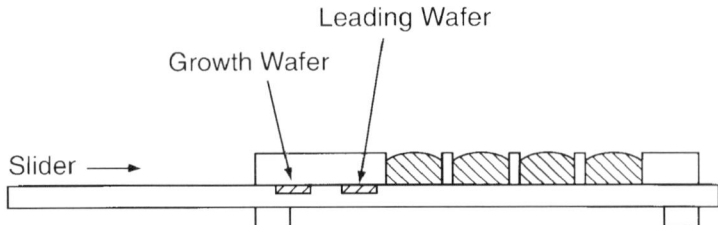

FIG. 6. Schematic of a horizontal graphite slider arrangement used for growing multiple-layer devices. A leading wafer is sometimes used to ensure that the melt is critically saturated when the growth wafer is introduced. (Reprinted from Dawson, 1974, *J. Crystal. Growth* **27**, 89; with the kind permission of The North-Holland Publishing Company.)

For the thick layers grown for the AlGaAs red LED structures, a significant variation in the AlAs mole fraction within the layer can occur during the course of the growth process due to depletion of Al from the melt. This variation can be calculated from the AlGaAs phase diagram (Ilegems and Pearson, 1969) and is shown in Fig. 7 for layers grown from different initial compositions starting at 1000°C. The depletion of Al from the melt is somewhat offset by the fact that the Al distribution coefficient increases at lower temperatures. While fairly uniform composition layers can be grown

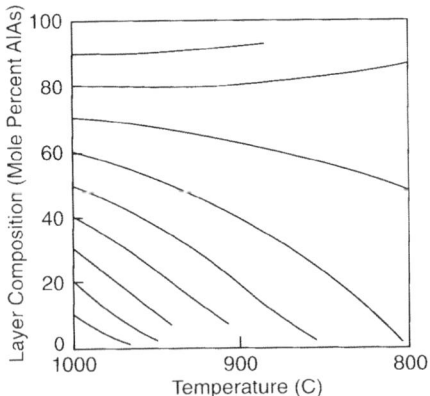

FIG. 7. The variation in the aluminum arsenide (AlAs) mole fraction of the solid grown as an AlGaAs melt is cooled from 1000°C for different starting compositions. The composition grades very little for $x > 0.7$ $Al_xGa_{1-x}As$ layers but grades rapidly for lower AlAs mole fraction starting compositions. AlAs, aluminum arsenide; AlGaAs, aluminum gallium arsenide. (Reprinted from Ilegems and Pearson, 1969, *Conf. Ser.–Inst. Phys.* **7**, 3; with the permission of the Institute of Physics Limited.)

in the 70% to 80% AlAs-mole-fraction region, the growth efficiency (microns of material grown per degree change in melt temperature) is approximately 1.5 times higher for $Al_{0.60}Ga_{0.40}As$ than for $Al_{0.80}Ga_{0.20}As$. Thus, especially for the growth of the very thick layers needed for DH-TS AlGaAs, lower Al-mole-fraction starting compositions are often chosen. The only constraint is that, at the end of the growth, the layer composition must be such that it is still transparent to the light generated in the active layer.

It is very important for the melts to be critically saturated when the substrate is introduced. If the melts are undersaturated, they will dissolve material from the substrate instead of depositing material onto it. If the melts are supersaturated, AlGaAs layers will grow quickly, resulting in poor thickness control and poor crystal quality. Two different techniques have been employed to ensure that the melts are critically saturated when the substrate is introduced: the "floater" method and the "leading-wafer" method.

In the floater method, excess GaAs is added to the melt and, because of its lower density, it floats on the surface of the melt. The melt will dissolve only enough GaAs to reach equilibrium at a given temperature. As long as the melts are not cooled too quickly, they will be critically saturated when the substrate is introduced. One drawback to this method, however, it that AlGaAs grows on the excess GaAs in addition to the substrate as the temperature is lowered. This accentuates the Al-depletion effects mentioned earlier and causes larger composition variations in the layers.

In the leading-wafer method, only the amount of GaAs needed to critically saturate the melt at the growth temperature (determined from the phase diagram) is added. To compensate for slight weighing errors or temperature drifts in the growth chamber, a sacrificial substrate is introduced into the melt prior to the growth substrate (Dawson, 1974), as shown in Fig. 6. Only after the melt has reached equilibrium on the leading wafer (either by depositing material or by etching some away) is the growth wafer introduced. The obvious drawback to this method is that it doubles the substrate cost.

To cost-effectively produce large volumes of AlGaAs red LEDs using the slow-cooling process, LPE growth reactors have been designed to grow simultaneously on a large number of wafers.

3. TEMPERATURE-DIFFERENCE LIQUID-PHASE EPITAXY

In temperature-difference LPE (Nishizawa et al., 1975; Ishimatsu and Okuno, 1989), a 10 to 20°C temperature gradient is established in the melt itself by differentially heating the top and bottom of the melt, as shown in

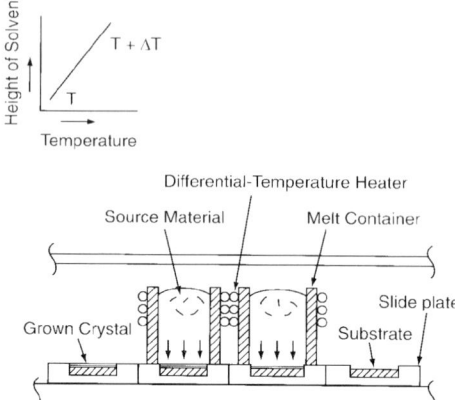

FIG. 8. Schematic of the reactor configuration used in the temperature-difference method. A separate heater for each melt is used to create a temperature difference between the top and bottom of the melt, which drives the deposition. (Reprinted from Ishimatsu and Okuno, 1989 *Optoelectron. Devices Technol.* **4**, 22; with the permission of Mita Press.)

Fig. 8. Excess GaAs is put into the melt to supply the As needed for the crystal growth. The saturation solubility of As in the melt is much higher at the higher temperture in the top of the melt, and it diffuses toward the lower-temperature region because of the concentration and kinetic energy difference. This causes the melt in the lower temperature region to become supersaturated, and epitaxial material is grown on the substrate. Because the growth takes place at a constant temperature, very thick layers can be grown without the compositional grading problems present in the slow-cooling LPE method. As the furnace temperature stays fixed, wafers can continually be sent under the melts until the source materials (Al, GaAs, dopants) are depleted. The growth rate is a function of the temperature difference across the melt; however, morphology problems appear if the growth rate is too high (Nishizawa *et al.*, 1975).

A schematic of a reactor design for continuous growth on many substrates is shown in Fig. 9. Two melts are shown—one p-type and one n-type—and therefore this configuration would be appropriate for growing an SH AlGaAs device structure. Growth substrates are stacked in trays on one side of the furnace, and the device wafers are collected on the other side after they have been pushed through the melts. In this type of reactor design, only two wafers can be grown upon simultaneously, and therefore many such reactors are necessary to produce high volumes of AlGaAs red LEDs.

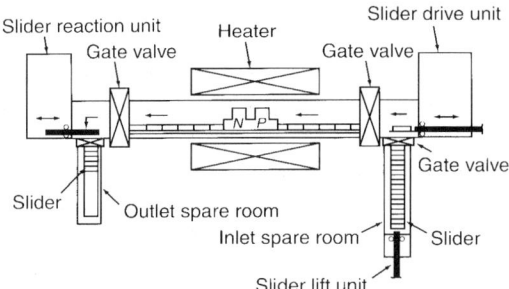

FIG. 9. Automated temperature-difference LPE reactor design. The substrates are loaded in a stack on one side of the reactor and are collected on the other side after passing under the melts. (Reprinted from Ishimatsu and Okuno, 1989 *Optoelectron. Devices Technol.* **4**, 22; with the permission of Mita Press.)

IV. Device Fabrication

To turn the epitaxial material into LED chips, contact metalization must be applied to the top and bottom of the wafer and it must be diced into discrete chips. In the case of DH-TS AlGaAs, the GaAs substrate must also be chemically removed. Brief descriptions of these processes follow.

1. Contact Formation

When a metal is deposited onto a semiconductor, a Schottky barrier is formed. However, if the doping level at the semiconductor surface is very high (over 10^{19} cm^{-3}), the space-charge layer associated with the Schottky barrier is thin enough for electrons to easily tunnel through the barrier under an applied voltage and the contact becomes "ohmic." It is very difficult to achieve such high electrically active doping levels in LPE growth (especially for AlGaAs layers with high Al content). Thus, such high surface concentrations must be diffused into the semiconductor. This can be accomplished either as a separate step in the wafer fabrication process or by including dopant atoms in the contact metallization, which can then diffuse into the semiconductor during a contact-alloying (heat-treatment) step. This second approach is typically employed for contact formation to n-type material because n-type diffusions are quite difficult to perform successfully. However, this approach is also frequently employed to make p-type contacts.

For n-type contacts to both GaAs and AlGaAs, a gold (Au)–germanium (Ge) (0.5% to 12% Ge) alloy contact is most frequently employed, whereas for p-type contacts, either Au–zinc (Zn) (1% to 6% Zn) or simply Al is used. Before depositing the contact metalization, the wafer surface must first be cleaned to remove oxides and other contaminants that can inhibit adhesion of the contact metal. This is accomplished by dipping the wafers in an oxide etch or a weak material etch. The contact metal is then deposited by either evaporation or sputtering. The bottom contacts on AS chips are deposited in sheet form. However, because the contact area can be light absorbing after the alloying step, the area covered by the back-contact metal needs to be as small as possible to avoid an excessive loss of efficiency in TS chips. For this reason, the back-contact metalization on TS chips is typically deposited in a pattern of small dots covering only 25% to 35% of the back surface of the wafer. The dots must be small enough that each chip has several of them, and the pattern is created by either evaporating through a metal mask that has small holes in it or by photolithographically patterning and etching a sheet-deposited film. The back-contact films are typically several hundred nanometers thick.

To make low-resistance contacts, the metal films are given short heat treatments (400 to 500°C for 10 to 30 min) in a forming-gas (85% nitrogen, 15% hydrogen) atmosphere. During this alloying step the dopant atoms (either Ge or Zn) diffuse into the semiconductor, creating the highly doped regions required to form ohmic contacts.

The top-contact metallization must be suitable for high-speed wire bonding. For high-quality Au wire bonds to be formed, the diameter of the top contact must be at least 100 μm, and the contact surface must be a relatively pure Au or Al film with no oxides or other contaminants present. Because AlGaAs is such a brittle material, the top contact metallization must typically be quite thick (2 to 3 μm) in order to avoid cracking the chips or "cratering" the contact area during the wire bonding process. To avoid having Ga diffuse to the top surface of the contact (where it can oxidize), the thick Al or Au contact pads are typically deposited after the ohmic-metal alloying step, and sometimes a barrier metal such as titanium tungsten (TiW) is deposited between the ohmic-contact metal and the thick contact pad. The top contacts are typically spaced 300- to 400-μm apart on the wafer surface, and roughly 10 to 20 thousand LED chips can be made from one 2-inch wafer.

2. DICING

To cut the wafer into individual LEDs, the wafers either are sawn with a thin diamond-impregnated blade or are scribed and broken. The advantage of the sawing process is that it can be used to separate the p-n junctions of the individual chips so that they can be tested while still in wafer form. This is accomplished by cutting grooves in the wafer that are deep enough to cut through the p-n junction but not so deep that the wafer cracks or breaks. After such a partial sawing step, the wafers are etched to remove residual saw damage prior to testing. LED chips that do not have acceptable light-output efficiency or electrical characteristics are covered with ink by the tester. The ink causes the automated pick-and-place machines to skip over such chips, and they are not assembled into final-product form. This increases the final-product yield and reduces the overall manufacturing cost.

3. SELECTIVE ETCHING

To form DH-TS AlGaAs chips, the epitaxial layers are grown very thick (greater than 150 μm) and the original GaAs substrate is removed, leaving only the AlGaAs layers. The GaAs substrate is usually removed in a two-step process. First, it is thinned either by lapping or chemical etching (see Kern, 1978, for a review of etching technology). Then it is etched in a selective etch—one that etches GaAs much more rapidly than it does AlGaAs. There are several types of such etches; however the commonly used versions are mixtures of hydrogen peroxide (H_2O_2) and ammonium hydroxide (NH_4OH). Such etches must either be sprayed (LePore, 1980) or mechanically agitated (Logan and Reinhart, 1973) to be effective. They work by growing an oxide film that cracks into small pieces and floats away from the surface, allowing new oxide sheets to form. Spraying or mechanical agitation helps to remove the oxides from the crystal surface. Differences in etch rates between GaAs and $Al_xGa_{1-x}As$ vary with the Al mole fraction x; however, factors of over 30 times are attainable with such etches.

V. Comparison with Other Types of Red Light-Emitting Diodes

The first visible LEDs were red LEDs made of $GaAs_{0.6}P_{0.4}$ grown on GaAs substrates (Holonyak and Bevacqua, 1962). $GaAs_{0.6}P_{0.4}$ has a direct energy gap; however, unlike AlGaAs, its lattice constant differs appreciably from that of GaAs (*see* Fig. 2). As a result, such films have very high defect densities and efficient heterostructure devices cannot be made. The $GaAs_{0.6}P_{0.4}$ LEDs are formed by selectively diffusing Zn into a thick constant-composition n-type layer, and lamps made with such chips have luminous efficiencies of approximately 0.3 lm/A (Fig. 10).

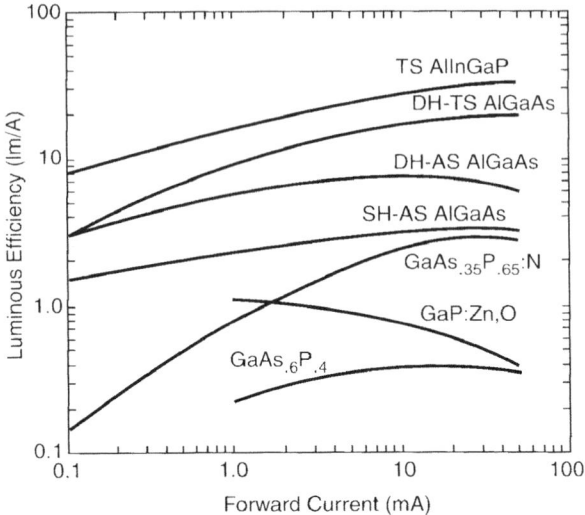

FIG. 10. Performance comparison of different red light-emitting diode (LED) technologies as a function of drive current. TS AlInGaP, transparent substrate aluminum indium gallium phosphide; DH-TS AlGaAs, double heterostructure–transparent substrate aluminum gallium arsenide; DH-AS AlGaAs, double heterostructure–absorbing substrate aluminum gallium arsenide; SH-AS AlGaAs, single heterostructure–absorbing substrate aluminum gallium arsenide; GaP:Zn–O, zinc-oxygen-doped gallium phosphide; GaAsP, gallium arsenide phosphide; GaAsP:N, nitrogen-doped gallium arsenide phosphide.

Zinc–oxygen (Zn–O)-doped GaP red LEDs followed in the late 1960s (Saul et al., 1969). They are formed by introducing Zn–O deep recombination centers into GaP. Such centers have very high radiative recombination efficiencies, and because GaP is transparent to the approximately 700-nm light emitted by this center, the optical extraction efficiency is also high. However, the sensitivity of the human eye is fairly poor at 700 nm, and the Zn–O centers rapidly become saturated as the current density is increased, causing the efficiency to fall. This limits the luminous efficiency at typical drive currents (10 to 20 mA) to only about 0.5 lm/A.

In the early 1970s it was discovered that nitrogen forms a very efficient isoelectronic recombination center in GaAsP (Groves et al., 1971), and red LEDs made of nitrogen-doped $GaAs_{0.4}P_{0.6}$ layers grown on GaP substrates were soon developed. These red LEDs have luminous efficiencies of just under 3 lm/A at typical drive currents and are comparable in performance to SH AlGaAs LEDs. (Their internal efficiency is not as high; however, because they are TS chips, their extraction efficiency is higher.) As has been mentioned earlier, DH-AS AlGaAs LEDs are roughly twice as bright as are the SH versions, and DH-TS AlGaAs LED chips are still another factor or two brighter. The only red LED technology having a higher performance than does DH-TS AlGaAs is the recently developed TS AlInGaP technology (Kish et al., 1994). The internal efficiency of the TS AlInGaP material is comparable to that of TS AlGaAs; however, it emits at shorter wavelengths (630 nm instead of 650 to 660 nm for DH-TS AlGaAs) to which the human eye is more sensitive. Because each red LED technology has its own unique cost-performance trade-off, red LEDs made with all of these different types of materials and structures are still commercially available.

VI. Temperature-Dependent Properties

In addition to being useful for LED applications design (Chapter 6), knowledge of the temperature-dependent properties of an LED can provide insight into the radiative and nonradiative recombination mechanisms that are active in the device. In this section, a summary of many of the temperature-dependent properties of DH AlGaAs LEDs is presented along with an analysis of the electroluminescent (EL) decay time data. This analysis provides an estimate of the radiative recombination coefficient in the p-type active layer of these devices and the temperature dependence of the nonradiative lifetime (Steranka et al., 1995).

1. I-V Curves and Forward Voltage

Log I-V curves of a DH-AS red AlGaAs LED at 89 and 298 K are shown in Fig. 11. The current passing through an LED as a function of the applied voltage can be expressed as

$$I(V) = I_1 \exp\left(\frac{q(V - I\mathbb{R}_s)}{kT}\right) + I_2 \exp\left(\frac{q(V - I\mathbb{R}_s)}{2kT}\right) \quad (1)$$

where q is the charge of an electron, V is the voltage across the device, I_1 is the saturation current due to diffusion, I_2 is the saturation current due to nonradiative recombination, and $\mathbb{R}_s$ is the series resistance of the device. The two components of the current are often referred to as the kT and $2kT$ currents. The kT current is due to diffusion of minority carriers across the depletion layer, and it gives rise to radiative recombination in the LED. The $2kT$ current in DH AlGaAs LEDs has been shown to be primarily due to nonradiative surface recombination at the perimeter of the p-n junction (Henry et al., 1978). From Fig. 11, we see that at low temperatures, the $2kT$

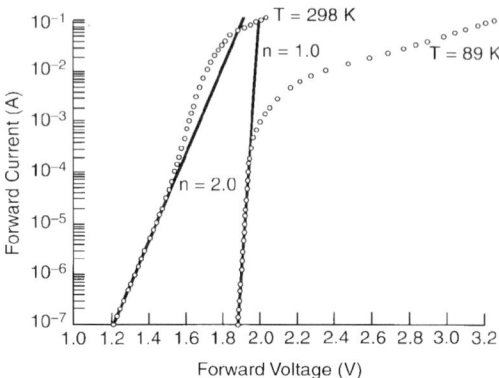

FIG. 11. Log I-V curves for a double heterostructure–absorbing substrate (DH-AS) AlGaAs red light-emitting diode (LED) at 298 and 89 K. At low temperatures, the $2kT$ current is frozen out and 100% of the current is kT current ($n = 1.0$). At room temperature, a significant amount of $2kT$ ($n = 2.0$) current is present. AlGaAs, aluminum gallium arsenide.

current is frozen out and 100% of the forward current is kT current. Between 89 K and room temperature, the $2kT$ current gradually grows and is approximately 10% of the total current under typical LED drive conditions (10 to 20 mA).

At a given forward current, the forward voltage across AlGaAs red LEDs decreases linearly with increasing temperature over the -55 to $+100°C$ temperature range that is important for LED applications. The temperature coefficient is $-1.7\,\mathrm{mV/°C}$.

2. Electroluminescent Spectra

Electroluminescent (EL) spectra at a constant bias current density of $27.4\,\mathrm{A/cm^2}$ at various temperatures between 89 and 390 K are shown in Fig. 12. At temperatures below approximately 180 K, distinct band-to-band (BB) and band-to-acceptor (BA) peaks are present in the spectra. At 89.9 K, the BB and BA peaks are comparable in magnitude, which means that roughly half of the total radiative recombination takes place through BA transitions at this temperature.

In the -55 to $+100°C$ temperature range, the spectra shift to longer wavelengths and increase in spectral width as the temperature is increased. The peak position moves linearly with temperature by $+0.16\,\mathrm{nm/°C}$.

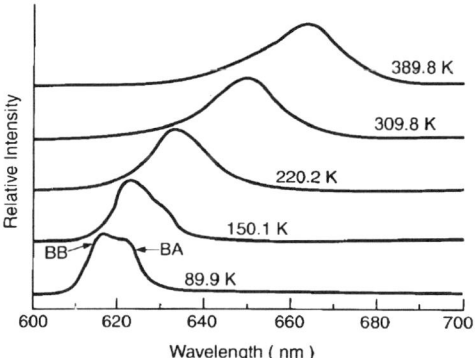

FIG. 12. Electroluminescent (EL) spectra at various temperatures for a double heterostructure–absorbing substrate (DH-AS) AlGaAs red light-emitting diode (LED) at a constant bias-current density of 27.4 A/cm². At temperatures below approximately 180 K, both band-to-band (BB) and band-to-acceptor (BA) transitions are observed. AlGaAs, aluminum gallium arsenide. (Reprinted from Steranka et al., 1995, J. Electron. Mater. **24**, 1411; with the permission of the Minerals, Metals and Materials Society.)

3. EFFICIENCY

Results of relative radiometric efficiency measurements at a constant bias current taken as a function of temperature from 90 to 400 K are shown in Fig. 13. These results show that the efficiency increases as the temperature is decreased but saturates at a little less than two times the efficiency at room temperature. The absolute external quantum efficiency at low temperatures is approximately 7%, which for LEDs with this structure is known to correspond to roughly 100% internal quantum efficiency (Steranka et al., 1995). If we assume that internal absorption does not vary significantly with temperature (which is reasonable for these DH-AS AlGaAs LEDs), then we can conclude that the internal quantum efficiency is slightly over 50% at room temperature in these visible red LEDs.

For visible LED applications, the variation in photometric (eye-response-normalized) efficiency with temperature is a more important parameter. It differs from the radiometric efficiency data in Fig. 13 as the EL peak position shifts to longer wavelengths with increasing temperature where the response of the human eye is weaker. Photometric efficiency data from -55 to $+100°C$ normalized to the room temperature value are shown in Fig. 14. The data are fit with the empirical relationship

$$\frac{\text{Efficiency } (T_1)}{\text{Efficiency } (T_2)} = \exp\left(\frac{-(T_1 - T_2)}{T'}\right) \qquad (2)$$

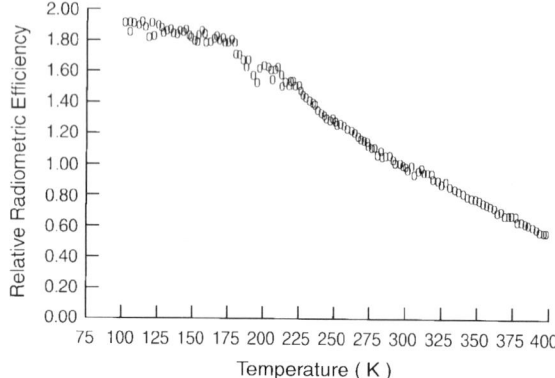

FIG. 13. Radiometric efficiency of a double hetrostructure–absorbing substrate (DH-AS) AlGaAs light-emitting diode (LED) as a function of temperature normalized to the room temperature value. The efficiency saturates at low temperatures at roughly twice the room temperature value. AlGaAs, aluminum gallium arsenide. (Reprinted from Steranka et al., 1995, J. Electron. Mater. **24**, 1409; with the permission of the Minerals, Metals and Materials Society.)

and T' is determined from the fit to be $T' = 105$ (photometric). A similar fit to the radiometric data in Fig. 13 yields $T' = 225$ (radiometric) for DH-AS AlGaAs. Because of internal absorption losses that vary somewhat with temperature, the efficiency of DH-TS AlGaAs changes a bit more rapidly with temperature. Similar fits to data for DH-TS AlGaAs yield $T' = 83$ (photometric) and $T' = 160$ (radiometric).

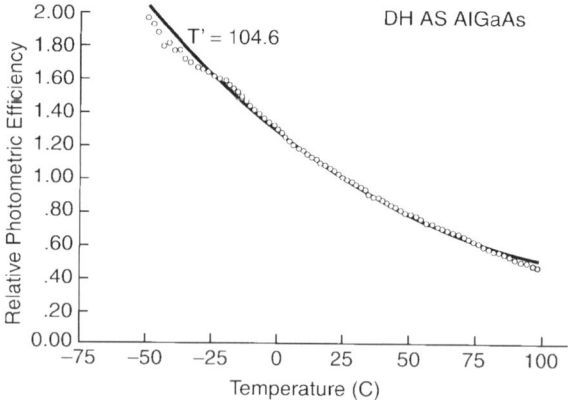

FIG. 14. Photometric efficiency of a double heterostructure–absorbing substrate (DH-AS) AlGaAs light-emitting diode (LED) as a function of temperature normalized to its room temperature value. The fit to the data is from Eq. (2). AlGaAs, aluminum gallium arsenide.

4. MINORITY-CARRIER LIFETIME AND ANALYSIS

Measurements of the electroluminescent (EL) decays at 100, 240, and 380 K are shown in Fig. 15. The EL decay times were measured using a small-signal approach (Su and Olshansky, 1982) at a bias current density of 20 A/cm². The minoritycarrier lifetimes obtained from fits to the data decrease by over an order of magnitude as temperature is lowered from 400 to 90 K.

The minority-carrier lifetime (τ_{mc}) in the p-type active layer of these near-crossover LEDs can be expressed as

$$\frac{1}{\tau_{mc}} = \frac{1}{\left[\frac{n_\Gamma}{n}\right]^{-1} \tau_{r\Gamma}} + \frac{1}{\tau_{nr}} \quad (3)$$

where τ_{nr} is the nonradiative lifetime, n_Γ/n is the fraction of electrons in the Γ conduction band, and $\tau_{r\Gamma}$ is the radiative lifetime for electrons in the Γ conduction band minimum. In the low-injection limit, n_Γ/n is given by

$$\frac{n_\Gamma}{n} = \frac{1}{1 + \left(\frac{m_x}{m_\Gamma}\right)^{3/2} \exp(-(E_x - E_\Gamma)/kT) + \left(\frac{m_L}{m_\Gamma}\right)^{3/2} \exp(-(E_L - E_\Gamma)/kT)} \quad (4)$$

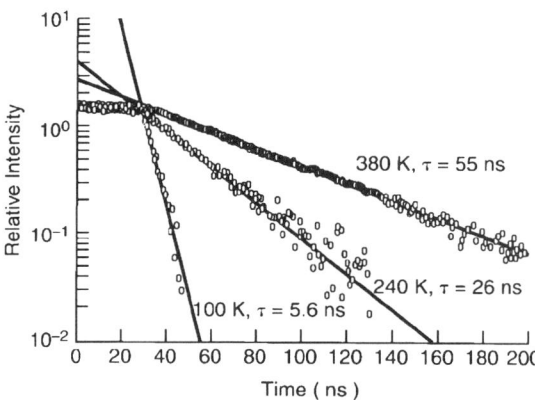

FIG. 15. Electroluminescent (EL) decay measurements at three temperatures for a double heterostructure–absorbing substrate (DH-AS) AlGaAs red light-emitting diode (LED). The minority-carrier lifetimes determined from the fits to the data decrease by over an order of magnitude as the temperature is decreased from 400 to 89 K. AlGaAs, aluminum gallium arsenide. (Reprinted from Steranka et al., 1995, J. Electron. Mater. 24, 1409; with the permission of the Minerals, Metals and Materials Society.)

where m_Γ, m_x, and m_L are the density of states masses for the Γ, X, and L conduction bands respectively, and E_Γ, E_x, and E_L are the corresponding energy minima. Values for these parameters can be found in Casey and Panish (1978). In the same notation, the internal quantum efficiency can be expressed as

$$\eta_i = \frac{1}{1 + \dfrac{\left[\dfrac{n_\Gamma}{n}\right]^{-1} \tau_{r\Gamma}}{\tau_{nr}}} \tag{5}$$

If we assume that the internal efficiency at low temperatures is 100%, then the data in Fig. 13 can be used to calculate the internal efficiency as a function of temperature. Using these values, the results of the τ_{mc} measurements, and Eqs. (3) to (5), both the radiative and nonradiative lifetimes can be calculated as a function of temperature. The results are shown in Fig. 16. The nonradiative lifetime was found to be virtually independent of temperature from 220 to 400 K. Neglecting BA transitions and the effects of photon recycling, the radiative lifetime in the low-injection limit can be expressed as

$$\frac{1}{\tau_{r\Gamma}} = Bp \tag{6}$$

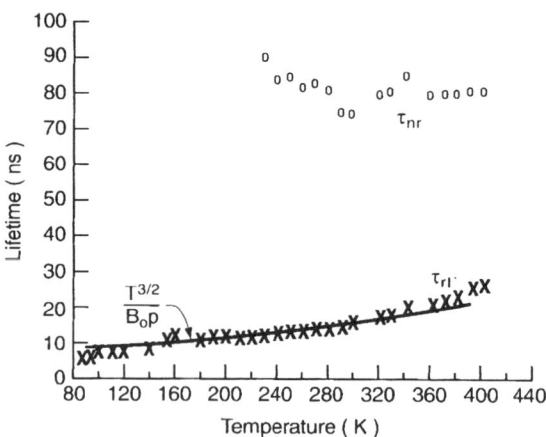

FIG. 16. Radiative and nonradiative lifetimes as a function of temperature for a double heterostructure–absorbing substrate (DH-AS) AlGaAs red light-emitting diode (LED). A fit to the data at temperatures greater than 180 K yields a value for the radiative combination coefficient of $B = 1.3 \times 10^{-10}$ cm^3 sec^{-1} at room temperature. AlGaAs, aluminum gallium arsenide. (Reprinted from Steranka et al., 1995, J. Electron. Mater. 24, 1410; with the permission of the Minerals, Metals and Materials Society.)

where $B = B_0 T^{-3/2}$ is the radiative recombination coefficient and p is the free-hole concentration in the active layer. By using measured values of the free-hole concentration as a function of temperature (Steranka *et al.*, 1995) the data in Fig. 16 were fit at tempertures above 180 K (where BA transitions are unimportant). The fit yields a room temperature value for B of 1.3×10^{-10} cm^3 sec^{-1}, which is close to the value reported by Nelson and Sobers (1978) and identical to the values reported by 'tHooft (1981) and Hwang and Dymet (1973) for GaAs.

The fact that the nonradiative lifetime was independent of temperture implies that the capture cross section of the dominant nonradiative recombination process must decrease slowly with temperature. While traps with such capture cross sections have been observed in both GaP and GaAs, the process responsible for the nonradiative recombination is unknown (Henry and Lang, 1977). If all of the nonradiative recombination is ascribed to interface recombination, then an upper bound for the interface-recombination velocity of approximately 1000 cm^2/sec^{-1} is obtained for these heterostructures with high AlAs mole fraction AlGaAs layers.

VII. Reliability Characteristics

As is typical for most LEDs, AlGaAs LEDs can be operated for many years under conditions of typical use with little loss in efficiency of light output. The data in Fig. 17 show the results of change in light-output efficiency measurements after stress of 47 A/cm^2 at +55°C on DH-AS AlGaAs LEDs at times of up to 10,000 hours (approximately one year). Even after 10,000 hours of stress, most of the LEDs are brighter than they were initially. This increase in brightness may be the result of recombination-enhanced annealing of point defects or other nonradiative recombination centers in the active layers of these devices (Kimerling and Lang, 1975).

The general reliability performance characteristics of AlGaAs red LEDs are quite good; however, operation at very low temperatures or in high-temperature high-humidity environments can cause problems for some of the AlGaAs device structures.

1. Low-Temperature Operation

At low temperatures, the clear epoxy encapsulants typically used to make LED lamps can exert significant stresses on the LED chips. This is primarily

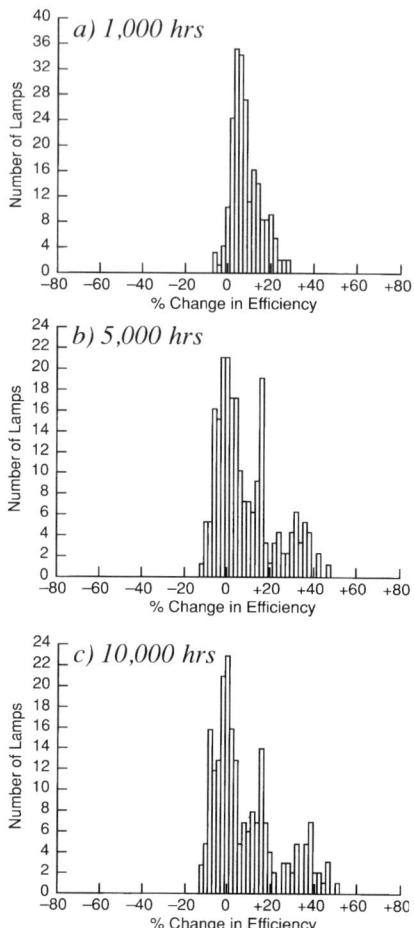

FIG. 17. Histograms of change in efficiency values for 215 double heterostructure–absorbing substrate (DH-AS) AlGaAs red light-emitting diodes (LEDs) after (1) 1000 hours, (b) 5000 hours, and (c) 10,000 hours of 55°C, 47 A/cm^2 stress. Even after 10,000 hours of stress, most lamps are brighter than they were initially. AlGaAs, aluminum gallium arsenide.

the result of the difference in the thermal expansion coefficient between the epoxy and the semiconductor. The modulus of elasticity and the linear expansion characteristics as a function of temperature for a typical clear epoxy are shown in Fig. 18. At the high temperatures employed to cure epoxies, the epoxy is soft (low modulus of elasticity), and little stress is applied to the LED chip. As the temperature is lowered below the glass

3 AlGaAs Red Light-Emitting Diodes

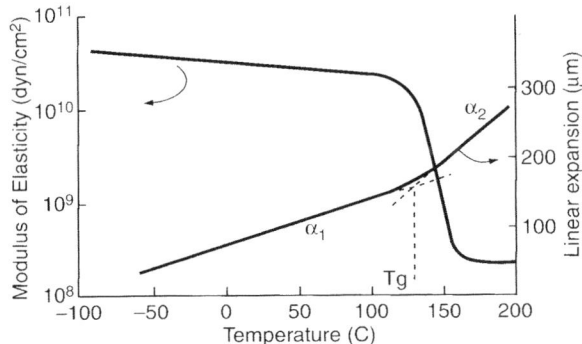

FIG. 18. Modulus of elasticity and linear expansion of a typical clear epoxy used for light-emitting diode (LED) encapsulation. Above the glass transition temperature (T_g) of the epoxy, it is soft (low modulus of elasticity) and has a high thermal expansion coefficient, $\alpha_2 = 180$ ppm/°C. Below the glass transition temperature, the epoxy becomes much harder and has a lower thermal expansion coefficient, $\alpha_1 = 60$ ppm/°C. However, α_1 still is approximately 10 times greater than is the expansion coefficient of AlGaAs, and this leads to large stresses on the semiconductor at low tempertures. AlGaAs, aluminum gallium arsenide.

transition temperature (T_g) of the epoxy, the epoxy becomes hard (high modulus of elasticity), and it continues to increase in hardness as the temperature is further decreased. The thermal expansion coefficient of the epoxy below T_g (α_1) typically ranges from 50 to 80 ppm/°C, whereas that of AlGaAs ranges between 5.7 and 6.4 ppm/°C (Adachi, 1985). Thus, the epoxy contracts more rapidly with temperature than does the semiconductor. This difference in thermal expansion coefficient as well as the increase in the modulus of elasticity at low temperatures of the epoxy lead to very high stresses on the LED chips at low temperatures.

Operating LEDs while mechanical stress above a certain threshold is applied has been shown to cause degradation in GaAs LEDs (Zaeschmar and Speer, 1979), and it appears to happen in AlGaAs red LEDs as well. The average change in efficiency as a function of stress time at $-20°$C for a group of 20 DH-AS AlGaAs LEDs is shown in Fig. 19. The efficiency falls rapidly in the first 24 hours and is down to 25% of the initial value at 1000 hours. A photograph of the light-up pattern of one of the chips from this evaluation that has gone through 1000 hours of low-temperature operation in shown in Fig. 20. Many dark-line defects (DLDs) have formed in the active layer of the device, dramatically reducing the light-output efficiency.

Light-emitting diode lamps made with soft epoxies or epoxies with low thermal-expansion coefficients exhibit dramatically reduced rates of light-output degradation and DLD formation after low-temperature stress. In

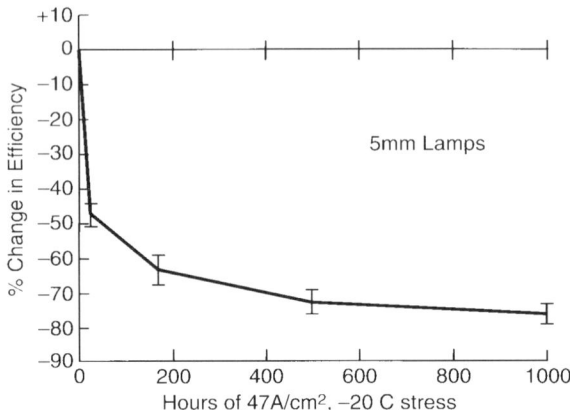

FIG. 19. Change in efficiency of 20 5-mm double heterostructure–absorbing substrate (DH-AS) AlGaAs light-emitting diode (LED) lamps as a function of hours of operation of 47 A/cm^2 at $-20°C$. The chips were unshaped, 10-mils (254 μm) tall, and 10.5×10.5 mils^2 (7.11×10^{-4} cm^2) in area. AlGaAs, aluminum gallium arsenide.

FIG. 20. Picture of the light-up pattern of a double heterostructure–absorbing substrate (DH-AS) AlGaAs red light-emitting diode (LED) chip that was stressed at a current density of 47 A/cm^2 in an epoxy-encapsulated lamp for 1000 hours' duration at $-20°C$. Many dark-line defects (DLDs) are visible in the active layer of the chip and the efficiency of the lamp decreased by over 70% due to the epoxy-stress–induced DLD formation. AlGaAs, aluminum gallium arsenide.

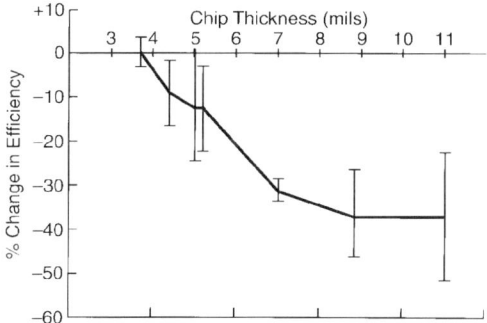

FIG. 21. Change in efficiency of groups of 20 double heterostructure–absorbing substrate (DH-AS) AlGaAs red light-emitting diode (LED) lamps after 24 hours of operation at a current density of 47 A/cm^2 and an ambient temperature of $-20°C$ as a function of LED chip thickness. Thin chips exhibit significantly less loss of light output than do thicker chips. AlGaAs, aluminum gallium arsenide.

addition, three changes in LED chip geometry impact the low-temperature reliability performance: chip size, thickness, and shape. Figures 21 and 22 show the results of change in light-output efficiency measurements made after 24 hours of operation at an ambient temperature of $-20°C$ on DH-AS AlGaAs red LED chips of different thicknesses, sizes, and shapes. Thinner chips, larger chips, and chips with beveled top corners all typically exhibit better performance. All these changes in geometry have been shown to result

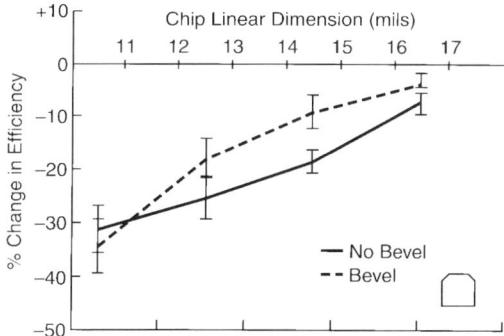

FIG. 22. Change in efficiency of groups of 20 double heterostructure–absorbing substrate (DH-AS) AlGaAs red light-emitting diode (LED) lamps after 24 hours of operation at an applied current of 20 mA and an ambient temperature of $-20°C$ as a function of LED chip size. Larger chips and chips with beveled top corners exhibit less loss of light output than do smaller unshaped chips. AlGaAs, aluminum gallium arsenide.

in lower stresses being applied to the semiconductor by finite-element analysis (FEA) modeling of LED lamps. By using large chips or by carefully shaping them, and using the right epoxies, stress-induced degradation problems can be minimized. SH AlGaAs and DH-TS AlGaAs chips are less sensitive to epoxy stress, and low-temperature operation is usually not an issue for these types of AlGaAs chips.

2. HIGH-TEMPERATURE, HIGHT-HUMIDITY OPERATION

One of the typical environmental stress tests run on semiconductor components is forward-biased operation under high-temperature (85°C) and high-humidity (85% R.H.) conditions. Such stressing of AlGaAs red LEDs results in a rapid loss of light-output efficiency, particularly for DH-TS AlGaAs structures. The light-output efficiency of such LEDs can decrease by more than 50% after less than 500 hours of 85% R.H. forward-biased stress at 85°C. This loss of efficiency is due to the formation of a dark absorbing oxide on the surface of the LEDs (Fig. 23). Under more typical operating conditions, this oxide forms very slowly; however it can still result in a significant loss of light-output efficiency after several years of operation. Recent work on wet atmospheric oxidation of AlGaAs has shown that a stable clear oxide can be formed on AlGaAs devices, which results in a

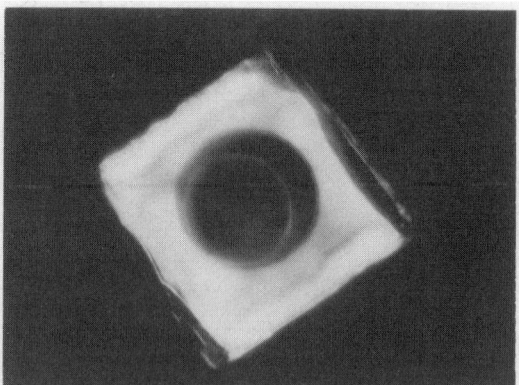

FIG. 23. Picture of a forward-biased double heterostructure–absorbing substrate (DH-AS) AlGaAs light-emitting diode (LED) chip after 1500 hours of operation at a current density of 27.4 A/cm^2 in an 85°C, 85% relative humidity (R.H.) ambient. A light-absorbing oxide formed on the surface of the chip causing a significant loss of efficiency in light output. The oxide is thickest on the sides of the chip where the high aluminum content layers are exposed. AlGaAs, aluminum gallium arsenide.

significant improvement in performance under 85°C, 85% R.H. testing (Richard et al., 1995). This approach and other types of passivating films are under investigation and may lead to the elimination of this reliability issue for DH-TS AlGaAs red LEDs.

Acknowledgments

The author thanks his colleagues at Hewlett-Packard who contributed to the work reported in this chapter. In particular, I thank the AlGaAs engineering development team: Dennis DeFevere, Michael Camras, Lou Cook, Chin-Wang Tu, Serge Rudaz, David McElfresh, Dan Steigerwald, Kwang Park, Dick Nelson, Karen Griswold, Jeff Rogers, and Wayne Snyder. Excellent technical assistance was provided by Charlotte Balassa, Anne Davis, Rick Pettit, Susan Wu, Paul Chow, Andrea Seymour, Cassandra Archer, Nora Felmoca, and Pat Tang. I especially thank Dimokratia Patterakis who made many of the characterization and reliability performance measurements reported in this chapter.

References

Adachi, S. (1985). *J. Appl. Phys.* **58**, R1.
Ahrenkiel, R. K., Keyes, T. C., Shen, T. C., Chyi, J. I., and Morkoç, H. (1991). *J. Appl. Phys.* **69**, 3094.
Alferov, Zh. I., Amosov, V. I., Garbuzov, D. Z., Zhilyaev, Yu. V., Konnikov, S. G., Kop'ev, P. S., and Trofim, V. G. (1973). *Sov. Phys.—Semicond. (Engl. Transl.)* **6**, 1620.
Alferov, Zh. I., Andreev, V. M., Garbuzov, D. Z., and Rumyantsev, V. D. (1975). *Sov. Phys.—Semicond. (Engl. Transl.)* **9**, 305.
Casey, H. C., Jr., and Panish, M. B. (1978). "Heterostructure Lasers." Academic Press, New York.
Cho, A. Y., and Arthur, J. R. (1975). *Progr. Solid State Chem.* **10**, 157.
Cook, L. W., Camras, M. D., Rudaz, S. L., and Steranka, F. M. (1988). In "Proceedings of the International Symposium on GaAs and Related Compounds" (A. Christon and H. S. Rupprecht, eds.), Vol. 91, p. 777. Inst. Phys., Bristol.
Craford, M. G., and Steranka, F. M. (1994). *Encycl. Appl. Phys.* **8**, 485.
Dawson, L. R. (1974). *J. Cryst. Growth* **27**, 86.
Dawson, P., Wilson, B. A., Tu, C. W., and Miller, R. C. (1986). *Appl. Phys. Lett.* **48**, 541.
Dupuis, R. D., and Dapkus, P. D. (1977). *Appl. Phys. Lett.* **31**, 466.
Groves, W. O., Herzog, A. H., and Craford, M. G. (1971). *Appl. Phys. Lett.* **19**, 184.
Henry, C. H., and Lang, D. V. (1977). *Phys. Rev. B.* **15**, 989.
Henry, C. H., Logan, R. A., and Merritt, F. R. (1978). *J. Appl. Phys.* **49**, 3530.
Holonyak, N., Jr., and Bevacqua, S. F. (1962). *Appl. Phys. Lett.* **1**, 82.
Hwang, C. J., and Dymet, J. C. (1973). *J. Appl. Phys.* **44**, 3240.
Ilegems, M. and Pearson, G. L. (1969). *Conf. Ser.—Inst. Phys.* **7**, 3.

Ishiguro, H., Sawa, K., Nagao, S., Yamanaka, H., and Koike, S. (1983). *Appl. Phys. Lett.* **43**, 1034.
Ishimatsu, S., and Okuno, Y. (1989). *Optoelectron.-Devices Technol.* **4**, 21.
Kellert, F. G., and Moon, R. L. (1986). *J. Electron. Mater.* **15**, 13.
Kern, W. (1978). *RCA Rev.* **39**, 278.
Kimerling, L. C., and Lang, D. V. (1975). "Lattice Defects in Semiconductors 1974," p. 589. Inst. Phys., London.
Kish, F. A. (1995). *Encycl. Chem. Technol.* **15**, 217.
Kish, F. A., Steranka, F. M., Defevere, D. C., Vanderwater, D. A., Park, K. G., Kuo, C. P., Osentowski, T. D., Peanasky, M. J., Yu, J. G., Fletcher, R. M., Steigerwald, D. A., and Craford, M. G. (1994). *Appl. Phys. Lett.* **64**, 2839.
Kressel, H., and Butler, J. K. (1977). "Semiconductor Lasers and Heterojunction LEDs." Academic Press, New York.
Kumagai, O., Kawai, H., Mori, Y., and Kaneko, K. (1984). *Appl. Phys. Lett.* **45**, 1322.
LePore, J. J. (1980). *J. Appl. Phys.* **51**, 6441.
Logan, R. A., and Reinhart, F. K. (1973). *J. Appl. Phys.* **44**, 4172.
Manasevit, H. M., and Simpson, W. I. (1969). *J. Electrochem. Soc.* **116**, 1725.
Nelson, H. (1963). *RCA Rev.* **24**, 603.
Nelson, R. J., and Sobers, R. G. (1978). *J. Appl. Phys.* **49**, 6103.
Nishizawa, J., and Suto, K. (1977). *J. Appl. Phys.* **48**, 3484.
Nishizawa, J., Shinozaki, S., and Ishida, K. (1973). *J. Appl. Phys.* **44**, 1638.
Nishizawa, J., Okuno, Y., and Tadano, H. (1975). *J. Cryst. Growth*, **31**, 215.
Panish, M. B., and Ilegems, M. (1972). *Prog. Solid State Chem.* **7**, 39.
Richard, T. A., Holonyak, N., Jr., Kish, F. A., Keever, M. R., and Lei, C. (1995). *Appl. Phys. Lett.* **66**, 2972.
Rupprecht, H., Woodall, J. M., and Pettit, G. D. (1967). *Appl. Phys. Lett.* **11**, 81.
Saul, R. H., Armstrong, J., and Hackett, W. H., Jr. (1969). *Appl. Phys. Lett.* **15**, 229.
Steranka, F. M., DeFevere, D. C., Camras, M. D., Tu, C. W., McElfresh, D. K., Rudaz, S. L., Cook, W. L., and Snyder, W. L. (1988). *Hewlett-Packard J.* **39**, 84.
Steranka, F. M., DeFevere, D. C., Camras, M. D., Rudaz, S. L., McElfresh, D. K., Cook, L. W., Snyder, W. L., and Craford, M. G. (1995). *J. Electron. Mater.* **24**, 1407.
Stringfellow, G. B. (1979). *In* "Crystal Growth: A Tutorial Approach" (W. Bardsley, D. T. J. Hurle, and J. B. Mullin, eds.), p. 217. North-Holland Publ., Amsterdam.
Stringfellow, G. B. (1981). *J. Cryst. Growth* **53**, 42.
Stringfellow, G. B., and Greene, P. E. (1971). *J. Electrochem. Soc.* **119**, 1780.
Su, C. B., and Olshansky, R. (1978). *Appl. Phys. Lett.* **32**, 761.
t'Hooft, G. W. (1981). *Appl. Phys. Lett.* **39**, 389.
Varon, J., Mahieu, M., Vandenberg, P., Boissy, M. C., and Lebailly, J. (1981). *IEEE Trans. Electron. Devices* **ED-28**, 416.
Watanabe, M. O., Yoshida, J., Mashita, M., Nakanishi, T., and Hojo, A. (1985). *J. Appl. Phys.* **57**, 5340.
Zaeschmar, G., and Speer, R. S. (1979). *J. Appl. Phys.* **50**, 5686.

CHAPTER 4

OMVPE Growth of AlGaInP for High-Efficiency Visible Light-Emitting Diodes

C. H. Chen, S. A. Stockman, M. J. Peanasky, and C. P. Kuo

Optoelectronics Division
Hewlett-Packard Company
San Jose, California

I.	INTRODUCTION	97
II.	GENERAL OVERVIEW OF ORGANOMETALLIC VAPOR-PHASE EPITAXY OF AlGaInP .	99
	1. *Source Materials*	99
	2. *Growth Conditions*	102
	3. *Growth of Optoelectronic Devices*	107
III.	ORGANOMETALLIC VAPOR-PHASE EPITAXY OF AlGaInP—CRITICAL ISSUES . . .	112
	1. *Dopant Incorporation Behavior*	112
	2. *Direct–Indirect Crossover*	117
	3. *Ordering*	119
	4. *Hydrogen Passivation of Acceptors*	122
	5. *Oxygen Incorporation*	127
IV.	HIGH-VOLUME MANUFACTURING ISSUES	136
	1. *Reactor Issues: Delivery and Exhaust Treatment*	136
	2. *Importance of Uniformity*	137
	3. *Source Quality Issues*	140
	4. *Color Control Issues*	141
	5. *Yield Loss Categories*	143
V.	SUMMARY	144
	References	144

I. Introduction

The quaternary alloy AlGaInP is an important material system for visible wavelength lasers and light-emitting diodes (LEDs). When grown lattice-matched on gallium arsenide (GaAs) substrate, the $(Al_xGa_{1-x})_{0.5}In_{0.5}P$ alloy has a direct bandgap from 1.9 to 2.26 eV ($x = 0$, $x \sim 0.5$, respectively), which covers the red to green portion of the visible spectrum. (Fig. 1) This property makes it an attractive material for making high-efficiency double heterostructure (DH) devices with a wide color range. However, this

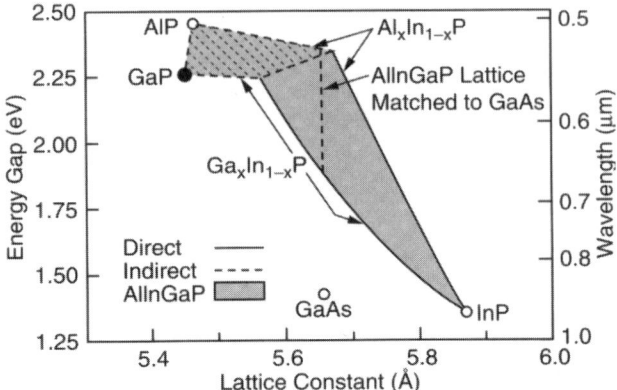

FIG. 1. The energy gap of the AlGaInP alloy and its corresponding wavelength versus lattice constant.

material cannot be grown by conventional high-volume growth processes for LEDs such as liquid-phase epitaxy (LPE) and hydride vapor-phase epitaxy (VPE). The difference in thermodynamic stability of aluminum phosphide (AlP) and indium phosphide (InP) makes compositional control extremely difficult by LPE. Additionally, the problem of forming a stable aluminum chloride (AlCl) compound during hydride or chloride vapor-phase epitaxy (HVPE and ClVPE, respectively) has prevented the successful growth of Al-containing phosphides by VPE. These technical problems had kept AlGaInP materials from being used in making LEDs or laser diodes until the late 1980s. Device-quality AlGaInP alloys were only reported by using kinetically controlled growth processes such as organometallic vapor-phase epitaxy (OMVPE) (Ikeda et al., 1985; Ishikawa et al., 1986; Kobayashi et al., 1985), molecular beam epitaxy (MBE) (Tanaka et al., 1987), and gas source molecular beam epitaxy (GSMBE) (Kikuchi et al., 1989). Much progress in the preparation of this material was made in the 1980s when OMVPE reactor technology improved and became established as a high-volume technique for the growth of III-V compound semiconductors including AlGaInP. Today, high-power AlGaInP visible semiconductor laser diodes that emit in the spectrum range from 680 to 625 nm have been realized by OMVPE (see, e.g., Bour and Shealy, 1987; Hatakoshi et al., 1991). Significant advances in the performance of LEDs have also been demonstrated in the wavelength range from about 660 to 560 nm. (Kuo et al., 1990; Fletcher et al., 1991; Sugawara et al., 1992a). These high-brightness AlGaInP LEDs are now produced commercially in high volume in both absorbing substrate (AS) (Fletcher et al., 1991) and transparent substrate

(TS) (Kish et al., 1994) form. They significantly outperform existing nitrogen-doped gallium phosphide (GaP) and GaAsP LEDs in the red-orange to yellow-green region and provide a reliable alternative to AlGaAs for red emitters. The AlGaInP LED technology and the recently developed AlInGaN LEDs (Amano et al., 1989; Nakamura et al., 1994, 1995) are now dominating new LED applications such as moving message panels, interior and exterior automotive lighting, traffic control signs, and low-power applications.

This chapter briefly describes the OMVPE growth of AlGaInP for high-efficiency LED applications. First, a broad overview of growth conditions and the requirements for growth of high-quality optoelectronic devices is provided. The subsequent sections focus on a variety of issues that are central to the growth process. These include n- and p-type doping, ordering effects, unintentional hydrogen passivation, and residual oxygen incorporation. A variety of manufacturing issues, such as uniformity, run-to-run reproducibility, and safety are also discussed.

II. General Overview of Organometallic Vapor-Phase Epitaxy of AlGaInP

1. SOURCE MATERIALS

A number of source material combinations may be used for the growth of AlGaInP by OMVPE. Some of the key issues revolve around the controllable delivery of the alkyl sources. The group III organometallic sources are generally the trimethyl-based or triethyl-based sources or both. Triethyl-based sources have been reported to produce less carbon incorporation in Al-containing epilayers such as AlGaAs (Kuech et al., 1987). However, it is necessary to use all triethyl-based precursors in order to achieve a reduction in carbon incorporation. If one source remains methyl-based, carbon from the methyl ligands can still be incorporated into the solid. The carbon incorporation can occur through radical exchange reaction in the gas phase between the trimethyl-based source and the Al source (Agnello and Ghandi, 1989) or the methyl ligand can undergo simple homolytic fission and bind to Al on the surface since the Al–C bond is strong (Stringfellow, 1989, Chapter 2). Thus, replacing trimethylaluminum (TMAl) alone with triethylaluminum (TEAl) without replacing other trimethyl-based sources will not solve the carbon contamination problem. (C. H. Chen and G. B. Stringfellow, unpublished results, 1988). However, exclusive use of all triethyl-based sources has generally been found to be impractical because triethylindium (TEIn) and triethylgallium (TEGa) sources have very low vapor pressures and do not provide sufficient growth rates in large-scale OMVPE reactors. Fortunately, C incorporation is very inefficient in In-containing alloys. For AlGaInP, carbon incorporation is also limited because In is the major constituent while Al never takes up more than 50% of the solid composition. Thus, the use of ethyl-based sources is not required for AlGaInP growth. For these reasons, the most commonly used group III sources are TMAl, trimethylgallium (TMGa), and trimethylindium (TMIn).

Another method to suppress carbon contamination in AlGaAs is the use of high V-III ratios. The atomic hydrogen from the hydrides can effectively recombine with the methyl radicals released from TMAl and TMGa thermal decomposition to form stable methane, which escapes from the thermal boundary layer rather than being incorporated into the solid (Kuech and Veuhoff, 1984). A high V-III ratio is also desirable for growth of AlGaInP, but for different reasons discussed later in this chapter.

A generic schematic of a gas delivery system for AlGaInP epilayer growth is shown in Fig. 2. For the alkyl delivery, individual pressure-controlled alkyl bubblers in temperature-controlled baths provide reliable and constant delivery. It is always critical in OMVPE that the gas plumbing provides rapid, efficient, and controllable switching of gases into the reactor chamber without perturbing the reactor pressure or the delivery pressure from the alkyl bubbler. For this reason it is typical to see a run–vent arrangement of injecting a set-up gas flow into the reactor. Many schemes have been developed through the years in attempts to minimize unswept "dead volume" and pressure transients that occur during switching, and the "best" approach typically depends on the specific application.

The delivery rate of a liquid or solid organometallic source depends on its temperature, the flow rate of the carrier gas through the organometallic bubbler, and the pressure of the bubbler. The flow of a source can be expressed as

$$Q_s = Q(\text{carrier gas})[P_s(T)/P_{\text{total}}] \quad (1)$$

where Q_s is the molar flow rate of the organometallic source, Q(carrier gas) is the molar flow rate of the carrier gas through the bubbler, $P_s(T)$ is the vapor pressure of the organometallic source at temperature T, and P_{total} is the total pressure inside the organometallic bubbler.

Following the delivery of organometallic and hydride sources through the plumbing manifold into the reactor chamber, chemical reactions occur within the thermal boundary near the GaAs substrate and on the surface for epitaxial growth of AlGaInP. Typically, the OMVPE growth of AlGaInP using TMAl, TMGa, TMIn, and phosphine (PH_3) occurs at growth

4 OMVPE Growth of AlGaInP for High-Efficiency Visible LEDs

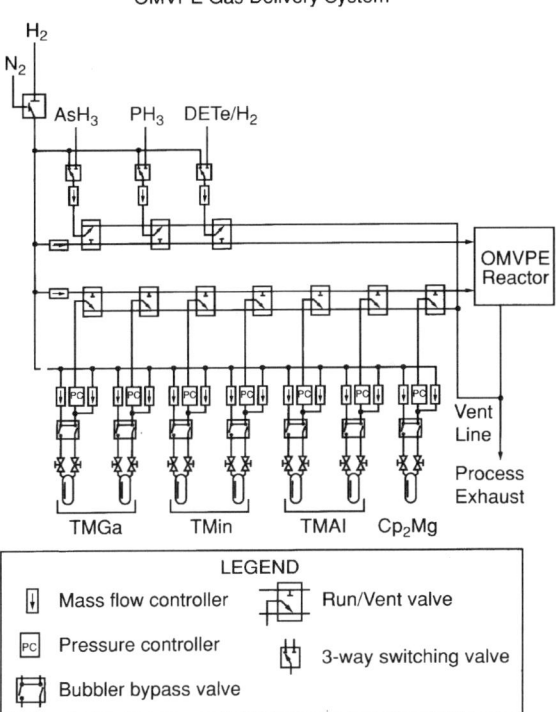

Fig. 2. Schematic of the gas delivery system for AlGaInP organometallic vapor-phase epitaxy (OMVPE) reactor.

temperatures of 700 to 800°C and at low reactor pressures of 0.02 to 0.2 atm. At these high growth temperatures, the TMAl, TMGa, and TMIn are likely to be partially decomposed in the gas phase by successive loss of methyl ligands (Stringfellow, 1989, Chapter 4). The PH_3 hydride, on the other hand, is quite stable (Larsen *et al.*, 1987) and is mainly decomposed heterogeneously on the AlGaInP surface. To first order, the AlGaInP epitaxial growth on the substrate surface may be described as follows:

$$x\,Al(CH_3)_n + y\,Ga(CH_3)_n + z\,In(CH_3)_n + PH_3 \Rightarrow Al_x Ga_y In_z P + CH_4 \quad (2)$$

where n can be 0, 1, 2, or 3.

The use of TMAl and TMGa sources is relatively trouble-free. However, the use of TMIn requires special attention because it is a solid source. The effective vapor pressure of a solid source depends on the surface area of the solid inside the bubbler. It is well known that the effective TMIn vapor pressure can vary in different bubblers and for the same bubbler at different times (Stringfellow, 1989, p. 28). Thus, it is useful to have some kind of *in-situ* monitoring for the TMIn vapor pressure to ensure the correct amount of TMIn is delivered into the reactor. Recently, alternative approaches have been directed at solving this problem. One approach is to use ethyldimethylindium, which is a liquid at room temperature. (Fry *et al.*, 1986; Knauf *et al.*, 1988). Alternatively, TMIn has been dissolved into a solvent so that more consistent carryout may be achieved (Air Products, 1995).

Due to the extreme toxicity of the arsine (AsH_3) and PH_3 hydride sources, great efforts and much progress have been made in the use of alternative group V sources (Stringfellow, 1988). For example, tertiarybutylarsine (TBAs) and tertiarybutylphosphine (TBP) have been found to be excellent replacements for AsH_3 and PH_3 (Chen *et al.*, 1987, 1988). However, the most commonly used hydride sources for the growth of AlGaInP continue to be AsH_3 and PH_3, because of their higher purity and lower costs than TBAs and TBP. High-purity AsH_3 and PH_3 may be obtained undiluted or in a mixture with H_2.

The most commonly used n-type dopants for AlGaInP are silicon (Si) and tellurium (Te). The Si source can be silane (SiH_4) or disilane (Si_2H_6), with the doping efficiency of the latter being less temperature-dependent. (Kuech *et al.*, 1984) The Te source can be delivered from a bubbler using H_2 as a carrier gas or from a ppm-level mixture in H_2 stored in a high-pressure cylinder. The most commonly used p-type dopants are zinc (Zn) and magnesium (Mg). The Zn source may be DEZn or DMZn in a liquid bubbler. The more common choice is DMZn because Zn incorporation efficiency is normally very low and the higher vapor pressure of DMZn helps in achieving a high Zn overpressure. The Mg source is generally bis(cyclopentadienyl)magnesium (Cp_2Mg) stored inside a bubbler. The choice of dopants significantly impacts not only doping efficiency, but also the dopant profile due to source memory effects and thermal diffusion of dopants. These issues are discussed in Part III, Section 1.

2. GROWTH CONDITIONS

The growth of high-quality $(Al_xGa_{1-x})_{0.5}In_{0.5}P$ can be quite difficult from the process point of view. Because AlGaInP is a quaternary system, careful control is required to ensure lattice-matching of AlGaInP to the GaAs substrate. This is crucial for obtaining high-quality material. Consequently, the issue of controlling lattice-matching from wafer-to-wafer and run-to-run dictates the design and the scale of a production OMVPE reactor. Second, aluminum is very reactive and binds easily to oxygen. Much work has been done in the past to minimize the incorporation of oxygen. It was found that one of the most effective ways to suppress oxygen incorporation is to increase the growth temperature (Yuan et al., 1985). However, high growth temperature is not ideal for In-containing alloys, which are typically grown at much lower temperatures. In addition, In re-evaporation can be a problem if the growth temperature is too high. Thus, the optimal growth temperature "window" for $(Al_xGa_{1-x})_{0.5}In_{0.5}P$ can be narrow.

The growth behavior of AlGaInP generally follows trends observed for other materials systems with a volatile group V element, such as GaAs, InP, or AlGaAs. Numerous review articles are available on the OMVPE technique (Stringfellow, 1985, 1989). The crystal growth occurs in kinetically limited regime when the growth temperature is low. At moderate growth temperatures (typically 500 to 800°C), the growth occurs in the diffusion-limited regime. At high temperature, growth rates decrease as a result of gas-phase decomposition and deposition of precursors on the reactor walls. The AlGaInP growth temperature generally falls between 650 and 800°C, and therefore the growth occurs in the diffusion-limited regime. In such a regime, the column III sublattice composition is approximately linearly proportional to the input gas-phase composition, except for some In re-evaporation at higher growth temperatures.

The V-III ratio is generally maintained at values higher than 200 for the growth of AlGaInP, similar to the case of other phosphides grown by OMVPE. This is due to the inefficient decomposition of PH_3 (Larsen et al., 1987). One problem with the use of high V-III is that the exhaust system can accumulate heavy phosphorus deposits in a short period of time. Extreme care must be taken when exposing the reactor's exhaust to the air since phosphorus burns instantly with oxygen. This is discussed in more detail in Part IV, Section 1.

Figure 3 shows the constant energy and lattice-constant contours for the AlGaInP materials system. The dashed lines show the lattice-constant contour and the solid lines the energy bandgap contour. These contour lines are generated from linear interpolation without considering bowing among the three ternary alloys: GaInP, AlInP, and AlGaP. The material has a direct energy bandgap on the left-hand side (below the bold line) and an indirect energy bandgap on the right (above the bold line). For lattice-matching to GaAs, one can simply fix the In/(Al + Ga + In) ratio at 0.5, as is seen in Fig. 3. Since OMVPE growth of AlGaInP occurs in the diffusion-limited regime, as mentioned earlier, one can first change the gas-phase TMIn/(TMAl + TMGa) ratio to achieve lattice-matching of AlGaInP on GaAs. Then, one can simply vary the TMAl/TMGa ratio (while keeping constant the total TMAl plus TMGa molar flow rate) to

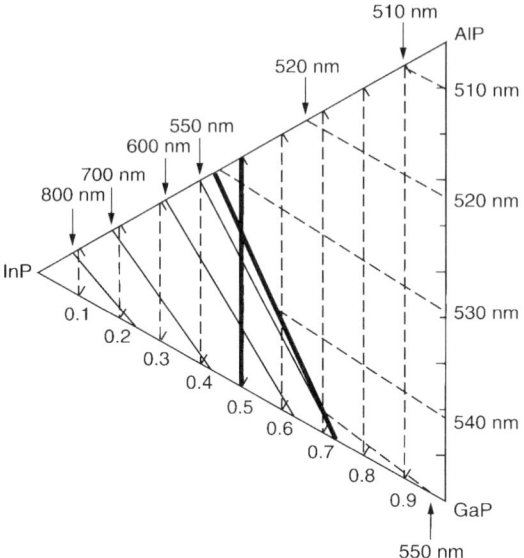

FIG. 3. The constant energy bandgap and lattice constant contours for the AlGaInP material system.

obtain different energy bandgaps of AlGaInP materials for the active layer and the cladding layers.

Because of the difference in thermal expansion coefficients between the GaAs substrate and the AlGaInP epilayer, the epilayer will be compressively strained at room temperature by the GaAs substrate when the lattice-matching condition is met at growth temperature. Table I shows the thermal expansion coefficients for the binary compounds (Bour, 1987) and examples of lattice constants are drawn in Figs. 4 and 5 for the ternary alloys of

TABLE I
LATTICE PARAMETERS AT ROOM TEMPERATURE (300 K) AND GROWTH TEMPERATURE (975 K) OF IMPORTANT BINARY COMPOUNDS

Compound	a(Å) at 300 K	$\alpha(10^{-6}/K)$	a(Å) at 975 K
Gallium arsenide	5.6533	6.86	5.6795
Indium phosphide	5.8686	4.75	5.8874
Gallium phosphide	5.4512	5.91	5.4729
Aluminum phosphide	5.4511	4.50	5.4677

Reprinted from Bour, 1987, Ph.D Thesis, Cornell University, Ithaca, NY; with permission.

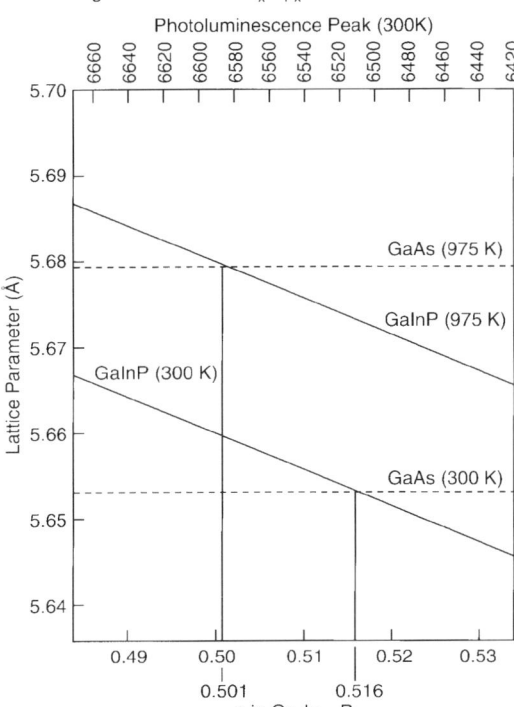

FIG. 4. Growth temperature and room temperature lattice matching conditions for $Ga_xIn_{1-x}P$ on gallium arsenide (GaAs) substrates.

GaInP and AlInP at room temperature and 975 K, respectively. The results suggest that it may be best to grow AlGaInP under compression (Indium-rich condition) as measured by room temperature X-ray rocking curves. Of course, the optimal compressive strain depends on the specific growth temperature and alloy composition.

Another important issue for the growth of AlGaInP epilayers is the switching from AsH_3 gas for the growth of GaAs to PH_3 for the growth of phosphide-based materials. Similar issues exist for the growth of GaInAs on an InP substrate for which switching of group V precursors is required. The difficulty with group V source switching is that the group V elements are volatile in most III-V materials. During the group V switching, re-evaporation of the group V element can occur. This problem is especially acute when the growth temperature and evaporation rates of the group V

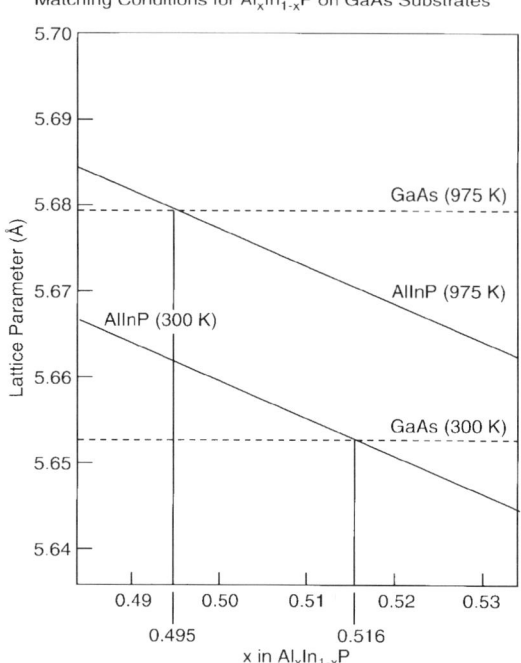

Fig. 5. Growth temperature and room temperature lattice matching conditions for $Al_xIn_{1-x}P$ on GaAs substrates.

elements are high. This can lead to poor material quality if not done properly. On the other hand, if the PH_3 and AsH_3 gases overlap too much, an intermediate layer with the presence of both As and P can occur, leading to a layer with large lattice mismatch and poor material quality. The subject has been under investigation in the metalorganic chemical vapor deposition (MOCVD) community for a long time (Ota et al., 1993). In general, rapid switching of the group V gases is necessary and the precise switching sequence and its timing are extremely critical.

The use of low-pressure conditions (0.02 to 0.2 atm) has been employed to resolve some issues relative to gas switching, uniformity, and minimizing gas phase pre-reactions. These factors have lead OMVPE equipment manufacturers to develop systems for the low-pressure growth of AlGaInP. The advantages of using low pressure vary considerably depending on the reactor geometry. The high-speed rotating-disk vertical reactor geometry requires low pressure or very high gas flow to compensate for the pumping action of the disk. That is, low pressure is required for stable flow characteristics in a rotating-disk reactor. For horizontal flow reactors, low pressure coupled with a low reactor chamber volume can provide efficient gas switching, and stable flow with a short entrance length. If the reactor geometry could allow for growth at atmospheric pressure, a vacuum system to maintain the chamber pressure with the precision required for stable and reproducible results would still be required.

Gas switching issues can be resolved at low pressure with a gas delivery system that uses pressure balancing and the proper sequence of gas injection valves, and an efficient (low residence time) reactor chamber. If gases are allowed to reside in a large volume that has a slow flow field, the effects of diffusion will dominate the composition of the gas. If the gas velocity is higher and the reaction chamber designed for a very short residence time, the effects of the valve switching in the gas plumbing system will then determine the composition of the gas stream and the epilayer film growth. Uniformity is generally improved by the increase in gas velocity over the wafer, which minimizes the alkyl depletion in the gas phase, especially for indium. Coupling low pressure and efficient chamber design with rotation in horizontal reactors can provide a means of achieving high alkyl utilization and good thickness and compositional uniformity. Excellent uniformity may also be achieved in vertical flow rotating-disk reactors, depending on the reactor pressure.

3. Growth of Optoelectronic Devices

For the application of high-brightness visible LEDs at Hewlett-Packard, the device structure of interest (Fig. 6) consists of a double heterostructure grown lattice-matched to a GaAs substrate. The GaAs substrates were (100) misoriented 2 degrees toward the $\langle 110 \rangle$ direction. The growth rate is kept at about $2\,\mu$m/h with a V-III ratio of 250. The growth temperatures were typically 700 to 800°C for the DH structure that consists of AlInP as the cladding layers and $(Al_xGa_{1-x})_{0.5}In_{0.5}P$ as the active layer. The composition of the active layer ranges from $x = 0.2$ to $x \sim 0.43$, which produces LEDs with emission wavelength from red-orange to green. The active layer is undoped and is 0.5- to 1.0-μm thick. The p- and n-type dopants for the AlInP layers are Cp_2Mg and DETe. The Mg-doped p-type AlInP layer is 0.5 μm, and the typical measurement of doping concentration is 0.5–1×10^{18} cm^{-3}. The Te-doped n-type AlInP has a 1.0-μm-thick layer, and the doping concentration is typically 1×10^{18} cm^{-3}. A p-type GaP window layer is deposited on top of the DH structure to serve as a transparent current spreading layer, which has been discussed elsewhere (Fletcher et al., 1991).

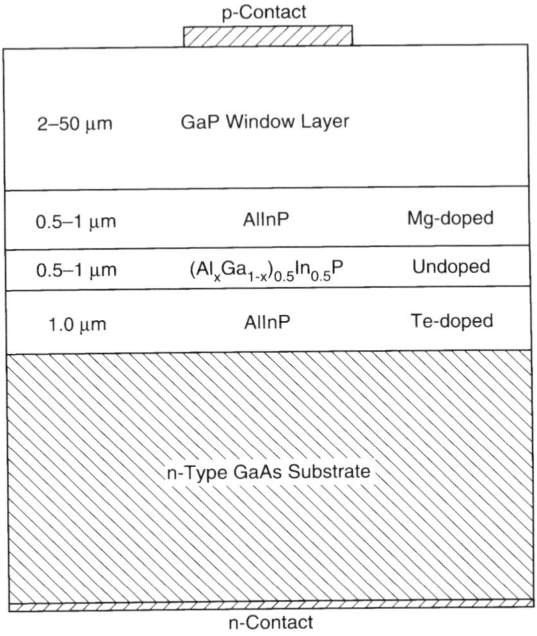

FIG. 6. Schematic diagram of an AlGaInP light-emitting diode (LED) structure.

The optical properties of $(Al_xGa_{1-x})_{0.5}In_{0.5}P$ are strongly dependent on the Al–Ga ratio of the alloy. As shown in Fig. 7, the bandgap energy of the $(Al_xGa_{1-x})_{0.5}In_{0.5}P$ increases from 1.9 to about 2.3 eV as the Al composition of the layer increases from 0 to 0.6. The energy bandgap of the material can be described by (Cao et al., 1990):

$$E_g = 1.91 + 0.61x \quad \text{(eV)} \tag{3}$$

The photoluminescence (PL) efficiency of the material decreases significantly as the Al composition increases from 0 to 0.6. The decrease in PL intensity with increasing Al composition is mainly due to the carrier overflow into the indirect (X) band. Impurities such as oxygen in the higher Al-content materials may also contribute somewhat to the drop in optical efficiency.

Figure 8 shows a double-crystal X-ray diffraction rocking curve of a typical 1-μm-thick AlInP layer. The full-width at half-maximum (FWHM) of the AlInP layer is 28 arc-sec compared with the 15 arc-sec measured for the GaAs substrate (the resolution limit of the X-ray diffractometer). The

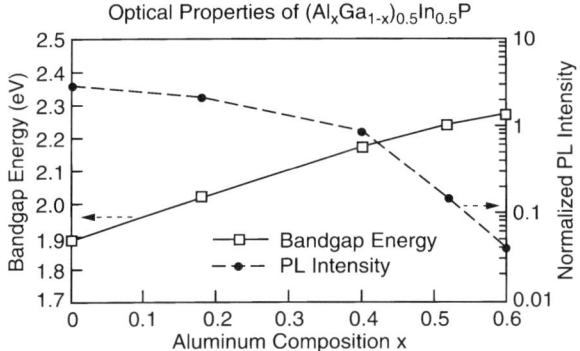

FIG. 7. Photoluminescent (PL) intensity and bandgap energy of AlGaInP alloy as a function of aluminum composition x.

ability of OMVPE to grow high-quality AlInP is clearly demonstrated. And the performance of the DH structure is strongly dependent on the quality of the AlInP cladding layers.

The background concentration of an undoped layer is typically measured in low 10^{15}'s for GaInP and near 1×10^{16} cm^{-3} for AlInP, with the dominant residual donor often being Si from the TMAl source. N-type AlGaInP material can be produced easily with low resisitivity by doping with Si, Se, or Te. The growth of highly p-doped AlGaInP is more difficult. The problem is shown in Fig. 9 when DMZn is used as p-type doping source. The doping behavior of GaInP ($x = 0$), $(Al_xGa_{1-x})_{0.5}In_{0.5}P$

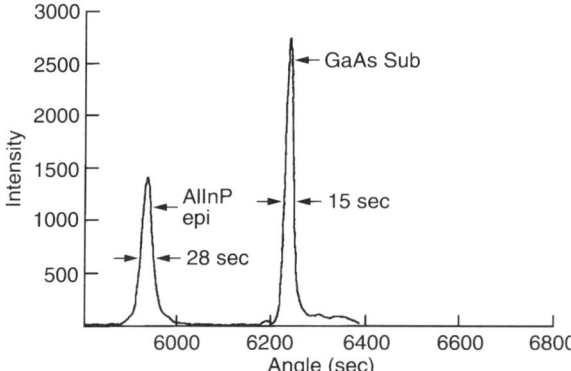

FIG. 8. Double-crystal X-ray diffraction rocking curve of an AlInP layer on a gallium arsenide substrate.

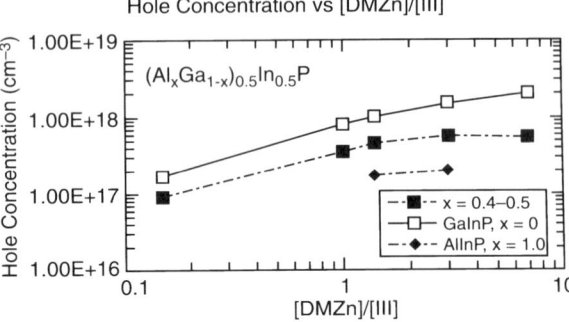

FIG. 9. Hole concentration of AlGaInP alloys ($x = 0$, 0.4–0.5, 1) as a function of [DMZn]/[III].

($x = 0.4$ to 0.5), and AlInP ($x = 1$) versus [DMZn]/[III] are shown, where [III] represents the total group III molar concentration. As the [DMZn]/[III] ratio is increased from 0.1 to nearly 10, the hole concentration increases only from the low 10^{17}'s to the low 10^{18}'s for GaInP. And the highest obtainable hole concentration decreases to the mid-10^{17}'s for the $(Al_xGa_{1-x})_{0.5}In_{0.5}P$ ($x = 0.4$ to 0.5). The highest p-doped AlInP is only $1-2 \times 10^{17}$ cm^{-3} using DMZn for p-doping. The situation is improved somewhat when Cp_2Mg is used as the p-type doping source. The typical Mg-doped p-type AlInP layer in this structure is in the $0.5-1 \times 10^{18}$ cm^{-3} range with a resistivity of 0.5 ohm-cm.

Two of the biggest problems in growing AlGaInP are the affinity of Al for oxygen and the low volatility of the resulting Al oxide. For example, Kuech et al. (1987) have calculated that the equilibrium O_2 concentration at 700°C with Al metal is less than 10^{-45} atm. Early attempts to grow AlGaInP mainly resulted in black epilayers with little or no PL because the background oxygen level was high and the epilayers contained a large amount of Al oxides (Yuan et al., 1985). The oxygen forms compensating deep-level states and the material becomes semi-insulating (McCalmont et al., 1992). The deep oxygen levels are also nonradiative centers that degrade PL and electroluminescence (EL) (Kondo et al., 1994a). It has been found that even a small amount of oxygen in the vapor phase can lead to 1×10^{19} cm^{-3} of oxygen in the solid of an Al-containing material (Kisker et al., 1982) Thus, reducing the oxygen incorporation is critical for growth of Al-containing materials.

The first place to reduce oxygen contamination is the OMVPE system. It is imperative to maintain a leak-tight OMVPE system. This may be verified by inboard He leak-testing. In addition, the H_2 carrier gas is typically purified by diffusion through a heated palladium (Pd) membrane. A glove box or ante-chamber for loading and unloading wafers is commonly used to isolate the reactor chamber from the atmosphere.

Source materials are another common place for oxygen contamination. Early results showed that AsH_3 and PH_3 were often the dominant sources of moisture (Dapkus et al., 1981). Advances in hydride purification processes have greatly improved the hydride purity. Moreover, a molecular sieve or other type of point-of-use purification unit can be used to further purify any residual moisture or oxygen present in the hydrides.

The group III sources can also contain alkoxides, which may lead to oxygen contamination. Currently, low-oxygen TMGa and TMAl sources are commonly available, mainly because TMGa and TMAl have been used extensively in the growth of AlGaAs for optoelectronic applications, and great effort has been put forth to reduce oxygen-containing impurities in these two sources. On the other hand, TMIn often contains more oxygen than does TMAl or TMGa (Roberts et al., 1994). In the past, TMIn was mainly used in OMVPE for the growth of non-Al-containing materials such as InGaAsP–InP, where oxygen incorporation is very inefficient. Thus, oxygen contamination was not a primary issue and considerably less effort was directed toward reducing oxygen content. The more recent development of AlGaInAs–InP and AlGaInP–GaAs, which utilize both In and Al and are much more sensitive to the presence of residual oxygen, has provided a new impetus for minimizing the concentration of oxygen-containing species in TMIn.

It is a good idea to screen both the hydride and the organometallic sources for the growth of AlGaInP in order to keep the level of oxygen to a minimum. Unfortunately, the commonly employed source analysis techniques do not allow the detection of oxygen down to a sub-ppm level. The most effective technique for screening of source materials is the actual growth of AlGaInP epilayers and devices and monitoring of the device performance.

There have been many investigations to find innovative ways to reduce oxygen contamination for the growth of Al-containing materials. One attempt is to use graphite baffles, which adsorb TMAl. The adsorbed TMAl then reacts with the oxygen in the vapor and removes it from the gas stream (Stringfellow and Hom, 1979). Another approach is to grow a high-Al buffer layer, which removes any residual oxygen in the reactor before the active layer of the device structure is grown (Hersee et al., 1981).

For the growth of high-quality AlGaInP device structures, a similar approach can be taken. Because AlGaInP is grown on a GaAs substrate, one can take advantage of the fact that AlGaAs tolerates oxygen incorporation much better than does AlGaInP. In other words, before the growth of the AlGaInP device structure, one can grow a high-Al AlGaAs layer on GaAs to getter out the residual oxygen in the reactor. In this way, the AlGaInP material quality can be improved, especially when the growth temperature is relatively low.

Careful optimization of process parameters may also be done to reduce oxygen incorporation. It has been universally observed that higher growth temperature and higher V-III ratio can lead to reduced oxygen incorporation for Al-containing materials (Kuech *et al.*, 1992), including AlGaInP (Kondo *et al.*, 1994a). In addition, the incorporation of oxygen is also highly dependent on substrate orientation (Kondo *et al.*, 1994a). These trends are discussed in detail below.

III. Organometallic Vapor-Phase Epitaxy of AlGaInP — Critical Issues

1. DOPANT INCORPORATION BEHAVIOR

For LED applications, successful doping of both n-type and p-type $(Al_xGa_{1-x})_{0.5}In_{0.5}P$ cladding layers ($0.6 \leqslant x \leqslant 1$) is required. Early attempts to dope AlGaInP with high Al content were quite difficult, especially for the p-type AlGaInP (Bour, 1987, pp. 109–111). This is mainly due to the following reasons. First, early AlGaInP epilayers likely contained high level of oxygen, as was discussed in Part II, Section 3. Because oxygen generates deep levels and compensates shallow acceptors (McCalmont *et al.*, 1992), p-type doping would be difficult when the background oxygen level is high. Second, unintentional hydrogen passivation of acceptors (Pearton *et al.*, 1987a; Hobson, 1994) can render the p-type dopant electrically neutral. This is especially severe for materials with high-energy bandgaps such as AlGaInP with high Al composition. It was later found that annealing the p-type materials can reduce the H passivation and more fully activate p-type dopants (Hobson, 1994). The H passivation of acceptors in AlGaInP is discussed in detail in Part III, Section 4. Third, higher Al composition leads to higher energy bandgaps and corresponding larger ionization energies for the acceptors. Thus, p-doping for high Al-content AlGaInP is more difficult, especially using Zn as the acceptor impurity.

After reduction of the background oxygen level and introduction of postgrowth annealing, both n- and p-type doping of AlGaInP can be routinely achieved without much difficulty. The characteristics of residual impurities and dopants have been reviewed by Stringfellow (1986), in which impurities have been conveniently categorized into two main groups: volatile and nonvolatile impurities. For volatile impurities, the incorporation decreases as the growth temperature is increased, due to re-evaporation of the impurity from the surface. The incorporation rate of volatile impurities does not depend on the growth rate. For nonvolatile impurities, the incorporation is not affected by the growth temperature because every impurity atom will be incorporated into the solid. The incorporation rate of nonvolatile impurities depends inversely on the growth rate. The incorporation of dopants for AlGaInP also follows these general trends. In addition, the dopant concentration in the solid generally increases as the concentration of the dopant in the vapor phase increases. However, as the dopant concentration increases in the vapor phase, the carrier concentration increases initially, then saturates and eventually drops again because too much dopant can generate compensating defects in the crystal.

The n-type doping of AlGaInP is relatively straightforward. Early OMVPE AlGaInP laser structures were fabricated using H_2Se as the n-type dopant (Bour, 1987). But H_2Se has become less popular because it has a severe memory effect (Lewis et al., 1984). The n-type dopants of choice for OMVPE AlGaInP are either Te or Si (Ishikawa et al., 1986; Kobayashi et al., 1985). The advantage of using SiH4 as a dopant is that SiH_4 has essentially no memory effect and Si exhibits relatively low diffusivity in the solid. Thus, control of the doping profile and junction position can be much easier with Si as the n-type dopant. However, SiH_4 is hydroscopic and its use can lead to poor surface morphology if moisture is present.

The incorporation of Te behaves as a volatile impurity in AlGaInP: less Te is incorporated when the growth temperature is raised. (C. H. Chen, T. Osentowski, and C. P. Kuo, unpublished results, 1994). However, the Si doping using SiH_4 as the source belongs to a special case. The Si atom itself can be categorized as a nonvolatile impurity (Stringfellow, 1986). However, Si doping with SiH_4 is kinetically limited by the decomposition of the SiH_4 source (Stringfellow, 1986). Thus, higher growth temperatures lead to increased decomposition of SiH_4 and increased Si incorporation. The results of Si doping using SiH_4 for different materials are shown in Fig. 10 as a function of growth temperature, where the distribution coefficient is defined as the solid dopant concentration divided by the gas-phase dopant concentration (normalized to the group III concentration in solid and gas phases, respectively). It is clear that higher growth temperature leads to higher electron concentration in the materials, including AlGaInP. Because of this temperature sensitivity, there have been efforts to replace SiH_4 with Si_2H_6, which exhibits a smaller doping dependence on growth temperature due to its lower thermal stability (Kuech et al., 1984).

The p-type doping is much more problematic for AlGaInP as compared with the n-type doping. The dopants of choice have been DMZn or Cp_2Mg, as discussed in Part II, Section 1. Zinc behaves like a volatile impurity

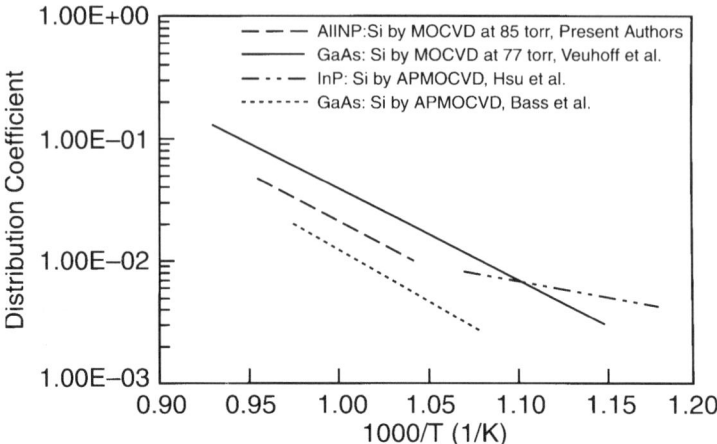

FIG. 10. Distribution coefficient of silicon (Si) doping in organometallic vapor-phase epitaxy (OMVPE) as a function of growth temperature.

(Stringfellow, 1986). Figure 9 shows the results for Zn doping at different DMZn flow rates for AlGaInP with different Al solid composition. It is interesting to note that it becomes harder to obtain a high hole concentration for high-Al AlGaInP. As the Al solid composition is increased, the energy bandgap and the ionization energy of Zn become larger. Therefore, it is more difficult to ionize the acceptor. (Nishikawa et al., 1989). In addition, as more Al is incorporated, more oxygen might be incorporated as well, leading to compensation (Nishikawa et al., 1992b).

Several problems exist for the use of Zn as a p-type dopant. First, Zn is extremely volatile and requires a high DMZn partial pressure in the vapor phase to achieve a high doping level. In fact, Fig. 9 shows that the DMZn molar flow rate is comparable to or higher than the group III molar flow rate in order to achieve a hole concentration in the range of 10^{17} cm^{-3}. In other words, only one Zn in a million is incorporated into the solid and the rest of the Zn atoms are re-evaporated back into the vapor phase. Thus, it is necessary to change the DMZn bubbler frequently, especially in a large-volume production reactor. The frequent change in dopant source can lead to fluctuation of material and device quality because source purity may not be constant from bubbler to bubbler. The second problem with Zn acceptor dopant is that Zn is known to diffuse extremely quickly in III-V materials (Kondo et al., 1994c), making it difficult to control the final p-type doping profile.

Another choice of acceptor for AlGaInP p-type doping is Cp_2Mg. In general, the Mg acceptor is expected to have a smaller ionization energy than that of Zn. Thus, it may be possible to achieve a higher hole concentration for AlGaInP with Mg, especially for high-Al AlGaInP (Nishikawa et al., 1992a). In addition, several studies have shown that Mg has a lower diffusion coefficient than that of Zn in III-V materials (Kondo et al., 1994c; Nelson and Westbrook, 1984; Kozen et al., 1986; Landgren et al., 1988; Veuhoff et al., 1989; Abernathy et al., 1993; Wu et al., 1994). Figure 11 (Wu et al., 1994) shows the hole concentration as a function of Cp_2Mg molar flow rate (normalized to group III). As a comparison, Fig. 11 also shows the dependence of hole concentration as a function of DMZn molar flow. It is seen that the molar flow rate of Cp_2Mg is at least two orders of magnitude less than that of DMZn, because Mg is much less volatile than Zn. This is consistent with the difference in the vapor pressures of Zn and Mg over their elemental solid or liquid, as shown in Fig. 12. Although less volatile than Zn, Mg still exhibits a temperature-dependent doping efficiency, with less Mg doping at higher growth temperature, as shown in Fig. 13.

The biggest problem with the use of a Cp_2Mg source is its notoriously strong memory effect (Kondo et al., 1994c; Wu et al., 1994; Roberts et al., 1984; Kuech et al., 1988). Nishikawa et al. (1993) showed that the Mg memory effect is not manifested as a slow and gradual rise in the doping profile. Instead, the doping profile is delayed with respect to the turn-on time of the Cp_2Mg source. The delay has been attributed to the adhesive

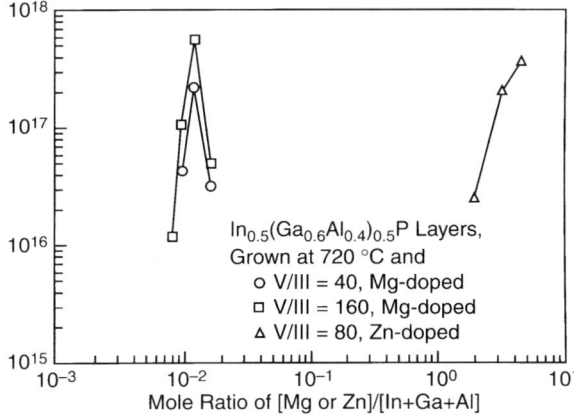

FIG. 11. Hole concentration as a function of dopant molar flow for AlGaInP.

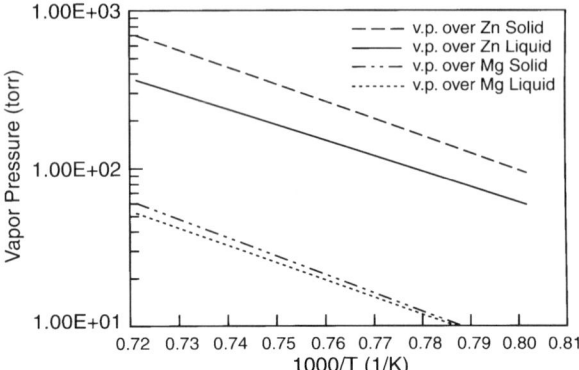

FIG. 12. Vapor pressure of zinc or magnesium over its liquid or solid as a function of temperature.

nature of the Cp_2Mg precursor on the internal surface of the gas handling system (Kuech *et al.*, 1988), including both stainless steel tubing and the reactor walls (Rask *et al.*, 1988). It was found that when TMAl or TEAl is added in addition to the Ga source (Kondo *et al.*, 1994c; Nishikawa *et al.*, 1992a), the Mg delay can be reduced, due to the competitive adsorption of TMAl to fill up the surface sites in the gas handling system and the reactor wall. It has also been demonstrated that the purging of the Cp_2Mg source during growth interruption can help reduce the delay time (Rask *et al.*, 1988); however, this involves precise timing because impurity accumulation on the epilayer surface can occur if the exposure time is too long.

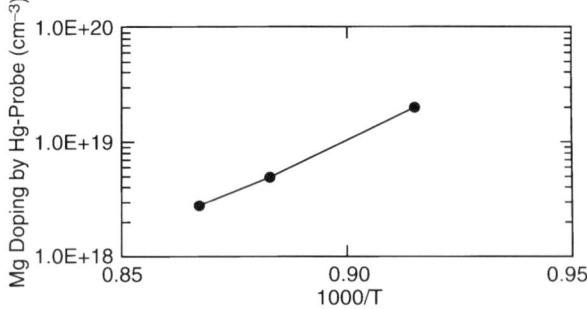

FIG. 13. Temperature dependence of magnesium (Mg) doping in gallium phosphide as a function of growth temperature.

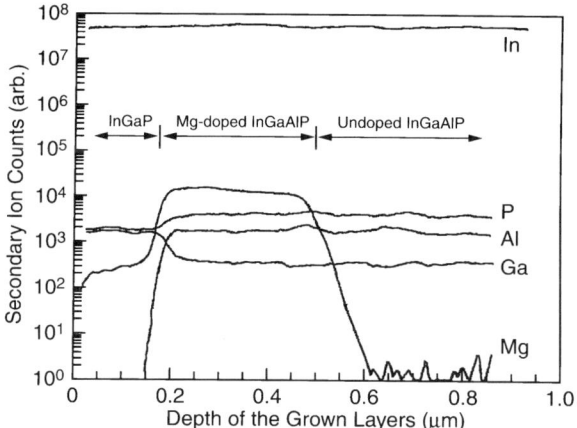

FIG. 14. Magnesium (Mg) doping profile of AlGaInP with a doping level of $5.6 \times 10^{17}\,\mathrm{cm}^{-3}$. Zero depth implies the top of the grown surface.

The Mg memory effect is also quite severe after the Cp_2Mg source is turned off. Figure 14 also shows the Mg doping tail (Wu et al., 1994). It is seen that the Mg concentration is decreased relatively rapidly by about two orders of magnitude when the Mg is turned off. However, the residual Mg level lingers on and is well above the background secondary ion mass spectroscopy (SIMS) detection limit for a long time. In fact, Mg may even be present in the reactor after overnight purging. This residual Mg can then deposit on the surface of the GaAs substrate during the heat-up process to form a doping spike at the substrate–epilayer interface in the next growth.

2. DIRECT–INDIRECT CROSSOVER

One important reason for the wide use of AlGaInP is that it has the highest direct energy bandgap among conventional III-V materials (excluding the III-N materials) (Stringfellow, 1978). Thus, AlGaInP can be used to make red, amber, yellow, and even green LEDs. Even with the successful demonstration of high-brightness nitride LEDs (Nakamura et al., 1995), AlGaInP remains to be the material of choice for yellow-green and longer wavelength colors because it is extremely difficult to incorporate enough In to lower the energy bandgap of GaInN (Nakamura et al., 1995). In other words, the nitride LEDs are currently best suited for blue, blue-green, and green applications, and AlGaInP LEDs are best suited for yellow-green, amber, orange, and red.

As the wavelength becomes shorter, more Al is needed in the AlGaInP solid. This can lead to more oxygen incorporation and degradation of material quality if the growth ambient contains oxygen, as discussed above. Thus, it is critical to reduce the background oxygen level in the active layer of the devices by using the approaches outlined in Part III, Section 5. These include the use of load-lock glove box, high growth temperatures (Yuan *et al.*, 1985) and high V-III ratios (Kuech *et al.*, 1992), and mis-orientated substrates. (Suzuki *et al.*, 1993) For state-of-the-art AlGaInP, the oxygen incorporation is very low ($<1 \times 10^{16}$ cm^{-3}). The lifetime of the minority carriers for high-quality AlGaInP has been determined to be the same for low and high Al-containing AlGaInP materials.

Under optimum growth conditions, the luminescence efficiency of AlGaInP is mainly limited by the direct–indirect crossover. The precise crossover composition for AlGaInP varies because of the ordering effect to be discussed in Part III, Section 3. Recently, several studies (Mowbray *et al.*, 1994; Prins *et al.*, 1995) have shown that the disordered AlGaInP crossover aluminum composition is lower than that previously determined using linear interpolation between ternany alloys (Cao *et al.*, 1990). After adjusting the low-temperature (2 K) high-pressure data of Prins *et al.* (1995) to account for bandgap shrinkage, the room temperature (Al$_x$Ga$_{1-x}$)$_{0.5}$In$_{0.5}$P energy bandgaps for the direct and indirect valleys are found to be

$$E_\Gamma = 1.91 + 0.61\,x \tag{4}$$

$$E_x = 2.19 + 0.085\,x \tag{5}$$

As the solid composition x increases, the direct bandgap E_Γ approaches the indirect bandgap E_x and eventually crosses the E_x for the AlGaInP to become an indirect material. The crossover solid composition is predicted to be at $x \cong 0.53$ from Eqs. (4) and (5). Archer (1972) has developed a model to predict the relationship between composition and the light output when this direct–indirect transition occurs. The internal quantum efficiency η_{in}, can be written as

$$\eta_{in} = 1/[1 + (\tau_r/\tau_n)(1 + n_X/n_\Gamma)] \tag{6}$$

where τ_r is the radiative lifetime for the direct valley, τ_n is the nonradiative lifetime for both the direct and the indirect valleys, and n_X and n_Γ are the carrier concentrations in indirect and direct valleys of the conduction band, respectively. The ratio of carrier populations in the direct and indirect valleys can be written as follows:

$$n_\Gamma/n_X = (N_\Gamma/N_X)\exp((E_X - E_\Gamma)/kT) \qquad (7)$$

where N_Γ and N_X are the density of states for the direct and indirect valleys in the conduction band. (Archer, 1972). The difference in energy between the x and Γ minima in the conduction band, $E_X - E_\Gamma$ can be calculated by using Eqs. (4) and (5). Because only the electrons in the direct valley may participate in radiative transitions, it is important to keep n_Γ/n_X as large as possible. However, as the solid composition x increases, $E_X - E_\Gamma$ will decrease, resulting in a corresponding decrease in n_Γ/n_X. This decrease is especially severe when x approaches the crossover solid composition. In fact, because the density of states for the indirect band is much larger than that for the direct band, the majority of the carriers occupy the indirect valley when $E_X = E_\Gamma$. This severe carrier overflow leads to a dramatic decrease in luminescence efficiency due to nonradiative recombination of carriers in the indirect valley, as is indicated by Eqs. (6) and (7). This principle has been applied to AlGaInP theoretically (Stringfellow, 1978) and confirmed experimentally (Cao et al., 1990).

The room-temperature AlGaInP PL intensity as a function of Al composition for AlGaInP epilayers lattice-matched to GaAs was shown in Fig. 7. Note that the intensity scale is logarithmic, and therefore the intensity decreases quite drastically when the Al composition x is increased beyond 0.5. A similar dependence of light output on wavelength is also seen in state-of-the-art green LEDs, as shown in Fig. 15. In Figure 15, the AS stands for the absorbing substrate (GaAs). It is seen that the light output (LOP, external luminous efficiency for a bare LED chip, expressed in lm/A) decreases by about 0.4 lm/A as the wavelength is decreased by one nanometer. Thus, the light output is highly sensitive to the wavelength in the green region in which the composition is close to the crossover solid composition.

3. ORDERING

Another important issue for LEDs is the tendency of AlGaInP material to be ordered. OMVPE GaInP grown at many laboratories was found to have an energy bandgap lower than the corresponding materials grown using liquid-phase epitaxy (LPE) (Kuo et al., 1985). In some cases, the difference can be as large as 50 meV even though the GaInP epilayers grown by both OMVPE and LPE techniques are lattice-matched to the GaAs substrate (Gomyo et al., 1986). This mysterious "50 meV" problem was

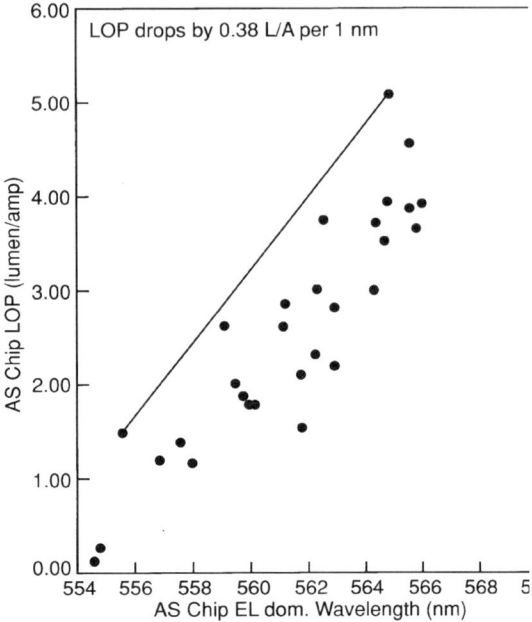

FIG. 15. Light output for a bare light-emitting diode (LED) chip as a function of electroluminescence (EL) dominant peak wavelength.

studied in detail by Gomyo et al. (1986) and has been discovered to be due to ordering of Ga and In atoms along the (111) planes (Gomyo et al., 1987). This change in energy bandgap by ordering was later shown to be as large as 160 meV under certain growth conditions (Su et al., 1993). A number of models have also been proposed to explain the formation of ordering (Stringfellow, 1993; Stringfellow and Chen, 1991; Philips et al., 1994; Ogale and Madhukar, 1992). The key to ordering is the surface reconstruction of the group V–rich III-V semiconductor surface into a (2 × 4) configuration (Stringfellow et al., 1995), which successfully explains the appearance of two of the four possible variances for ordering along the (111) planes for both the III-V-V and III-III-V alloys.

Transmission electron microscope (TEM) study of GaInP has shown that the ordered material typically contains ordered domains rather than a continuous ordered phase (Su et al., 1994b). These domains give superspots in TEM micrographs because the periodicity of the lattice for the ordered domains is twice the lattice constant for a random alloy. The material is typically called ordered even though it has compositional nonhomogeneity with ordered domains randomly imbeded in the random (disordered) alloy. When ordering occurs, the energy bandgap is lowered (Stringfellow, 1993). Since the excited carriers thermalize to the lowest energy bandgap, the PL peak typically shows the ordered domains. Thus, PL has also been used extensively to study ordering, with the lowering in PL peak energy used to gauge the degree of ordering.

Ordering is regarded as a kinetically controlled process occurring on the surface (Stringfellow *et al.*, 1995). The ordered phase is unstable in the bulk (Barnard *et al.*, 1988) and can be destroyed by dopant diffusion and annealing (Su *et al.*, 1994a; Gavrilovič *et al.*, 1988; Suzuki *et al.*, 1988). The degree of ordering was found to depend on numerous growth parameters (Gomyo *et al.*, 1987; Kurtz *et al.*, 1994). Substrate misorientation from (001) (Kurtz *et al.*, 1994; Gomyo *et al.*, 1989) and other substrate orientations (Gomyo *et al.*, 1988) have been found to have a major influence on the ordering because the formation of the dimer rows and the motion of the steps on the surface change when the substrate orientation is changed. The growth temperature (Gomyo *et al.*, 1986; Su *et al.*, 1994b) and V-III ratio (Gomyo *et al.*, 1986; Chun *et al.*, 1996) also affects the degree of ordering. The combined effect has been studied by Gomyo *et al.* (1987). At high growth temperatures, the degree of ordering is reduced (Gomyo *et al.*, 1987), possibly due to the annealing effect the bulk material undergoes (Kurtz *et al.*, 1990). In addition, growth rate (Cao *et al.*, 1991; Su *et al.*, 1994a) has also been found to affect the degree of ordering. At low growth temperatures, higher growth rates produced more disordered materials because the atoms do not have time to reach the ordered state on the surface. On the other hand, lower growth rate at high temperatures would favor more disordering because the material has more time to be annealed and become disordered (Kurtz *et al.*, 1990).

Ordering also occurs readily in AlGaInP. (Hamada *et al.*, 1991; Valster *et al.*, 1991). When GaInP or AlGaInP is ordered, its energy bandgap is lowered as compared with the random alloy. As a result, the energy bandgap or the color of the device varies as the degree of ordering changes. Thus, the elimination of ordering is important from the viewpoint of wavelength control. For AlGaInP green device applications, ordering is especially detrimental because it lowers the direct–indirect crossover energy (Sugawara *et al.* 1992b). Figure 16 shows the energy bandgap for AlGaInP at two growth temperatures. It is seen that when the growth temperature is higher, the energy bandgap is also higher, indicating less ordering in the material. Since every nanometer in wavelength is important in the green color, as shown in Fig. 15, it is imperative to use a combination of growth parameters discussed earlier to minimize ordering in AlGaInP for green device applications.

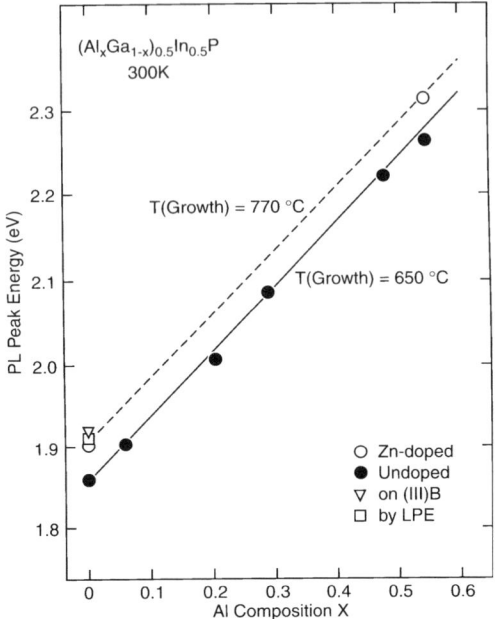

FIG. 16. Photoluminescence (PL) peak energy as a function of aluminum (Al) composition x at growth temperatures of 770 and 650°C.

4. HYDROGEN PASSIVATION OF ACCEPTORS

Controllable p-type doping of wide-bandgap $(Al_xGa_{1-x})_{0.5}In_{0.5}P$ is complicated by a number of factors. As discussed earlier, the incorporation of Zn is very inefficient at the high growth temperatures required for growth of high-quality material, and Mg is known to have a severe memory effect, depending on reactor geometry. In addition, the incorporation of both Mg and Zn is highly dependent on temperature, making reproducibility difficult. The high diffusivity of Mg and Zn results in redistribution of the p-type dopant during growth and subsequent high-temperature ($\geqslant 600°C$) processing or epitaxial regrowth. Another factor that has a significant impact on p-type doping is the unintentional H incorporation and resulting acceptor passivation which occurs during OMVPE growth and during the post-growth cool-down.

The importance of H incorporation in OMVPE-grown III-V materials has only recently been recognized (Hobson, 1994). For a III-V semiconductor doped with a column II acceptor that resides on the column III sublattice, the H is believed to reside between the acceptor (e.g., Zn) and a nearest-neighbor column V atom (phosphorus). The H forms a bond with the phosphorus, leaving the Zn atom tricoordinated and thus electrically inactive (Pajot *et al.*, 1989). Although no evidence for a direct bond between hydrogen and the column II acceptor has been found, the stability of the acceptor-hydrogen complex has been shown to depend on both the acceptor species (Szafranek and Stillman, 1990) and the column V element (e.g., P and As).

The effect of the cooling ambient on the passivation of Zn acceptors in InP grown by atmospheric-pressure OMVPE was first described by Antell *et al.* (1988) and Cole *et al.* (1988), and the general trends they reported were later shown to apply to the AlGaInP material system as well. Antell *et al.* (1988) grew Zn-doped InP using TMIn, PH_3, and DMZn ([Zn] $\sim 2 \times 10^{18}$ cm^{-3}). They found that in a p-type InP layer capped with p-InGaAs and exposed to AsH_3 during postgrowth cooling, the hole concentration could be significantly reduced (by about 80%) compared with a Zn-doped InP layer with an n-type cap. The hole concentration could be restored to the expected value simply by annealing the sample in N_2. This behavior was attributed to partial H passivation of Zn acceptors, with the source of H being the AsH_3 present during sample cooling. (The solubility of H in p-InP is significantly higher during cooling than at the growth temperature of 650°C.) Passivation is minimized when an n-type cap layer is employed due to the lower solubility of H in n-type materials (Pearton *et al.*, 1987b).

Cole *et al.* (1988) studied the effect of various cooling ambients on acceptor passivation in InP. They observed measurable H incorporation and acceptor passivation in p-type InP (Zn- or Cd-doped) cooled in a PH_3-containing ambient, although at a much lower level than when a comparable amount of AsH_3 was present. These authors concluded that the concentration of atomic hydrogen generated at the semiconductor surface in the presence of AsH_3, PH_3, and H_2 was related to the relative stability of these molecules (no passivation was detected when cooling in H_2 alone). They also reported that no effect of the cooling ambient could be observed for p-type GaAs (Zn-doped), n-type GaAs (Si-doped), p-type InGaAs (Zn- or Cd-doped), or n-type InP (S- or Sn-doped).

The role of cooling the ambient and capping layers in Zn activation in OMVPE grown $(Al_{0.7}Ga_{0.3})_{0.5}In_{0.5}P$ was first reported by Minagawa and Kondo (1990) and Minagawa *et al.* (1992). The hole concentration, measured by electrochemical capacitance-voltage profiling, is shown for several cooling and capping conditions in Fig. 17. For the samples grown with a 0.25-μm n-type GaAs cap, the cap layer was selectively removed before profiling. Secondary ion mass spectrometry (SIMS) analysis confirmed that the Zn concentration was the same and that some H was present in all

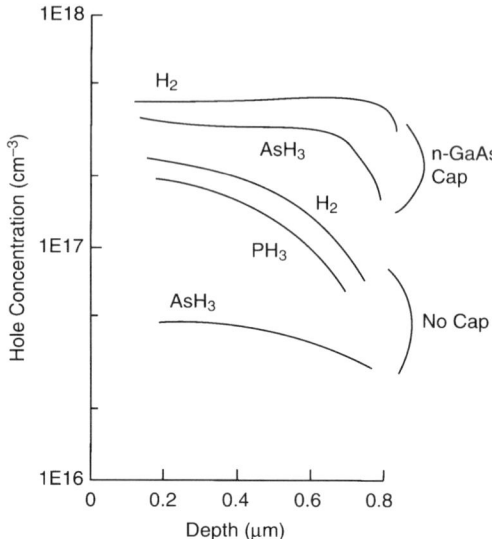

FIG. 17. Hole concentration profiles of p-AlGaInP cooled from the growth temperature (750°C) to 150°C in various ambients. (Reprinted from Minagawa et al., 1992, J. Cryst. Growth **118**, 425, with the kind permission of Elsevier Science – NL.)

layers. No surface degradation due to insufficient As or phosphorus overpressure was evident in any of the samples. Figure 17 clearly shows that the Zn activation is highest when cooling in H_2 alone, and is lowest when AsH_3 is present due to enhanced H passivation. In addition, the n-type GaAs cap is shown to result in maximum Zn activiation, presumably due to the low solubility of H in n-type layers. This layer then acts as a barrier to H in-diffusion during cooling. Minagawa and Kondo also showed that use of a p-type GaAs cap gives an intermediate result, which is likely due to an intermediate solubility or diffusivity of H. The role of H incorporation was more directly demonstrated in a similar experiment using SIMS by Nishikawa et al. (1992b) in Fig. 18. The H concentration in the AlGaInP is clearly higher when the cap layer is p-type.

The re-activation of Zn by annealing in H_2 is shown in Fig. 19. The activation occurs due to dissociation of the acceptor-H complex and subsequent out-diffusion of atomic H, and is observed to occur between 400 and 500°C. Minagawa and Kondo (1990) and Minagawa et al. (1992) pointed out that p-type doping is sensitive not only to the cooling ambient and cap layer structure, but also the temperature at which the hydride source is switched and the thermal history to which the samples are subjected.

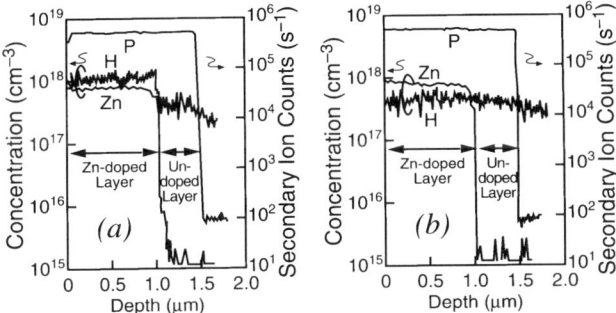

FIG. 18. Secondary ion mass spectrometry (SIMS) profiles of hydrogen, zinc, and phosphorus for AlGaInP layers grown with a p-type GaAs cap layer (a) and an n-type GaAs cap layer (b). The samples were cooled in an arsine-containing ambient and the GaAs cap layers were moved before the SIMS measurements.

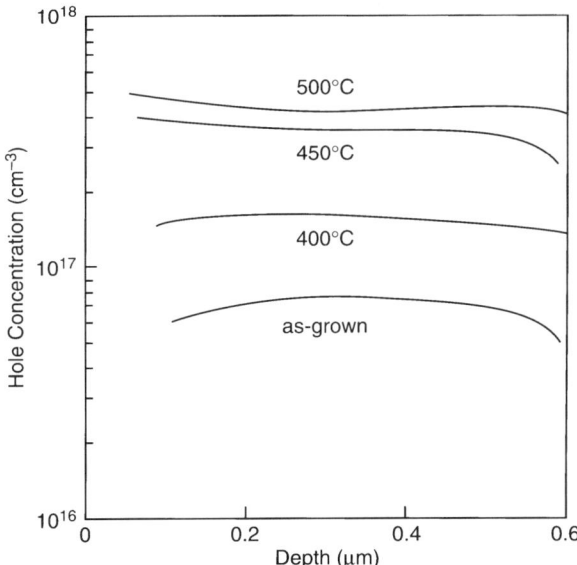

FIG. 19. Hole concentration profiles versus postgrowth annealing temperature (30 sec annealing in H_2) for a zinc-doped AlGaInP sample originally cooled in an arsine-containing ambient. (Reprinted from Minagawa et al., 1992, J. Cryst. Growth **118**, 425; with the kind permission of Elsevier Science – NL.)

The effect of substrate orientation and Al content (x) on activation of Zn in OMVPE-grown $(Al_xGa_{1-x})_{0.5}In_{0.5}P$ was investigated by Hamada et al. (1992). They showed that the degree of passivation was independent of the orientation for substrates misoriented from (100) toward the (011) direction by up to 7 degrees. They also found that the degree of passivation increased with Al composition over the range of $x = 0$ to $x = 0.65$, as shown in Fig. 20. SIMS analysis showed a decrease in H content after annealing at 500°C, and therefore the degree of passivation can be inferred from the increase in hole concentration on annealing. Thus, proper thermal treatment to maximize the p-type conductivity is especially important for high Al-content ($x = 0.6$ to 1) confining layers in laser and LED structures.

Ishibashi et al. (1994) reported that H passivation of Zn acceptors in AlGaInP may suppress Zn diffusion and column-III interdiffusion (disordering) during thermal annealing. They annealed strongly ordered p-type AlGaInP samples with a GaAs cap layer at 750°C in an AsH_3-H_2 ambient, and observed that disordering occurred only if the GaAs cap was of n-type. In the case of a p-type cap layer, which allows H in-diffusion, the degree of ordering remained unchanged. However, when the annealing ambient was switched to N_2 so that no free hydrogen was present, disordering occurred in the sample with the p-type cap layer. It was speculated that hydrogen may terminate the dangling bonds of native defects, slowing the diffusion of Zn and the interdiffusion of column III atoms. Thus, the presence of hydrogen during growth or processing may affect not only the acceptor

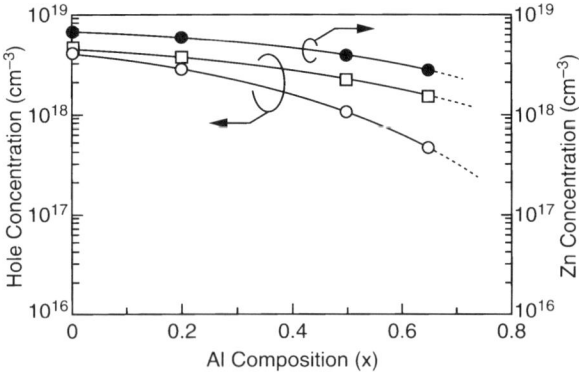

FIG. 20. Hole concentration, as-grown and after annealing at 500°C, and zinc concentration versus aluminum composition in $(Al_xGa_{1-x})_{0.5}In_{0.5}P$. (Reprinted from Hamada et al., 1992, Electron. Lett. 28, 585; with the permission of the IEEE.)

activation but also p-type doping profiles, and possibly the emission wavelength of AlGaInP-based lasers and LEDs.

The intentional hydrogenation of GaInP (Dallesasse et al., 1990a) and AlGaInP (Gorbylev et al., 1994) by exposure to a hydrogen plasma has also been studied. The behavior is qualitatively quite similar to that observed in other III-V semiconductors, that is, hydrogen can passivate deep-level defects as well as shallow donors and acceptors. This technique may be used in device processing. Dallesasse et al. (1990b) used masked hydrogenation to define resistive current blocking regions in stripe-geometry AlGaInP quantum well laser diodes. They produced devices with better performance than conventional oxide-defined diodes due to reduced thermal resistance. Plasma hydrogenation might also lead to less damage than other isolation techniques such as proton implantation. One aspect of hydrogenation that remains largely unexplored is device stability over long periods of operation. This is an especially critical issue because hydrogen-dopant and hydrogen-defect complexes in GaAs have been reported to be unstable under conditions of minority-carrier injection (Tavendale et al., 1985).

5. Oxygen Incorporation

One of the primary challenges in the growth of high-quality AlGaInP LEDs by OMVPE is control over the incorporation of residual oxygen. Oxygen has long been known to be detrimental to PL efficiency and to the performance of AlGaAs-based LEDs and lasers (Terao and Sunakawa, 1984; Mihashi et al., 1994) due to the introduction of deep-level states that may act as nonradiative recombination centers. Oxygen incorporation is most efficient, and thus most problematic, in III-V materials such as $Al_xGa_{1-x}As$ and AlGaInP with a high Al content due to the strong affinity of Al to react with oxygen. The inability to effectively minimize oxygen contamination in the manufacture of AlGaInP LEDs can result in poor reproducibility due to natural variations in the source of contamination, as well as low average internal quantum efficiency. The solution to this problem must involve a two-pronged approach. First, the various sources of oxygen contamination must be identified and controlled. These sources may include leaks or virtual leaks in the gas handling system or reaction chamber, alkoxide impurities in the source materials, water or O_2 in the carrier gas, and reactor components that are contaminated. Second, growth conditions must be chosen such that the oxygen incorporation efficiency is sufficiently minimized. These conditions may include temperature, V-III ratio, and substrate orientation.

Although the microstructure of oxygen-related defects in AlGaInP is not known, the case of oxygen in GaAs has been the subject of several studies and provides some insight into the nature of oxygen in other III-V materials (Schneider et al., 1989; Skowronski et al., 1990). Infrared absorption data acquired from bulk GaAs doped with ^{16}O show local vibrational mode (LVM) spectra consistent with two distinct types of oxygen-related defects, as shown in Fig. 21 (Schneider et al., 1989). The first is interstitial oxygen that is bonded to both a Ga atom and an As atom, and is believed to be electrically inactive. The second is a quasi-substitutional oxygen (V_{As}–O), in which oxygen occupies an off-center As site and is bonded to two nearest-neighbor Ga atoms. The electrical nature of this type of defect is quite complex, especially in alloys such as AlGaAs and AlGaInP. It is generally observed that substitutional oxygen can lead to a number of defects and deep states in III-V materials can be responsible for electrical compensation in either n-type or p-type materials, as well as decreased radiative efficiency.

The problem of unintentional oxygen incorporation during OMVPE growth of $Al_xGa_{1-x}As$ is especially severe because of the high Al content. Terao and Sunakawa (1984) showed that intentional introduction (doping) of O_2 during growth resulted in a reduction in carrier concentration, PL efficiency, and degraded surface morphology. The primary effect of water, however, was a decrease in Al content (x) due to a strong gas-phase reaction with TMAl leading to formation of non-volatile aluminum hydroxide

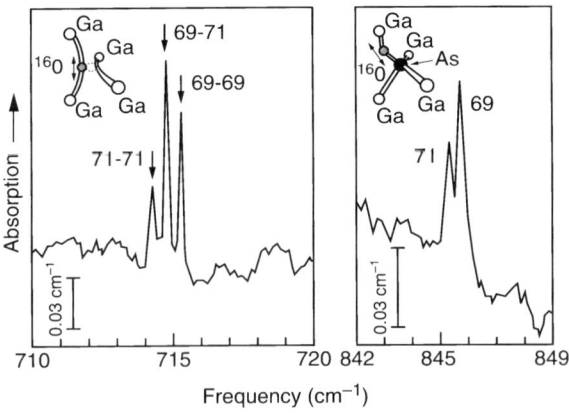

FIG. 21. Fine structure of the local vibrational mode absorption of oxygenated gallium arsenide (GaAs:O) taken at 4.2 K with 0.1 cm^{-1} resolution. Suggested models that explain the ^{69,71}Ga ligand isotope splittings are shown as inserts. (Reprinted from Schneider et al., 1989. Appl. Phys. Lett. **54**, 1442; with the permission of the American Institute of Physics.)

deposits. Metal alkoxides such as $(CH_3)_2AlO(CH_3)$ are common contaminants in the column III organometallic sources, and have been found to result in very efficient oxygen incorporation (Terao and Sunakawa, 1984; Huang et al., 1994). The most commonly employed technique for reducing oxygen levels in AlGaAs is to increase the growth temperature — hence temperatures above 700°C are used for growth of AlGaAs optoelectronic devices (Kuech et al., 1987). In most cases a high V-III ratio (AsH_3 flow) is also beneficial, although this may not be the case if the AsH_3 cylinder is the source of the oxygen-bearing impurity or if adduct formation occurs between AsH_3 and the oxygen-bearing species (Huang et al., 1994).

Several authors have intentionally doped AlGaInP with a dilute O_2 source in order to study oxygen incorporation trends and oxygen-related defects. McCalmont et al. (1992) grew semi-insulating $(Al_{0.4}Ga_{0.6})_{0.5}In_{0.5}P$ by OMVPE and found a square root dependence of oxygen incorporation (measured by SIMS) on O_2 pressure, $[O] \propto F(O_2)^n$ with $n = \frac{1}{2}$ which is consistent with a simple incorporation mechanism involving heterogeneous dissociation of O_2 at the growing surface. C-V analysis of metal-insulator-semiconductor (MIS) capacitors showed that the trap concentration decreased with increasing oxygen content. The authors speculated that the dominant trap was a DX center related to residual Si donor incorporation, and that the Si donor concentration could be reduced by formation of a Si–O complex, thereby reducing the trap concentration.

Oxygen-doping of AlGaInP by OMVPE was also studied by Kondo et al. (1994a). They found a much stronger dependence of oxygen incorporation on O_2 flow. For $(Al_xGa_{1-x})_{0.5}In_{0.5}P$ with $x = 0.7$ grown at $T_g = 690°C$ they observed that $[O] \propto F(O_2)^n$ where $n = 4$, and T_g is the substrate temperature as shown in Fig. 22. They attributed the superlinear dependence to gettering of oxygen by the quartz reactor surfaces. However, a similar strong dependence has been observed in alkoxide-doped AlGaAs (Huang et al., 1994) and suggests an interaction of multiple oxygen-bearing species as part of a complex oxygen incorporation process. Another study has revealed a superlinear dependence for AlInP grown by OMVPE and doped with O_2 or DEAlO, with the order n ranging from 2.5 to 5 depending on the growth conditions (S. A. Stockman, T. D. Ostenowski, B. Liang, J. Tarn, C. H. Chen, and M. J. Peanasky, unpublished data, 1995).

Kondo et al. (1994a) also employed isothermal capacitance transient spectroscopy to identify three deep levels in these samples: D1 ($E_T = 0.37$ eV), D2 (0.46 eV) and D3 (1.0 eV). D1 was found to be related to Si donors, while D2 and D3 are correlated with oxygen as is evident from Fig. 22. Kondo proposed that D2 and D3 were charge-state dependent levels related to off-center substitutional oxygen ($O-V_P$). They also pointed out that the total concentration of deep levels (D2 and D3) was always less than 10% of

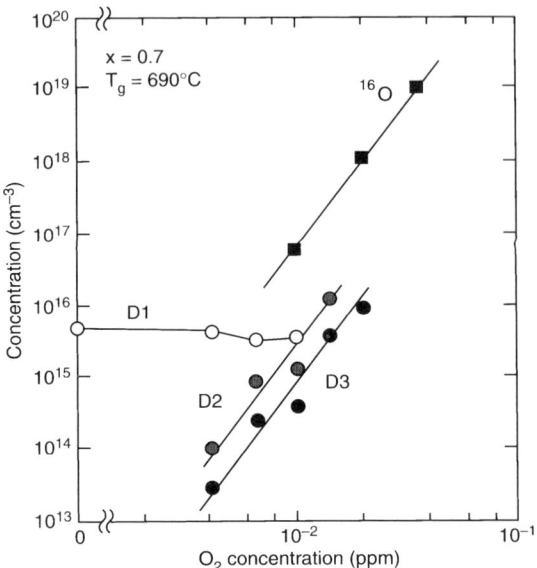

FIG. 22. Dependence of oxygen ($[^{16}O]$) and deep level (D1, D2, D3) concentrations on oxygen concentration in the gas phase for $(Al_{0.7}Ga_{0.3})_{0.5}In_{0.5}P$ grown at 690°C. (Reprinted from Kondo et al., 1994a, J. Electron. Mater. **23**, 355; with the permission of the Minerals, Metals and Materials Society.)

the total oxygen content over a wide range of growth conditions. This suggests that most of the oxygen is in the form of electrically inactive interstitial oxygen or some other neutral oxygen-related complex.

The incorporation of oxygen and oxygen-related defects in $(Al_xGa_{1-x})_{0.5}In_{0.5}P$ is also greatly enhanced as the Al content (x) is increased, as shown in Fig. 23. This is consistent with the behavior of oxygen in AlGaAs, and may be related to a gas-phase reaction between O_2 and TMAl to form DMAlO. The desorption of oxygen-containing species may also be reduced as x is increased due to the increased probability that an oxygen atom will encounter a stable surface adsorption site involving two neighboring Al atoms. In an LED or laser structure, oxygen incorporation is very efficient in the high Al content (x ~ 0.6 to 1) confining layers. In the active layer, oxygen incorporation is much more efficient in the green and amber portion of the spectrum (λ ~ 560 to 590 nm, x ~ 0.5 to 0.4) than it is in the red (λ ~ 630 to 650 nm, x ~ 0.1 to 0). Thus, increased [O] works together with reduced carrier confinement and the approaching direct–indirect transition to make realization of high-efficiency devices more difficult as the emission is shortened from the red toward the green.

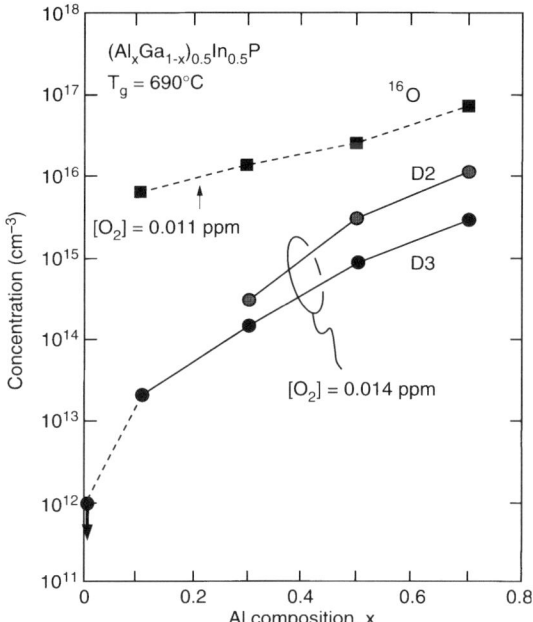

FIG. 23. Dependence of oxygen (O_2) and deep level (D2 and D3) concentrations on aluminum (Al) composition. AlGaInP, ??? . (Reprinted from Kondo et al., 1994a. *J. Electron. Mater.* **23**, 355; with the permission of the Minerals, Metals and Materials Society.)

Oxygen incorporation is also strongly dependent on substrate temperature T_g, as illustrated for O_2-doped $(Al_{0.7}Ga_{0.3})_{0.5}In_{0.5}P$ in Fig. 24. For $T_g < 690°C$, the measured oxygen concentration is temperature-independent, indicating that the incorporation efficiency is high and that incorporation is essentially mass-transport-limited. This behavior has been observed for residual oxygen from TMAl by other authors as well. (Nishikawa et al., 1993). For higher temperatures ($T_g > 690°C$ in Fig. 24) the oxygen concentration is observed to decrease rapidly with increasing T_g. This suggests that desorption of oxygen-containing species has become the rate-limiting step in the incorporation process. This dependence is quite strong, with an associated activation energy greater than 5 eV. Growth of high-quality AlGaInP optoelectronic devices is typically performed at $T_g \geq 700°C$, primarily for the purpose of minimizing [O] and oxygen-related defect levels. Growth at lower temperatures ($T_g < 700°C$) may have other advantages such as suppressing dopant diffusion. However, low-T_g growth of AlGaInP is very sensitive to the presence of low-level oxygen-containing contaminants and thus is unacceptable in a high-yield manufacturing environment.

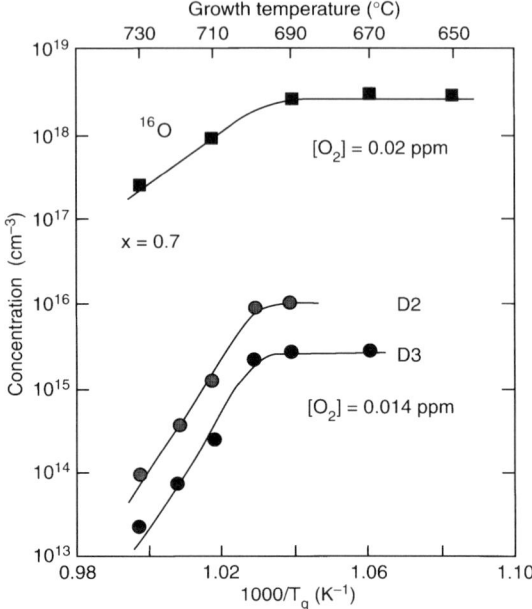

FIG. 24. Dependence of oxygen (O_2) and deep level (D2 and D3) concentrations on growth temperature, T_g. (Reprinted from Kondo et al., 1994a. J. Electron. Mater. **23**, 355; with the permission of the Minerals, Metals and Materials Society.)

The phosphorus overpressure during growth can strongly influence the incorporation of oxygen. For a constant growth rate we can describe relative changes in this quantity using the input V-III ratio, which is defined as the molar flow rate of PH_3 divided by the combined column III molar flow rates. Figure 25 shows the dependence of residual [O] on V-III for AlInP grown at two different T_g's. The source of oxygen in this case is believed to be the column III organometallics. This suggests that at higher T_g, [O] decreases with increasing V-III ratio in both cases. However, for $T_g = 750°C$, [O] is much less dependent on the V-III ratio than for growth at 785°C. In fact, the optimal set of growth conditions for minimizing oxygen incorporation involves high T_g and high V-III ratio. Unfortunately, the use of very high PH_3 flows is often limited by practical considerations related to trapping and disposal of excess phosphorus in the reactor exhaust.

There are two possible explanations for the strong dependence of [O] on the V-III ratio. One is a site competition model, in which phosphorus and oxygen "compete" for colume V surface adsorption sites. Thus, a high V-III ratio reduces the column V vacancy concentration at the surface and

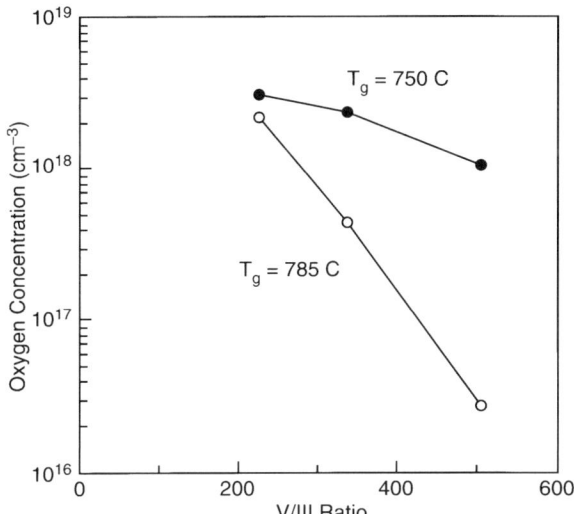

FIG. 25. Dependence of oxygen incorporation on V-III ratio for AlInP growth at $T_g = 750$ and 785°C. (S. A. Stockman, T. D. Ostenowski, B. Liang, J. Tarn, C. H. Chen, and M. J. Peanasky, unpublished data, 1995.)

suppresses oxygen incorporation. A second model is that hydrogen released from decomposition of PH_3 heterogeneously or in the gas phase may react with oxygen to form highly volatile compounds such as water, which do not lead to efficient oxygen incorporation. In this model, higher T_g may also accelerate the formation or evaporation, or both, of the volatile oxygen-containing species.

High levels of oxygen in AlGaInP can result in electrical compensation of shallow donors or acceptors. This is illustrated for the case of p-type AlInP in Fig. 26 (S. A. Stockman, T. D. Ostentowski, B. Liang, J. Tarn, C. H. Chen, and M. J. Peanasky, unpublished data, 1995), where the hole concentration ($p = N_A - N_D$, measured at room temperature by C-V) is plotted versus the concentration of oxygen (measured by SIMS). The Mg doping level is held roughly constant at $[Mg] = 5-8 \times 10^{17}\,cm^{-3}$ in all cases. As can be seen, the degree of compensation is relatively small for $[O] \leq [Mg]$, and the layers do not become fully depleted (insulating) until $[O] > 10^{19}\,cm^{-3}$. In addition, the compensation behavior is not strongly dependent on the oxygen source. Nearly identical behavior is observed whether the oxygen level is varied intentionally using O_2 or DEAlO as a dopant source, or unintentionally due to changes in organometallic source purity. This result is consistent with the observation by Kondo *et al.* (1994a) that the ratio of

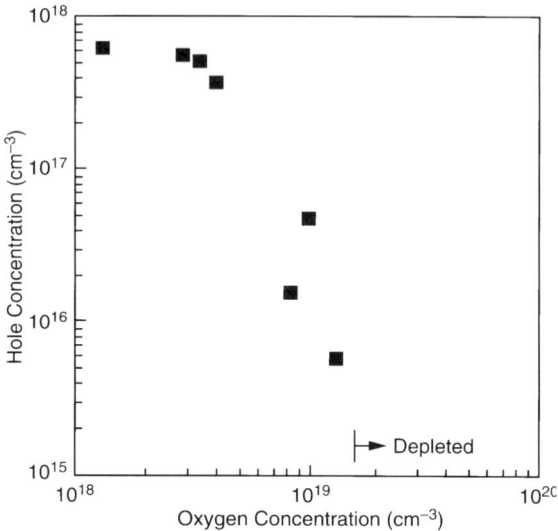

FIG. 26. Room temperature hole concentration ($p = N_A - N_D$) versus oxygen concentration for magnesium-doped AlInP with [Mg] = $5-9 \times 10^{17}$ cm^{-3}. The layers are fully depleted ($p < 10^{14}$ cm^{-3}) for [O] $> 1.5 \times 10^{19}$ cm^{-3}. (S. A. Stockman, T. D. Ostenowski, B. Liang, J. Tarn, C. H. Chen, and M. J. Peanasky, unpublished data, 1995.)

the deep-level concentration to the total oxygen concentration is less than 0.1 under a wide range of growth conditions, and suggests that most of the oxygen is incorporated as electrically inactive interstitial oxygen or in some other type of inactive complex.

Orientation of the GaAs substrate also has a dramatic impact on oxygen incorporation. This can be seen for O_2-doped $(Al_{0.7}Ga_{0.3})_{0.5}In_{0.5}P$ in Fig. 27 (Kondo et al., 1994a,b). The oxygen incorporation efficiency decreases rapidly for miscut toward (111)A, and [O] is below the SIMS detection limit for the (311)A orientation. Kondo et al. (1994b) have pointed out the similarity of this behavior to that of column VI donors such as S, Se, and Te, which incorporate substitutionally on the column V sublattice. For a (111)A surface, group V sites encountered by oxygen are weak adsorption sites with only one bond to an underlying group III atom. Thus, tilting of the surface orientation toward (111)A increases the density of weak adsorption sites, leading to decreased oxygen incorporation. Suzuki et al. (1993) have demonstrated that for a fixed set of growth conditions, AlGaInP LED performance may be enhanced by tilting the GaAs substrate 15 degrees toward the (111)A direction. This enhancement in LED quantum efficiency was attributed to a reduction in the oxygen-related trap concentration.

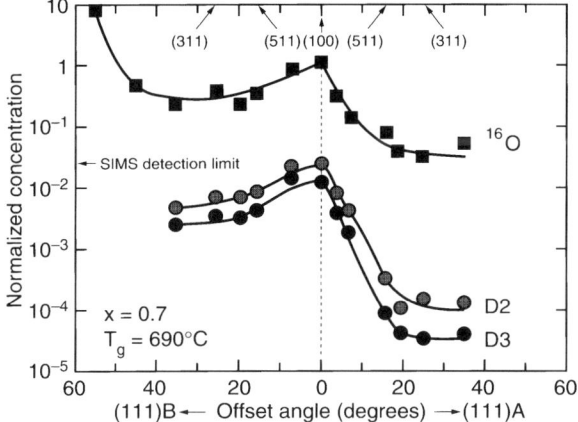

FIG. 27. Dependence of normalized oxygen and deep level (D2 and D3) concentrations on gallium arsenide substrate misorientation with respect to the (100) surface. (Reprinted from Kondo et al., 1994a, J. Electron. Mater. **23**, 355; with the permission of the Minerals, Metals and Materials Society.)

The deleterious effects of incorporation of oxygen directly into the active region of AlGaInP LEDs and lasers have been noted by many groups. It has also been reported that oxygen-related deep levels in the high-Al content confining layers may reduce the radiative efficiency. Figure 28 shows the dependence of interface recombination velocity and luminescence intensity

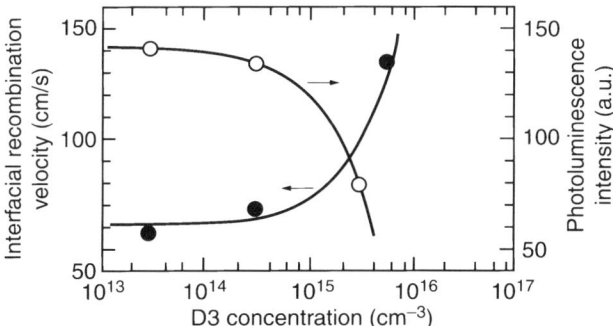

FIG. 28. Interface recombination velocity and photoluminescent intensity of AlGaInP–GaInP double heterostructure samples as a function of the D3 trap concentration in the $Al_{0.7}Ga_{0.3})_{0.5}In_{0.5}P$ confinement layers. (Reprinted from Domen et al., 1991, J. Cryst. Growth **115**, 529; with the kind permission of Elsevier Science – NL.)

on the D3 trap concentration in the $(Al_{0.7}Ga_{0.3})_{0.5}In_{0.5}P$ confinement layers of AlGaInP–GaInP DH samples (Domen et al., 1991). The increase in interface recombination velocity and corresponding decrease in PL intensity can be attributed to nonradiative recombination at the interfaces via oxygen-related deep levels in the adjacent confining layers. Thus, even devices with an Al-free GaInP active layer may be quite sensitive to oxygen contamination. As a result, great care must be taken to ensure that oxygen incorporation is minimized.

IV. High-Volume Manufacturing Issues

There are many issues that manifest themselves in the process of scaling the epitaxial reactor from a single wafer, research size to a larger, multiwafer (more than 10 wafers) reactor. There are issues of gas delivery to the chamber. The exhaust system must be able to handle large quantities of phosphorus efficiently and safely. Uniformity of the deposition is paramount for high-volume manufacturing of LEDs. Consumption rates of the reactants should be considered. Yield loss must be minimized and reactor through-put and up-time must be optimized.

1. Reactor Issues: Delivery and Exhaust Treatment

One of the critical and practical issues in III-V OMVPE is the safe use and disposal of the group V hydride gas and the associated As- and phosphorus-containing deposits. This issue dominates gas delivery and exhaust system design and maintenance for AlGaInP OMVPE reactors because of the properties and hazards associated with hydride gases and phosphorus. One must carefully consider many aspects of laboratory design and safety in the manufacture of OMVPE materials. Several reviews exist on these issues (Johnson et al., 1984; Messham and Tucker, 1986; Kaufman et al., 1988; Grodzinski et al., 1995).

Arsine and PH_3 gases are highly toxic (Steere, 1967; OSHA Publication, 1989; Hess and Riccio, 1986). The permissible exposure level (PEL) for AsH_3 has been set at 50 ppb in air. Toxic gas monitoring is absolutely required. Phosphine is slightly less toxic with a PEL of 300 ppb, but nonetheless is considered highly toxic. The use of tertiarybutylarsine (TBAs) and tertiarybutylphosphine (TBP) greatly reduces the hazards associated with the group V precursor. However, even the toxicity of these organic substitutes are reported to be high (American Cyanamid, 1988). In addition to the hydride gas toxicity, the toxicity of the elemental and the oxide forms of the group V precursor need to be addressed because these compounds will be encountered in the gas delivery and exhaust systems during maintenance. The elemental phosphorus is toxic and flammable! Extreme care must be exercised in the handling of phosphorus deposits.

Whether utilizing a vertical flow or horizontal flow multiwafer geometry, the increase in deposition area and the increase in PH_3 flow required poses a much larger set of problems than for a single-wafer, research and development–scale reactor. This results in very high loading of phosphorus in the exhaust system if the gas temperature falls below the vapor-phase equilibrium curve for the given reactor pressure. To prevent phosphorus condensation, heating of the exhaust system may be required. Heating of the lines allows the phosphorus to be carried down the exhaust system. At some point the phosphorus needs to be dealt with by either condensation or chemical conversion.

The treatment of the exhaust after the reactor is termed abatement. Careful consideration of the residual hydride gas abatement is needed for a safe and environmentally sound OMVPE operation. The conventional method of activated charcoal may require large canisters and high disposal costs for the given PH_3 use rate in large multiwafer reactors. Resin bed scrubbers may be very expensive as well. Liquid chemical scrubbers are labor intensive and messy. Combustion and decomposition methods generate fine particles of the group V oxide that must be collected. The chemical conversion of phosphorus and PH_3 is exothermic, which causes a need for thermal management of the abatement system. Figure 29 shows an example of an exhaust system and an effluent abatement system utilizing various phosphorus condensers followed by hydride gas combustion and oxide particle collection. The available facility support system often defines the abatement method that is right for the user.

2. IMPORTANCE OF UNIFORMITY

It may seem intuitively obvious that highly uniform epitaxial layers are required for the successful manufacture of LEDs. Whole-wafer, single-color binning (2.5 nm/color bin) is the goal of any LED process. However, with the AlGaInP material system spanning well into the most wavelength-sensitive region of visible spectrum (amber and green), the task of mass-producing epilayers wafers with high uniformity is a real challenge.

Early OMVPE crystal growers were well aware of the difficulties of producing highly uniform layers over the entire wafer. One standard approach on horizontal flow reactors has been to increase flow rates to

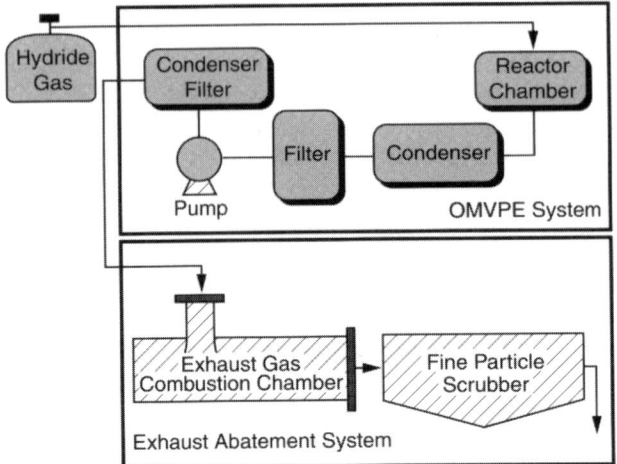

FIG. 29. An example of an exhaust system and organometallic vapor-phase epitaxy (OMVPE) reactor effluent abatement system.

achieve a high flow velocity across the wafer allowing less depletion of the gas stream to occur. While successful for single-wafer reactors, the concept does have the drawback of high reactant use rates. Low pressure is another method of increasing the gas velocity over the wafer without increasing the use of column III reactants. The drawback of low pressure is that it lowers the partial pressure of the group V species, necessitating an increase of the hydride gas flow in order to maintain the same effective V-III ratio.

Uniformity is the result of the reactor design and careful process optimization. Over the past decade the field of OMVPE has crossed a threshold with commercial reactors capable of producing multiple wafer runs with excellent compositional, thickness, and doping uniformity. Modern production-scale OMVPE reactors now utilize some form of rotation and low-pressure growth to generate uniformity. High-speed rotating-disk reactors reduce the boundary layer thickness and pump the vertically injected gas across the wafers, yielding high uniformity and efficient alkyl use rates. A similar technique is using vertical stagnation flow reactors without high-speed rotation in which highly uniform gas injection results in highly uniform deposition. A competitive approach is to rotate the wafer in a horizontal flow field at a rate comparable to a monolayer of film growth. In this approach, efficient alkyl use rates can be achieved along with highly uniform film growth.

Good compositional uniformity within a wafer is critical for a high yielding process as well as wafer-to-wafer uniformity and run-to-run repeatability. Nonuniformity within a wafer may be the result of temperature nonuniformity on the wafer. Good heater design is required to be able to tune the reactor temperature profile to provide the best uniformity. The hardware design, combined with creative process development to overcome remaining design deficiencies, are the tools of the epilayer grower for providing highly uniform epitaxial wafers.

Uniformity studies using GaInP lattice-matched to GaAs have been performed in various production-scale reactors. The results for a stagnation flow reactor geometry are presented by Vernon et al. (1994). Rotating-disk reactor data are shown by McKee et al. (1992). Multiwafer horizontal flow reactor geometry is presented by Schmitz et al. (1992). The compositional uniformity as measured by PL of the GaInP layer can be well within the single color bin needed in the LED industry (± 1.25 nm) using 2-mm edge exclusion. Figure 30 shows an example of a wavelength contour map showing ± 0.35 nm. The compositional uniformity can be examined by

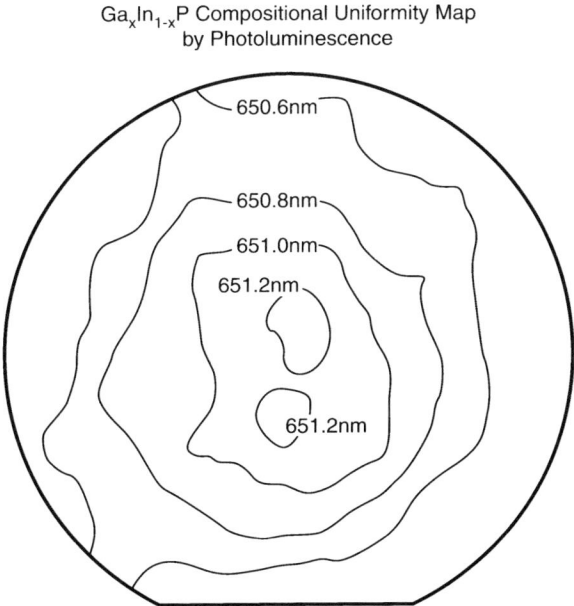

FIG. 30. GaInP (tellurium-doped, $n = 5 \times 10^{17}$ cm^{-3}) photoluminescent peak wavelength contour map with a 2-mm edge exclusion on a 2-inch wafer. (M. J. Peanasky, C. P. Kuo, T. D. Ostenowski, D. A. Steigerwald, S. A. Stockman, and A. J. Moll, unpublished, 1995.)

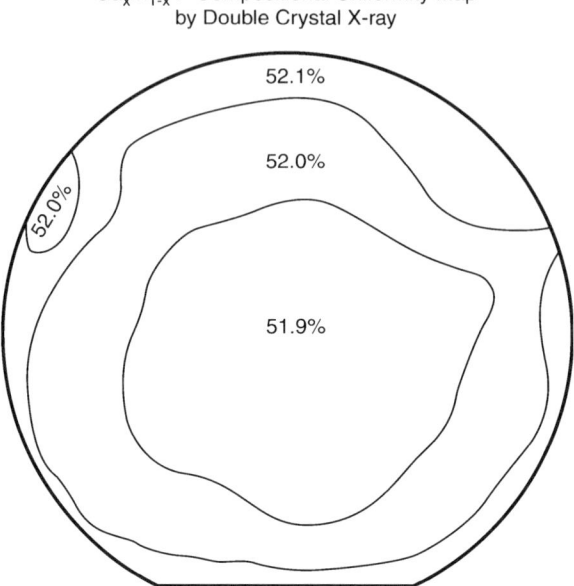

FIG. 31. GaInP (tellurium doped, $n = 5 \times 10^{17}\,\text{cm}^{-3}$) double-crystal X-ray diffractometry, compositional uniformity map. The composition is presented as the percent gallium (x in $Ga_xIn_{1-x}P$). (M. J. Peanasky, C. P. Kuo, T. D. Ostenowski, D. A. Steigerwald, S. A. Stockman, and A. J. Moll, unpublished, 1995.)

double crystal X-ray maps. Figure 31 shows the range of composition variation on a 2-inch $Ga_xIn_{1-x}P$ wafer of 0.25% Ga (approximately 2-mm edge exclusion). These are truly impressive results!

3. SOURCE QUALITY ISSUES

Source purity has improved dramatically from the mid-1980s to the mid-1990s for both the hydride gases and the typical alkyl sources. However, as discussed in Part II, Section 3, oxygen contamination from source materials remains a primary concern.

For the hydride gases, typical source purity levels previously were around 1 to 3 ppm for oxygen and moisture for the best grades, but now one can commonly find impurity levels less than 500 ppb. Because PH_3 is used in excess of the stochiometric ratios with the group III source, the impurity flux from the hydride gas can be high.

Likewise, the source purity for alkyls is improving rapidly. The predominant oxygen-related impurities present themselves as alkoxides (such as $(CH_3)_2AlO(CH_3)$) making it difficult for the manufacturer to detect. With a sufficient understanding of the incorporation of the alkoxide species into the epilayer, it is possible to evaluate the purity of the alkyl source. Trimethylindium is more difficult to purify than are the other conventional liquid alkyls trimethylaluminum and trimethylgallium. Contacting and purifying schemes are inherently easier for liquids than for solids. This provides a secondary motivation for the development of liquid alternatives to solid TMIn. (The primary motivation is improved control over In delivery.)

Metallic and other residual dopant impurities may also significantly affect device performance. It is not uncommon to find Si as an alkyl impurity. Without careful screening of the alkyl sources, good quality and consistency are difficult to achieve from cylinder to cylinder.

4. Color Control Issues

A stable process for AlGaInP LED manufacture requires that the color of the LED be tightly controlled. Color is a fundamental property of an LED. Many hardware issues affect the control of color. The calibration drift of mass flow controllers and pressure controllers and variation in process temperature all have an impact on color control.

In the LED industry, PL from the epilayer just grown is often used in some manner to predict the color that the LED will produce. The timely PL feedback usually provides effective control over the product color. However, the fact that the AlGaInP active layer may be ordered or disordered (Part III, Section 3) complicates the accurate determination of color based on PL, because the degree of ordering may change with DH structure growth conditions, GaP window layer regrowth and device processing.

Fortunately, the color variation from run to run is the result of the lack of control over only one group III element, In. Hence, the primary focus is placed on the control of In delivery and consumption in the reactor. One feature of the AlGaInP growth structure is that the ratio of In to Ga plus Al in each layer is approximately the same. Hence, In incorporation is the dominant variable to control the LED color. The active layer In composition easily can be obtained using double-crystal X-ray diffractometry.

There are several reasons why variation in the In solid composition far exceeds that of Al or Ga. First, the control of Al and Ga is more stable than that of In because the sources are liquids and TMIn is a solid. Delivery rates of liquids from bubblers can be accurately controlled simply by controlling the bubbler temperature, the carrier gas pressure, and the flow rate. On the other hand, the vapor pressure of the solid TMIn changes over time as the surface area of TMIn in the bubbler changes with sublimation. This leads to variation in the TMIn delivery rate even when the bubbler temperature, pressure, and carrier gas flow rate are maintained constant. To ensure accurate and precise control of the concentration of TMIn being dispensed from the bubbler from run to run, an alkyl monitor is needed to measure and actively control TMIn flux from the bubbler. Alternatively, efforts have been made to replace TMIn with a liquid source.

Another reason that In incorporation is more problematic is that the TMIn is less stable to thermal decomposition than the TMGa or TMAl (Stringfellow, 1989, Chapter 2). This leads to a higher tendency for parasitic decomposition of TMIn on reactor chamber surfaces rather than the wafers. The nonsteady losses of indium are possibly due to the nonsteady wall temperatures and effective surface roughness of the walls. The TMIn decomposition may be unsteady with time as the reactor walls come to a stable temperature, or as the cooling water temperature changes, or by any other nonsteady temperature effects inherent to the reactor. This wall temperature effect has been reported to not only influence composition through the epilayer growth, but also the dopant incorporation across the wafer (Li and Giling, 1995). The nonsteady delivery of In to the wafers and the nonsteady losses of In in the reactor chamber may be the dominant problem for achieving color control. These nonsteady losses in the chamber are commonly called the "liner" effect in a horizontal tube reactor.

The third reason for the variation in the In solid composition is the high substrate temperature used to grow the AlGaInP device structures. The incorporation of Al and Ga into the solid is determined by the diffusion through the boundary layer near the substrate; Al and Ga re-evaporation is limited. Thus, Al and Ga solid compositions show minimal dependence on growth temperature. On the other hand, at high growth temperatures typically used for AlGaInP growth, In re-evaporation from the surface is observed so that In incorporation depends on growth temperature. Thus, variation in process temperature will result in a variation in In solid composition and the color of the device.

The thermocouple calibration drift or slight movement of the thermocouple position will certainly effect wafer temperature (i.e., color). Temperature drift resulting from a change in the heater coupling efficiency to the susceptor may affect color dramatically. Temperature control on the wafer is fundamental to OMVPE processes but difficult to achieve. Optical pyrometry techniques may offer a tool to help minimize wafer temperature variation.

In summary, process stability is mainly dominated by variation in the Indium incorporation. Indium incorporation can only be achieved by careful control of the indium flux out of the bubbler and by the control of nonsteady In losses on the reactor walls. Tight control on growth temperature is also necessary. In addition, active monitoring of the In solid composition using X-ray rocking curves will enable corrective actions to be taken once the drift in the In incorporation occurs.

5. Yield Loss Categories

Yield loss categories are typically morphology, color control, and lattice-matching criteria, electrical properties, light output and the stability of the light output of the device. Yield may be strongly dependent on the consistency of the substrate supply, morphology, and wafer handling. Substrate preparation for large quantities of wafers can become a major task in the laboratory, and excessive or improper wafer handling may lead to marring of the wafer surface.

Considering the growth rates and device thickness requirements, epilayer growth for AlGaInP will likely remain a large batch process with more than 10 wafers. The nondestructive testing of an epilayer run typically is limited to PL, X-ray rocking curves, and surface doping. Destructive testing is required for the measurement of carrier concentration profiles and the measurement of other impurities. The destructive testing may be accomplished on a test piece that is smaller than a normal wafer. Although the correlation between the test piece and the wafer may not be exact, it is often possible to control a process with this method.

Typically in OMVPE, test structures are grown to maintain and control some more difficult epilayer characteristics. In the production environment, excessive use of test structures reduces the device through-put of the reactor. Obviously, a balanced strategy is required.

Device processing will generate a whole new set of yield loss categories. The electrical characteristics can yield information about dopant levels or interface problems. By the time poor electrical data is available it is often too late and many runs will have been lost. Likewise, the real color of the device is measured after processing. The device electroluminescent color must be correlated with the predicted color from the epilayer film. In addition, stable device performance (both optical and electrical) must be guaranteed to the customer and verified in the manufacturing process. Sampling techniques are required to provide statistical process control. At the present state of OMVPE, the manufacturing process is optimized by efficient epilayer wafer screening and a short cycle time for wafer evaluation.

V. Summary

The progress in the growth of AlGaInP alloys using OMVPE has made possible the commercialization of new high-brightness visible LEDs. These AlGaInP green, yellow, orange and red LEDs, with their high luminous performance, provide a viable alternative in applications that have traditionally employed incandescent lamps. This chapter describes the OMVPE growth process for these AlGaInP alloys. The properties of the epilayer depend strongly on the growth conditions as well as the ability to control background impurity incorporation, especially oxygen. Inability to effectively minimize oxygen contamination in the manufacture AlGaInP LEDs can result in poor reproducibility due to variation in source purity and other factors. The production yield for these AlGaInP LEDs is expected to increase each year as manufacturers continue to improve the growth process and increase the productivity of the OMVPE reactors. Control charts and statistical process control methods are also found to be essential to a high-yield OMVPE process. All of these improvements are expected to lower the cost of AlGaInP LEDs further, and this will be a dominant factor in accelerating the penetration of AlGaInP devices into new applications.

Acknowledgments

The authors acknowledge the full support of Hewlett-Packard Company, Hewlett-Packard Laboratories, and the Optoelectronics Division. Without the support from a company such as HP the task would be made even more challenging than it already is. Credit is also given to the ever-present encouragement and coaching from George Craford. Finally, special recognition also goes to all the engineers, technicians, and operators who have contributed to the innumerable tasks required to be successful in introducing the world's brightest LEDs.

References

Abernathy, C. R., Wisk, P. W., Pearton, S. J., and Ren, F. (1993). *Appl. Phys. Lett.* **62**, 258.
Agnello, P. D. and Ghandi, S. K. (1989). *J. Cryst. Growth* **94**, 311.
Air Products (1995). "The Cutting Edge: Organometallics Update," June issue.
Amano, H., Kato, M., Hiramatsu, K., and Akasaki, I. (1989). *Conf. Ser. Inst. Phys.* **106**, 725.
American Cyanamid (1988). Appl. No. 2.
Antell, G. R., Briggs, A. T. R., Butler, B. R., Kitching, S. A., Stagg, J. P., Chew A., and Sykes, D. E. (1988). *Appl. Phys. Lett.* **53**, 758.

Archer, R. J. (1972). *J. Electron. Mater.* **1**, 1.
Barnard, J. E., Ferreira, L. G., Wei, S.-H., and Zunger, A. (1988). *Phys. Rev. B* **38**, 6338.
Bour, D. P. (1987). Ph.D. Thesis, Cornell University, Ithaca, NY.
Bour, D. P., and Shealy, J. R. (1987). *Appl. Phys. Lett.* **51**, 1658.
Cao, D. S., Kimble, A. W., and Stringfellow, G. B. (1990). *J. Appl. Phys.* **67**, 739.
Cao, D. S., Reihlen, E. H., Chen, G. S., Kimbal, A. W., and Stringfellow, G. B. (1991). *J. Cryst. Growth* **109**, 279.
Chen, C. H., Larsen, C. A., and Stringfellow, G. B. (1987). *Appl. Phys. Lett.* **50**, 218.
Chen, C. H., Cao, D. S., and Stringfellow, G. B. (1988). *J. Electron. Mater.* **17**, 67.
Chun, Y. S., Murata, H. Hsu, T. C., Ho, I. H., Su, L. C., Hosokawa, Y., and Stringfellow, G. B. (1996). *J. Appl. Phys.* (to be published).
Cole, S., Evans, J. S., Harlow, M. J., Nelson, A. W., and Wong, S. (1988). *Electron. Lett.* **24**, 929.
Dallesasse, J. M., Szafranek, I., Baillargeon, J. N., El-Zein, N., Holonyak, N. Jr., Stillman, G. E., and Cheng, K. Y. (1990a). *J. Appl. Phys.* **68**, 5866.
Dallesasse, J. M., El-Zein, N., Holonyak, N. Jr., Fletcher, R. M., Kuo, C. P., Osentowski, T. D., and Craford, M. G. (1990b). *J. Appl. Phys.* **68**, 5871.
Dapkus, P. D., Manasevit, H. M., Hess, K. L., Low, T. S., and Stillman, G. E. (1981). *J. Cryst. Growth* **55**, 10.
Domen, K., Sugiura, K., Anayama, C., Kondo, M., Sugawara, M., Tanahashi, T., and Nakajima, K. (1991). *J. Cryst. Growth* **115**, 529.
Fletcher, R. M., Kuo, C. P., Osentowski, T. D., Huang, K. H., Craford, M. G., and Robbins, V. M. (1991). *J. Electron. Mater.* **20**, 1125.
Fry, K. L., Kuo, C. P., Larsen, C. A., Cohen, R. M., Stringfellow, G. B., and Melas, A. (1986). *J. Electron. Mater.* **15**, 91.
Gavrilovic, P., Dabkowski, F. P., Meehan, K., Willians, J. W., Stutius, W., Hsieh, K. C., Holonyak, N., Jr., Shahid, M. A., and Mahajan, S. (1988). *J. Cryst. Growth* **93**, 426.
Gomyo, A., Kobayashi, K., Kawata, S., Hino, I., and Susuki, T. (1986). *J. Cryst. Growth* **77**, 367.
Gomyo, A., Kobayashi, K., Kawata, S., Hino, I., and Yuasa, T. (1987). *Appl. Phys. Lett.* **50**, 673.
Gomyo, A., Suzuki, T., Iijima, S., Hotta, H., Fujii, H., Kawata, S., Kobayashi, K., Uenno, Y., and Hino, I. (1988). *Jpn. J. Appl. Phys.* **27**, L2370.
Gomyo, A., Kawata, S., Suzuki, T., Iijima, S., and Hino, I. (1989). *Jpn. J. Appl. Phys.* **28**, L1728.
Gorbylev, V. A., Chelniy, A. A., Polyakov, A. Y., Pearton, S. J., Smirnov, N. B., Wilson, R. G., Milnes, A. G., Cnekalin, A. A., Govorkov, A. V., Leiferov, B. M., Borodina, O. M., and Balmashnov, A. A. (1994). *J. Appl. Phys.* **76**, 7390.
Grodzinski, P., DenBaars, S. P., and Lee H. C. (1995). *Miner. Met. Mater. Soc. — JOM* **47**, 12.
Hamada, H., Shono, M., Hondo, S., Hiroyama, R., Yodoshi, K., and Yamaguchi, T. (1991). *IEEE. J. Quantum Electron.* **27**, 1483.
Hamada, H., Honda, S., Shono, M., Hiroyama, R., Yodoshi, Y., and Yamaguchi, T. (1992). *Electron. Lett.* **28**, 585.
Hatakoshi, G., Itaya, K., Ishikawa, M., Okajima, M., and Uematsu, Y. (1991). *IEEE. J. Quantum Electron.* **27**, 1476.
Hersee, S. D., DiForte-Poisson, M. A., Baldy, M., and Duchemin, J. P. (1981). *J. Cryst. Growth* **55**, 53.
Hess, K. L., and Riccio, R. J. (1986). *J. Cryst. Growth* **77**, 95.
Hobson, W. S. (1994). *Mater. Sci. Forum* **148–149**, 27.
Huang, J. W., Gaines, D. F., Kuech, T. F., Potemski, R. M., and Cardone, F. (1994). *J. Electron. Mater.* **23**, 659.
Ikeda, M., Mori, Y., Sato, H., Kaneko, K., and Watanabe, N. (1985). *Appl. Phys. Lett.* **47**, 1027.
Ishibashi, A., Mannoh, M., Kidoguchi, I., Ban, Y., and Ohnaka, K. (1994). *Appl. Phys. Lett.* **65**, 1275.

Ishikawa, M., Ohba, Y., Sugawara, H., Yamamoto, M., and Nakanisi, T. (1986). *Appl. Phys. Lett.* **48**, 207.
Johnson, E., Tsui, R., Convey, D., Nellen, N., and Curless, J. (1984). *J. Cryst. Growth* **68**, 497.
Kaufman, L. M. F., Heuken, M., Tilders, R., Heime, K., Jurgensen, H., and Heyen, M. (1988). *J. Cryst. Growth* **93**, 279.
Kikuchi, A., Kishino, K., and Kaneko, Y. (1989). *J. Appl. Phys.* **66**, 4557.
Kish, F. A., Steranka, F. M., DeFevere, D. C., Vanderwater, D. A., Park, K. G., Kuo, C. P., Osentowski, T. D., Peanasky, M. J., Yu, J. G., Fletcher, R. M., Steigerwald, D. A., Craford, M. G., and Robbins, V. M. (1994). *Appl. Phys. Lett.* **64**, 2839.
Kisker, D., Miller, J. N., and Stringfellow, G. B. (1982). *Appl. Phys. Lett.* **40**, 614.
Knauf, J., Schmitz, D., Strauch, G., Jurgensen, H., Heyen, M., and Melas, A. (1988). *J. Cryst. Growth* **93**, 34.
Kobayashi, K., Kawara, S., Gomyo, A., Hino, I., and Suzuki, T. (1985). *Electron. Lett.* **21**, 931.
Kondo, M., Okada, N., Domen, K., Sugiura, K., Anayama, C., and Tanahashi, T. (1994a). *J. Electron. Mater.* **23**, 355.
Kondo, M., Anayama, C., Okada, N., Sekiguchi, H., Domen, K., and Tanahashi, T. (1994b). *J. Appl. Phys.* **76**, 914.
Kondo, M., Anayama, C., Sekiguchi, H., and Tanahashi, T. (1994c). *J. Cryst. Growth* **141**, 1.
Kozen, A., Nojima, S., Tenmyo, J., and Asahi, H. (1986). *J. Appl. Phys.* **59**, 1156.
Kuech, T. F. (1986). *Mater. Sci. Rep.* **2**, 1.
Kuech, T. F. and Veuhoff, E. (1984). *J. Cryst. Growth* **68**, 148.
Kuech, T. F., Veuhoff, E., and Meyerson, B. S. (1984). *J. Cryst. Growth* **68**, 48.
Kuech, T. F., Wolford, D. J., Veuhoff, E., Deline, V., Mooney, P. M., Potemsky, R., and Bradley, J. (1987). *J. Appl. Phys.* **62**, 632.
Kuech, T. F., Wang, P. J., Tischler, M. A., Potemski, R., Scilla, G. J., and Cardone, F. (1988). *J. Cryst. Growth* **93**, 624.
Kuech, T. F., Potemski, R., Cardone, F., and Scilla, G. (1992). *J. Electron. Mater.* **21**, 341.
Kuo, C. P., Vong, S. K., Cohen, R. M., and Stringfellow, G. B. (1985). *J. Appl. Phys.* **57**, 5428.
Kuo, C. P., Fletcher, R. M., Osentowski, T. D., Lardizabal, M. C., Craford, M. G., and Robbins, V. M. (1990). *Appl. Phys. Lett.* **57**, 27.
Kurtz, S. R., Olsen, J. M., and Kibbler, A. (1990). *Appl. Phys. Lett.* **57**, 1922.
Kurtz, S. P., Olsen, J. M., Arent, D. J., Bode, M. H., and Bertness, K. A. (1994). *J. Appl. Phys.* **75**, 5110.
Landgren, G., Rask, M., Anderson, S. G., and Lundberg, A. (1988). *J. Cryst. Growth* **93**, 646.
Larsen, C. A., Buchan, N. I., and Stringfellow, G. B. (1987). *J. Cryst. Growth* **85**, 148.
Lewis, C. R., Ludowise, M. J., and Dietze, W. T. (1984). *J. Electron. Mater.* **13**, 447.
Li, Y., and Giling, L. J. (1995). *J. Cryst. Growth* **156**, 177.
McCalmont, J. S., Casey, H. C., Jr., Wang, T. Y., and Stringfellow, G. B. (1992). *J. Appl. Phys.* **71**, 1046.
McKee, M. A., McGivney, T., Walker, D., Capuder, K., Norris, P. E., Stall, R. A., and Rose, B. C. (1992). *J. Electron. Mater.* **21**, 289.
Messham, R. L., and Tucker, W. K. (1986). *J. Cryst. Growth* **77**, 101.
Mihashi, Y., Miyashita, M., Kaneno, N., Tsugami, M., Fujii, N., Takamiya, S., and Mitsui, S. (1994). *J. Cryst. Growth* **141**, 22.
Minagawa, S., and Kondo, M. (1990). *J. Electron. Mater.* **19**, 597.
Minagawa, S., Kondow, M., Yanagisawa, H., and Tanaka, T. (1992). *J. Cryst. Growth* **118**, 425.
Mowbray, D. J., Kowalski, O. P., Hopkinson, M., Skolnick, M. S., and David, J. P. R. (1994). *Appl. Phys. Lett.* **65**, 213.
Nakamura, S., Mukai, T., and Senoh, M. (1994). *J. Appl. Phys.* **76**, 8189.
Nakamura, S., Senoh, M., Iwasa, N., and Nagahama, S. (1995). *Jpn. J. Appl. Phys.* **34**, L797.

Nelson, A. W., and Westbrook, L. D. (1984). *J. Cryst. Growth* **68**, 102.
Nishikawa, Y., Ishikawa, M., Tsuburai, Y., and Kokubun, Y. (1989). *Jpn. J. Appl. Phys.* **28**, L2092.
Nishikawa, Y., Sugawara, H., and Kokubun, Y. (1992a). *J. Cryst. Growth* **119**, 292.
Nishikawa, Y., Suzuki, M., Ishikawa, M., Kokubun, Y., and Hakakoshi, G. (1992b). *J. Cryst. Growth* **123**, 181.
Nishikawa, Y., Suzuki, M., and Okajima, M. (1993). *Jpn. J. Appl. Phys.* **32**, 498.
Ogale, S. B., and Madhukar, A. (1992). *Appl. Phys. Lett.* **60**, 2095.
OSHA Final Rule Air Contaminants Permissible Exposure Limits (1989). Title 29, Code of Federal Regulations, Part 1910. 1000 Fed. Regist., Jan. 19, 1989.
Ota, T., Otake, S., and Iwasa, I. (1993). *J. Cryst. Growth* **133**, 207.
Pajot, B., Chevallier, J., Jalil, A., and Rose, B. (1989). *Semicond. Sci. Technol.* **4**, 91.
Pearton, S. J., Corbett, J. W., and Shi, T. S. (1987a). *Appl. Phys. A* **43**, 153.
Pearton, S. J., Dautremont-Smith, W. C., Lopata, J., Tu, C. W., and Abernathy, C. R. (1987b). *Phys. Rev. B* **36**, 4260.
Philips, B. A., Norman, A. G., Seong, T. Y., Mahajan, S., Booker, G. R., Skowronski, M., Harbison, J. P., and Keramidas, V. G. (1994). *J. Cryst. Growth* **140**, 249.
Prins, A. D., Sly, J. L., Meney, A. T., Dunstan, D. J., O'Reilly, E. P., Adams, A. R., and Valster, A. (1995). *J. Phys. Chem. Solids* **56**, 349.
Rask, M., Landgren, G., Anderson, S. G., and Lundberg, A. (1988). *J. Electron. Mater.* **17**, 311.
Roberts, J. S., Mason, N. J., and Robinson, M. (1984). *J. Cryst. Growth* **68**, 422.
Roberts, J. S., Button, C. C., and Chew, A. (1994). *J. Cryst. Growth* **135**, 365.
Schmitz, D., Lengeling, G., Strauch, G., Hergeth, J., and Jurgensen, H. (1992). *J. Cryst. Growth* **124**, 278.
Schneider, J., Dischler, B., Seelewind, H., Mooney, P. M., Lagowski, J., Matsui, M., Beard, D. R., and Newman, R. C. (1989). *Appl. Phys. Lett.* **54**, 1442.
Skowronski, M., Neild, S. T., and Kremer, R. E. (1990). *Appl. Phys. Lett.* **57**, 902.
Steere, N. V., ed. (1967). "CRC Handbook of Laboratory Safety." Chemical Rubber Co., Cleveland, OH.
Stringfellow, G. B. (1978). *Annu. Rev. Mater.* **8**, 73.
Stringfellow, G. B. (1985). In "Semiconductors and Semimetals" (R. K. Willandson and A. K. Beer, eds.), Vol. 22A, p. 209. Academic Press, New York.
Stringfellow, G. B. (1986). *J. Cryst. Growth* **75**, 91.
Stringfellow, G. B. (1988). *J. Electron. Mater.* **17**, 327.
Stringfellow, G. B. (1989). "Organometallic Vapor Phase Epitaxy: Theory and Practice." Academic Press, Boston.
Stringfellow, G. B. (1993). In "Common Themes and Mechanisms of Epitaxial Growth" (P. Fuoss, J. Tsao, D. W. Kisker, A. Zangwill, and T. Kuech, eds.), Mater. Res. Soc. Pittsburg. pp. 35–46.
Stringfellow, G. B., and Chen, G. S. (1991). *J. Vac. Sci. Technol., B*[2] **9**, 2182.
Stringfellow, G. B., and Hom, G. (1979). *Appl. Phys. Lett.* **34**, 794.
Stringfellow, G. B., Su, L. C., Strausser, Y. E., and Thorton, J. T. (1995). *Appl. Phys. Lett.* **66**, 3155.
Su, L. C., Pu, S. T., Stringfellow, G. B., Christen, J., Selber, H., and Bimberg, D. (1993). *Appl. Phys. Lett.* **62**, 3496.
Su, L. C., Ho, I. H., and Stringfellow, G. B. (1994a). *J. Appl. Phys.* **75**, 5135.
Su, L. C., Ho, I. H., and Stringfellow, G. B. (1994b). *Appl. Phys. Lett.* **65**, 749.
Sugawara, H., Itaya, K., Ishikawa, M., and Hatakoshi, G. (1992a). *Jpn. J. Appl. Phys.* **31**, 2446.
Sugawara, H., Ishikawa, M., Kokubun, Y., Nishikawa, Y., Naritruka, S., Itaya, K., Hatakoshi, G., and Suzuki, M. (1992b). U.S. Pat. 5,153,889.

Suzuki, M., Itaya, K., Nishikawa, Y., Sugawara, H., and Okajima, M. (1993). *J. Cryst. Growth* **133**, 303.
Suzuki, T., Gomyo, A., Hino, I., Kobayashi, K., Kawata, S., and Iijima, S. (1988). *Jpn. J. Appl. Phys.* **27**, L1549.
Szafranek, I., and Stillman, G. E. (1990). *J. Appl. Phys.* **68**, 3554.
Tanaka, H., Kawamura, Y., Nojima, S., Wakita, K., and Asahi, H. (1987). *J. Appl. Phys.* **61**, 1713.
Tavendale, A. J., Alexiev, D., and Williams, A. A. (1985). *Appl. Phys. Lett.* **47**, 316.
Terao, H., and Sunakawa, H. (1984). *J. Cryst. Growth* **68**, 157.
Valster, A. Liedenbaum, C. T. H. F., Finke, M. N., Severens, A. L. G., Boermans, M. J. B., Vandenhoudt, D. E. W., and Bulle-Lieuwma, C. W. T. (1991). *J. Cryst. Growth* **107**, 403.
Vernon, S. M., Colter, P. C., McNulty, D. D., Hogan, S. J., Weyburne, D. W., and Ahern, B. S. (1994). *Int. Conf. InP Relat. Mater. 6th*, p. 137.
Veuhoff, E., Baumeister, H., Treichler, R., and Brandt, O. (1989). *Appl. Phys. Lett.* **55**, 1017.
Wu, C. C., Chang, C. Y., Chen, P. A., Chen, H. D., Lin, K. C., and Chan, S. H. (1994). *Appl. Phys. Lett.* **65**, 1269.
Yuan, J. S., Hsu, C. C., Cohen, R. M., and Stringfellow, G. B. (1985). *Appl. Phys. Lett.* **57**, 1380.

CHAPTER 5

AlGaInP Light-Emitting Diodes

F. A. Kish and R. M. Fletcher

OPTOELECTRONICS DIVISION
HEWLETT-PACKARD COMPANY
SAN JOSE, CALIFORNIA

I. INTRODUCTION	149
II. ACTIVE LAYER DESIGN	153
1. *Band Structure*	154
2. *Double Heterostructure Devices*	159
3. *Quantum Well Devices*	169
III. CURRENT-SPREADING IN DEVICE STRUCTURES	170
1. *Ohmic Contact Modifications*	171
2. *Growth on P-Type Substrates*	172
3. *Current-Spreading Window Layers*	172
4. *Indium Tin Oxide and Other Approaches*	176
IV. CURRENT-BLOCKING STRUCTURES	178
V. LIGHT EXTRACTION	180
1. *Upper Window Design*	181
2. *Substrate Absorption*	183
3. *Distributed Bragg Reflector Light-Emitting Diodes*	187
4. *Wafer-Bonded Transparent-Substrate Light-Emitting Diodes*	195
VI. WAFER FABRICATION TECHNIQUES	206
VII. DEVICE PERFORMANCE CHARACTERISTICS	208
1. *Quantum Efficiency*	208
2. *Luminous Efficiency*	211
3. *Current–Voltage Characteristics*	213
4. *Electroluminescence Spectra*	215
5. *Reliability*	217
VIII. CONCLUSIONS AND OUTLOOK	219
References	220

I. Introduction

Progress in the development of new materials for light-emitting diodes (LEDs) has been ongoing since the first red light emitting gallium arsenide phosphide (GaAsP) devices were introduced in the early 1960s in low volumes (Holonyak and Bevaqua, 1962) and later that decade in high volumes (Herzog et al., 1969). The materials first developed included p-n homojunction diodes in $GaAs_{1-x}P_x$ and zinc-oxygen–doped gallium phosphide (GaP) for red-spectrum devices; nitrogen-doped $GaAs_{1-x}P_x$ for red, orange, and yellow devices; and nitrogen-doped GaP for yellow-green devices. A milestone was reached in the mid-1980s with the development and introduction of aluminum gallium arsenide (AlGaAs) LEDs that made use of a direct bandgap material system and a highly efficient double heterostructure (DH) active region. These AlGaAs devices raised the luminous efficiency of LEDs to levels comparable to red-filtered incandescent lamps. Following in 1990, Hewlett-Packard Company (Kuo et al., 1990) and Toshiba Corporation (Sugawara et al., 1991a) independently developed and introduced a new family of LEDs based on the aluminum gallium indium phosphide (AlGaInP) material system. With AlGaInP, highly efficient LEDs emitting light from the red to the yellow-green part of the spectrum were made available. New record light-efficiency levels were achieved over this spectral regime, and as a consequence, new applications for LEDs are in the process of being developed. Some of these applications include automotive lighting (e.g., for tail lights and turn signal lights), outdoor variable message signs, highway information signs, outdoor large-screen video displays, and traffic signal lights.

Simple homojunction devices are produced by growing layers of GaAsP using hydride vapor-phase epitaxy (HVPE) or GaP using liquid-phase epitaxy (LPE). A p-n junction is created by introducing dopants, such as zinc (Zn) and tellurium (Te), into the growth process or by performing a sealed ampoule Zn diffusion into an n-type epitaxial layer. For GaAsP devices, epitaxial growth is performed on both GaAs and GaP substrates where grading techniques are used to address the problem of lattice mismatch between epilayer and substrate. Heterostructure devices are not practical with these materials because of the lack of suitable lattice-matched confining layer alloys. On the other hand, both AlGaAs and AlGaInP are well-suited for the growth of lattice-matched heterostructure devices (grown on GaAs substrates). Both material systems are direct bandgap semiconductors over a large range of their alloy compositions, and by changing the aluminum (Al) alloy composition, suitable lattice-matched confining layers can be grown. Examples of various types of LED epitaxial structures from homojunctions to heterojunctions in various material systems are illustrated in Fig. 1 (Fletcher et al., 1993).

Conventional vapor-phase epitaxy (VPE) is not possible for the growth of Al-bearing alloys due to the attack of the hot-wall quartz growth chamber by the Al-containing gases. In the case of AlGaAs LEDs, the usual growth technique is LPE, especially for visible (red) emitting devices. The oxygen

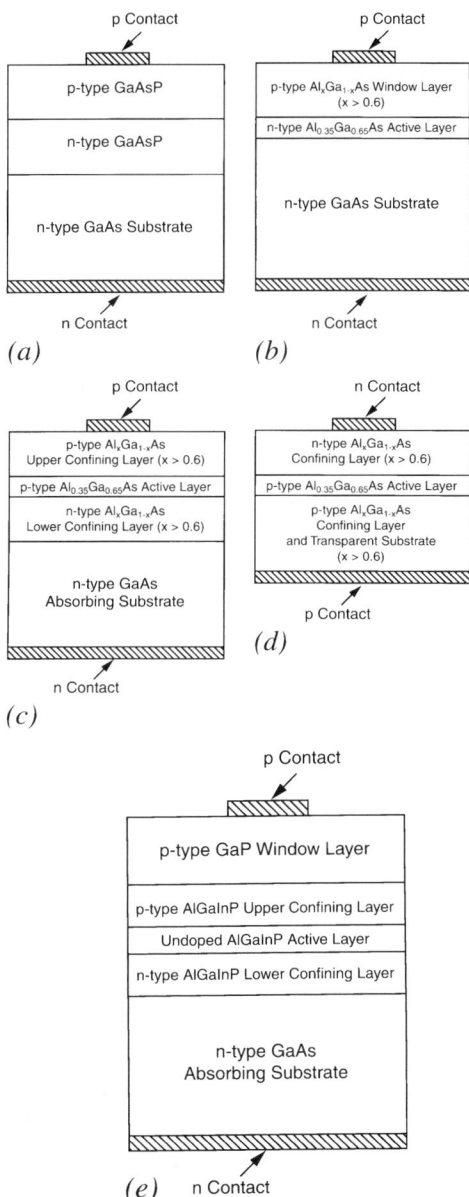

FIG. 1. Examples of various light-emitting diode (LED) devices using different material systems. (a) A typical GaAsP device. (b) A single heterostructure AlGaAs device. (c) An AlGaAs absorbing-substrate (AS) double heterostructure (DH) LED. (d) An AlGaAs transparent-substrate DH LED. (e) An AS AlGaInP DH LED. (© 1993 Hewlett-Packard Company. Reproduced with permission.)

gettering effect that occurs in the Ga melts used in the LPE process makes possible the growth of the high Al-composition layers enabling the fabrication of ~650-nm emission LEDs with internal quantum efficiencies of ~50% (Cook et al., 1988). In the case of AlGaInP LEDs, LPE is not a suitable growth technique because of Al segregation in the melt (Kazamura et al., 1983). For the $(Al_xGa_{1-x})_{0.5}In_{0.5}P$ material system (lattice-matched to a GaAs substrate), and for infrared-emitting AlGaAs devices for which lower Al compositions are used, the organometallic vapor-phase epitaxy (OMVPE) technique is used (Manasevit, 1968; Dupuis and Dapkus, 1977). Molecular beam epitaxy (MBE) has also been used for the growth of high-quality AlGaInP materials and devices (Arthur, 1968; Cho, 1971; Tsang, 1980), especially laser diodes; however, OMVPE is the dominant crystal-growth technique used for the fabrication of AlGaInP LEDs and laser diodes. Organometallic vapor-phase epitaxial growth has resulted in the highest efficiency AlGaInP LEDs and the only commercial devices realized to date. (For a complete discussion of OMVPE growth of AlGaInP, see Chapter 4.) These facts are attributed to the improved material quality, high growth rates, and economy of scale afforded by the OMVPE process. In fact, current production of AlGaInP LEDs by OMVPE marks the first time that this growth technique has been used for truly high-volume manufacture of optoelectronic devices.

Organometallic vapor-phase epitaxy is a highly controllable thin-film process. Layer compositions, doping levels, and thicknesses can all be manipulated more or less independently in order to produce a complex heterostructure device. The flexibility of the process allows for the growth of current-blocking layers that impede the current flow to certain areas of the device, and distributed Bragg reflectors (DBRs) that direct light away from an absorbing substrate and back out the top of the chip. For increased light extraction and current-spreading, complementary techniques such as VPE can be coupled with OMVPE to produce thick window layers. Compound semiconductor wafer-bonding techniques also can be used to replace the original opaque GaAs substrate with a transparent GaP substrate to completely eliminate substrate absorption. These innovations are being used to make AlGaInP LEDs as efficient as possible.

This chapter focuses on a number of topics important to the development and production of high-performance AlGaInP LED devices. Active layer design of AlGaInP LEDs is described, with emphasis on the materials properties of the AlGaInP alloys. Next, specific parts of the LED structure that enhance light-output performance are discussed, including current-spreading layers, current-blocking layers, and window layers. Light extraction technologies that enhance performance, such as DBRs and wafer bonding of transparent substrates are then described, followed by a brief section on the AlGaInP LED wafer fabrication process. A discussion of AlGaInP device performance, including efficiency, color, electrical, and reliability data follow. The final section contains concluding remarks and a discussion of the future outlook for AlGaInP LEDs.

II. Active Layer Design

The AlGaInP alloy system possesses the largest direct bandgap of any III-V semiconductor with the exception of the nitrogen-based materials, with corresponding light emission spectra extending from the red to green. As a result, the AlGaInP alloy is an attractive system for the construction of lasers and direct bandgap LEDs. Work on AlGaInP LEDs was motivated by the potential of such direct bandgap emitters to substantially exceed the efficiency of the more conventional indirect bandgap emitters in this spectral regime (e.g., GaAsP, nitrogen-doped GaAsP (GaAsP:N), GaP, and nitrogen-doped GaP (GaP:N)), (Archer, 1972; Loebner, 1973; Duke and Holonyak, 1973). The earliest light emitters fabricated in this material system consisted of bulk and heteroepitaxial InGaP emitters (Burnham *et al.*, 1970, 1971; Macksey *et al.*, 1971; Nuese *et al.*, 1971, 1972; Scifres *et al.*, 1972). The ability to substitute Al for Ga as in $Al_xGa_{1-x}As$ (Alferov *et al.*, 1967; Rupprecht *et al.*, 1967) leads to a larger bandgap quarternary alloy $(Al_xGa_{1-x})_yIn_{1-y}P$ and the additional ability to form lattice-matched heterostructures. Because both aluminum phosphide (AlP) and GaP have almost identical lattice constants (5.4510 and 5.4512 Å, respectively) (Sze, 1981), lattice-matching can be achieved by simply adjusting the indium (In) mole fraction of the quaternary alloy.

As in virtually every III-V semiconductor light-emitting compound, low crystalline defect densities are essential for the high-efficiency operation of AlGaInP LEDs. The sole exception to the aforementioned trend are the InGaN-based blue and green emitters (Lester *et al.*, 1995). Gallium arsenide is the only binary compound semiconductor substrate (5.6533 Å lattice constant) suitable for lattice-matching to the $(Al_xGa_{1-x})_yIn_{1-y}P$ system. Lattice-matching occurs for compositions of $y \sim 0.5$ over the entire compositional range (Casey and Panish, 1978). In addition, the thermal expansion coefficients of the $(Al_xGa_{1-x})_{0.5}In_{0.5}P$ alloy and GaAs are sufficiently alike to facilitate thermal cycling (e.g., from room-temperature to growth temperature, less than 800°C) without the generation of deleterious crystalline defects (Minagawa and Kakibayashi, 1985; Bour and Shealy, 1988). A number of workers have attempted to fabricate InGaP LEDs from bulk material or on lattice-mismatched substrates. Lattice-mismatched homojunction InGaP LEDs have been fabricated on GaP (Nuese et al., 1972, 1973), Si (Kondo et al., 1989), GaAs (Nuese et al., 1971; Masselink and Zachau, 1992), and GaAsP pseudo-substrates (consisting of "thick" compositionally graded heteroepitaxial GaAsP films on GaAs substrates) (Macksey et al., 1973; Lin et al., 1993a). Furthermore, lattice-mismatched DH $(Al_xGa_{1-x})_{0.65}In_{0.35}P$ LEDs have been fabricated on $GaAs_{0.7}P_{0.3}$ pseudo-substrates (Lin et al., 1993b). Despite utilizing a variety of crystal-growth technologies, compositional grading schemes, and device designs, these devices contain sufficient densities of crystalline dislocations and defects to limit their performance substantially compared with $(Al_xGa_{1-x})_{0.5}In_{0.5}P$ LEDs grown lattice-matched on GaAs. The most efficient lattice-mismatched devices reported to date consist of InGaP homojunction LEDs grown by VPE on transparent GaP substrates with peak efficiencies of $\sim 10\,lm/A$ ($\sim 0.9\%$ external quantum efficiency) at $\lambda \sim 590\,nm$ (Stinson et al., 1991). Although these devices are three to four times as efficient as commercial 590 nm GaAsP:N–GaP LEDs (Craford, 1992), they are still approximately an order of magnitude less efficient than the best lattice-matched $(Al_xGa_{1-x})_{0.5}In_{0.5}P$ devices (Kish et al., 1996a).

1. BAND STRUCTURE

a. Bandgap Energies

The $(Al_xGa_{1-x})_{0.5}In_{0.5}P$ alloy lattice-matched to GaAs results in the highest crystal quality and, thus, the most efficient LED material. Knowledge of the band structure of the $(Al_xGa_{1-x})_{0.5}In_{0.5}P$ alloy is essential for the optimal design of LED structures. Anomalous behavior (reduction by up to 190 meV) in the relation between alloy composition and bandgap energy has been observed in the $(Al_xGa_{1-x})_{0.5}In_{0.5}P$ alloy. This phenomena is attributed atomic ordering of In and Ga (or Al) on the group III sublattice, resulting in a monolayer indium phosphide (InP)–GaP (or InP–AlP) superlattice along the {111} crystal planes (Gomyo et al., 1986, 1987; Ohba et al., 1986; Bellon et al., 1988; T. Suzuki et al., 1988; Nozaki et al., 1988; Kondow et al., 1988; Schneider et al., 1992; Buchan et al., 1992; Kobayashi et al., 1995). Ordering in the AlGaInP system is discussed in detail Chapter 4. Because the primary application of AlGaInP LEDs is for short-wavelength ($<650\,nm$) operation, the disordered random alloy is preferred as a result of its higher energy gap. For equivalent emission wavelengths, the disordered random alloy contains less Al than that of the ordered alloy. Lower Al compositions are generally considered advantageous for the growth of higher quality material with less nonradiative recombination centers (Sugiura et al., 1991; Naritsuka et al., 1991; Sugawara et al., 1992a; Kondo et al., 1994). In addition, disordered AlGaInP alloys have been shown to exhibit narrower photoluminescence (PL) linewidths than ordered alloys (Fouquet et al., 1990; Schneider et al., 1992). As a result of such considerations, the highest performance AlGaInP LEDs are generally fabricated in disordered material (Kuo et al., 1990; Fletcher et al., 1991a; Huang et al., 1992; Sugawara et al., 1992a,b; Itaya et al., 1994; Kish et al., 1994a). The disordered alloy is generally obtained by growth on misoriented substrates, by the modification of the epitaxial growth conditions, or both. Because the disordered alloy is most useful for visible LED applications, the remainder of this chapter focuses on AlGaInP LED devices fabricated from the disordered random alloy.

The 300 K energy gap versus alloy composition (x) for the disordered $(Al_xGa_{1-x})_{0.5}In_{0.5}P$ alloy is shown in Fig. 2. These relationships are derived from the low-temperature (2 K) measurements of $(Al_xGa_{1-x})_{0.5}In_{0.5}P$ crystals under hydrostatic pressure (Prins et al., 1995a). The data for the Γ band are decreased by 70 meV to account for the change in bandgap from 2 to 300 K (Bour and Shealy, 1988). Accordingly, the direct (Γ) bandgap variation with composition is given by

$$E_\Gamma(x) = 1.91 + 0.61x \quad (eV) \quad (1)$$

Similarly, decreasing the indirect (X) bandgap by 70 meV results in a 300 K bandgap compositional dependence given by

$$E_X(x) = 2.19 + 0.085x \quad (eV) \quad (2)$$

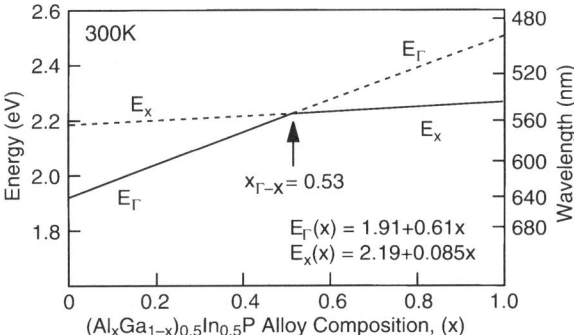

FIG. 2. Room temperature (300 K) bandgap energies and corresponding emission wavelengths of the Γ and X bands for the unordered $(Al_xGa_{1-x})_{0.5}In_{0.5}P$ alloy as a function of alloy composition x adopted from low-temperature (2 K) pressure measurements (Prins et al., 1995a). At room-temperature the direct–indirect crossover occurs for an alloy composition of $x = 0.53$, corresponding to an emission wavelength of 555 nm.

The energy gap relationships of Fig. 1 indicate that the $(Al_xGa_{1-x})_{0.5}In_{0.5}P$ alloy is direct from 555 to 650 nm. Furthermore, these relationships show a linear variation of the Γ and X energy gaps with composition. The absence of bowing is consistent with much of the experimental data reported to date. This behavior has been attributed to the similarity of the AlP and GaP lattice constants, the small effect of chemical disorder, and the linear variation of the potential form factors of the alloy (Mowbray et al., 1994; Baldereschi et al., 1977). The E_Γ dependence agrees with that measured by various workers (Asahi et al., 1982; Nozaki et al., 1988; Cao et al., 1990). The E_X dependence results in a $\Gamma - X$ crossover at 2.23 eV (555 nm) for an alloy composition of $x = 0.53$. This 300 K crossover composition is in close agreement with that determined by high-pressure bandgap data, $x = 0.58$ (Meney et al., 1995) and optical spectroscopic measurements, $x = 0.50 \pm 0.02$ (Mowbray et al., 1994). Furthermore, the 555 nm crossover wavelength agrees well with the shortest wavelength LEDs fabricated in the $(Al_xGa_{1-x})_{0.5}In_{0.5}P$ material system (Fletcher et al., 1991a; Huang et al., 1992). Analogously, the shortest wavelength stimulated emission observed in the $(Al_xGa_{1-x})_{0.5}In_{0.5}P$ alloy occurs for devices with active regions compositions of $x \sim 0.56$ at 77 K (Kuo et al., 1988). Thus, consistent with Eqs. (1) and (2), a significant body of data indicates that the crossover occurs for an alloy composition in the range $x = 0.5$ to 0.6.

The E_X dependence in Eq. (2) differs significantly from that previously determined by other workers (Onton and Chicotka, 1970; Nelson and Holonyak, 1976; Asahi et al., 1982; Bour et al., 1987) and the application of Vegard's law for AlP, GaP, and InP (Cao et al., 1990). Such data combined with the E_Γ dependence of Eq. (1) indicate a $\Gamma - X$ crossover energy of 2.3 eV (540 nm) for a composition of $x = 0.7$. The implications of the lower energy gap of the X conduction band minima in Eq. (2) and corresponding reduced crossover energy (by ~ 70 meV) are significant. The 2.23 eV indirect crossover energy drastically limits the efficiency of emitters in the deep green portion of the visible spectrum. As a result, the efficiency of LEDs degrades drastically as the alloy composition is tuned from the yellow ($\lambda \sim 590$ nm) to the yellow-green ($\lambda \sim 565$ nm), as discussed in Part VII, Sections 1 and 2 herein. Furthermore, the lower energy of the energy gap of the X minima results in increased difficulty in obtaining sufficient electron confinement in heterostructure devices (see Part II, Section 2.c).

The position of the L minima in the $Ga_{0.5}In_{0.5}P$ alloy has been theoretically determined to be 100 to 200 meV above that of the Γ minimum at 300 K (Campbell et al., 1974; Chen and Sher, 1981; Bugajski et al., 1983). Further theoretical work (Brennan and Chiang, 1992) placed the L band ~ 40 meV above the Γ band for $x = 0.52$. Little experimental evidence exists for the position of the L minima in the $(Al_xGa_{1-x})_{0.5}In_{0.5}P$ alloy. Hydrostatic high-pressure measurements indicate that the L minima lie above energy gap over the entire alloy range, with at least a 120 meV separation for $x = 0$ (Prins et al., 1995b).

b. *Carrier Effective Masses*

The carrier effective masses are important in the design of LEDs because they determine the density of states (DOS), and hence the carrier distribution in the direct and indirect bands in both the active layer and confining layers of heterostructure LEDs. These values become especially important as the direct–indirect transition is approached in determining the radiative efficiency, carrier injection efficiency, and carrier confinement in the device. The dependence of the DOS electron effective masses in the Γ band as a function of alloy composition is given by (Rennie et al., 1993)

$$m_{e\Gamma}(x) = (0.11 + 0.00915x - 0.0024x^2)m_0 \qquad (3)$$

No data for the DOS electron effective mass currently exist for $(Al_xGa_{1-x})_{0.5}In_{0.5}P$, InP, or AlP in the X band. However, the X-band effective mass is relatively independent of cation species, and thus can be approximated by the X band DOS electron effective mass in GaP (Sze, 1981):

$$m_{eX}(x) = m_{eX}(\text{GaP}) = 0.82m_0 \qquad (4)$$

The heavy hole (Honda et al., 1985) and light hole (Lawaetz, 1971) effective masses can be interpolated from the values at the ternary endpoints and are given by

$$m_{hh}(x) = (0.62 + 0.05x)m_0 \qquad (5)$$

$$m_{lh}(x) = (0.11 + 0.03x)m_0 \qquad (6)$$

From these values, the DOS effective hole mass can be calculated from the relation $m_h(x) = [m_{hh}(x)^{3/2} + m_{lh}(x)^{3/2}]^{2/3}$. Note that the effective DOS carrier masses in the $(Al_xGa_{1-x})_{0.5}In_{0.5}P$ system are significantly higher than those in the $Al_xGa_{1-x}As$ system (Adachi, 1985). These parameters affect a number of device design considerations important in LED devices, including the DOS of the direct and indirect energy bands as well as carrier injection and confinement in heterostructure devices.

c. Band Offsets

Knowledge of the heterojunction band offsets is important for proper design of optoelectronic device structures. Band offsets have been measured in the $(Al_xGa_{1-x})_{0.5}In_{0.5}P$ system by a variety of techniques, including low-temperature PL analysis (Tanaka et al., 1986, 1987b; Liedenbaum et al., 1990), absorption measurements (Tanaka et al., 1987a), capacitance-voltage profiling (Watanabe and Ohba, 1987), hydrostatic high-pressure measurements (Patel et al., 1993; Prins et al., 1995b), and internal photoemission (Yow et al., 1995). The band offsets are determined by measuring the conduction or valence band offset and subsequently determining the other band offset by knowledge of the energy gap dependence. Figure 3 shows the conduction band offsets (ΔEc) and valence band offsets (ΔEv) from a number of workers wherein the data have been adjusted for the energy gap

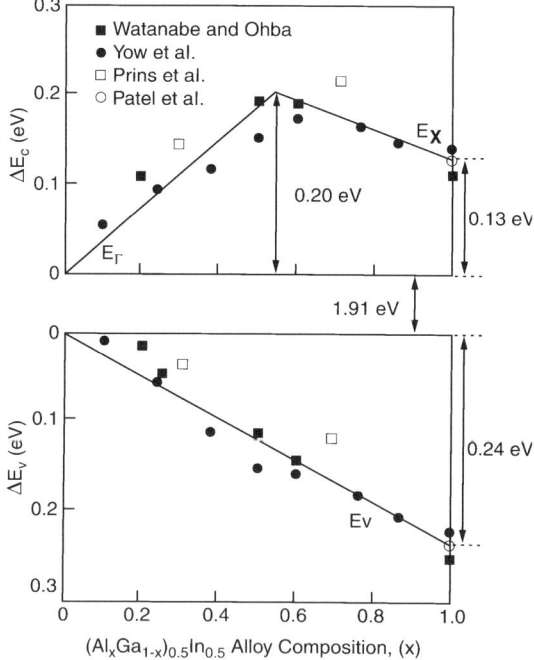

FIG. 3. Energy band offsets for the conduction and valence bands for unordered $(Al_xGa_{1-x})_{0.5}In_{0.5}P$ relative to $Ga_{0.5}In_{0.5}P$ as a function of alloy composition x. The maximum conduction band offset (ΔEc) is approximately 0.2 eV and occurs in the range of $x \sim 0.5$ to 0.7. The maximum valence band offset (ΔEv) occurs for $Al_{0.5}In_{0.5}P$ and is approximately 0.24 eV.

dependence given by Eqs. (1) and (2). A least squares fit to these data points results in expressions for the conduction band offsets given by

$$\Delta Ec(x) = 0.369x \quad (eV) \quad x \leqslant 0.53 \quad (7)$$

$$\Delta Ec(x) = 0.285 - 0.157x \quad (eV) \quad x > 0.53 \quad (8)$$

and valence band offsets given by

$$\Delta Ev(x) = 0.241x \quad (eV) \quad (9)$$

Although there is a fair amount of scatter in the data, Fig. 3 indicates that the maximum conduction band offset is $\sim 0.2\,eV$ for alloy compositions in the range $x = 0.5$ to 0.7. The conduction band offset is reduced for larger Al mole fractions due to the crossover of the Γ and X bands. The maximum valence band offset is $\sim 0.24\,eV$, which occurs for an alloy composition of $x = 1.0$. Furthermore, the conduction band offset of $Ga_{0.5}In_{0.5}P$ relative to GaAs has been measured to be 0.20 to 0.25 eV (Watanabe and Ohba, 1987; Feng et al., 1993).

2. Double Heterostructure Devices

The advantages of utilizing heterostructures for improved performance in light-emitting devices has been known for several decades (Kroemer, 1963; Alferov et al., 1968; Alferov et al., 1969a,b). Accordingly, heterojunction LEDs offer inherent advantages, including the following: enhanced injection efficiency of electrons or holes, confinement of injected carriers in the active layer, and formation of transparent window and substrate layers for improved current-spreading and light extraction. Devices based on the AlGaAs system were the first practical devices to employ heterostructures for improved carrier confinement and enhanced injection efficiency (Nishizawa et al., 1977, 1983). The earliest devices consisted of single heterostructure (SH) active regions. However, the potential benefits afforded by heterojunctions have been shown to be maximized by employing a DH active layer (Nishizawa et al., 1983; Ishiguro et al., 1983). Accordingly, the most efficient red (650 nm) AlGaAs LEDs produced to date employ DH active regions and exhibit external quantum efficiencies of 18% at 300 K (Cook et al., 1988).

Leveraging the extensive knowledge of AlGaAs DH LEDs, $(Al_xGa_{1-x})_{0.5}In_{0.5}P$ LEDs (grown lattice-matched to GaAs) were first developed with DH active regions (Kuo et al., 1990; Sugawara et al., 1991a).

The band diagram of the active region of a typical $(Al_xGa_{1-x})_{0.5}In_{0.5}P$ DH LED is shown schematically in Fig. 4. Such devices consist of $Al_{0.5}In_{0.5}P$ or $(Al_{0.7}Ga_{0.3})_{0.5}In_{0.5}P$ wide bandgap N- and P-type injecting–confining regions (~ 0.5- to 1.0-μm thick) and an $(Al_xGa_{1-x})_{0.5}In_{0.5}P$ lower bandgap active region (~ 0.3 to 1.0 μm) whose composition (x) is tuned for emission in the 650 nm red ($x = 0$) to 555 nm green ($x \sim 0.53$) spectral regime. The largest bandgap injection–confining layers are generally preferred to maximize injection efficiency and carrier confinement within the device. However, difficulty in obtaining high p-type doping levels and reasonable conductivity in $Al_{0.5}In_{0.5}P$ has resulted in the employment of $(Al_{0.7}Ga_{0.3})_{0.5}In_{0.5}P$ confining layers in many LED structures (Ohba et al., 1986; Suzuki et al., 1994). The doping of the wide bandgap confining layers is chosen to maximize carrier injection efficiency and confinement as well as current-spreading in the LED. The doping must further be sufficiently high to minimize the resistance of these layers. However, care must be taken to minimize the effects of nonradiative centers introduced by the dopants near the active layer. Generally, the optimal doping for the $(Al_xGa_{1-x})_{0.5}In_{0.5}P$ LED confining layers ($x = 0.7$ or 1.0) has been reported to be $\sim 1 \times 10^{16}$ to 5×10^{18} cm^{-3} for the N-type layers and $\sim 4 \times 10^{17}$ to 2×10^{18} cm^{-3} for the P-type layers (Kuo et al., 1991; Sugawara et al., 1991b, 1992c). The donor species Te and silicon (Si) have been employed successfully as n-type dopants in $(Al_xGa_{1-x})_{0.5}In_{0.5}P$ LEDs whereas the p-type dopant typically is Mg or Zn (Kuo et al., 1990; Fletcher et al., 1991a; Sugawara et al., 1991a, 1992a). Most workers describe the $(Al_xGa_{1-x})_{0.5}In_{0.5}P$ active layer as unintentionally doped. However, the optimal doping level for the active region has been reported to be lightly n- or p-doped $\lesssim 1 \times 10^{17}$ cm^{-3} (Sugawara et al., 1992c). Thus, the doping of the DH active region may be

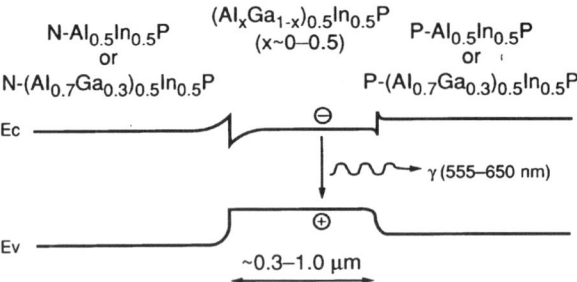

FIG. 4. Schematic diagram of the energy band structure of a typical double-heterostructure $(Al_xGa_{1-x})_{0.5}In_{0.5}P$ light-emitting diode.

N-p-P or N-n-P, or N-n-p-P, wherein the p-n junction is located within the active region in the last case.

a. Minority Carrier Injection

Light-emitting diodes in the $(Al_xGa_{1-x})_{0.5}In_{0.5}P$ system are generally operated at relatively low current densities (< 150 A/cm^2) and hence low injected carrier densities ($< 1 \times 10^{17}$ cm^{-3}). Anisotype heterojunctions (e.g., N-p or n-P) offer enhanced injection of minority carriers into the lower bandgap material, and consequently increased LED efficiency for heterostructure devices. For an ideal N-p heterojunction wherein the N-layer is the wide bandgap injection layer and the p-layer is the lower bandgap active layer, the electron/hole injection ratio is given by (Kressel and Butler, 1977)

$$\frac{J_e}{J_h} = \left(\frac{D_e^p N_e^N}{L_e^p} \frac{L_h^N}{D_h^N N_h^p}\right) \left(\frac{m_e^p m_h^p}{m_e^N m_h^N}\right)^{3/2} e^{\Delta E_g/kT} \quad (10)$$

where $J_{e,h}$ are the injected electron and hole current densities; $D_{e,h}^{p,N}$, $L_{e,h}^{p,N}$, $N_{e,h}^{N,p}$, are the minority electron (hole) diffusion coefficient, minority electron (hole) diffusion length, and electron (hole) doping density, respectively; and $m_e^{N,p}$, $m_h^{N,p}$ are the DOS electron (hole) effective masses in the wide bandgap N-confining layer and lower bandgap p-active layer, respectively. The energy gap difference (ΔEg) is given by the difference in the confining layer and active layer bandgap energies

$$\Delta Eg = E_g(\text{Confining layer}) - E_g(\text{Active layer}) \quad (11)$$

For DH devices, the thickness of the active layer is generally less than that the minority carrier diffusion length in the active layer. In this situation, the active layer thickness is substituted for the active layer minority carrier diffusion length in Eq. (6). Analogously, a similar expression can be derived for the hole/electron injection ratio of a P-n heterojunction consisting of a wide bandgap P-injecting layer and a lower bandgap n-active layer. In this case, the hole/electron injection ratio is given by

$$\frac{J_h}{J_e} = \left(\frac{D_h^n N_h^P}{L_h^n} \frac{L_e^P}{D_e^P N_e^n}\right) \left(\frac{m_e^n m_h^n}{m_e^P m_h^P}\right)^{3/2} e^{\Delta E_g/kT} \quad (12)$$

The hole/electron injection ratio is also dominated by the exponential term containing the energy gap difference, as given precisely in Eq. (11). Equations (10) and (12) assume an ideal heterojunction and neglect the effects of interfacial recombination and any conduction and valence band offset spikes (ΔEc, ΔEv) at the heterointerface. These equations indicate that the injection ratio is predominated the energy gap difference; consequently, this difference should be maximized for optimal injection efficiency.

Red (630 nm) and red-orange (615 nm) $(Al_xGa_{1-x})_{0.5}In_{0.5}P$ LEDs possess a sufficiently large energy gap difference ($\Delta Eg > 0.2\,eV$) between the active layer and confining–injecting layers ($x = 1.0$ or $x = 0.7$) to ensure essentially single-sided injection for electrons (holes) at an ideal N-p (P-n) heterojunction. Table I shows the electron and hole injection efficiencies for N-p and P-n $(Al_xGa_{1-x})_{0.5}In_{0.5}P$ heterojunctions LEDs emitting at 590 and 570 nm wherein the energy gap difference between the active layer and $x = 1.0$ or $x = 0.7$ injecting layers is less than 0.2 eV. Accordingly, the electron and hole injection efficiencies are defined as $J_e/(J_e+J_h)$ and $J_h/(J_e+J_h)$, respectively. The carrier injection efficiencies are calculated for both P-n and N-p heterojunctions typically employed in $(Al_xGa_{1-x})_{0.5}In_{0.5}P$ LED applications (Kuo et al., 1991; Sugawara et al., 1991b, 1992c). The calculations also assume a DH device with a 1 μm active layer. Because no data are available for the minority carrier mobilities, majority carrier mobilities (Ohba et al., 1986; Suzuki et al., 1994) are used in their place. Similarly, minority carrier diffusion lengths of 4 and 1 μm for electrons and holes are assumed, respectively. As a result of these uncertainties, the values calculated in Table I should be considered trends; hence, only one significant digit is shown. The calculated values indicate that in short-wavelength 570 nm yellow-green LEDs, single-sided injection (injection efficiency = 1) can be achieved for electrons in a N-p heterojunction for

TABLE I

CALCULATED CARRIER INJECTION EFFICIENCIES FOR $(Al_xGa_{1-x})_{0.5}In_{0.5}P$ HETEROJUNCTIONS

$(Al_xGa_{1-x})_{0.5}In_{0.5}P$ Active Layer	$Al_{0.5}In_{0.5}P$ Injection Layer		$(Al_{0.7}Ga_{0.3})_{0.5}In_{0.5}P$ Injection Layer	
Emission Wavelength (nm)	N-p Electron Injection Efficiency[a,b]	P-n Hole Injection Efficiency[a,b]	N-p Electron Injection Efficiency[a,b]	P-n Hole Injection Efficiency[a,b]
590	1	1	1	1
570	1	0.8	1	0.6

[a]Calculations employ standard doping levels for the active layer and injection layer (Kuo et al., 1991; Sugawara et al., 1991b, 1992c).
[b]Assumes $L_e = 4\,\mu m$, $L_h = 1\,\mu m$, and 1-μm active layer thickness. Majority carrier mobilities (Ohba et al., 1986; Suzuki et al., 1994) used in place of minority carrier mobilities in above calculations.

both $x = 1.0$ and $x = 0.7$ wide bandgap injecting layers. Hole mobilities are 5 to 50 times lower than those of electrons in the $(Al_xGa_{1-x})_{0.5}In_{0.5}P$ system (Ohba et al., 1986; Suzuki et al., 1994). Consequently, lower hole injection efficiencies are observed at a P-n heterojunction. In the shorter wavelength regimes, the hole injection efficiency at P-n heterojunctions can be improved by utilizing the wider bandgap ($x = 1.0$) injection layers, which provide an additional ~ 25 meV energy gap difference compared with $x = 0.7$ injection layers. For these devices, optimization of the material quality and junction parameters (e.g., mobility, minority carrier diffusion lengths, and doping) is essential to obtain the highest hole injection efficiency. Note that the aforementioned calculations and discussion are for ideal heterojunctions. Other factors, such as interfacial recombination, band offset discontinuities, and carrier recombination, or generation in the space charge region, may affect or predominate performance depending on the device design and material quality.

b. Carrier Recombination

After minority carrier injection occurs at the p-n junction or heterojunction, the injected carriers may recombine radiatively or nonradiatively within the LED active layer or escape ("leak") to the adjoining confining layers in a DH device. A DH device provides enhanced efficiency by virtue of reduction of the recombination volume through confinement of the carriers to the active layer, yielding an increased injected carrier density. Higher injected carrier densities generally result in increased radiative efficiency due to the saturation of nonradiative recombination centers. The effects of the DH are realized provided the active layer thickness is less than the injected minority carrier diffusion length, placing an upper constraint on the active layer thickness. This requirement varies as a function of active layer doping type and level, injected carrier density, and active layer compostion. The minority carrier diffusion length is proportional to the square root of the minority carrier mobility. Consequently, thicker active regions can be realized for p-type $(Al_xGa_{1-x})_{0.5}In_{0.5}P$ active layers (wherein electrons are injected into the active layer) because the electron mobility substantially exceeds that of holes in this alloy. For $(Al_xGa_{1-x})_{0.5}In_{0.5}P$ LEDs, active layers with thickness less than 1 to 2 μm generally satisfy the requirement that the active layer is less than the minority carrier diffusion length. Substantially thicker active regions result in a reduction in LED efficiency caused by lower injected carrier densities (increasing the effect of nonradiative recombination) absorption of light within the active layer, or both.

The internal quantum efficiency of the LED is determined by the ratio of the active layer radiative recombination rate to the total recombination rate within the LED. Nonradiative recombination may occur within the bulk active layer and confining layers as well as at heterojunction interfaces. Nonradiative centers have been identified in the $(Al_xGa_{1-x})_{0.5}In_{0.5}P$ alloy and have been shown to degrade radiative efficiency in this material system (Sugiura et al., 1991; Naritsuka et al., 1991). Specifically, in $(Al_xGa_{1-x})_{0.5}In_{0.5}P$ deep levels have been shown to be related to the presence of the p-type dopant Zn (Nozaki and Ohba, 1989), the n-type dopants Si (Nojima et al., 1986; Suzuki et al., 1991) and Se (Watanabe and Ohba, 1986), as well as the contaminant oxygen (Suzuki et al., 1993; Kondo et al., 1994). Time-resolved photoluminescence (TRPL) measurements of Zn-doped $(Al_xGa_{1-x})_{0.5}In_{0.5}P$ DHs have shown that bulk nonradiative recombination rates increase with increasing Zn concentration in the active layer and upper confining layers (Domen et al., 1993). In addition, the deep levels associated with the n-type dopants have been shown to exhibit characteristics similar to the D-X center (Suzuki et al., 1993; Kondo et al., 1994; Watanabe and Ohba, 1986), which is known to degrade the radiative characteristics of LEDs (Craford et al., 1991). These data indicate the importance of selecting the dopant species as well as controlling and optimizing the dopant concentration and position in $(Al_xGa_{1-x})_{0.5}In_{0.5}P$ LED structures.

Minimization of oxygen contamination has been identified as a critical parameter in producing high-efficiency $(Al_xGa_{1-x})_{0.5}In_{0.5}P$ LEDs. Oxygen contamination of $(Al_xGa_{1-x})_{0.5}In_{0.5}P$ has been correlated to the presence of two deep electron traps with thermal activation energies (E_T) of ~ 0.46 to $0.64\,eV$ (denoted trap D2) and ~ 0.9 to $1.3\,eV$ (denoted trap D3) (Suzuki et al., 1993; Kondo et al., 1994). The correlation between oxygen concentration, trap D2 and D3 density, and $(Al_xGa_{1-x})_{0.5}In_{0.5}P$ LED efficiency for devices with $x \sim 0.4$ active layers is shown in Fig. 5 (Suzuki et al., 1993). Accordingly, an order of magnitude increase in oxygen concentration results in approximately two orders of magnitude increase in the density of the deep-level traps. These increases are associated with a dramatic decrease in LED external quantum efficiency. As a result of these trends, control of oxygen in $(Al_xGa_{1-x})_{0.5}In_{0.5}P$ DH LEDs is essential. Oxygen incorporation in $(Al_xGa_{1-x})_{0.5}In_{0.5}P$ alloys occurs even in the presence of very low levels of oxygen-containing contaminants during epitaxial growth. Consequently, a variety of techniques have been employed to minimize the concentration of background oxygen in the $(Al_xGa_{1-x})_{0.5}In_{0.5}P$ alloy, including the growth on misoriented substrates (Sugawara et al., 1992a, b; Suzuki et al., 1993; Itaya et al., 1994; Kondo et al., 1994), optimization of the epitaxial growth parameters (e.g., V-III ratio and growth temperature), and improvement of OMVPE source purity (Nishikawa et al., 1993; Kondo et al., 1994).

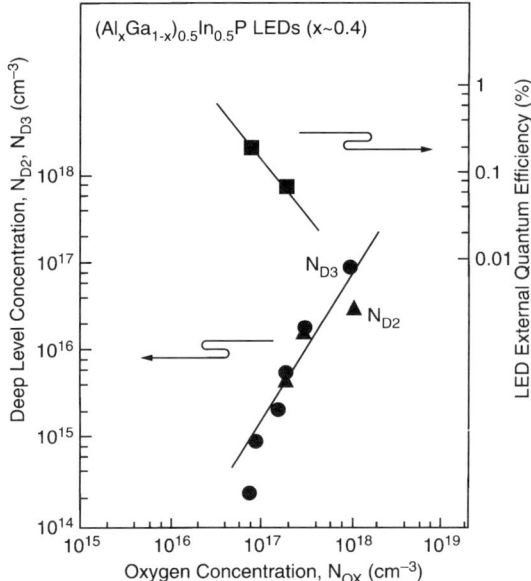

FIG. 5. Relationship between oxygen concentration (N_{ox}) and associated D2 and D3 deep-level concentrations (N_{D2}, N_{D3}) in $(Al_xGa_{1-x})_{0.5}In_{0.5}P$ light-emitting diodes (LEDs), $x \sim 0.4$ (left). The presence of increased oxygen and its associated deep levels is correlated with dramatic decreases in $(Al_xGa_{1-x})_{0.5}In_{0.5}P$ light-emitting diode (LED) external quantum efficiency (right) (Reprinted from Suzuki et al., 1993, J. Cryst. Growth 133, 303–308; with the kind permission of Elsevier Science – NL.)

Contributions to the total nonradiative lifetime also occur from interfacial recombination at the DH active layer–confining layer interfaces and within the confining layers. For relatively thin (less than 0.4 μm) undoped $(Al_xGa_{1-x})_{0.5}In_{0.5}P$ DH active regions, TRPL measurements indicate that interfacial recombination has a significant effect on the total nonradiative recombination rate (Domen et al., 1991, 1994). In addition, TRPL studies indicate Zn-doped heterointerfaces exhibit an increase in interfacial recombination velocity as the Zn doping level increases (Domen et al., 1993). Deep-levels within the confining layers of the DH also contribute to nonradiative recombination. An increased density of oxygen-related deep levels (trap D3) in the confining layers has been shown to increase the nonradiative recombination rate in DH samples (Domen et al., 1991). This behavior is speculated to result from the tunneling of carriers from the active layer into the adjoining deep levels of the confining layers (Tsang, 1978). Consequently, these data indicate that the minimization of nonradiative sites at the DH heterointerfaces and within the confining layers is important for the realization of high-efficiency $(Al_xGa_{1-x})_{0.5}In_{0.5}P$ LEDs.

c. *Carrier Confinement*

As discussed in Part II, Section 2.b, DH LEDs generally exhibit enhanced radiative efficiency as a result of the confinement of carriers to the active layer of the device. The benefits of this confinement are realized provided that the active layer thickness is approximately less than the injected minority carrier diffusion length. In order for the carriers to be confined to the active layer, the injected carriers must lack sufficient thermal energy to leak over the potential barrier at the active-confining layer interface. For a fixed injected current density, both the active layer thickness (hence the injected carrier density) and the electron and hole heterobarrier heights directly affect the confinement of carriers. In $(Al_xGa_{1-x})_{0.5}In_{0.5}P$ laser diodes, lack of sufficient electron confinement in DH and separate confinement heterostructure (SCH) quantum well (QW) devices leads to increased threshold currents and reduced characteristic and operating temperatures (Hagen *et al.*, 1990; Ishikawa *et al.*, 1991; Hatakoshi *et al.*, 1991; Bour *et al.*, 1993). However, laser diodes differ from LEDs in that laser diodes typically possess thinner active layers and are operated at much higher injected current densities. Consequently, carrier leakage is typically less problematic in LED heterostructure devices than in laser diodes; however, this problem cannot be totally neglected in LEDs.

Carrier leakage in a DH device can consist of either drift or diffusion of carriers that possess sufficient thermal energy to be transported from the active layer to the confining layers. The drift leakage component increases with injected current density (Anthony and Schumaker, 1980; Dutta, 1981). Thus, drift leakage is a significant component of the total leakage in devices that operate at relatively high current densities (e.g., laser diodes). Light-emitting diodes typically operate at one to two orders of magnitude lower injected current densities; consequently, diffusive leakage is primarily of concern for LED structures. Analysis of $(Al_xGa_{1-x})_{0.5}In_{0.5}P$ SCH QW lasers indicates that drift leakage is comparable to diffusive leakage for an injected current density of 400 A/cm^2 for an $\sim$0.3-μm total active layer thickness (QW and SCH) (Bour *et al.*, 1993). This injected current density is much larger than that of $(Al_xGa_{1-x})_{0.5}In_{0.5}P$ DH LEDs, which typically operate between 5 and 150 A/cm^2 and generally possess thicker active layers (leading to lower injected carrier densities). In a DH device, the electron diffusive leakage at a p-P active–confining layer heterointerface is given by (Kressel and Butler, 1977)

$$J_L = \frac{q\mu_e^P kT}{L_e^P} 2\left(\frac{2\pi m_e^P kT}{h^2}\right)^{3/2} \exp((E_{fn}^{act} - \Delta\varepsilon)/kT) \qquad (13)$$

where μ_e^P, L_e^P, m_e^P are the electron minority carrier mobility, diffusion length, and DOS effective mass in the upper confining layer, respectively; E_{fn}^{act} is the electron quasi-Fermi level in the active layer (measured from the conduction band edge); and $\Delta\varepsilon$ is the electron confining potential. The electron quasi-Fermi level in the active layer is a function of the injected minority carrier density (current density) in the active layer. The electron confining potential is the effective barrier for electrons at the active–confining layer interface and is given by

$$\Delta\varepsilon = (E_{gap}^{conf} - E_{gap}^{act}) + (E_{fp}^{act} - E_{fP}^{conf}) \qquad (14)$$

wherein the first term represents the difference in bandgap energy between the confining layer and active layer and the second term is the difference in hole quasi-Fermi levels in the active and confining layers (measured from the edge of the valence band). Note that Eqs. (13) and (14) assume an ideal heterojunction with no interface states and that any conduction band offset spike present is negligible or sufficiently narrow to allow electrons to tunnel easily through it. These equations indicate that the electron diffusive leakage is minimized by maximizing the difference in energy gaps between the confining layer and active layer and by heavy doping of the P-confining layer. The latter ensures the maximum difference between the Fermi levels in the active and confining layers. Consequently, heavily p-doped $Al_{0.5}In_{0.5}P$ confining layers are optimal for minimizing electron leakage. Higher maximum operating temperatures and characteristic temperatures (indicative of lower electron leakage) in $(Al_xGa_{1-x})_{0.5}In_{0.5}P$ DH laser diodes have been shown to result from increased doping of the p-type confining layer (Hatakoshi et al., 1991). However, high p-type doping is difficult to achieve in high Al-content ($x > 0.7$) $(Al_xGa_{1-x})_{0.5}In_{0.5}P$ layers unless optimized conditions for p-type doping are employed (Chapter 4). This difficulty often results in the choice of lower composition ($x = 0.7$) confining layers in $(Al_xGa_{1-x})_{0.5}In_{0.5}P$ DH LEDs.

Similar equations can also be derived for hole diffusive leakage at an n-N heterojunction. However, hole leakage generally is small compared with electron leakage due to the fact that the hole mobility is typically an order of magnitude less than that of electrons. Furthermore, the DOS effective mass of holes is significantly larger than that of electrons in the direct bandgap active layer. As a result, the density of holes with sufficient thermal energy to surmount the same confining layer potential barrier is significantly less than that of electrons. For example in $(Al_xGa_{1-x})_{0.5}In_{0.5}P$ DH laser diodes, the hole leakage current has been calculated to be 1/100th of the electron leakage current (Ishikawa et al., 1991). Consequently, electron diffusive leakage is the primary carrier confinement issue for $(Al_xGa_{1-x})_{0.5}In_{0.5}P$ DH LEDs.

The exponential term containing the electron confinement potential in Eq. (13) dominates the electron diffusive leakage current. For 630 and 615 nm $(Al_xGa_{1-x})_{0.5}In_{0.5}P$ DH LEDs with 1 μm active layers, the electron confining potential is sufficiently large to result in negligible electron leakage for $(Al_xGa_{1-x})_{0.5}In_{0.5}P$ confining layers ($x = 0.7$ or 1.0) with doping levels greater than $10^{17}\,cm^{-3}$ and typical injected carrier densities (less than $10^{17}\,cm^{-3}$). However, optimization of electron confinement becomes potentially more critical in shorter wavelength devices with smaller electron confining potentials. Table II shows calculated electron diffusive leakage parameters for 590 and 570 nm $(Al_xGa_{1-x})_{0.5}In_{0.5}P$ LEDs with 1 μm active layers, $x = 0.7$ and $x = 1.0$ confining layers, and typical doping levels employed in LED applications (Kuo et al., 1991; Sugawara et al., 1991b, 1992c) In Table II, the electron confinement efficiency is defined as $1 - (J_L/J)$, where J is the total injected current density. These calculations indicate that electron diffusive leakage does not degrade LED performance for devices operating at $\lambda \sim 590$ nm in the yellow spectral regime. However, the efficiency of shorter wavelength ($\lambda \sim 570$ nm) yellow-green devices may be significantly reduced by electron leakage. In this situation, the utilization of higher bandgap $x = 1.0$ confining layers results in a 1.5 times improvement in electron confinement efficiency compared with devices with $x = 0.7$ confining layers provided sufficiently high p-type doping can be achieved. These calculations are in close agreement with experimental data reported for $\lambda \sim 566$ nm devices that exhibit a 1.5 times improvement in external quantum efficiency as the confining layer composition is increased from $x = 0.7$ to 1.0 (Suzuki et al., 1994). Furthermore, the aforementioned

TABLE II

CALCULATED ELECTRON CONFINMENT PARAMETERS IN p-P $(Al_xGa_{1-x})_{0.5}In_{0.5}P-(Al_yGa_{1-y})_{0.5}In_{0.5}P$ HETEROJUNCTIONS ($y = 0.7, 1.0$)

$(Al_xGa_{1-x})_{0.5}In_{0.5}P$ Active Layer	P-$Al_{0.5}In_{0.5}P$ Confining Layer		P-$(Al_{0.7}Ga_{0.3})_{0.5}In_{0.5}P$ Confining Layer	
Emission Wavelength (nm)	Electron Confinement Potential[a] $\Delta\varepsilon$ (eV)	Electron Confinement Efficiency[b] $1 - (J_L/J)$	Electron Confinement Potential[a] $\Delta\varepsilon$ (eV)	Electron Confinement Efficiency[b] $1 - (J_L/J)$
590	0.21	0.99	0.18	0.97
570	0.13	0.83	0.11	0.55

[a] Calculations assume standard active layer and confining layer doping levels (Kuo et al., 1991; Sugawara et al., 1991b, 1992c) and injected electron density of $2.9 \times 10^{16}\,cm^{-3}$ (40 A/cm² injected current density).
[b] Calculations assume the parameters in footnote a, a minority carrier diffusion length L_e^p of 4 μm, and a minority carrier mobility μ_e^p of 200 cm²/V·sec in the confining layer.

calculations are for 1 μm active regions. Devices with thinner active regions possess higher injected carrier densities for the same injected current, resulting in increased electron leakage as the thickness of the active layer is reduced. Consequently, these data indicate that electron confinement can be a performance limiting factor in some $(Al_xGa_{1-x})_{0.5}In_{0.5}P$ LEDs.

3. QUANTUM WELL DEVICES

Quantum wells (QWs) have been employed extensively in $(Al_xGa_{1-x})_{0.5}In_{0.5}P$ laser diodes to effect numerous improvements in device performance, including the following: reduced threshold currents, shortened wavelength cw operation, increased characteristic and operating temperatures, and increased power conversion efficiencies (Bour, 1993). Most of the work to date on $(Al_xGa_{1-x})_{0.5}In_{0.5}P$ LEDs has concentrated on DH devices. However, "thin" QW LED devices offer the potential of higher injected carrier densities, and hence more efficient radiative recombination. A thin active layer also serves to minimize self-absorption in LEDs with relatively low internal quantum efficiencies, which is important in device structures wherein photons make multiple passes through the active layer before escaping from the LED (Part V, Sections 2 to 4). Furthermore, quantum size effects (QSE) can be employed to shorten the emission wavelength without a corresponding increase in Al compostion.

Quantum well structures utilizing QSE to shorten the emission wavelength are potentially advantageous in $(Al_xGa_{1-x})_{0.5}In_{0.5}P$ LEDs because the radiative efficiency decreases anomalously faster than that predicted by band structure calculations with increasing Al content (Part VII, Section 1). Accordingly, multiple QW structures have been employed to fabricate 600 nm $(Al_xGa_{1-x})_{0.5}In_{0.5}P$ LEDs, wherein QSEs are employed to shorten the emission wavelength by 12 nm from that of bulk DH active layer LEDs (Sugawara et al., 1994a). The active region of these QW devices consists of 50 Å $(Al_{0.2}Ga_{0.8})_{0.5}In_{0.5}P$ wells separated by 40 Å $(Al_{0.5}Ga_{0.5})_{0.5}In_{0.5}P$ barriers. Note that the quantized energy levels in the wells are inversely proportional to the carrier effective masses, resulting in the requirement of thinner wells for AlGaInP devices compared to similar AlGaAs structures. The efficiency of these $(Al_xGa_{1-x})_{0.5}In_{0.5}P$ QW LEDs increases as the number of wells is increased from 3 to 40 in the device. This behavior is believed to result from increased carrier capture and confinement as the number of wells increases. Furthermore, such an increase in light output with an increasing number of wells (and interfaces) indicates that recombination at the quantum well interfaces does not limit the performance of these devices. Multiple QW LEDs with 20 wells exhibit enhanced external quantum efficiencies compared with DH LEDs operating at the *same* emission wavelength. However, the maximum external quantum efficiency reported in this multiple QW LED structure is 1.6% at 600 nm (fabricated on an absorbing GaAs substrate). This performance is significantly lower than the best-reported DH absorbing substrate LEDs at 600 nm which exhibit external quantum efficiencies of $\sim 6\%$ (Huang et al., 1992). Furthermore, the growth of thin (50 Å) QW layers presents serious control and uniformity issues for the very high volume OMVPE growth required to manufacture these LEDs. As a result of these considerations, QW active layers have not been employed in commercial $(Al_xGa_{1-x})_{0.5}In_{0.5}P$ LEDs to date.

III. Current-Spreading in Device Structures

Current-spreading is of major importance for obtaining a high-efficiency surface-emitting LED chip used for display applications. In the case of the most common LED chip structure, such as the one shown in Fig. 1(e), the top surface of the device is partially covered by a circular top contact metallization to which a bond wire is attached. Current supplied to the chip flows from the top contact through the p-type layers and down through the junction where light generation occurs. However, if the resistivity of these p-type layers is too high, the current will not spread out appreciably from the contact and will remain confined beneath the metal. Light generated only under the contact is trapped and absorbed within the LED chip. (In contrast, in a stripe laser diode structure this current-confining behavior is desirable, and the current is intentionally limited to the area beneath the ohmic contact.)

During the early development of AlGaInP LEDs, the problem of current-spreading was quickly recognized as a fundamental limitation of the AlGaInP material system. Typical structures investigated were grown on n-type substrates making a p-side-up heterostructure necessary. As a result, three characteristics of the AlGaInP material system contribute to the current-spreading problem. First, the carrier mobility in p-type $(Al_xGa_{1-x})_{0.5}In_{0.5}P$ is low. The mobility is on the order of 10 cm²/V·sec for the high Al-content alloys used in the upper confining layers (Ohba et al., 1986). Second, doping levels in $(Al_xGa_{1-x})_{0.5}In_{0.5}P$ saturate in p-type material around 10^{18} cm^{-3}. Attempts to increase the doping level result in serious degradation of the epitaxial crystal quality. Third, increasing the thickness of the AlGaInP upper confining layer beyond 2 to 5 μm in order to increase its sheet conductance results in significant degradation of the crystal quality of the layer.

1. OHMIC CONTACT MODIFICATIONS

Several approaches exist to solve the current-spreading problem and circumvent these material limitations. A simple technique is to modify the ohmic contact pattern to extend outward from the center of the chip. Examples of these patterns are illustrated in Fig. 6. The first contact configuration shown in Fig. 6(a) is that just described above in which a circular contact is placed in the center of the top surface of the chip. A modification of the circular contact is to add finger projections, as shown in Fig. 6(b); a further modification, shown in Fig. 6(c), adds an additional ring of contact metallization around the periphery of the die. This last configuration is often used in large LED chips, such as high-power infrared emitters. Other configurations are also possible.

Ohmic contact modifications such as those just described are effective for enhancing current-spreading if the sheet resistivity is not very high. However, for AlGaInP epitaxial layers, the sheet resistance is sufficiently high that such techniques only marginally improve performance. Furthermore, any benefit achieved by the enhanced current-spreading is partially offset by the blocking effect of the additional metal covering the surface of the chip. Some of the first AlGaInP devices made by Hewlett-Packard Company (San Jose, CA) consisted of a DH where the upper confining layer was p-type $Al_{0.5}In_{0.5}P$, and there was no separate current-spreading window or ohmic contact layer. Consequently, current-spreading was negligible. The maximum luminous performance obtained from these devices using a simple circular ohmic contact was 0.4 lm/W, with the light emission occurring at the very edges of the contact itself (most of the light being generated under and blocked by the contact metal). By using a more complex metallization pattern such as that shown in Fig. 6(c), the light output increased to 1.0 lm/W; however, no further significant improvements were achieved by these techniques.

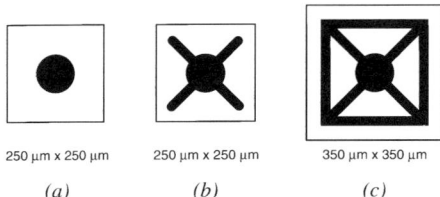

FIG. 6. Various top ohmic contact geometries used for light-emitting diodes. The circular contact (a) is simplest and most common. Finger projections (b) and more complex patterns (c) are used to improve current-spreading, especially for larger die.

2. Growth on P-Type Substrates

Another approach toward improving material conductivity is to invert the doping in the structure such that the top layer is n-type instead of p-type. The mobility in n-type alloys of AlGaInP with high Al content are around 400 cm^2/V·sec (Ohba et al., 1986), approximately 5 to 10 times that of the p-type alloys. This would indeed improve current-spreading; however, the problem of growing thick layers of high crystal quality still exists. Furthermore, the substrate must now be p-type. Zinc-doped p-type substrates have higher defect densities than do n-type substrates. These higher defect densities may lead to the formation of dislocations and nonradiative defects in the epitaxial device layers. Also, Zn is a relatively mobile impurity, diffusing easily into the epitaxial layers during growth and compensating the n-type layers. Despite these difficulties, AlGaInP LEDs have been fabricated on p-GaAs substrates with n-type wide bandgap AlGaInP current-spreading window layers (Lin et al., 1995a,b). However, the current performance of these structures is significantly worse than that of more conventional AlGaInP LEDs grown on n-type substrates with p-type current-spreading layers.

3. Current-Spreading Window Layers

The current approach used to maximize performance in commercially available AlGaInP LEDs is to grow a thick p-type window layer of a material other than $(Al_xGa_{1-x})_{0.5}In_{0.5}P$ on top of the DH. The requirements for such a layer, in addition to the growth being compatible with $(Al_xGa_{1-x})_{0.5}In_{0.5}P$ alloys, are first that it possess a high sheet conductance, and second that it be transparent to the emitted light. High sheet conductance can arise from high mobility and carrier concentration, large layer thickness, or a combination of both. Transparency depends on the window material used and the wavelength of light being generated in the device.

Two materials have been successfully used as a window layer for AlGaInP LEDs: GaP and AlGaAs. The research groups that independently developed these window structures were at the Hewlett-Packard Company (GaP window structure) (Kuo et al., 1990), and Toshiba Corporation (AlGaAs window structure) (Sugawara et al., 1991a). Each of these structures has unique characteristics both in terms of the geometry of the layers as well as the physical properties of the window material itself.

The compound $Al_xGa_{1-x}As$ is a likely choice as a window material for an AlGaInP LED for several reasons. First, p-type $Al_xGa_{1-x}As$ can be doped heavily while maintaining relatively high carrier mobility to achieve the current-spreading needed for an efficient device. Second, for compositions where $x \geq 0.7$, $Al_xGa_{1-x}As$ is transparent to the light emitted in the red to yellow-green wavelength range and also possesses an indirect bandgap, minimizing absorption losses. Third, AlGaAs is intrinsically lattice-matched for all Al compositions (x) to the GaAs substrate and to the $(Al_xGa_{1-x})_{0.5}In_{0.5}P$ structure; therefore, AlGaAs is relatively easy to grow by OMVPE in the same epitaxial growth run as the AlGaInP DH.

Gallium phosphide also is an excellent choice for a window material. As with AlGaAs, p-GaP can be doped heavily to achieve sufficient current-spreading. It also is an indirect semiconductor and is transparent to the wavelengths of interest. If fact, GaP is even more transparent than AlGaAs at wavelengths below 600 nm. The unusual feature of GaP, however, is that it has a 3.6% lattice mismatch to the GaAs substrate and the $(Al_xGa_{1-x})_{0.5}In_{0.5}P$ layers. Nevertheless, GaP can be grown on top of the heterostructure without affecting the internal quantum efficiency or reliability of the LED device, despite the presence of a dense network of dislocations at the interface between the upper confining layer and the window layer (Fletcher *et al.*, 1991a). Fortunately, the dislocations remain confined to the first several microns of the GaP window layer and do not propagate downward into the device structure where they would cause nonradiative recombination sites in the active layer. The transparency and conductivity of the GaP layer are not appreciably affected by the dislocations as well. The GaP layer can be grown in the OMVPE reactor after the growth of the $(Al_xGa_{1-x})_{0.5}In_{0.5}P$ double heterostructure. In the HewlettPackard structure, however, a thick layer of GaP ($\sim 50\,\mu m$) is grown in a separate reactor using hydride VPE. The separate growth process takes advantage of the economy of the VPE process and the ability to quickly grow GaP layers with thicknesses in excess of 20 μm.

The current-spreading characteristics of these window layers can be understood and compared by using a model for current-spreading in stripe lasers (Thompson, 1980). Figure 7 illustrates a simplified schematic cross section of an AlGaInP LED with a window layer. Because of its symmetry, the device can be bisected, and only the right half of a whole device is shown. The left edge of the contact shown in the diagram is actually at the center of the chip. The model assumes a constant potential (V_0) and current density (J_0) underneath the metal. The potential throughout the substrate (V_{0s}) is also assumed to be a constant. The current density $J(x)$ extending away from the contact can be described by the expression

$$J(x) = 2\frac{J_0}{(x/l_s + 2^{1/2})^2} \tag{15}$$

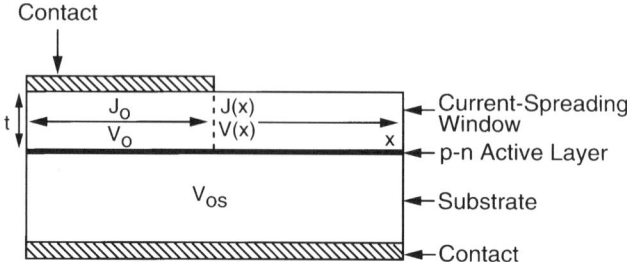

FIG. 7. Cross section of an AlGaInP die used to model current-spreading in the window layer. Because of symmetry only half the chip is needed; the right half is shown here.

where l_s is a spreading length given by

$$l_s = \left(\frac{gtnkT}{J_0 e}\right)^{1/2} \tag{16}$$

where g is the conductivity of the window layer, t is the window thickness, and n is the diode ideality factor (assumed to be 1.3 for all calculations herein). Typical parameters for various materials and both doping types are shown in Table III. Current-spreading curves derived from Eqs. (15) and (16) using the parameters in Table III are shown in Fig. 8 for both p-type (a) and n-type (b) GaP, $Al_{0.8}Ga_{0.2}As$, and $Al_{0.5}In_{0.5}P$. The thickness values for GaP and AlGaAs are typical of commercially available devices. The value of 1 μm for $Al_{0.5}In_{0.5}P$ is typical for a DH, and the result for a 5 μm layer is included for comparison and to illustrate the difficulty of using $Al_{0.5}In_{0.5}P$ as a current-spreading layer.

TABLE III

TYPICAL WINDOW LAYER MATERIAL PARAMETERS

Window Composition	p-Type		n-Type	
	Doping (cm^{-3})	Mobility (cm^2/V·sec)	Doping (cm^{-3})	Mobility (cm^2/V·sec)
GaP	1×10^{18}	62	1×10^{18}	125
$Al_{0.8}Ga_{0.2}As$	1×10^{18}	36	1×10^{18}	151
$Al_{0.5}In_{0.5}P$	5×10^{17}	10	1×10^{18}	211

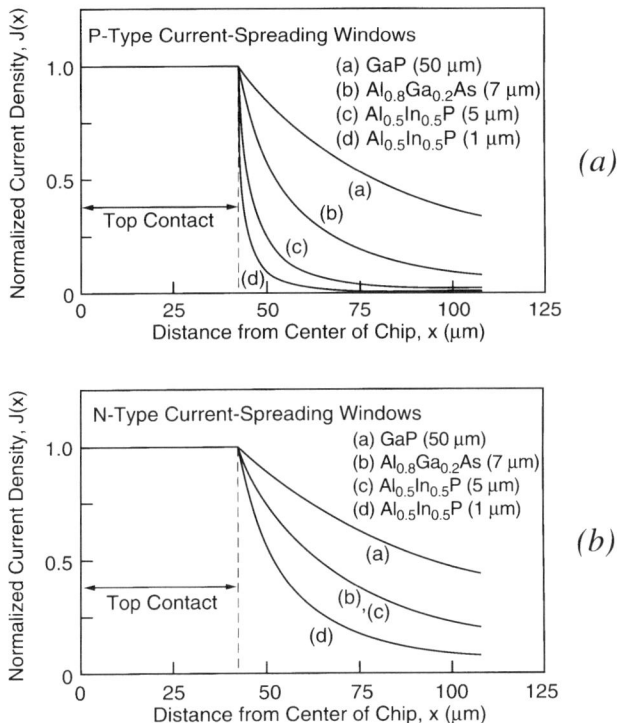

FIG. 8. Current densities for various window materials and thicknesses are shown as functions of the distance from the ohmic contact in an AlGaInP light-emitting diode (LED) based on the model described in the text. Results for p-type current-spreading windows are shown in (a), and results for n-type current-spreading windows in (b).

The data graphed in Fig 8(a) show that with a 1 μm thickness of p-type $Al_{0.5}In_{0.5}P$, the current density falls precipitously within a few micrometers of the top contact. A device operating under these conditions will appear to light up only at the very edge of the metal contact. Increasing the thickness of the $Al_{0.5}In_{0.5}P$ current-spreading layer to 5 μm results in some improvement; however, present epitaxial growth technology does not allow for an increase in thickness much beyond this point. Significant improvement is obtained with the use of $Al_{0.8}Ga_{0.2}As$ at a 7 μm thickness, but the best result is with the 50 μm GaP layer. In this case, the initial current density falls by only ~50% at a distance halfway to the chip edge. Figure 8(b) illustrates the results for n-type window layers, with dramatic improvement in the current-spreading behavior, especially for the $Al_{0.5}In_{0.5}P$ layers. Note that

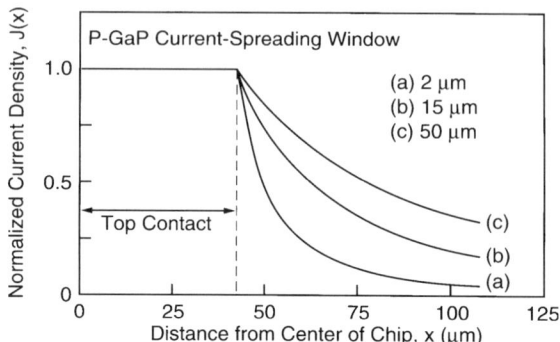

FIG. 9. Current density calculations for p-GaP current-spreading windows of various thicknesses.

the current-spreading for GaP does not show much difference, owing to the similarity in mobility for both n- and p-type materials. However, the limitations on the growth of an n-side-up structure described earlier still hold.

Figure 9 shows the effect of thickness on current-spreading for p-GaP only. Even a 2 µm layer yields significant spreading. An increase in thickness to 15 µm and then to 50 µm suggests that little improvement would be gained by increasing the thickness beyond 50 µm. This observation is supported by the data in Fig. 10, where the light emission from the surface of three LED chips with different GaP window thicknesses are measured as a function of position across the chip (Fletcher *et al.*, 1991a). This data was obtained by imaging the surface of each chip with a video camera and plotting the intensity of the emission using a video analyzer. The dotted line in the inset of the figure indicates the position on the chip at which the intensity measurement was obtained. For a 2 µm window thickness, current crowding is significant, and the emission intensity falls sharply close to the ohmic contact. A 5 µm window thickness improves the emission considerably, and with 15 µm of GaP the light emission approaches maximum intensity almost to the edge of the chip.

4. INDIUM TIN OXIDE AND OTHER APPROACHES

Another recently reported technique for improving current-spreading in AlGaInP LEDs is to apply a transparent ohmic contact to the surface of the double heterostructure. This is similar in concept to the current-

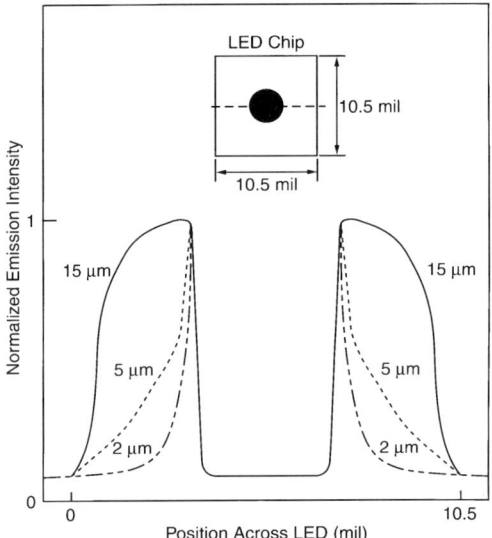

FIG. 10. Using a video camera and a video signal analyzer, the surface emission intensity was measured for three AlGaInP light-emitting diode (LED) chips with GaP window thicknesses of 2, 5, and 15 μm. The dotted line in the inset indicates the position on each LED at which the data in the lower part of the figure were taken. The 2 μm window device shows the effect of severe current crowding, whereas the 15 μm device shows good light output uniformity from the improved current-spreading. (Reprinted from Fletcher et al., 1991a J. Electron. Mater. 20, 1125, with the permission of the Minerals, Metals and Materials Society.)

spreading window; however, rather than growing an epitaxial window layer, a conductive film of indium tin oxide (ITO) is deposited by reactive electron beam evaporation on top of the DH. The sheet conductivity of the ITO films is very high (on the order of 5Ω/□) and the transparency is greater than 90% over the wavelength range of interest. Unfortunately, it is not possible to make direct ohmic contact between the ITO and AlGaInP, and lower bandgap contact layers must be grown on top of the AlGaInP, resulting in some absorption of the emitted light. Lin et al. (1994) describe a 500 Å GaAs contact layer in between the ITO and the $(Al_{0.7}Ga_{0.3})_{0.5}In_{0.5}P$ upper confining layer of their device. Aliyu et al. (1995) use a 300 to 500 Å GaAs layer to make contact with the ITO; however, they also add a transition layer of $Ga_{0.5}In_{0.5}P$ between the AlGaInP upper confining layer and the GaAs. The effect of the $Ga_{0.5}In_{0.5}P$ is to minimize the potential barrier between the high bandgap AlGaInP and the lower bandgap GaAs, resulting in a lower forward turn-on voltage for the device. In fact, using such techniques, Aliyu et al. (1995) report a forward voltage of 1.71 V at 20 mA, as compared with 2.0 V at 20 mA reported by Lin et al. (1994).

The light-output performance of AlGaInP LEDs using ITO so far has not been described in sufficient detail to facilitate comparison with commercially available devices that have a GaP or AlGaAs window layer. Lin et al. (1994) reported 0.5 mW of light emitted from a 620 nm device at 20 mA. However, they did not specify whether the device was encapsulated in clear epoxy or measured in air. (Currently, commercially available lamps with the chips encapsulated in epoxy routinely produce over 2.0 mW at 620 nm peak wavelength.) The question of encapsulation is important because the epoxy improves the coupling of the light out of the chip, often by as much as a factor of three or four, by providing refractive index matching between the high index semiconductor and the air. On a bare chip, ITO can potentially only serve as an anti-reflection coating. The reliability of devices with ITO has also not been investigated in detail, although Lin et al. (1994) described stable operation at 20 mA out to 3000 hours at room temperature. Clearly, more research on the ITO technique is needed and may yield a device that compares favorably with the other window structures.

Other methods for improving current-spreading may be developed in the future. At present, however, the use of AlGaAs and GaP as a current-spreading window provides excellent device performance that will be difficult to improve upon.

IV. Current-Blocking Structures

The LED chip structure described so far incorporates a top metal contact that partially covers the surface of the device. This configuration has the advantage of simplicity; however, the light generated under the contact where the current density is highest is either blocked and absorbed by the contact or reflected back into the chip. The reflected light has a high probability of being absorbed within the chip itself. This problem is especially severe if the current-spreading window layer is relatively thin. Under such conditions, there is little chance for the light generated under the contact to escape by multiple reflections within the structure or through the sides of the chip. These considerations are discussed in more detail in the next section. Several techniques exist, however, that minimize the contact shadowing problem by blocking the flow of current directly underneath the contact and forcing it to flow through the device outside the contact region, minimizing the light generated under the contact.

Current-blocking can be accomplished by a number of methods. One method used for selective area current-blocking is to introduce a heterobarrier or a p-n junction barrier within the epitaxial structure. These techniques are not new to the AlGaInP material system. Many AlGaInP laser structures incorporate such blocking layers in order to confine the current flow to a narrow stripe region of the device (Itaya et al., 1989; Kobayashi et al., 1985; Okuda et al., 1989).

To fabricate a p-n blocking structure it is necessary to perform a two-step growth process in which first the complete p-side-up double heterostructure is grown and then topped with a thin (~ 500 Å) n-type layer, which could be $Al_{0.5}In_{0.5}P$, $Ga_{0.5}In_{0.5}P$, or GaAs, for example. The wafer is then removed from the OMVPE reactor and processed using photolithography and etching to define islands in the top n-type layer where the ohmic contacts will later be deposited. The wafer is then reintroduced into the OMVPE reactor, and a final p-type current-spreading window layer is grown over the n-type islands. Final processing of the wafer involves aligning the top ohmic contacts over the n-type islands. The presence of this extra p-n junction under the contact forces the current to flow laterally through the window layer and out from under the contact before it can flow down through the p-n junction in the double heterostructure. The composition of the blocking layer depends on the practical consideration of being able to grow the material successfully and being able to etch it without disturbing the other $(Al_xGa_{1-x})_{0.5}In_{0.5}P$ epitaxial layers. Ideally, a material transparent to the emitted light is desired to minimize absorption; however, thin layers of GaAs and $Ga_{0.5}In_{0.5}P$ do not add appreciably to absorption and have the added benefit that they can be selectively etched.

The heterobarrier blocking layer structure also involves a two-step growth process; however, instead of using an n-type layer on top of the DH, a 500 Å thick p-type layer of GaAs is grown. In this case, the heterobarrier created between the high bandgap $Al_{0.5}In_{0.5}P$ upper confining layer and the lower bandgap GaAs is sufficient to provide some current blocking. Processing and regrowth steps are similar to the p-n blocking structure.

Another technique for creating a p-n blocking structure takes advantage of a selective area diffusion process (Sugawara et al., 1992c). In this process, a normal $(Al_xGa_{1-x})_{0.5}In_{0.5}P$ double heterostructure is grown until it is approximately halfway through the Zn-doped p-type upper confining layer. At that point, a 0.5-μm-thick layer of moderately Si-doped n-type $(Al_{0.7}Ga_{0.3})_{0.5}In_{0.5}P$ is inserted, and then the rest of the Zn-doped upper confining layer is completed. Next, the p-type AlGaAs current-spreading window layer is grown and then capped by a 0.5-μm-thick layer of heavily Si-doped GaAs. Once the epitaxial growth is complete, photolithography is used to define and etch holes in the GaAs cap layer where the ohmic contacts will later be deposited. The wafer is then annealed, and Zn in the split upper confining layer diffuses into the sandwiched thin layer of n-type

$(Al_{0.7}Ga_{0.3})_{0.5}In_{0.5}P$ and converts it to p-type. However, this diffusion and type conversion only occurs in the regions in which the Si-doped GaAs cap was not removed. The result is that buried islands of n-type $(Al_{0.7}Ga_{0.3})_{0.5}In_{0.5}P$ remain within the p-type upper confining layer and act as a current-blocking region. Final processing includes removing the rest of the GaAs cap and patterning ohmic contacts over the current-blocking regions.

Although the actual mechanism for this selective area diffusion is not understood the technique appears to work. Green AlGaInP LEDs with such a structure have been described with luminous performance more than twice that of conventional GaP LEDs. One limitation of this technique is that the diffusion process may not be easy to control and reproduce. Also, the diffusion mechanism depends on the proximity of the heavily Si-doped GaAs cap to the layer to be diffused and converted. Because the GaAs cap is on top of the current-spreading window layer, the window must be made sufficiently thin to still permit the diffusion process to occur. This necessarily puts a limit on the thickness, and therefore the effectiveness, of the current-spreading window layer, leading to a trade-off between optimizing current-spreading and current-blocking. Nevertheless, the process is attractive for its simplicity and low cost.

V. Light Extraction

We have described the essential features of an AlGaInP LED, namely, the double heterostructure design for efficiently generating the light and the current-spreading window layer for ensuring that current flow and light production are distributed throughout the device. The formation of current-blocking structures to inhibit current flow in parts of the device to improve light-output efficiency has also been described. We now turn the discussion to the subject of extracting light from the LED chip once it has been generated. This area of research is important for all types of LEDs; however, in the case of AlGaInP devices, light extraction techniques are being applied to the fullest degree possible to take advantage of the high internal quantum efficiency and color range of the material.

The light generated in the active layer of an LED chip can be divided in two: the half of the light that propagates up toward the top of the chip has a good chance of escaping; the other half that propagates down toward the substrate side of the chip has a good chance of being absorbed. Because of this division, it is reasonable to address the issue of light extraction of each separately. First, in the case of upward-directed light, we describe the design of the window layer. Second, in the case of downward-directed light, we describe the use of distributed Bragg reflectors (DBRs) to reflect the light back toward the top of the chip and also the replacement of the GaAs absorbing substrate (AS) with a transparent substrate (TS).

1. UPPER WINDOW DESIGN

Because the index of refraction of semiconductors in general is high, the critical angle of reflection is small for rays of light originating inside the chip; therefore, the fraction of light that can escape is limited by total internal reflection. Consider the simplified AlGaInP LED chip shown in Fig. 11 (Huang et al., 1992). The light-emitting DH is represented by a thin layer on top of the GaAs absorbing substrate and is covered by a transparent GaP window layer of finite thickness. The top contact pad has been omitted from this diagram for the sake of clarity. From an arbitrary point near the center of the chip, light is generated and travels in every direction. Virtually all the light directed downward is absorbed by the substrate. A ray of light directed upward can escape from the chip only if it strikes the surface of the chip at an angle less than the critical angle of reflection. Otherwise, total internal reflection occurs, and the ray tends to get trapped in the chip and eventually absorbed. The region for rays of less than the critical angle can be represented by so-called cones of acceptance (or escape cones), as illustrated in Fig. 11. One full cone is directed straight up, and there are four half cones directed toward each side of the chip (the other halves are

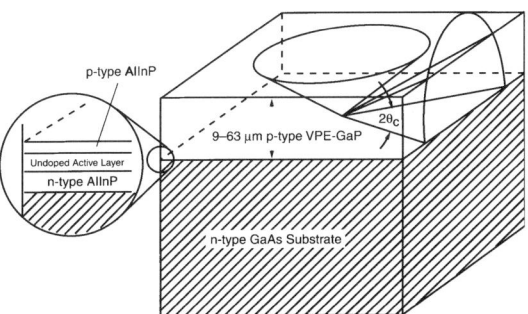

FIG. 11. Simplified diagram of an AlGaInP LED chip with a GaP window layer. The circular top ohmic contact has been omitted for clarity. For an arbitrary emission point near the center of the device, the cones of acceptance are shown for rays striking the surface of the chip at less than the critical angle. Only one of four side-directed cones is shown here. (After Huang, et al., 1992, Appl. Phys. Lett. **61**, 1045–1047; courtesy of the American Institute of Physics.)

absorbed by the substrate). Only one of the side-directed half cones is shown in the figure to keep the diagram uncluttered. All rays falling in the directions outside these cones of acceptance suffer total internal reflection and tend to be trapped and absorbed within the chip. Of course, some rays may experience multiple reflections and may eventually escape; however, to a first approximation, these can be ignored. The critical angle, which determines the size of the cones, is defined by Snell's law as

$$\theta_c = \sin^{-1}(n_1/n_2) \tag{17}$$

where n_1 and n_2 are the index of refraction of the medium surrounding the chip and that of the GaP window, respectively. If the surrounding medium is air, then $n_1 = 1$, $n_2 \approx 3.4$ for GaP, and the critical angle is about 17.1 degrees, making the apex of the cone 34.2 degrees. In most cases, however, the chip is encapsulated in epoxy, such that $n_2 = 1.5$, which widens the critical angle to 26.2 degrees and the cone apex to 52.4 degrees, an increase of 53%. This calculation clearly illustrates one advantage of encapsulating an LED chip in epoxy: The amount of light that can escape the chip is automatically increased by a factor of ~ 2 to 4 times.

Referring back to Fig. 11, one can understand the role that the window thickness plays on the light extraction efficiency by using the cones of acceptance. It is clear that the thickness of the window has no effect on rays within the upward-directed cone; the same amount of light within this cone will escape, regardless of thickness. However, for the side-directed rays, if the window is thin, a fraction will strike the surface of the chip and be totally reflected into the substrate before reaching the sidewall. For an infinitesimal emitting region and a given sidewall, the proportion of transmitted light can be estimated by the solid angle of intersection between the side wall of the window layer and the cone of acceptance for the side-directed light. The total light extraction efficiency is then the integration of this solid angle over the entire light-emitting area for each of the four sidewalls, plus the contribution of the upward cone (Huang et al., 1992). Figure 12 shows the relative light output for a bare chip in air and for a chip encapsulated in epoxy (a lamp) as a function of GaP window layer thickness. The theoretical curves were calculated using the model just described. The total light output at zero thickness begins at 1 because of the upward-directed cone, increases with window thickness, and then levels off as the window accommodates the entire half cones of the side-directed rays. For an encapsulated chip, the cone angle of acceptance is larger; therefore, a greater window thickness is required before the light output begins to level off. The data points plotted with the curves show close agreement with theory and support the validity of the model.

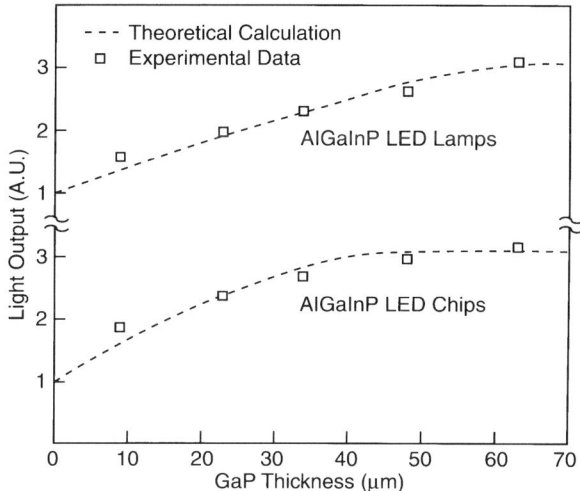

FIG. 12. Theoretic data for the light output of an AlGaInP light-emitting diode (LED) chip with an absorbing substrate for various window layer thicknesses. Actual measured data for chips with a GaP window layer agree with the theoretical model. (After Huang, et al., 1992, Appl. Phys. Lett. **61**, 1045–1047; courtesy of the American Institute of Physics.)

The theoretical and measured results show that a thick window can improve light extraction efficiency significantly. For example, an AlGaInP LED with a 50-μm-thick window emits roughly twice as much light flux as does an LED with a 10-μm-thick window. Thick layers and high growth rates are easily achieved using the VPE process for the growth of GaP. Consequently, the VPE process is an economically attractive technique for forming thick GaP window layers. On the other hand, thick AlGaAs windows are generally not produced because of the cost of the required OMVPE process.

2. Substrate Absorption

As discussed in Part II, lattice-matched growth is essential to the realization of the highest crystal quality AlGaInP, and hence highest quality LED material. Consequently, the $(Al_xGa_{1-x})_{0.5}In_{0.5}P$ alloy lattice-matched to GaAs is the material of choice for visible AlGaInP LED applications in the green-red (555 to 650 nm) spectral regime. However, GaAs substrates are problematic for such devices because they are optically absorbing across the entire visible spectrum. Light emission from the LED active layer is nearly isotropic. Consequently, the GaAs substrate absorbs at least half of the light emitted by the active layer. To circumvent absorption of light by the GaAs substrate, many attempts have been made to grow the AlGaInP alloy lattice-mismatched on transparent substrates (e.g., GaP) (Nuese et al., 1972, 1973; Stinson et al., 1991; V. M. Robbins, C. P. Kuo, R. M. Fletcher, T. D. Ostenouski, and M. G. Craford, personal communication, 1990). However, the benefits of enhanced crystal quality afforded by lattice-matched growth on an absorbing substrate (AS) supersede the significant improvements in light extraction efficiency (by two times or more) facilitated by lattice-mismatched growth on a transparent substrate (TS). As a result, two alternative approaches have been employed to minimize or eliminate the effects of the absorbing GaAs substrate, while allowing the growth of DH $(Al_xGa_{1-x})_{0.5}In_{0.5}P$ devices lattice-matched to GaAs. These device structures are shown schematically in Fig. 13 and consist of either a highly reflective lattice-matched semiconductor DBR placed between the absorbing GaAs substrate and the active layer (Fig. 13(a)), or a semiconductor wafer-bonded transparent GaP substrate, which is substituted in place of the GaAs substrate subsequent to the lattice-matched growth of the $(Al_xGa_{1-x})_{0.5}In_{0.5}P$ device layers (Fig. 13(b)). Both device structures result in improved performance relative to devices grown on GaAs. However, the TS device architecture results in significantly higher light extraction efficiency.

The light extraction efficiencies of the DBR and TS LED structures can be evaluated by considering the total number of escape cones of light from the LED, as depicted schematically in Fig. 14 (Kish, 1995). These cones are defined by the critical angle (θ_c) for total internal reflection given by Snell's law. As discussed previously, the critical angles for emission into air ($n_2 = 1$) and epoxy ($n_2 = 1.5$) are 17.1 and 26.2 degrees, respectively (assuming a semiconductor index (n_2) of 3.4). The extraction efficiency (C_{ex}) is defined as the ratio of the external to internal quantum efficiency, and for the devices of Fig. 14 can be estimated by the expression:

$$C_{ex} \approx N \frac{\int_0^{2\Pi}\int_0^{\theta_c} [1 - R(\theta)] \sin\theta \, d\theta \, d\phi}{\int_0^{2\Pi}\int_0^{\Pi} \sin\theta \, d\theta \, d\phi} \qquad (18)$$

where θ, ϕ are the azimuthal angle perpendicular to the active layer and radial angle in the plane of the active layer, respectively; θ_c is the critical angle; $[1 - R(\theta)]$ is the average power transmission coefficient of the transverse electric (TE) and transverse magnetic (TM) polarizations for a plane wave incident on a dielectric interface (Marcuse, 1972); and N is the

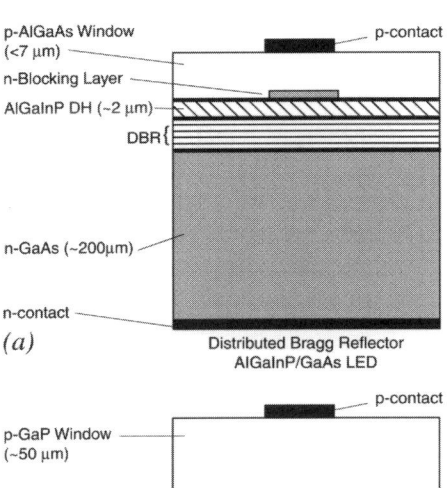

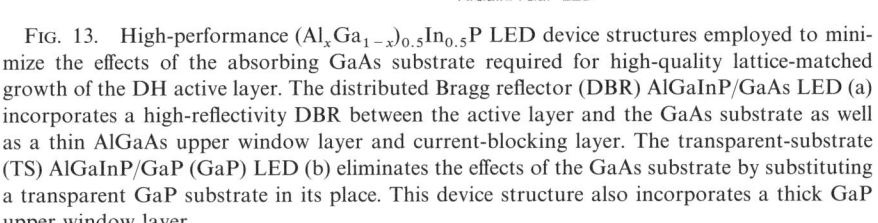

FIG. 13. High-performance $(Al_xGa_{1-x})_{0.5}In_{0.5}P$ LED device structures employed to minimize the effects of the absorbing GaAs substrate required for high-quality lattice-matched growth of the DH active layer. The distributed Bragg reflector (DBR) AlGaInP/GaAs LED (a) incorporates a high-reflectivity DBR between the active layer and the GaAs substrate as well as a thin AlGaAs upper window layer and current-blocking layer. The transparent-substrate (TS) AlGaInP/GaP (GaP) LED (b) eliminates the effects of the GaAs substrate by substituting a transparent GaP substrate in its place. This device structure also incorporates a thick GaP upper window layer.

total number of light extraction cones present in the device. Calculating the light extraction into epoxy ($n \sim 1.5$) from a transparent GaP or AlGaAs window substrate ($n \sim 3.4$) layers employed in the $(Al_xGa_{1-x})_{0.5}In_{0.5}P$ LED architectures of Fig. 14 yields an approximate extraction efficiency of $C_{ex} \approx N \times (4.1\%)$. The DBR LED is generally grown entirely by OMVPE and thus possesses a relatively "thin" (less than 7 μm) top window layer, resulting in emission essentially only from the top surface of the chip. The DBR mirror reflects most of the downward-reflected light, resulting in a

Light Extraction in AlGaInP Light-Emitting Diodes

(a) Thin Window Layer + Bragg Reflector + GaAs Absorbing Substrate

(b) Wafer-Bonded GaP Transparent Substrate

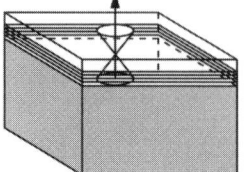
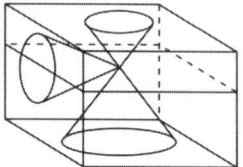

Top Cone + Bottom Cone

Top Cone + Bottom Cone + 4 Side Cones

FIG. 14. Schematic depiction of the light extraction properties, (escape cones of light emission defined by the critical angle) for the high-performance $(Al_xGa_{1-x})_{0.5}In_{0.5}P$ LED structures of Fig. 13. The absorbing substrate (AS) DBR LED possesses a maximum of two cones (a), whereas the transparent substrate (TS) LED possesses a maximum of six cones (b). (Reprinted from Kish, 1995. *Encycl. Chem. Technol.* **15**, 217–246, with the permission of John Wiley and Sons.)

maximum of $N = 2$ escape cones present in the device ($C_{ex} \approx 2 \times 4.1\% = 8.2\%$). In reality, the DBR is an imperfect mirror and does not reflect all of the downward-directed light, resulting in $N < 2$ for emission into epoxy (Part V, Section 3). Thus, the extraction efficiency of the DBR LED is somewhat less than that of the AS LED with a "thick" upper GaP window described in Part V, Section 1, wherein $N \sim 2.5$ to 3 cones. The TS LED of Figs. 13(b) and 14(b) is generally grown by a combination of epitaxial growth technologies, thin-film OMVPE growth for the $(Al_xGa_{1-x})_{0.5}In_{0.5}P$ active layer and thick-film VPE growth for the $\sim 50\,\mu m$ upper transparent GaP window layer. Thus, $N \sim 6$ cones of light extraction are present in this device. Absorption of top and bottom cones of light by the alloyed contacts plays a greater role in the TS LED structure, compared with the DBR LED structure wherein a wide bandgap transparent n-type $(Al_xGa_{1-x})_{0.5}In_{0.5}P$ current-blocking layer beneath the top contact minimizes its effect. As a result, the number of effective escape cones is reduced in the TS structure to $N \sim 5$ to 6 ($C_{ex} \approx 5$–$6 \times 4.1\% = 20.5$ to 24.6%). Consequently, the TS LED structure possesses a light extraction efficiency approximately twice that of an AS LED with a thick transparent window and roughly 2.5 to 3 times that of a DBR LED structure with a thin transparent window. Note that the aforementioned discussion of extraction efficiency neglects light absorption within the chip and does not fully capture the effects of randomization and extraction of multiple pass light from the chip. The latter effects are most prevalent in TS LED structures. For example, red 650 nm

TS AlGaAs LED lamps possess an estimated extraction efficiency of $\sim 30\%$ (Craford and Steranka, 1994), which is significantly higher than the maximum calculated value of $\sim 25\%$ (6 escape cones) predicted by Eq. (18). This difference can be attributed to the extraction of multiple pass light from the LED.

3. Distributed Bragg Reflector Light-Emitting Diodes

The utilization of semiconductor DBR mirror stacks for improving the light output performance of LEDs has been known since at least the early 1980s (Burnham *et al.*, 1982). Shortly thereafter, workers demonstrated the ability to grow high reflectivity mirror stacks by OMVPE (Thornton *et al.*, 1984). Utilizing such techniques, AlGaAs–GaAs DBR LEDs emitting in the infrared (and grown on absorbing GaAs substrates) were demonstrated with significantly enhanced light output (Kato *et al.*, 1991). This DBR device structure was not particularly useful for AlGaAs devices operating in the red visible spectral regime due to previously developed and mature TS AlGaAs LED technology (Ishiguro *et al.*, 1983; Cook *et al.*, 1988). Furthermore, to realize the highest efficiency devices, material growth considerations require that red AlGaAs devices be grown by LPE, a technique that is not amenable to the growth of the thin layered structures required to form a $\lambda/4$ semiconductor DBR mirror stack. However, the DBR LED structure is well-suited to minimize the absorption of light by the GaAs substrate in high-efficiency $(Al_xGa_{1-x})_{0.5}In_{0.5}P$ visible LEDs grown by thin-film OMVPE techniques. Consequently, $(Al_xGa_{1-x})_{0.5}In_{0.5}P$ LEDs have been fabricated with a variety of DBR mirror structures to minimize the substrate absorption effects (Sugawara *et al.*, 1992b, 1994b; Itaya *et al.*, 1994).

a. Distributed Bragg Reflector Design Considerations

A highly reflective DBR mirror is generally formed from a repeated periodic stack of alternating high and low index quarter-wavelength layers. The design thickness L of these high (h) and low (l) index layers is governed by the equation

$$L_{h,l} = \frac{m\lambda_0}{4n_{h,l}\cos\theta_{h,l}} \qquad (19)$$

where m is an odd integer, λ_0 is the free space design wavelength, n is the index of refraction of the layer, and θ is the incident angle of light relative the normal. In order to minimize cost, it is desirable to minimize the total thickness of epitaxial growth required for LEDs, resulting in the utilization of $\lambda/4$ stacks ($m = 1$).

For a fixed number of DBR periods, the reflectivity of the DBR mirror stack is maximized when the refractive index difference (Δn) between the high and low index $\lambda/4$ layers is maximized. The semiconductor alloys $Al_xGa_{1-x}As$ and $(Al_xGa_{1-x})_{0.5}In_{0.5}P$ are lattice-matched to GaAs and provide sufficient differences in refractive indices to allow the formation of high reflectivity DBRs. The refractive index of GaAs, AlAs, and $(Al_xGa_{1-x})_{0.5}In_{0.5}P$ are shown in Fig. 15 for the photon energies

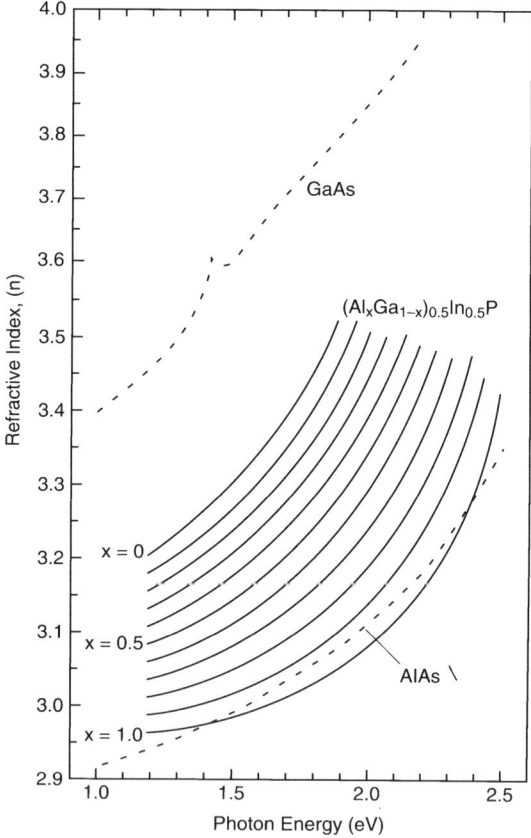

FIG. 15. Refractive index dispersion for the $(Al_xGa_{1-x})_{0.5}In_{0.5}P$ alloy (solid lines) and AlAs and GaAs (dashed lines) (Adachi, 1990) as a function of photon energy. (Reprinted from Adachi et al., 1994, J. Appl. Phys. **78**, 478–480, with the permission of the American Institute of Physics.)

of interest for $(Al_xGa_{1-x})_{0.5}In_{0.5}P$ LEDs. The refractive indices of $(Al_xGa_{1-x})_{0.5}In_{0.5}P$ shown in Fig. 15 were determined by the fitting of spectroscopic ellipsometry data to theory (Adachi et al., 1994), and the indices typically range from ~ 2.9 to ~ 3.8 in the 1.9 to 2.3 eV range. Also shown in Fig. 15 are the refractive indices of AlAs and GaAs (Adachi, 1990) which span a larger range of refractive indices.

A variety of materials have been employed in $(Al_xGa_{1-x})_{0.5}In_{0.5}P$ LED DBR mirrors. Fabrication of high-performance LEDs necessitates that the materials be lattice-matched to GaAs and possess a low resistivity. Table IV lists some relevant properties of DBR mirrors employed in $(Al_xGa_{1-x})_{0.5}In_{0.5}P$ LEDs. Generally, the DBRs may be classified into two categories: "lossy" and "transparent." In the case of the lossy mirrors, the high index $\lambda/4$ layers possesses a smaller bandgap than the emission energy of the LED, leading to substantially increased absorption within the DBR stack. All layers in the transparent mirrors possess a bandgap greater than the emission wavelength of the LED. The lossy mirrors employed to date consist of $Al_{0.5}In_{0.5}P$–GaAs or $Al_{0.5}In_{0.5}P$–$Ga_{0.5}In_{0.5}P$ $\lambda/4$ stacks. Both of these mirrors are lossy over the entire range of $(Al_xGa_{1-x})_{0.5}In_{0.5}P$ LED emission wavelengths. Furthermore, the lossy stacks typically possess fairly large differences in refractive index between the DBR layers. Despite difficulties that occur in OMVPE growth of alternating As- and P-based layers, $Al_{0.5}In_{0.5}P$–GaAs DBRs are generally preferred for $(Al_xGa_{1-x})_{0.5}In_{0.5}P$ LED applications due to their larger refractive index difference (Sugawara et al., 1992a, 1993). The transparent mirrors employed in DBR $(Al_xGa_{1-x})_{0.5}In_{0.5}P$ LEDs generally consist of $Al_{0.5}In_{0.5}P$–$(Al_xGa_{1-x})_{0.5}In_{0.5}P$ $\lambda/4$ stacks wherein the Al mole fraction in the DBR is chosen to possess a wider bandgap than the emission energy of the LED (Sugawara et al., 1993). As shown in Table IV, these DBRs

TABLE IV

PROPERTIES OF SEMICONDUCTOR DISTRIBUTED BRAGG REFLECTOR (DBR) MIRRORS UTILIZED IN $(Al_xGa_{1-x})_{0.5}In_{0.5}P$ LIGHT-EMITTING DIODES

DBR Materials	Refractive Index Difference (Δn) at Design Wavelength[a]	Mirror Type
$Al_{0.5}In_{0.5}P$–GaAs	0.85 (590 nm)	Lossy
$Al_{0.5}In_{0.5}P$–$Ga_{0.5}In_{0.5}P$	0.61 (590 nm)	Lossy
$Al_{0.5}In_{0.5}P$–$(Al_{0.3}Ga_{0.7})_{0.5}In_{0.5}P$	0.37 (615 nm)	Transparent ($\lambda \gtrsim 592$ nm)
$Al_{0.5}In_{0.5}P$–$(Al_{0.4}Ga_{0.6})_{0.5}In_{0.5}P$	0.35 (590 nm)	Transparent ($\lambda \gtrsim 576$ nm)
$Al_{0.5}In_{0.5}P$–$(Al_{0.5}Ga_{0.5})_{0.5}In_{0.5}P$	0.33 (570 nm)	Transparent ($\lambda \gtrsim 560$ nm)

[a]Adachi et al. (1994) and Adachi (1990).

are characterized by significantly smaller differences in refractive index, thus requiring more periods to achieve the same reflectivity as the lossy DBRs. Furthermore, because the refractive index difference decreases as the design wavelength is shortened for the transparent DBRs, more periods are required to achieve the same DBR reflectivity. Note that the peak emission wavelength of the LEDs must be chosen significantly longer than the transparency wavelength (Table IV, column 3) to accommodate the full emission bandwidth (Part VII, Section 4). In addition to those materials listed in Table IV, other materials are potentially viable for utilization in DBR mirrors for $(Al_xGa_{1-x})_{0.5}In_{0.5}P$ LEDs. For example, $AlAs-Al_xGa_{1-x}As$ DBRs have been shown to be capable of exhibiting relatively high reflectivities at 550 nm (Young et al., 1992). Workers have avoided using high Al-composition $Al_xGa_{1-x}As$ layers ($x \geqslant 0.6$) in $(Al_xGa_{1-x})_{0.5}In_{0.5}P$ LED structures to date due to a concern for the hydrolization degradation of these compounds under high-humidity conditions (Dallesasse et al., 1990a,b; Richard et al., 1995). However, experiments indicate that the hydrolization degradation is significantly different for thin ($\lesssim 200$ Å) AlGaAs layers than for thick (~ 1000 Å) layers (Dallesasse et al., 1990a,b), making it unclear at this time how problematic thin high-composition $Al_xGa_{1-x}As$ layers are for LED DBR applications.

The reflectivity of the DBR mirror stacks increases with an increasing number of periods. This effect is shown in Fig. 16, which shows calculated reflectivity versus wavelength for both (a) transparent $Al_{0.5}In_{0.5}P-(Al_{0.4}Ga_{0.6})_{0.5}In_{0.5}P$ and (b) lossy $Al_{0.5}In_{0.5}P-GaAs$ DBRs with a design wavelength of $\lambda_0 = 590$ nm. These and the following calculations are performed using standard matrix techniques (Born and Wolf, 1989) incorporating the effects of absorption within the DBR layers, and utilize the refractive index values of Fig. 15 (Adachi, 1990; Adachi et al., 1994). The calculations assume unpolarized light incident from a wide bandgap $Al_{0.5}In_{0.5}P$ lower confining layer. The calcuations of Fig. 16 show marked differences for the transparent and lossy DBRs. The reflectivity of both DBR types increases with an increasing number of periods. However, the reflectivity of the lossy DBR (b) tends to saturate at a value less than 60% wherein reflectivities greater than 95% can readily be achieved for the transparent DBR (20-period stacks) (a). This difference is ascribed to the increased absorption in the lossy DBR. However, the lossy DBR tends to possess a wider spectral bandwidth (greater than 70 nm full width at half maximum (FWHM), 20-period mirror) compared with that of the transparent DBR (less than 40 nm FWHM, 20-period mirror) as a result of the larger refractive index difference between the DBR layers.

The benefits of lossy and transparent DBRs may be combined by employing a hybrid reflector stack wherein a transparent DBR is placed on top of a lossy DBR grown on the GaAs substrate (Sugawara et al.,

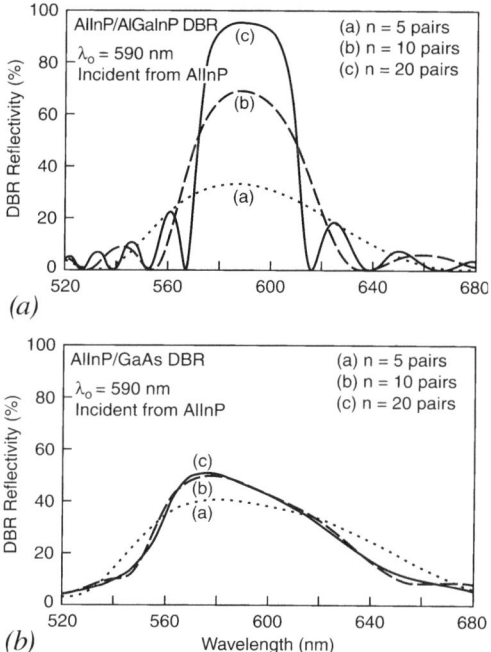

FIG. 16. Calculated reflectivity versus wavelength for (a) transparent AlInP–AlGaInP DBRs and (b) lossy AlInP–GaAs DBRs. The reflectivity of the lossy DBRs (b) tends to saturate at a peak reflectivity less than 60% with an increasing number of DBR periods; however, the reflectivity of the transparent DBRs (a) increases significantly with an increasing number of periods, resulting a peak reflectivity of over 95% for a 20-period mirror structure.

1994b). The potential benefits of such a hybrid DBR are shown in Fig. 17, which shows calculated reflectivity versus wavelength for (a) a lossy $Al_{0.5}In_{0.5}P$–GaAs DBR (20 periods), (b) a transparent $Al_{0.5}In_{0.5}P$–$(Al_{0.4}Ga_{0.6})_{0.5}In_{0.5}P$ DBR (20 periods), and (c) a hybrid $Al_{0.5}In_{0.5}P$–$(Al_{0.4}Ga_{0.6})_{0.5}In_{0.5}P$ (10-period) plus $Al_{0.5}In_{0.5}P$–GaAs (10-period) DBR. The FWHM of the reflectivity spectrum of the hybrid DBR is increased by ~25% over that of the transparent DBR, while maintaining a reflectivity of greater than 90%. Furthermore, the enhanced reflectivity bandwidth of the hybrid DBR results in an enhanced ability to reflect light of grazing incidence. This characteristic is shown in Fig. 18, wherein the calculated reflectivity versus the incident angle (measured from the normal) is shown for the (a) lossy, (b) transparent, and (c) hybrid DBRs of Fig. 17. These data indicate that the hybrid DBR provides an increased angular bandwidth of ~20%. Consequently, the hybrid DBR is a very good compromise between the high reflectivity of the transparent DBR and large bandwidth of the lossy DBR.

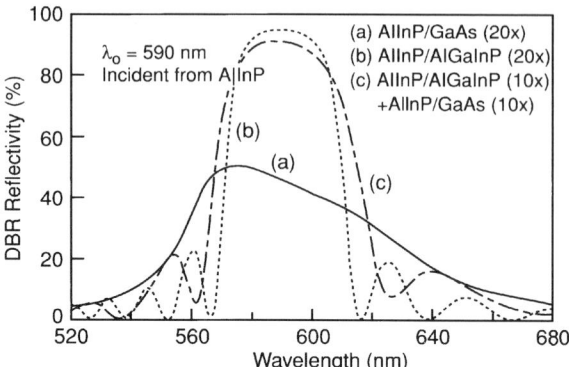

FIG. 17. Calculated reflectivity versus wavelength for a (a) lossy 20-period AlInP–GaAs DBR, a (b) transparent 20-period AlInP–AlGaInP DBR, and a (c) hybrid 10-period transparent AlInP–AlGaInP plus 10-period AlInP–GaAs lossy DBR. The hybrid DBR (c) combines the benefits of high reflectivity and wide spectral bandwidth of the transparent and lossy DBRs.

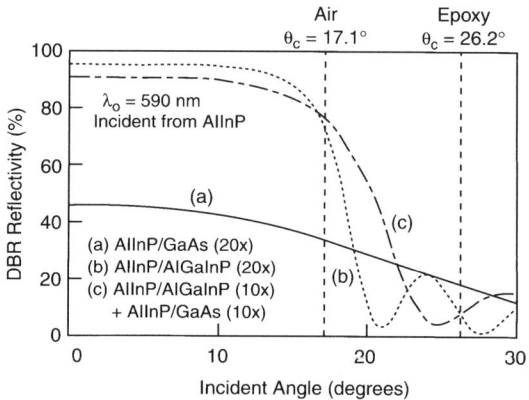

FIG. 18. Calculated reflectivity versus incident angle for a (a) lossy 20-period AlInP–GaAs DBR, a (b) transparent 20-period AlInP–AlGaInP DBR, and a (c) hybrid 10-period transparent AlInP–AlGaInP plus 10-period AlInP–GaAs lossy DBR. The hybrid DBR (c) possesses high reflectivity (>90%) over a large angular bandwidth; however, this bandwidth is less than the critical angle for total internal reflection for emission into epoxy (26.2 degrees).

b. Device Performance

The calculated reflectivity curves of Figs. 15 to 17 indicate that the semiconductor DBR mirror is an imperfect reflector, possessing less than 100% peak reflectivity over a finite band of wavelengths and incident angles. Consequently, DBR LEDs encapsulated in epoxy do not extract all of the light emitted toward the substrate from the active region. In Part V, Section 2, an analysis of the extraction efficiency of a DBR LED structure indicated that the light extraction efficiency (C_{ex}) of the device is $\sim 8.2\%$ ($2 \times 4.1\%$). A more precise analysis of the extraction efficiency considers the true reflectivity (including angular and spectral bandwidth) of the DBR mirror. In this case, the extraction efficiency is given by

$$C_{ex} \approx \frac{\int_0^{2\pi} \int_0^{\theta_c} [T_{DBR}(\theta)] \sin\theta \, d\theta \, d\phi}{\int_0^{2\pi} \int_0^{\Pi} \sin\theta \, d\theta \, d\phi} \quad (20)$$

wherein $T_{DBR}(\theta)$ is the effective power transmission coefficient of the DBR LED structure and the remainder of the terms are identical to those found in Eq. (18). The effective power transmission coefficient of the DBR structure can be derived by considering the simple ray tracing model of Fig. 19. Accordingly, the total power transmitted through the top surface of the DBR LED is given by

$$T_{DBR}(\theta) = (1 - R_T) \sum_{i=0}^{\infty} R_T^i R_B^i + (1 - R_T)R_b \sum_{i=0}^{\infty} R_T^i R_B^i$$

$$= \frac{(1 - R_T)(1 + R_B)}{1 - R_T R_B} \quad (21)$$

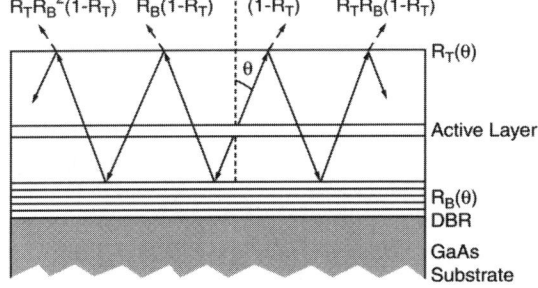

FIG. 19. Schematic ray-tracing diagram of a DBR light-emitting diode structure indicating the effects of multiple pass light and reflection at the DBR and ambient interfaces.

wherein $R_T = R_T(\theta)$ is the average TE and TM power reflection coefficient at the top window and $R_B = R_B(\theta)$ is the reflectivity of the DBR as a function of the incident angle. For purposes of calculation, an infinite DBR reflectivity spectral bandwidth and a step function reflectivity angular bandwidth are assumed of the form

$$R_B(\theta) = R \quad |\theta| \leq \theta_{\text{DBR}} \tag{22}$$

$$R_B(\theta) = 0 \quad |\theta| > \theta_{\text{DBR}} \tag{23}$$

wherein θ_{DBR} is the angular bandwidth of the DBR mirror. Using these relations, a DBR light-output gain (equivalent to N as defined in Eq. (18)) can be defined as the ratio of the extraction efficiency of the DBR LED structure to that of an LED with a single escape cone ($N = 1$). Figure 20 shows the calculated DBR light-output gain for (a) an LED chip in air ($n = 1$) and (b) an encapsulated LED ($n = 1.5$) as a function of DBR peak reflectivity R for emission from a transparent window ($n = 3.4$). These calculations assume a DBR angular bandwidth (θ_{DBR}) of ± 20.8 degrees which is equivalent to that of the hybrid DBR of Fig. 18(c). The DBR light-output gain is significantly higher for an LED chip in air compared with an encapsulated LED. This phenomenon is due to the fact that the DBR angular bandwidth exceeds the critical angle for total internal reflection for emission into air ($\theta_c = 17.1°$) but is less than the critical angle for

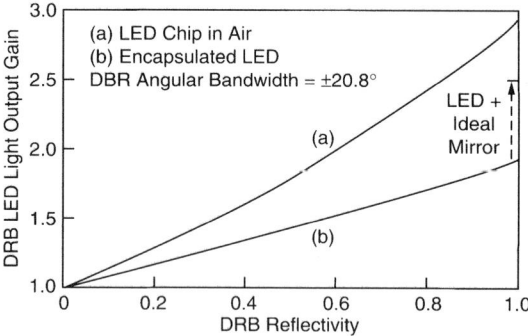

FIG. 20. Calculated light output gain for thin window DBR light-emitting diodes versus DBR reflectivity for emission into (a) air ($n = 1$) and (b) epoxy ($n = 1.5$). These calculations assume an angular reflectivity bandwidth of ± 20.8 degrees, an infinite spectral bandwidth, and neglect and absorption within the LED chip. For emission into epoxy (b), the DBR angular bandwidth is less than the critical angle (26.2 degrees), resulting in reduced light-output gains. An ideal mirror (100% reflectivity, infinite spectral bandwidth, ± 90 degrees) results in approximately a 2.5 times improvement light output.

emission into epoxy ($\theta_c = 26.2°$), as shown in Fig. 18. Consequently, the full downward escape cone is not captured for emission into epoxy and results in a DBR light-output gain (or N in Eq. (18)) of 1.7 to 1.9 as the DBR reflectivity varies from 80% to 100%. A "perfect" mirror (± 90-degree angular bandwidth, 100% reflectivity) captures all of the downward-directed light, resulting in a DBR light-output gain of ~ 2.5 for emission into epoxy. Note that this model neglects any absorption (or reflections) with the LED internal heterolayers and at the top contact that tend to reduce the extraction efficiency of the structure.

The incorporation of DBR mirror structures between the absorbing GaAs substrate and active layer of $(Al_xGa_{1-x})_{0.5}In_{0.5}P$ LEDs results in a significant improvement in light-output performance. As discussed previously, the choice of the DBR mirror design depends on a number of trade-offs in performance (peak reflectivity versus spectral and angular bandwidths). In addition, the number of periods (total mirror thickness) and complexity of the DBR (e.g., utilization of hybrid stacks) results in a cost–performance trade-off. High-performance $(Al_xGa_{1-x})_{0.5}In_{0.5}P$ LED lamps utilizing DBRs have been fabricated with external quantum efficiencies of $\sim 7\%$ at ~ 610 nm and $\sim 0.8\%$ at ~ 573 nm (Itaya et al., 1994). Furthermore, both n-type AlInP–GaAs and AlInP–AlGaInP DBR mirrors employed in these structures possess low resistance, adding less than 0.3 V to the forward voltage of typical LED devices at 50 mA (Sugawara et al., 1993) (see Part VII, Section 3). As a result, the DBR LED structures are also capable of achieving high power-conversion efficiencies. Distributed Bragg reflector $(Al_xGa_{1-x})_{0.5}In_{0.5}P$ LEDs with a current-blocking region beneath the top contact formed via diffusion (Part IV and Fig. 13) were commercially introduced in 1993 by Toshiba Corporation and are their latest generation of devices currently in production as of this writing.

4. Wafer-Bonded Transparent-Substrate Light-Emitting Diodes

A transparent substrate LED offers the highest light extraction efficiency of all conventional device architectures. Although material quality considerations dictate that the AlGaInP alloy must be grown lattice-matched on an absorbing GaAs substrate, a variety of means are conceivable for producing TS AlGaInP devices. The absorbing GaAs substrate may be removed after epitaxial growth of a thick transparent window layer (greater than 50 μm) on top of the device (e.g., GaP as in Part III, Section 3). However, removal of the substrate is not sufficient to realize enhanced light output as shown schematically in Fig. 21. After removal of the GaAs substrate, the exposed

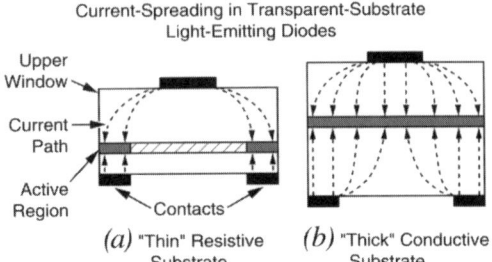

FIG. 21. Schematic diagram of current-spreading in transparent substrate (TS) LED structures. A thin resistive TS (a) results in current-crowding around the patterned bottom contacts of the device. The optimal design consists of a thick conductive TS (b) to facilitate adequate current-spreading within the device.

thin $(Al_xGa_{1-x})_{0.5}In_{0.5}P$ layers possess high resistivity, and results in devices with severe current-crowding (Fig. 21(a)) with limited improvement in light-output performance. Furthermore, the thin "substrate" of such an LED chip results in a reduced ability to extract the side cones of light (see Fig. 14(b)) in the device due to total internal reflection of light emitted from the active layer into the thin "substrate" and subsequent parasitic absorption within the LED chip (e.g., at the active layer and alloyed contacts). Consequently, a thick (greater than 50 μm) conductive substrate (Fig. 21(b)) is desirable for optimal TS LED performance. This thick substrate may be produced by epitaxial regrowth of a wide bandgap conductive layer after removal of the GaAs absorbing substrate (Fletcher et al., 1991b). Alternatively, a thick (greater than 50 μm) transparent buffer layer (e.g., consisting of AlGaAs) may be grown on the GaAs substrate prior to the growth of the device structure (Kuo et al., 1991; Sugawara et al., 1992c). Subsequent removal of the GaAs substrate results in a TS device structure. Both of the aforementioned techniques are problematic in that thick layers are costly, can be very difficult to grow, and often are not compatible with the realization of a high-quality LED active region. As a result of these difficulties, wafer-bonding a TS to an $(Al_xGa_{1-x})_{0.5}In_{0.5}P$ active layer to produce a thick conductive TS LED is an attractive alternative technology.

a. Semiconductor Wafer Bonding

Semiconductor wafer bonding, the attachment of a semiconductor substrate (or epitaxial layer) to another substrate (or epitaxial layer) facilitates the integration of dissimilar materials while generally preserving the bulk properties of the original materials. A variety of techniques have been employed to bond two surfaces together, the methods being distinguished by the characteristics of the bonded interface. Table V lists some of the more common wafer-bonding techniques and their associated properties. The fabrication of wafer-bonded TS LED structures requires a robust, transparent interface. In addition, cost considerations (wafer area utilization) and the relatively low sheet conductivity of the AlGaInP alloys dictate that $(Al_xGa_{1-x})_{0.5}In_{0.5}P$ LED device architectures be designed for vertical current flow (with contacts on the top and bottom of the device), resulting in the further requirement that the wafer-bonded interface exhibit low-resistance conduction. Consequently, van der Waals wafer-bonding techniques (Yablonovitch *et al.*, 1990) are not suitable because they do not possess the required mechanical strength or interfacial electrical conductivity necessary for such LED applications. Good electrical conductivity can be achieved across the wafer-bonded interface provided an alloyed metallic interlayer is employed; however, the bonded interface then becomes absorbing or moderately reflective ($R \lesssim 70\%$) (Schnitzer *et al.*, 1993). As a result, metallic or eutectic bonding techniques are not suitable for TS $(Al_xGa_{1-x})_{0.5}In_{0.5}P$ LEDs. Techniques of epoxy bonding or direct wafer bonding of oxide (or glass) to a semiconductor also are not viable because of the poor electrical properties of such wafer-bonded interfaces. However, direct wafer bonding (the fusion of surfaces without any adherence interlayers) of two semiconductor surfaces is a viable candidate for TS LED applications because it is capable of forming strong chemical bonds across a transparent and electrically conductive wafer-bonded interface.

Direct semiconductor wafer bonding of Si wafers is a relatively mature technology for the bonding of Si to Si and Si to SiO_2 and is well known in the art (Bengtsson, 1992; Mitani and Gosele, 1992). Direct bonding of compound semiconductors to glass to fabricate high-performance GaAs photocathodes on transparent substrates has been known for over 20 years (Antypas and Kinter, 1973; Antypas and Edgecumbe, 1975). More recently, direct bonding of InP to GaAs has been performed using elevated temperatures (520 to 830°C) and applied uniaxial pressures (Liau and Mull, 1990). This work led to the utilization of compound semiconductor wafer bonding for a variety of optoelectronic device applications, including: the integration of InGaAsP laser diodes on GaAs substrates (Lo *et al.*, 1991) and the integration InGaAs–GaAs laser diodes on Si substrates (Lo *et al.*, 1993); the fabrication of long-wavelength InGaAsP (1.3 to 1.5 μm) vertical-cavity surface emitting lasers (VCSELs) (Dudley *et al.*, 1992; Babic *et al.*, 1995) and resonant-cavity photodetectors (Tan *et al.*, 1994) using high-index contrast AlGaAs DBRs; and the fabrication of long-wavelegth InGaAs photodiodes on GaAs and Si substrates (Ejeckam *et al.*, 1995). In 1994, Hewlett-Packard

TABLE V
COMPARISON OF SEMICONDUCTOR WAFER-BONDING TECHNIQUES

Bonding Technique	Bonding Mechanism	Bond Strength	Thermal Cycling Capability	Interfacial Electrical Conductivity	Interfacial Optical Transparency
van der Waals	Electrostatic forces	Poor	Poor	Poor	Excellent
Metal–eutectic	Semiconductor or intermetallic alloy	Good	∼Eutectic temperature (<600°C)	Excellent	Poor
Epoxy	Adhesive	Good	Continuous use temperature (<200°C)	Poor	Good
Oxide (glass)	Chemical bond	Excellent	∼Bonding temperature	Insulating	Excellent
Direct Semiconductor-semiconductor	Chemical bond	Excellent	∼Bonding temperature	Process dependent	Process dependent

Company introduced the first commercial devices (high-efficiency visible-spectrum $(Al_xGa_{1-x})_{0.51}In_{0.5}P$ LEDs integrated on a transparent GaP substrate) using direct wafer bonding of two compound semiconductor wafers (Kish et al., 1994a,c, 1996b). This innovation required a number of technological advances at Hewlett-Packard Company that enabled formation of a robust, low-resistance, optically transparent, compound semiconductor wafer-bonded interface suitable for demanding LED applications and amenable to high-volume manufacturing techniques.

b. Wafer-Bonded Light-Emitting Diode Fabrication

Figure 22 shows a schematic diagram of the TS $(Al_xGa_{1-x})_{0.5}In_{0.5}P$/GaP LED fabrication process. The LED structures employed in this technique consist of a DH p-n $(Al_xGa_{1-x})_{0.5}In_{0.5}P$ LED grown latticematched by OMVPE on a GaAs substrate with a thick (~ 50-μm) GaP window layer grown on top of the structure by VPE. The GaP window serves to enhance current-spreading from the top contact (Part III, Section 3) and increases the light extraction from the top of the device (Part V, Section 1). After epitaxial growth, the absorbing GaAs substrate is removed by conventional chemical etching techniques (Adachi and Oe, 1983). At this point, the thick GaP window provides mechanical stability to facilitate handling of the epitaxial layers. The exposed n-type layers of the DH are subsequently wafer-bonded to an ~ 8 to 10-mil n-type GaP substrate at elevated tempera-

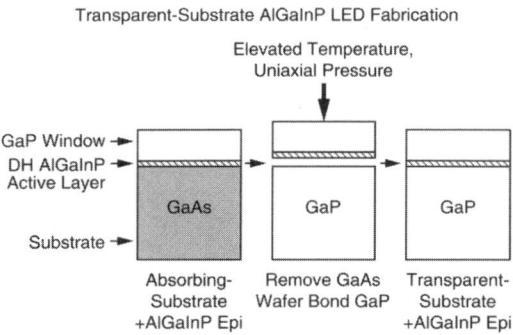

FIG. 22. Schematic diagram of the fabrication process for wafer-bonded transparent-substrate (TS) AlGaInP LEDs. After selective removal of the original GaAs growth substrate, elevated pressure and uniaxial pressure are applied to the GaP substrate and AlGaInP–GaP epitaxial (epi) film, resulting in formation of a single TS LED wafer.

ture and uniaxial pressure (Liau and Mull, 1990; Kish et al., 1994a). Patterned ohmic contacts are then applied to *both* GaP surfaces and the bonded wafers are diced into 8.5 × 8.5 mil^2 LED chips, which are typically encapsulated in standard 5-mm LED lamps.

The wafer bonding of compound semiconductors is an attractive alternative to integrating lattice-mismatched materials because the crystal quality adjacent to the wafer bonding interface is generally preserved. Transmission electron microscope (TEM) analysis of the wafer-bonded GaP/$(Al_xGa_{1-x})_{0.5}In_{0.5}P$ interfaces indicate that the misfit dislocations consist entirely of edge dislocations. Figure 23 shows a cross-sectional TEM image of a wafer-bonded TS GaP-$(Al_xGa_{1-x})_{0.5}In_{0.5}P$/GaP LED. Threading dislocations occur within the upper GaP window above the DH as a result of the lattice-mismatched epitaxial growth techniques employed to form the

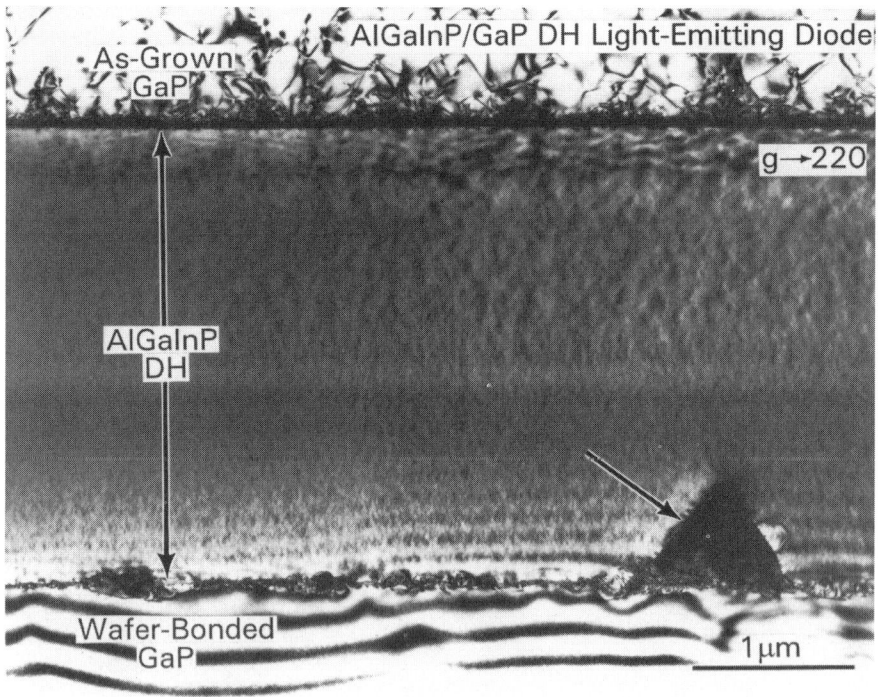

FIG. 23. Transmission electron microscope image of a wafer-bonded transparent-substrate GaP-AlGaInP GaP DH LED. Threading dislocations are observed in the upper epitaxial GaP window layer; however, no threading dislocations are observed at the lower wafer-bonded interface. The arrow (at bottom, right) indicates a crystalline defect formed during the wafer-bonding process. Such defects do not adversely affect device performance.

window. However, the image of Fig. 23 (as well as images along other directions) shows no signs of threading dislocations at the wafer-bonded interface, indicating that the as-grown crystal quality of the DH is preserved. These results are in agreement with the absence of threading dislocations reported at wafer-bonded interfaces of other compound semiconductors (Lo et al., 1991, 1993; Wada et al., 1993; Okuno et al., 1995). Figure 23 also shows a crystalline defect consisting of an AlGaInP alloy of different crystal composition and orientation than the surrounding layers present at the wafer-bonded interface (lower right arrow). These defects have been observed to be a function of the wafer-bonding conditions. However, such defects are considered benign in that their presence and density do not degrade any aspect of LED device performance (e.g., forward voltage, light output, or reliability). Thus, compound semiconductor wafer bonding facilitates the integration of lattice-mismatched materials for LED applications.

Although many analogies can be drawn between the techniques required to wafer-bond elemental Si and III-V compound semiconductors wafers, the processes are distinctly different. These differences arise from variations in the mechanical, chemical, and electrical properties of the materials. Table VI compares a number of properties of III-V compounds and Si that are important for direct wafer-bonding. Dissimilar III-V compound semiconductor wafers possess a significant mismatch in thermal expansion coefficients ($\sim 10^{-6}/°C$). Consequently, unlike the bonding of Si to Si, wafer bonding different compound semiconductor wafers requires the application of uniaxial pressure to ensure that the wafers remain in contact as the temperature is cycled throughout the bonding process. Furthermore, Si wafers can be prepared with a surface flatness on nearly an atomic scale across an entire wafer. The flatness of III-V compound semiconductor

TABLE VI

COMPARISON OF SILICON (Si) AND III-V DIRECT WAFER-BONDING TECHNOLOGIES

	Si to Si Wafer-Bonding	$III_A V_B$ to $III_C V_D$ Wafer-Bonding
Thermal expansion mismatch	None	$\sim 10^{-6}/°C$
Surface flatness	Excellent	Moderate
Surface volatility	Low	High
	Stable to $>1000°C$	Stable to 500–900°C Unequal III-V vapor pressures
Surface electrical reactivity	High	Moderate

wafers is grossly different, varying by as much as several microns across a 50-mm-diameter wafer. As a result, special techniques are required to accommodate this variation when bonding compound semiconductor wafers over large areas. Silicon also possesses a low equilibrium vapor pressure, being stable to temperatures in excess of 1000°C (the conditions under which direct Si wafer bonding is typically performed). However, the surface volatility of III-V compounds is fairly high, becoming unstable in the temperature range of ~500 to 900°C, with the column V element(s) exhibiting a much higher equilibrium vapor pressure compared with that of the column III element(s). This volatility occurs in the regime wherein compound semiconductor wafer bonding is performed, increasing the complexity of the bonding processes. Differing electrical characteristics of the surfaces of Si and compound semiconductor wafers also result in varying considerations for the preparation of low-resistivity interfaces. Thus, III-V compound semiconductor wafer bonding is substantially different from that of elemental Si.

The achievement of low-resistance conduction across wafer-bonded interfaces imposes different requirements for the bonding of III-V semiconductors compared with Si. Direct wafer bonding of Si wafers is known to result in low-resistance conduction across unipolar (p-p or n-n) bonded interfaces (Shimbo *et al.*, 1986; Bengtsson, 1992). However, a large variation has been observed in the electrical transport properties across compound semiconductor wafer-bonded interfaces (Liau and Mull, 1990; Lo *et al.*, 1993; Wada *et al.*, 1993; Dudley *et al.*, 1994; Kish *et al.*, 1994a) with most workers reporting high-resistance conduction. An essential element in achieving low-resistance conduction across unipolar compound semiconductor wafer-bonded interfaces is maintaining the proper relative surface orientation *and* relative rotational alignment of the bonded crystals (Kish *et al.*, 1995). Such effects are shown in Fig. 24, which depicts the current-voltage (I-V) characteristics of TS $(Al_xGa_{1-x})_{0.5}In_{0.5}P/GaP$ LEDs for (a) aligned bonded wafers (no relative misalignment of the crystallographic planes) and (b) misaligned bonded wafers (90 degree rotational misalignment of the crystallographic directions in the plane of the wafer surfaces and 2.8 degree misalignment of the wafer surface orientations). The I-V characteristics indicate that low forward voltages (2.0 V at 20 mA) are typically achieved in the aligned case ((a) in Fig. 24)), which are generally less than 0.2 V higher than devices fabricated from the same absorbing GaAs substrate $(Al_xGa_{1-x})_{0.51}In_{0.5}P$ LEDs. However, the misaligned case ((b) in Fig. 24) exhibits a forward voltage greater than 3.0 V at 20 mA. The increase in forward voltage is thought to be caused by an increased density of electrically charged defects (e.g., point defects, edge dislocations, antiphase domains) at misaligned wafer-bonded interfaces. Such defects result in an

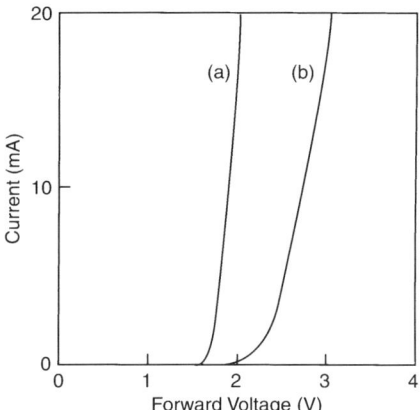
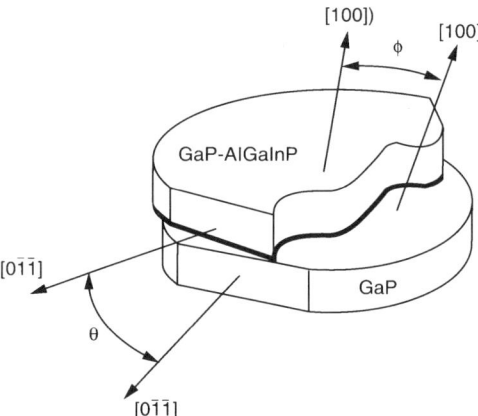

FIG. 24. Current-voltage characteristics for TS AlGaInP/GaP LEDs for wafers wherein the crystallographic planes of bonded wafers are (a) aligned and (b) misaligned. In the aligned case, low forward voltage operation (~2.0 V at 20 mA) is achieved. Significant crystallographic misalignment results in the substantial increase in the forward voltage of the LED.

increased potential barrier height at the wafer-bonded heterojunction. The reverse I-V characteristics also typically exhibit standard avalanche reverse breakdown at ~ −15 to −30 V in the aligned case (Part VII, Section 3). Consequently, the wafer-bonded interface does not adversely affect the LED I-V characteristics, provided the wafer bonding is performed properly. These trends are substantially different from those observed in Si direct wafer bonding wherein little dependence on crystallographic orientation is observed unless the bonded wafers are lightly doped ($\sim 10^{15}\,\text{cm}^{-3}$) (Laporte et al., 1994; A. Laporte, personal communication, 1995). These differences can be attributed to the different chemical and electrical properties of Si and III-V compound semiconductors.

As discussed in Chapter 1, LEDs are the highest volume compound semiconductor commercial devices. Consequently, the manufacture of wafer-bonded LEDs requires techniques commensurate with achieving very high wafer-bonding yields over large areas. Despite the previously described complexities required for compound semiconductor wafer bonding, high-yield wafer bonding of 50-mm diameter GaP substrates to AlGaInP-GaP epitaxial films has been demonstrated (Hofler et al., 1996). This feasibility is demonstrated by the transmitted light photograph of Fig. 25 (see color plate section) for a TS GaP-$(\text{Al}_x\text{Ga}_{1-x})_{0.5}\text{In}_{0.5}\text{P}$/GaP bonded LED wafer ($x \sim 0.3$, $\lambda \sim 590$ nm). The uniform red color indicates that no macroscopic bonding imperfections are present (such defects are easily observed due to changes in refractive index at a void). The red color results from absorption transmitted light above the bandgap of the ~ 0.75-μm-thick active region. Optical transmission measurements show no detectable absorption at the wafer-bonded interface across such wafers. Furthermore, full-wafer electrical mapping of mesa-defined LEDs indicate a device yield of $\sim 97\%$ on such wafers, with the only imperfections arising from wafer fabrication problems (e.g., missing metallization, and handling induced scratches). Such results indicate that compound semiconductor wafer bonding is a viable technique for the high-volume manufacturing of LEDs and other optoelectronic devices.

c. *Device Characteristics*

The benefits of improved light extraction that can be achieved with wafer bonding are shown in the photomicrographs of Fig. 26 (see color plate section) of (a) absorbing substrate (GaAs) and (b) wafer-bonded transparent substrate (GaP) $(\text{Al}_x\text{Ga}_{1-x})_{0.5}\text{In}_{0.5}\text{P}$ LED chips (Fig. 26(b)) (Kish et al., 1994a). In the AS LED (a), light escapes only from the thick GaP window and active layer. However, light appears to radiate relatively evenly from all surfaces of TS LED (b), indicative of uniform current flow across the wafer-bonded interface and throughout the LED chip. As discussed in Part V, Section 2, the TS LED possesses approximately two times the light extraction efficiency of an AS LED with a thick transparent window. Figure 27 shows the light-output versus current (L-I) curves for an (a) AS LED lamp (with a thick GaP upper window) and a (b) wafer-bonded TS LED lamp fabricated from the same $(Al_xGa_{1-x})_{0.5}In_{0.5}P$ LED wafer. The DC L-I curves show that the TS device is two times as efficient as the AS device, indicating that the wafer-bonded interface is optically transparent to the light emitted from the active layer (Kish et al., 1994a). The wafer-bonded TS LED lamp operating at $\lambda \sim 604$ nm emits 1.9 lm (4.7 mW) at 20 mA and 4.0 lm (10.2 mW) at 50 mA. Quasi-DC (10 ms pulse) characteristics are also shown for the TS LED lamp, which emits a maximum flux of 11.5 lm (29.2 mW) at 100 mA. Furthermore, the inset of Fig. 27 shows a typical electroluminescence (EL) spectrum for the wafer-bonded TS LED lamps, possessing a FWHM of 16.6 nm (56.6 meV), similar to that of equivalent AS LEDs.

The compound semiconductor wafer-bonding technique is fully capable of satisfying all requirements necessary for the high-volume manufacture and practical implementation of LEDs. As discussed in Part V, Section 4.b, low-resistance conduction across the bonded interface can be readily achieved, thus facilitating low operating voltages and high power conversion efficiencies. Furthermore, the high-quality of the wafer-bonded interface (e.g., absence of threading dislocations) results in excellent device operating life reliability characteristics. Thus, the reliability of wafer-bonded TS LEDs is essentially equivalent to that of their counterpart AS LEDs (Part VII, Section 5). In addition, the wafer-bonded interface is mechanically robust,

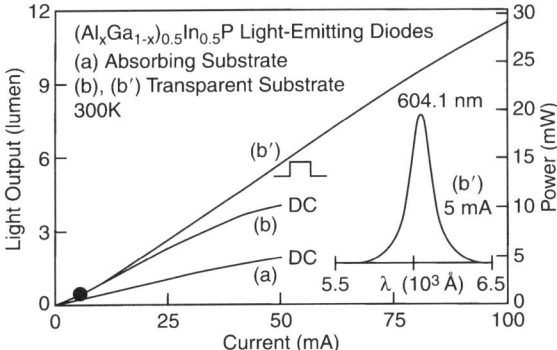

FIG. 27. Room temperature (300 K) light output versus current (L-I) curves for representative $\lambda \sim 604$ nm (a) absorbing substrate (AS) and (b) wafer-bonded transparent-substrate $(Al_xGa_{1-x})_{0.5}In_{0.5}P$ LED lamps fabricated from the same wafer. Under DC operation, the TS device exhibits a twofold improvement in light-output efficiency compared with the AS device. Quasi DC (10 ms pulse) operation of the TS AlGaInP LED lamp (c) results in emission of 11.5 lm (29.2 mW) at 100 mA. The electroluminescence spectrum of the TS AlGaInP/GaP LED (inset) indicates that the wavelength peaks at $\lambda \sim 604$ nm, with a typical full width half maximum of 56.6 meV (16.6 nm). (After Kish et al., 1994a, Appl. Phys. Lett. **64**, 2839–2841; courtesy of the American Institute of Physics.)

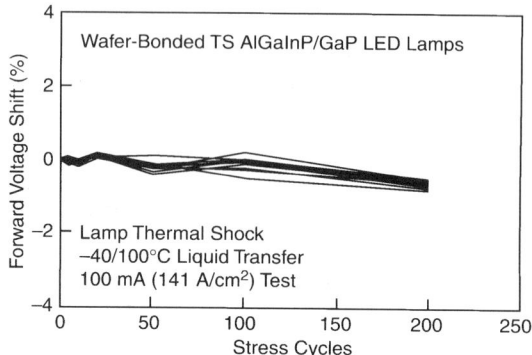

FIG. 28. Forward voltage shift (%) at 100 mA (141 A/cm^2) versus the number of stress cycles for wafer-bonded TS AlGaInP/GaP LED lamps subjected to thermal shock testing (-40 to 100°C, liquid–liquid transfer). The absence of an increase in forward voltage is indicative of the structural stability and integrity of the wafer-bonded interface. (Reprinted from Kish *et al.*, 1996a, *Electron Lett.* **32**, 132–134; with the permission of the IEEE.)

satisfying the expectation of LEDs to endure a myriad of harsh environmental operating conditions. For example, Fig. 28 shows the forward voltage shift at 100 mA as a function of the number of thermal shock stress cycles (-40 to 100°C) for wafer-bonded TS $(Al_xGa_{1-x})_{0.5}In_{0.5}P$ LED lamps (Kish *et al.*, 1996a). No significant increase occurs in the operating voltage of the devices indicating the high structural integrity of the wafer-bonded interface. Delamination of the wafer-bonded substrate from the LED device layers would be expected to be manifested by significant increases in the operating voltage of these devices. These properties indicate that wafer-bonding is a viable technique for the production of TS LEDs. In 1994, the Hewlett-Packard Company introduced TS $(Al_xGa_{1-x})_{0.5}In_{0.5}P$/GaP LEDs to the commercial marketplace, the highest performance AlGaInP LEDs currently available for sale. Furthermore, the TS devices have yielded the highest efficiencies reported to date for LEDs in the yellow-green to red spectral regime (Part VII, Sections 1 and 2), exhibiting an external quantum efficiency of 23.7% (636 nm) and a luminous efficiency of 50.3 lm/W (607.4 nm) (Kish *et al.*, 1996a).

VI. Wafer Fabrication Techniques

In comparison to other types of semiconductor devices, the fabrication of LEDs is a relatively simple process. Typically, there are only one or two levels of masking needed, and high-resolution photolithography is not required. Figure 29 illustrates a typical wafer fabrication and die fabrication process flow. Of course, this is only an example, and there can be other processing steps involved and numerous variations on the sequence of the steps.

In this case, the first fabrication step after the completion of the epitaxial growth is the formation of the top ohmic contacts. Metal layers for this contact are sheet-deposited on the top surface of the wafer, either by evaporation or sputtering. For p-type contacts, various compositions of gold and zinc are commonly used. Patterning of the top contact geometry is then accomplished by photolithography and etching of the metal layers. The shape of the top contact is circular or a pattern with finger projections. Possible examples have been discussed previously and are illustrated in Fig 6. A brief high-temperature alloying step is used to generate the actual ohmic contact at the metal–semiconductor interface. Typical alloy conditions are on the order of 400 to 500°C for a duration of 5 to 30 minutes, depending on the exact metal and window layer materials used. If AlGaAs is used for the window, there may also be a thin layer of GaAs on top of the AlGaAs to facilitate ohmic contact formation. If this is the case, an

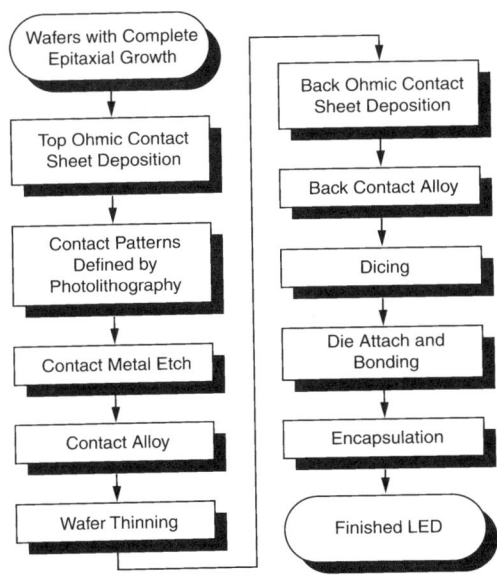

FIG. 29. Flow diagram for a typical AlGaInP LED fabrication process.

additional etching step is required to remove the GaAs outside of the metal contact area, as even a thin GaAs layer can absorb a significant amount of the visible light emitted by the device. Selective etches composed of hydrogen peroxide and either ammonium hydroxide or sulfuric acid can be used to etch the GaAs without affecting the rest of the epitaxial structure.

The next step in the process is to thin the substrate to the final die thickness. This step can be accomplished by standard mechanical lapping techniques. A final thickness of 10 to 12 mils is typically desired. A chemical etching step often follows the lapping to remove any microscopic damage induced in the substrate by the mechanical lapping process. Various formulations of etches consisting of hydrogen peroxide and either sulfuric acid or ammonium hydroxide have been developed for GaAs, and aqua regia is used for GaP. The last step in the wafer fabrication process is to form the back ohmic contact to the n-type substrate. This contact is commonly a sheet-evaporated layer of gold-germanium. Alloying of this contact is necessary and is performed at a temperature around 400 to 500°C for 5 to 30 minutes.

The die fabrication process consists of singulating the devices on the wafer into individual chips. Either scribe and break techniques or sawing can be used. However, because the die are small, anywhere from 8 to 14 mils square, scribe and break techniques often suffer from yield loss due to excessive edge chipping and incomplete singulation during the break process. Consequently, sawing is generally preferred. As in the silicon integrated circuit industry, high-speed dicing saws using diamond-impregnated dicing wheels are used for this process. A single saw may be able to process as many as 50 thousand die per hour, depending on the die size.

The completion of an LED device is accomplished by mounting the chip in a suitable lead frame or circuit board package. A silver-impregnated conductive die attach epoxy is used to cement the back contact of the chip to the lead frame or circuit board. Thermosonic bonding is then used to attach a wire to the top contact pad. Finally, the entire device is encapsulated in a transparent epoxy and is thoroughly tested. The final device might be an individual lamp ("thru-hole" or surface mount) or a multichip alphanumeric array package.

VII. Device Performance Characteristics

1. Quantum Efficiency

An important measure of LED performance is external quantum efficiency (η_{ext}). Figure 30 shows the best-reported values for $(Al_xGa_{1-x})_{0.5}$-$In_{0.5}P$ LED external quantum efficiency (300 K, DC, 20 mA)

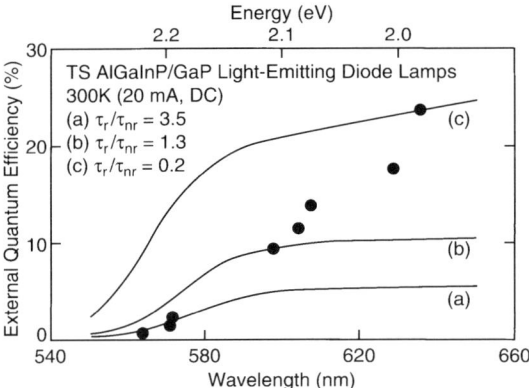

FIG. 30. External quantum efficiency versus peak electroluminescent wavelength for the best-reported transparent-substrate $(Al_xGa_{1-x})_{0.5}In_{0.5}P/GaP$ LED lamps reported to date under DC operation (300 K) at 20 mA (44 A/cm^2) (Kish et al., 1996a). The curves show theoretical fits to the data obtained by applying Boltzmann statistics for the $\Gamma - X$ transition for constant radiative-to-nonradiative lifetime ratios of (a) 3.5, (b) 1.3, and (c) 0.2.

as a function of peak EL wavelength (Kish et al., 1996a). All data points are for wafer-bonded TS $(Al_xGa_{1-x})_{0.5}In_{0.5}P/GaP$ LEDs consisting of an $\sim 1\,\mu m$ $(Al_xGa_{1-x})_{0.5}In_{0.5}P$ DH active layer with $Al_{0.5}In_{0.5}P$ confining layers. The TS LEDs exhibit the highest efficiencies as a result of their enhanced light extraction properties. The best-reported external quantum efficiencies of AS $(Al_xGa_{1-x})_{0.51}In_{0.5}P/GaAs$ LEDs with either a "thick" transparent GaP upper window (Huang et al., 1992) or a "thin" AlGaAs upper window with DBR below the active layer (Itaya et al., 1994) are approximately a factor of two lower than those shown in Fig. 30. A maximum η_{ext} of 23.7% is observed for devices emitting at 635.6 nm in the red portion of the visible spectrum. This performance significantly exceeds that of the best-reported 650 nm red TS AlGaAs LEDs (18% η_{ext}) (Cook et al., 1988). The external quantum efficiency decreases by more than 2.5 times as the wavelength is decreased to 597.7 nm (yellow-orange spectral regime) wherein the best devices exhibit $\eta_{ext} \sim 9.2\%$. The efficiency decreases even more rapidly in the yellow-green spectral regime, wherein $\eta_{ext} \sim 2.2\%$ for the best devices emitting at 571.4 nm.

Theoretical fits to the experimental data also are shown in Fig. 30. These curves are derived from calculating the relative population of carriers in the direct (Γ) and indirect (X) minima, as given by Boltzmann statistics. Accordingly, the external quantum efficiency is given by the product of the light extraction efficiency and internal quantum efficiency:

$$\eta_{ext} = C_{ex} \left\{ \frac{1}{1 + \left(\frac{\tau_r}{\tau_{nr}}\right)\left[1 + \left(\frac{m_{eX}}{m_{e\Gamma}}\right)^{3/2} \exp((E_\Gamma - E_X)/kT)\right]} \right\} \quad (24)$$

where C_{ex} is the light extraction efficiency, (τ_r/τ_{nr}) is the minority carrier radiative nonradiative lifetime ratio, m_{eX} and $m_{e\Gamma}$ are the electron DOS effective masses, and E_Γ and E_X are the bandgaps of the Γ and X minima, respectively. Equation (24) neglects the population of the L indirect valley as well as any radiative recombination from the X indirect valley. The latter is a valid assumption because the radiative lifetimes are generally orders of magnitude larger in the indirect valleys than those of the direct valley (Kaliski et al., 1985). In addition, Eq. (24) assumes that the nonradiative lifetimes of all bands (Γ and X) are equal ($\tau_{nr} = \tau_{nr\Gamma} = \tau_{nrX}$). The calculations assume a constant minority carrier lifetime ratio and use the band structure parameters given by Eqs. (1) to (4). Fits are shown in Fig. 30 for radiative nonradiative lifetimes of (a) 3.5, (b) 1.3, and (c) 0.2. These lifetime ratios correspond to internal quantum efficiencies of (a) 22%, (b) 43%, and (c) 82% for purely direct bandgap (650 nm) emission. As discussed in Part V, Section 2, multiple pass light-extraction in a TS LED can result in extraction efficiencies as high as 30%. However, parasitic absorption (e.g., by an LED active layer with low internal quantum efficiency), reduces the probability that multiple pass light can escape from the chip. Consequently, the high internal quantum efficiency curve (a) of Fig. 30, assumes that the light extraction efficiency varies linearly from 30 to 20.5% from 650 to 550 nm (6 to 5 cones of light extracted from the chip, as depicted in Fig. 14). The remaining lower internal quantum efficiency curves (b) and (c) assume a linear light extraction efficiency variation of 22.6 to 20.5% from 650 to 550 nm (5.5 to 5 cones).

The theoretical curves and experimental data shown in Fig. 30 indicate that $(Al_xGa_{1-x})_{0.5}In_{0.5}P$ LED efficiency decreases anomalously with decreasing wavelength (increasing Al content, x), faster than that predicted by the overflow of electrons from the Γ to X bands (Naritsuka et al., 1991; Sugawara et al., 1992a; Kish et al., 1996a). Calculations of the minority carrier injection efficiency in Part II, Section 2.a (Table I) and carrier confinement in the DH in Part II, Section 2.c (Table II) indicate that these mechanisms do not degrade $(Al_xGa_{1-x})_{0.5}In_{0.5}P$ LED efficiency for emission wavelengths over 590 nm. Furthermore, the calculations in Tables I and II indicate that the LED efficiency is decreased by no more than 40% in the 590 to 570 nm spectral regime as result of these phenomena. As a result, the anomalous decrease in external quantum efficiency with decreasing wavelength may be attributed to an increased concentration in nonradiative centers in the higher Al-content active layers. Deep traps have been identified in the $(Al_xGa_{1-x})_{0.5}In_{0.5}P$ alloy system whose density increases with increasing Al composition (Watanabe and Ohba, 1986; Suzuki et al., 1991, 1993; Naritsuka et al., 1991; Sugiura et al., 1991; Kondo et al., 1994). Such traps have been correlated with the presence of oxygen (Suzuki et al., 1993; Kondo et al., 1994) and donor impurities (Watanabe and Ohba, 1986; Suzuki et al., 1991, 1993). Specifically, the presence of increased oxygen in $(Al_xGa_{1-x})_{0.5}In_{0.5}P$ LEDs has been correlated directly with decreased LED external quantum efficiency (Suzuki et al., 1993). Another potential cause of this anomalous decrease may be the population of the L minima. As discussed in Part II, Section 1a, little data is available on the bandgap dependence of the L minima, and this effect has been neglected in the calculated curves in Fig. 30. Thus, a variety of phenomena may be responsible for the dramatic decrease in $(Al_xGa_{1-x})_{0.5}In_{0.5}P$ LED quantum efficiency with decreasing wavelength.

2. LUMINOUS EFFICIENCY

Luminous efficiency is the luminous flux emitted by the device per input power (lm/W). This parameter is an important measure of performance for visible LEDs and other light sources. Figure 31 shows a plot of the best-reported luminous efficiencies (300 K, DC, 20 mA) versus peak EL wavelength for TS $(Al_xGa_{1-x})_{0.5}In_{0.5}P$/GaP LEDs (solid line and circles) and AS $(Al_xGa_{1-x})_{0.5}In_{0.5}P$/GaAs LEDs (dashed line). Note that the curve for the AS AlGaInP LEDs is representative of both the best thick GaP window devices (Huang et al., 1992) and the best thin AlGaAs window devices incorporating a DBR beneath the active layer (Itaya et al., 1994). Also shown are the average performance for conventional commercial LEDs (Craford and Steranka, 1994) and the best-measured single quantum well InGaN–Al_2O_3 LEDs emitting in the blue and deep-green portions of the visible spectrum obtained commercially from Nichia Chemical Company (Nakamura et al., 1995). In the yellow-green to red spectral regime (570 to 640 nm), the performance of the TS AlGaInP/GaP LEDs exceeds that of all other technologies, being twice as efficient as the best-reported AS AlGaInP/GaAs LEDs. The TS AlGaInP/GaP LEDs exhibit a maximum luminous efficiency of 50.3 lm/W at 607.4 nm. Thus, AlGaInP LEDs (AS-TS) are approximately 25 to 50 times as efficient as conventional GaAsP:N/GaP devices that emit in the yellow and orange portion of the visible spectrum. The efficiency of the TS AlGaInP/GaP devices decreases in the

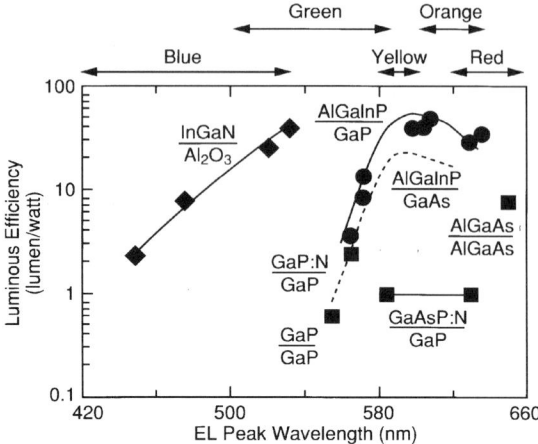

FIG. 31. Luminous efficiency versus electroluminescent (EL) peak wavelength for the best transparent-substrate $(Al_xGa_{1-x})_{0.5}In_{0.5}P/GaP$ (solid line and circles) and best absorbing-substrate $(Al_xGa_{1-x})_{0.5}In_{0.5}P/GaAs$ (dashed line) LED lamps operated at 20 mA, DC at 300 K. The TS AlGaInP LED lamps exhibit a twofold improvement in efficiency compared with AS AlGaInP LED lamps. The luminous efficiency of the AlGaInP emitters is superior to all existing commercial LED technologies in the 555-to-650-nm spectral regime. At $\lambda \sim 607$ nm, a peak performance of 50.3 lm/W is achieved by TS AlGaInP LEDs, exceeding that of all reported LED devices.

yellow-green regime, with the best LEDs exhibiting a luminous efficiency of 13.6 lm/W at 571.4 nm. This performance is greater than 5 times that of conventional GaP:N/GaP LEDs emitting at the same wavelength in the yellow-green spectral regime. In the red portion of the visible spectrum, the best TS AlGaInP/GaP devices exhibit an efficiency of 35.5 lm/W at 635.6 nm. Consequently, the performance of AS and TS AlGaInP devices exceeds that of longer wavelength conventional TS AlGaAs/AlGaAs 650 nm red-spectrum LEDs.

The efficiency of AlGaInP LEDs shown in Fig. 31 is also competitive with many conventional lighting sources. In the yellow-green to red spectral regime (570 to 640 nm), the performance of the best TS AlGaInP LEDs exceeds that of an unfiltered 60 W incandescent bulb (15 lm/W). Furthermore, in the yellow through red spectral regime, the performance surpasses that of an unfiltered 30 W halogen source (25 lm/W) and begins to approach that of 400 W mercury vapor sources (60 lm/W), 70 W metal halide sources (70 lm/W), and 50 W high-pressure sodium lamps (80 lm/W). Consequently, the performance of AlGaInP LEDs rivals that of a variety of conventional lighting sources for low-flux (power) applications. Currently, the average performance of commercial AS and TS AlGaInP LEDs is roughly half that of the curves shown in Fig. 31. This performance is expected to improve with continued developments in material growth techniques.

3. CURRENT–VOLTAGE CHARACTERISTICS

The current-voltage (I-V) characteristics of typical AS and TS $(Al_xGa_{1-x})_{0.5}In_{0.5}P$ LEDs ($\lambda \sim 590$ nm) with transparent thick GaP upper window layers are shown in Figs. 32 and 33, respectively. Both types of devices exhibit normal diode turn-on forward characteristics (~ 1.7 to 1.8 V) and avalanche reverse breakdown in the range of ~ -15 to -30 V. The resistance under forward-biased operation for the AS device is ~ 3 to $6\,\Omega$, resulting in a typical forward voltage of ~ 2.0 to 2.1 V at 20 mA (Fig. 32). The forward resistance of the TS device is somewhat higher, 10 to 14 Ω, resulting in ~ 0.1 V higher operating voltages than equivalent AS devices at 20 mA. The increased forward resistance in the TS device is primarily due to the higher resistivity GaP substrate and patterned back ohmic contacts that are present in TS devices. Resistance associated with the wafer-bonded TS interface does not contribute significantly to this resistance, provided the

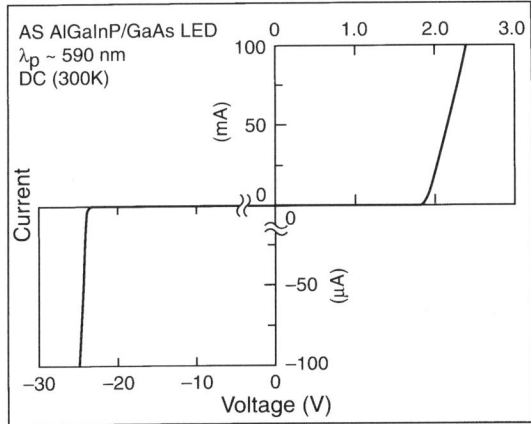

FIG. 32. Current-voltage (I-V) characteristics of a typical absorbing-substrate GaP-AlGaInP/GaAs light-emitting diode ($\lambda \sim 590$ nm) under 300 K DC operation. The LED exhibits a forward voltage of approximately 2.0 to 2.1 V at 20 mA. Standard avalanche reverse breakdown occurs in the range of approximately -15 to -30 V for these devices.

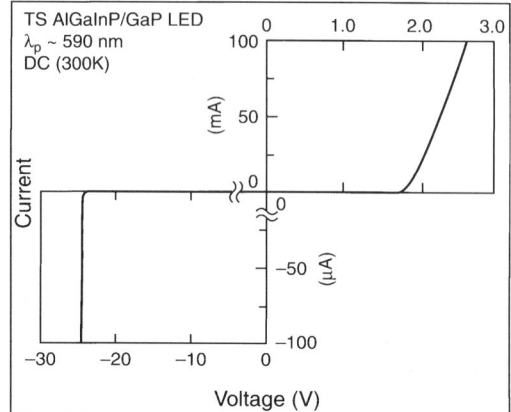

FIG. 33. Current-voltage (I-V) characteristics of a typical transparent-substrate GaP-AlGaInP/GaP light-emitting diode ($\lambda \sim 590$ nm) under 300 K DC operation. The LED exhibits a forward voltage of approximately 2.1 to 2.2 V at 20 mA. Standard avalanche reverse breakdown occurs in the range of approximately -15 to -30 V for these devices.

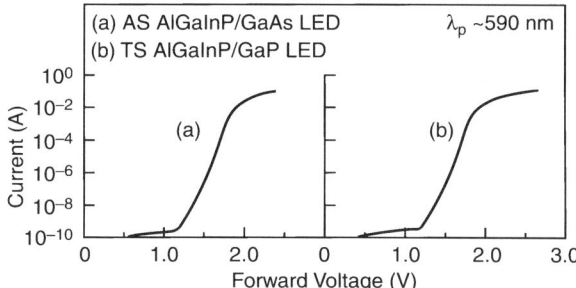

FIG. 34. Semi-logarithmic current-voltage (Log I versus V) plot of the forward DC characteristics of typical (a) absorbing-substrate GaP-AlGaInP/GaAs LEDs and (b) transparent-substrate GaP-AlGaInP/GaP LEDs ($\lambda \sim 590$ nm). Transition from the recombination dominated injection regime ($n = 2$ diode ideality factor) to the radiative diffusion injection regime ($n = 1$ diode ideality factor) typically occurs at approximately 100 μA (0.2 A/cm^2).

wafer-bonding is performed properly (Part V, Section 4.b). Figure 34 shows semilogarithmetic plots of the I-V characteristics for typical (a) AS and (b) TS devices with thick GaP upper window layers. Both devices exhibit similar low-current behavior, wherein the transition from recombination dominated injection ($n = 2$ diode ideality factor) to radiative diffusion injection ($n = 1$ diode ideality factor) occurs at roughly 100 μA (0.2 A/cm^2).

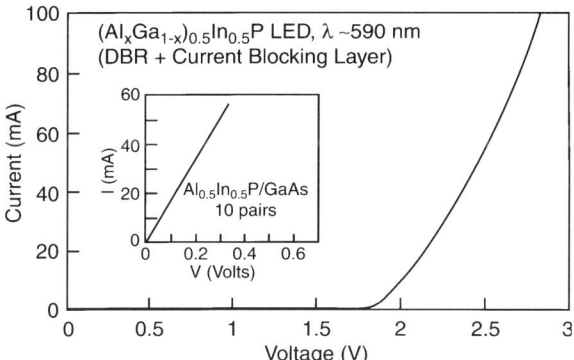

FIG. 35. Forward current-voltage (I-V) DC characteristics of a typical absorbing-substrate AlGaAs-$(Al_xGa_{1-x})_{0.5}In_{0.5}P$/GaAs light-emitting diode ($\lambda \sim 590$ nm) incorporating a distributed Bragg reflector (DBR) and current-blocking layer. The forward voltage of these devices are typically approximately 2.1 to 2.2 V at 20 mA. The I-V characteristics of a chip consisting solely of the AlInP/GaAs DBR are shown in the inset (Reprinted from Sugawara et al., 1993, J. Appl. Phys. **74**, 3189–3193; with the permission of the American Institute of Physics.)

The forward I-V characteristics of a typical AS $(Al_xGa_{1-x})_{0.5}In_{0.5}P$ LED ($\lambda \sim 590$ nm) with a thin ($\sim 4\,\mu$m) AlGaAs window, a current-blocking layer, and an AlInP/GaAs DBR are shown in Fig. 35. The device exhibits a forward resistance of 10 to 15 Ω after diode turn-on, resulting in a forward voltage of ~ 2.1 to 2.2 V at 20 mA. The contribution of the DBR layers to the forward resistance can be estimated from the I-V characteristics of an n-type AlInP/GaAs DBR grown on a GaAs substrate that was subsequently contacted with full-sheet ohmic metallization and diced into LED size chips (14×14 mil^2). The I-V characteristics of the DBR (Fig. 35, inset) indicate that the DBR contributes less than 5 Ω to the forward resistance of the DBR LED (Sugawara et al., 1993). The majority of the additional resistance most likely arises from current-crowding effects in the thin AlGaAs window layer.

4. ELECTROLUMINESCENCE SPECTRA

The EL spectrum of an AlGaInP LED is characteristic of a direct bandgap semiconductor with symmetrical line shape and a halfwidth on the order of 50 meV, or about 15 nm. Figure 36 illustrates a series of spectra taken from $(Al_xGa_{1-x})_{0.5}In_{0.5}P$ LEDs with the Al composition x adjusted in the active layer to give emission peaks ranging from 554 to 622 nm (Fletcher et al., 1991a). The spectral halfwidth is largely determined by the doping level in the active layer of the device. The spectra shown here become

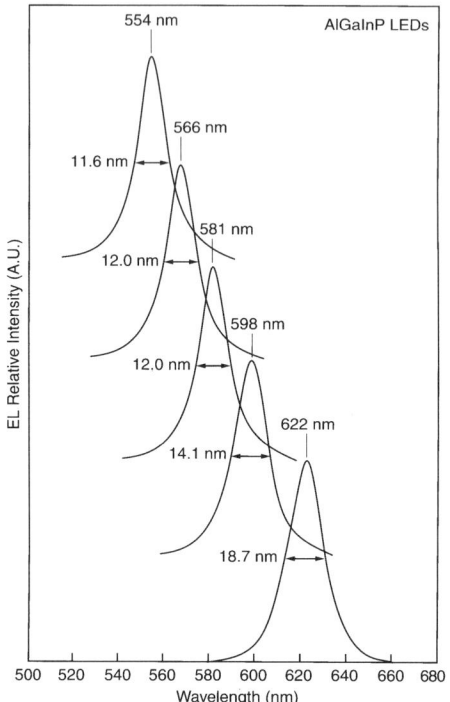

FIG. 36. A series of spectra taken from five $(Al_xGa_{1-x})_{0.5}In_{0.5}P$ light-emitting diodes with different aluminum (Al) compositions x in the active layer (Fletcher et al., 1991a). Peak intensities are normalized for each spectrum. Spectral halfwidths decrease with decreasing wavelength because of lower doping levels in the active layers of the higher Al-content devices. (Reprinted from Fletcher et al., 1991a, J. Electron. Mater. **20**, 1125, with the permission of the Minerals, Metals and Materials Society.)

narrower with decreasing wavelength as a result of reduced doping in the active layers as more Al is added.

Also characteristic of direct bandgap materials is a fairly well-defined change in emission wavelength with temperature, which is consistent with a change in the bandgap of the semiconductor with temperature. Temperature coefficients for emission peak wavelength in AlGaInP LEDs have been measured and found to be 0.078 nm/°C for peak emission around 590 nm and 0.096 nm/°C for peak emission around 620 nm.

Drive current can also influence color shift, mostly due to resistive heating in the device. In applications in which color control is important, limiting the drive current and proper heat sinking of the device may be necessary. Ambient temperature may also have to be considered, especially in outdoor applications where the operating temperature may approach 85°C.

5. Reliability

Light-emitting diodes based on the $(Al_xGa_{1-x})_{0.5}In_{0.5}P$ alloy typically exhibit excellent reliability characteristics. This reliability is a definite advantage for many applications wherein AlGaInP solid-state lamps are currently competing with incandescent sources. Such differences are a consequence of the fact that the light-generating electron-hole recombination event within the LED is not inherently a destructive process (in comparison to a high-temperature incandescent filament that tends to evaporate over time). Figure 37 shows the long-term high-temperature reliability behavior of eight wafer-bonded TS $(Al_xGa_{1-x})_{0.5}In_{0.5}P$/GaP LED lamps operating at 50 mA (70 A/cm^2) at an ambient temperature of 55°C. After 17,000 hours of stress, the devices exhibit an average degradation of 14% (Kish et al., 1996a). Extrapolating the data for times greater than 1000 hours (dashed line) results in an average projected lifetime of over 120,000 hours for degradation to 70% of the initial light output. Similar behavior is also observed for AS $(Al_xGa_{1-x})_{0.5}In_{0.5}P$ LEDs with thick lattice-mismatched GaP windows, indicating that the wafer-bonding technique or mismatched window does not adversely affect the reliability characteristics. These LED operating lives are substantially greater than those of traditional incandescent sources, which typically exhibit lifetimes in the range of 1000 to 10,0000 hours. Shorter term (1500 h),

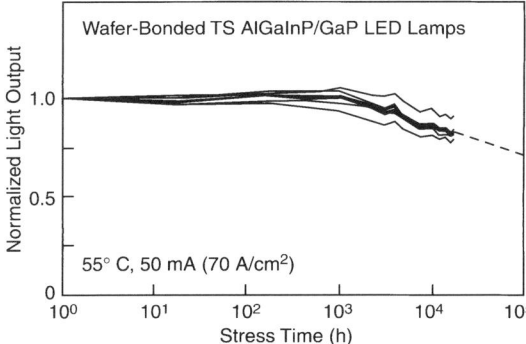

Fig. 37. Long-term, high-temperature (55°C ambient) reliability behavior of wafer-bonded TS AlGaInP/GaP LED lamps stressed at 50 mA (70 A/cm^2). After 17,000 hours of stress, the LEDs exhibit an average light-output degradation of 14%. Extrapolation of the degradation behavior for times over 1000 h (dashed line) results in a projected lifetime of over 120,000 hours for degradation to 70% of the initial light output of the devices. (Reprinted from Kish et al., 1996a, Electron. Lett., **32**, 132–134: with the permission of the IEEE.)

higher temperature (80°C) data for AS $(Al_xGa_{1-x})_{0.5}In_{0.5}P$ LEDs with thin AlGaAs windows also indicate excellent reliability behavior (Sugawara et al., 1992a). These data indicate the feasibility of utilizing $(Al_xGa_{1-x})_{0.5}In_{0.5}P$ LEDs for applications requiring extremely high-reliability operation.

The reliability of $(Al_xGa_{1-x})_{0.5}In_{0.5}P$ LEDs is very sensitive to the operational current density of the device, becoming progressively worse with increasing current density. For the present generation of devices, the current densities should be kept to 150 A/cm^2 or less to ensure high-reliability operation. Furthermore, in the -40 to 55°C temperature range, the reliability behavior *is not* a strong function of temperature, making it difficult to determine an accurate activation energy for device degradation. These current-density and temperature trends are in agreement with those observed for $(Al_xGa_{1-x})_{0.5}In_{0.5}P$ laser diodes (Ishikawa et al., 1989; Nitta et al., 1992).

High-reliability operation under 85°C, 85% relative humidity (WHTOL) conditions is characteristic of $(Al_xGa_{1-x})_{0.5}In_{0.5}P$ LEDs with GaP window layers (AS and TS). Figure 38 shows the WHTOL reliability characteristics of (a) TS AlGaInP LEDs wherein essentially no degradation occurs after 2000 hours of stress time. This behavior is in marked contrast with the behavior of TS AlGaAs LEDs in (b), which exhibit significant degradation under WHTOL stress due to hydrolization degradation of the high Al-

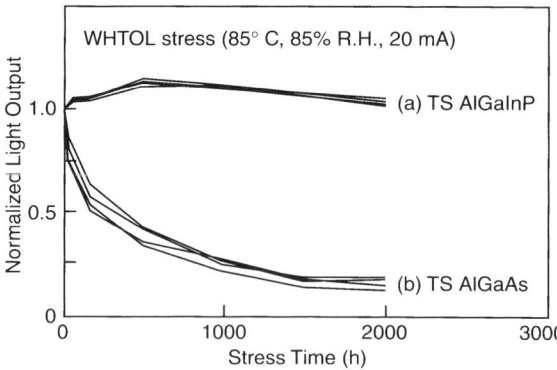

FIG. 38. Normalized light output versus stress time for (a) TS GaP-AlGaInP/GaP LED lamps and (b) TS AlGaAs/AlGaAs LED lamps operated at 85°C, 85% relative humidity (WHTOL) under 20 mA operation (Kish et al., 1996a). Unlike AlGaAs-based devices, no significant degradation occurs for GaP-AlGaInP-based LEDs. The degradation of the TS AlGaAs LEDs is attributed to hydrolization degradation of the high Al composition AlGaAs layers within the device.

composition AlGaAs layers within the device (Dallesasse et al., 1990a,b). Similar behavior is also observed for $(Al_xGa_{1-x})_{0.5}In_{0.5}P$ LEDs with high-composition AlGaAs window layers unless the AlGaAs is protected by a stable native oxide (Dallesasse et al., 1990c; Richard et al., 1995) or a passivating AlGaInP layer. Thus, high-reliability operation of $(Al_xGa_{1-x})_{0.5}In_{0.5}P$ LEDs can be achieved in very harsh environments.

VIII. Conclusions and Outlook

The development of $(Al_xGa_{1-x})_{0.5}In_{0.5}P$ emitters has resulted in a revolution in visible LED performance, with yellow and orange visible spectrum devices exhibiting an ~ 50 times improvement in efficiency compared with conventional LEDs. Likewise, the efficiency of red-spectrum $(Al_xGa_{1-x})_{0.5}In_{0.5}P$ LEDs now exceeds that of conventional high-brightness transparent-substrate AlGaAs LEDs by two to four times. Furthermore, the luminous performance of the best $(Al_xGa_{1-x})_{0.5}In_{0.5}P$ LEDs in the yellow-green to red spectral regime (570 to 640 nm) exceeds that of unfiltered incandescent lamps (15 lm/W). Luminous efficiencies as high as 40 to 50 lm/W have been realized in the yellow and orange spectral regimes. Such levels of performance have led to the emergence of new markets for LEDs, wherein LEDs have begun to replace incandescent solutions in a myriad of applications. These markets will continue to expand as the efficiency of this new LED technology continues to improve. In the future, significant demand will exist for both the lower performance, lower cost absorbing-substrate (AS) devices as well as the higher performance, higher cost transparent-substrate (TS) devices. Such demand will depend primarily on the requirements of the application (e.g., energy efficiency and brightness, fill factor of light, and sensitivity of the system price to the number of LED parts).

The development of $(Al_xGa_{1-x})_{0.5}In_{0.5}P$ LEDs has been enabled by OMVPE epitaxial growth techniques that facilitate the growth of high-quality, lattice-matched material. Furthermore, commercialization of $(Al_xGa_{1-x})_{0.5}In_{0.5}P$ LEDs has required advances in OMVPE growth techniques and large-chamber crystal growth reactors in order to allow cost-effective high-volume manufacturing of these devices. The enhanced control afforded by OMVPE growth also has resulted in the implementation of a number of sophisticated structures aimed at optimizing device performance, including double heterostructure (DH) active regions, current-blocking layers, current-spreading layers, and distributed Bragg reflectors (DBRs). The low growth rates of the OMVPE process make it inherently a thin film technique. Consequently, other technologies commensurate with the cost-effective realization of thick (greater than 50 μm) transparent windows for enhanced current-spreading and light extraction have been combined with OMVPE $(Al_xGa_{1-x})_{0.5}In_{0.5}P$ LED structures to maximize device performance. As a result, both vapor-phase epitaxy (VPE) and newly invented techniques of compound semiconductor wafer bonding have been employed to realize the most efficient devices. Thus, $(Al_xGa_{1-x})_{0.5}In_{0.5}P$ LEDs are amongst the most efficient and sophisticated visible LEDs available on the commercial market.

Visible $(Al_xGa_{1-x})_{0.5}In_{0.5}P$ LEDs have only been commercially available for less than 7 years. Future improvements in materials growth technology and the implementation of more sophisticated device structures should result in continued improvements in LED efficiency. It is very conceivable that such advances may ultimately result in the attainment of $(Al_xGa_{1-x})_{0.5}In_{0.5}P$ commercial devices with efficiencies in the range of 80 to 100 lm/W. Such improvements will lead to new markets that compete with a broader array of lighting sources, exploiting the enhanced efficiency and reliability offered by solid-state lamps. Currently, LEDs are well-suited for low-flux (low-power) applications. Consequently, the best commercial LEDs emit several lumens of visible flux, compared with 60 W incandescent lamps that emit ~ 1000 lm. In order to truly compete with such light sources, high-power LEDs will need to be developed. Monolithic large-area $(Al_xGa_{1-x})_{0.5}In_{0.5}P$ LEDs that emit 84 lm have been demonstrated in the laboratory (Kish et al., 1994b). However, considerable improvements in device efficiency and operating characteristics are required to commercialize high-power LED devices. Thus, many significant advances in $(Al_xGa_{1-x})_{0.5}In_{0.5}P$ LEDs can be expected in the future that will enable new applications for solid-state lamps.

References

Adachi, S. (1985). *J. Appl. Phys.* **58**, R1–R29.
Adachi, S. (1990). In "Properties of Gallium Arsenide," EMIS Datareview Ser., Vol. 2, pp. 513–528. INSPEC, Institute of Electrical Engineers, New York.
Adachi, S., and Oe, K. (1983). *J. Electrochem. Soc.* **130**, 2427.
Adachi, S., Kato, H., Moki, A., and Ohtsuka, K. (1994). *J. Appl. Phys.* **78**, 478–480.
Alferov, Z. I., Andreev, V. M., Korol'kov, V. I., Tretyakov, D. N., and Tuchkevich, V. M. (1967). *Sov. Phys. Semicond.* (*Engl. Transl.*) **1**, 1313–1316.
Alferov, Z. I., Andreev, V. M., Korol'kov, V. I., Tret'yakov, D. N., and Tuchkevick, V. M. (1968). *Sov. Phys. Semicond.* (*Engl. Transl.*) **1**, 1313–1314.
Alferov, Z. I., Andreev, V. M., Korol'kov, V. I., Portnoi, E. L., and Tret'yakov, D. N. (1969a). *Sov. Phys. Semicond.* (*Engl. Transl.*) **2**, 843–844.

Alferov, Z. I., Andreev, V. M., Korol'kov, V. I., Portnoi, E. L., and Yakovenko, A. A. (1969b). *Sov. Phys. Semicond. (Engl. Transl.)* **3**, 785–787.
Aliyu, Y. H., Morgan, D. V., Thomas, H., and Bland, S. W. (1995). *Electron. Lett.* **31**, 2210–2212.
Anthony, P. J., and Schumaker, N. E. (1980). *IEEE Electron. Device Lett.* **EDL-1**, 58–59.
Antypas, G. A., and Edgecumbe, J. (1975). *Appl. Phys. Lett.* **26**, 371–372.
Antypas, G. A., and Kinter, M. L. (1973). U.S. Pat. 3,769,536.
Archer, R. J. (1972). *J. Electron. Mater.* **1**, 1–26.
Arthur, J. R. (1968). *J. Appl. Phys.* **39**, 4032–4034.
Asahi, H., Kawamura, Y., and Nagai, H. (1982). *J. Appl. Phys.* **53**, 4928–4930.
Babic, D. I., Streubal, K., Mirin, R. P., Margalit, N. M., Bowers, J. E., Hu, E. L., Mars, D. E., Yang, L., and Carey, K. (1995). *IEEE Photon. Technol. Lett.* **7**, 1225–1227.
Baldereschi, A., Hess, E., Maschke, K., Neumann, H., Schulze, K.-R., and Unger, K. (1977). *J. Phys. C: Solid State Phys.* **10**, 4709–4717.
Bellon, P., Chevalier, J. P., Martin, G. P., Dupont–Nivet, E., Thiebaut, C., and Andre, J. P. (1988). *Appl. Phys. Lett.* **52**, 567–569.
Bengtsson, S. (1992). *J. Electron. Mater.* **21**, 841–862.
Born, M., and Wolf, E. (1989). "Principles of Optics," 6th ed. Pergamon, New York.
Bour, D. P. (1993). *In* "Quantum Well Lasers" (P. S. Zory, Jr., ed.), pp. 415–460. Academic Press, Boston.
Bour, D. P., and Shealy, J. R. (1988). *IEEE J. Quantum Electron.* **24**, 1856–1863.
Bour, D. P., Shealy, J. R., Wicks, G. W., and Schaff, W. J. (1987). *Appl. Phys. Lett.* **50**, 615–617.
Bour, D. P., Treat, D. W., Thornton, R. L., Geels, R. S., and Welch, D. F. (1993). *IEEE J. Quantum Electron.* **29**, 1337–1343.
Brennan, K. F., and Chiang, P.-K. (1992). *J. Appl. Phys.* **71**, 1055–1057.
Buchan, N., Heuberger, W., Jakubowicz, A., and Roentgen, P. (1992). *In* "Proceedings of the 18th International Symposium on GaAs and Related Compounds" (G. B. Stringfellow, ed.), Vol. 120, pp. 529–534. Inst. Phys., Bristol.
Bugajski, M., Kontkiewicz, A. M., and Mariette, H. (1983). *Phys. Rev. B* **28**, 7105–7114.
Burnham, R. D., Holonyak, N., Jr., Keune, D. L., Scifres, D. R., and Dapkus, P. D. (1970). *Appl. Phys. Lett.* **17**, 430–432.
Burnham, R. D., Holonyak, N., Jr., Keune, D. L., and Scifres, D. R. (1971). *Appl. Phys. Lett.* **18**, 160–162.
Burnham, R. D., Scifres, D. R., and Streifer, W. (1982). U.S. Pat. 4,309,670.
Campbell, J. C., Holonyak, N., Jr., Craford, M. G., and Keune, D. L. (1974). *J. Appl. Phys.* **45**, 4543–4555.
Cao, D. S., Kimball, D. S., and Stringfellow, G. B. (1990). *J. Appl. Phys.* **67**, 739–744.
Casey, H. C., Jr., and Panish, M. B. (1978). "Heterostructure Lasers. Part B: Materials and Operating Characteristics." Academic Press, New York.
Chen, A.-B. and Sher, A. (1981). *Phys. Rev. B* **23**, 5360–5373.
Cho, A. Y. (1971). *J. Vac. Sci. Technol.* **8**, S31-S38.
Cook, L. W., Camras, M. D., Rudaz, S. L. and Steranka, F. M. (1988). *In* "Proceedings of the 14th International Symposium on GaAs and Related Compounds" (A. Christou and H. S. Rupprecht, eds.), Vol. 91, pp. 777–780. Inst. Phys., Bristol.
Craford, M. G. (1992). *IEEE Circuits and Devices* **8**, 24–29.
Craford, M. G., and Steranka, F. M. (1994). In "*Encycl. Appl. Phys.*" (G. Trigy, ed.), pp. 485–514. VCH, New York.
Craford, M. G., Stillman, G. E., Holonyak, N., Jr., and Rossi, J. A. (1991). *J. Electron. Mater.* **20**, 3–12.

Dallesasse, J. M., Gavrilovic, P., Holonyak, N., Jr., Kaliski, R. W., Nam, D. W., and Vesely, E. J. (1990a). *Appl. Phys. Lett.* **56**, 2436–2438.
Dallesasse, J. M., El-Zein, N., Holonyak, N., Jr., Hsieh, K. C., Burnham, R. D., and Dupuis, R. D. (1990b). *J. Appl. Phys.* **68**, 2235–2238.
Dallesasse, J. M., Holonyak, N., Jr., Sugg, A. R., Richard, T. A., and El-Zein, N. (1990c). *Appl. Phys. Lett.* **57**, 2844–2846.
Domen, K., Sugiura, K., Anayama, C., Kondo, M., Sugawara, M., Tanahashi, T., and Nakajima, K. (1991). *J. Cryst. Growth* **115**, 529–532.
Domen, K., Kondo, M., and Tanahashi, T. (1993). In "Proceedings of the 19th International Symposium on GaAs and Related Compounds" (T. Ikegami, F. Hasegawa, and Y. Takeda, eds.), Vol. 129, pp. 447–452. Inst. Phys., Bristol.
Domen, K., Kondo, M., and Tanahashi, T. (1994). *Appl. Phys. Lett.* **64**, 3629–3630.
Dudley, J. J., Ishikawa, M., Babic, D. I., Miller, B. I., Mirin, R., Jiang, W. B., Bowers, J. E., and Hu, E. L. (1992). *Appl. Phys. Lett.* **61**, 3095–3097.
Dudley, J. J., Babic, D. I., Mirin, R., Yang, L., Miller, B. I., Ram, R. J., Reynolds, T., Hu, E. L., and Bowers, J. E. (1994). *Appl. Phys. Lett.* **64**, 1463–1465.
Duke, C. B., and Holonyak, N., Jr. (1973). *Phys. Today* **26**, 23–31.
Dupuis, R. D., and Dapkus, P. D. (1977). *Appl. Phys. Lett.* **31**, 466–468.
Dutta, N. K. (1981). *J. Appl. Phys.* **51**, 70–73.
Ejeckam, F. E., Chua, C. L., Zhu, Z. H., Lo, Y. H., Hong, M., and Bhat, R. (1995). *Appl. Phys. Lett.* **67**, 3936–3938.
Feng, S. L., Krynicki, J., Donchev, V., Bourgoin, J. C., Forte-Poisson, M. D., Brylinski, C., Delage, S., Blanck, H., and Alaya, S. (1993). *Semicond. Sci. Technol.* **8**, 2092–2096.
Fletcher, R. M., Kuo, C. P., Osentowski, T. D., Huang, K. H., Craford, M. G., and Robbins, V. M. (1991a). *J. Electron. Mater.* **20**, 1125–1130.
Fletcher, R. M., Kuo, C. P., Osentowski, T. D., and Robbins, V. M. (1991b). U.S. Pat. 5,008,718.
Fletcher, R. M., Kuo, C. P., Osentowski, T. D., Yu, J. G., and Robbins, V. M. (1993). *Hewlett-Packard Journal* **44**, 6–14.
Fouquet, J. E., Robbins, V. M., Rosner, S. J., and Blum, O. (1990). *Appl. Phys. Lett.* **57**, 1566–1568.
Gomyo, A., Kobayashi, K., Kawata, S., Hino, I., and Suzuki, T. (1986). *J. Cryst. Growth* **77**, 367–373.
Gomyo, A., Suzuki, T., Kobayashi, K., Kawata, S., Hino, I., and Yuasa, T. (1987). *Appl. Phys. Lett.* **50**, 673–675.
Hagen, S. H., Valster, A., Boermans, M. J. B., and van der Heyden, J. (1990). *Appl. Phys. Lett.* **57**, 2291–2293.
Hatakoshi, G., Itaya, K., Ishikawa, M., Okajima, M., and Uematsu, Y. (1991). *IEEE J. Quantum Electron.* **27**, 1476–1482.
Herzog, A. H., Groves, W. O., and Craford, M. G. (1969). *Appl. Phys. Lett.* **40**, 1830–1833.
Hofler, G. E., Vanderwater, D. A., DeFevere, D. C., Kish, F. A., Camras, M. D., Steranka, F. M., and Tan, I.-H. (1996). *Appl. Phys. Lett.* **69**, 803–805.
Holonyak, N., Jr., and Bevaqua, S. F. (1962). *Appl. Phys. Lett.* **1**, 82–83.
Honda, M., Ikeda, M., Mori, Y., Kaneko, K., and Watanabe, N. (1985). *Jpn. J. Appl. Phys.* **24**, L187–L189.
Huang, K. H., Yu, J. G., Kuo, C. P., Fletcher, R. M., Osentowski, T. D., Stinson, L. J., Craford, M. G., and Liao, A. S. H. (1992). *Appl. Phys. Lett.* **61**, 1045–1047.
Ishiguro, H., Sawa, K., Nagao, S., Yamanaka, H., and Koike, S. (1983). *Appl. Phys. Lett.* **43**, 1034–1036.
Ishikawa, M., Okuda, H., Itaya, K., Shiozawa, H., and Uematsu, Y. (1989). *Jpn. J. Appl. Phys.* **28**, 1615–1621.

Ishikawa, M., Shiozawa, H., Itaya, K., Hatakoshi, G., and Uematsu, Y. (1991). *IEEE J. Quantum Electron.* **27**, 23–29.
Itaya, K., Ishikawa, M., Watanabe, Y., Nitta, K., Hatakoshi, G., and Uematsu, Y. (1989). *Jpn. J. Appl. Phys.* **27**, L2414–L2416.
Itaya, K., Sugawara, H., and Hatakoshi, G. (1994). *J. Cryst. Growth* **138**, 768–775.
Kaliski, R. W., Epler, J. E., Holonyak, N., Jr., Peanasky, M. J., Hermannsfeldt, G. A., Drickamer, H. G., Tsai, M. J., Camras, M. D., Kellert, F. G., Wu, C. H., and Craford, M. G. (1985). *J. Appl. Phys.* **57**, 1734–1738.
Kato, T., Susawa, H., Hirotani, M., Saka, T., Ohashi, Y., Shichi, E., and Shibata, S. (1991). *J. Cryst. Growth* **107**, 832–835.
Kazamura, M., Ohta, I., and Teramoto, I. (1983). *Jpn. J. Appl. Phys.* **22**, 654–657.
Kish, F. A. (1995). *Encycl. Chem. Technol.* **15**, 217–246.
Kish, F. A., Steranka, F. M., DeFevere, D. C., Vanderwater, D. A., Park, K. G., Kuo, C. P., Osentowski, T. D., Peanasky, M. J., Yu, J. G., Fletcher, R. M., Steigerwald, D. A., Craford, M. G., and Robbins, V. M. (1994a). *Appl. Phys. Lett.* **64**, 2839–2841.
Kish, F. A., DeFevere, D. C., Vanderwater, D. A., Trott, G. R., Weiss, R. J., and Major, J. S., Jr. (1994b). *Electron. Lett.* **30**, 1790–1791.
Kish, F. A., Steranka, F. M., DeFevere, D. C., Robbins, V. M., and Uebbing, J. (1994c). U.S. Pat. 5,376,580.
Kish, F. A., Vanderwater, D. A., Peanasky, M. J., Ludowise, M. J., Hummel, S. G., and Rosner, S. J. (1995). *Appl. Phys. Lett.* **67**, 2060–2062.
Kish, F. A., Vanderwater, D. A., DeFevere, D. C., Steigerwald, D. A., Hofler, G. E., Park, K. G., and Steranka, F. M. (1996a). *Electron. Lett.* **32**, 132–134.
Kish, F. A., Steranka, F. M., DeFevere, D. C., Robbins, V. M., and Uebbing, J. (1996b). U.S. Pat. 5,502,316.
Kobayashi, K., Kawata, S., Gomyo, A., Hino, I., and Suzuki, T. (1985). *Electron. Lett.* **21**, 931–932.
Kobayashi, T., Kojima, H., Deol, R. S., Buchan, N., Heuberger, W., Jakubowicz, A., and Roentgen, P. (1995). *J. Phys. Chem. Solids* **56**, 311–317.
Kondo, M., Okada, N., Domen, K., Sugiura, K., Anayama, C., and Tanahashi, T. (1994). *J. Electron. Mater.* **23**, 355–358.
Kondo, S., Nagai, N., Itoh, Y., and Yamaguchi, M. (1989). *Appl. Phys. Lett.* **55**, 1981–1983.
Kondow, M., Kakibayashi, H., Miagawa, S., Inoue, Y., Nishino, T., and Hamakawa, Y. (1988). *J. Cryst. Growth* **93**, 412–417.
Kressel, H., and Butler, J. K. (1977). "Semiconductor Lasers and Heterojunction LEDs." Academic Press, New York.
Kroemer, H. (1963). *Proc. IEEE* **51**, 1782–1783.
Kuo, C. P., Fletcher, R. M., Osentowski, T. D., Craford, M. G., Nam, D. W., Holonyak, N., Jr., Hsieh, K. C., and Fouquet, J. E. (1988). *J. Cryst. Growth* **93**, 389–395.
Kuo, C. P., Fletcher, R. M., Osentowski, T. D., Lardizabal, M. C., Craford, M. G., and Robbins, V. M. (1990). *Appl. Phys. Lett.* **57**, 2937–2939.
Kuo, C. P., Fletcher, R. M., and Osentowski, T. D. (1991). U.S. Pat. 5,060,028.
Laporte, A., Benamara, M., Sarrabayrouse, G., Rocher, A., Lescouzeres, L., Peyrelavigne, A., and Claverie, A. (1994). *In* "Third International Symposium on Semiconductor Wafer Bonding: Physics and Applications" (C. E. Hunt, H. Baumgart, S. S. Iyer, T. Abe, and U. Gosele, eds.), Vol. 95–7. Electrochemical Society, Reno, NV.
Lawaetz, P. (1971). *Phys. Rev. B* **4**, 3460–3467.
Lester, S. D., Ponce, F. A., Craford, M. G., and Steigerwald, M. G. (1995). *Appl. Phys. Lett.* **66**, 1249–1251.
Liau, Z. L., and Mull, D. E. (1990). *Appl. Phys. Lett.* **56**, 737–739.

Liedenbaum, C. T. H. F., Valster, A., Severens, A. L. G. H., and 't Hooft, G. W. (1990). *Appl. Phys. Lett.* **57**, 2698–2700.

Lin, J.-F., Wu, M.-C., Jou, M.-J., Chang, C.-M., Chen, C.-Y., and Lee, B.-J. (1993a). *J. Appl. Phys.* **74**, 1781–1786.

Lin, J.-F., Wu, M.-C., Jou, M.-J., Chang, C.-M., and Lee, B.-J. (1993b). *Electron. Lett.* **29**, 1346–1347.

Lin, J.-F., Wu, M.-C., Jou, M.-J., Chang, C.-M., Lee, B.-J., and Tsai, Y.-T. (1994). *Electron. Lett.* **30**, 1793–1794.

Lin, J.-F., Wu, M.-C., Jou, M.-J., Chang, C.-M., and Lee, B.-J. (1995a). *Solid-State Electron.* **38**, 305–308.

Lin, J.-F., Wu, M.-C., Jou, M.-J., Chang, C.-M., and Lee, B.-J. (1995b). *J. Electrochem. Soc.* **142**, 1293–1297.

Lo, Y. H., Bhat, R., Hwang, D. M., Koza, M. A., and Lee, T. P. (1991). *Appl. Phys. Lett.* **58**, 1961–1963.

Lo, Y. H., Bhat, R., Hwang, D. M., Chua, C., and Lin, C.-H. (1993). *Appl. Phys. Lett.* **62**, 1038–1040.

Loebner, E. E. (1973). *Proc. IEEE* **61**, 837–861.

Macksey, H. M., Holonyak, N., Jr., Scifres, D. R., Dupuis, R. D., and Zack, G. W. (1971). *Appl. Phys. Lett.* **19**, 271–273.

Macksey, H. M., Lee, M. H., Holonyak, N., Jr., Hitchens, W. R., Dupuis, R. D., and Campbell, J. C. (1973). *J. Appl. Phys.* **44**, 5035–5040.

Manesevit, H. M. (1968). *Appl. Phys. Lett.* **12**, 156–159.

Marcuse, D. (1972). "Light Transmission Optics." Van Nostrand-Reinhold, New York.

Masselink, W. T., and Zachau, M. (1992). *Appl. Phys. Lett.* **61**, 58–60.

Meney, A. T., Prins, A. D., Phillips, A. F., Sly, J. L., O'Reilly, E. P., Dunstan, D. J., Adams, A. R., and Valster, A. (1995). *IEEE J. Sel. Top. Quantum Electron.* **1**, 697–706.

Minagawa, S., and Kakibayashi, H. (1985). *Jpn. J. Appl. Phys.* **24**, 1569–1570.

Mitani, K. and Gosele, U. M. (1992). *J. Electron. Mater.* **21**, 669–676.

Mowbray, D. J., Kowalski, O. P., Hopkinson, M., Skolnick, M. S., and David, J. P. R. (1994). *Appl. Phys. Lett.* **65**, 213–215.

Nakamura, S., Senoh, M., Iwasa, N., Nagahama, S., Yamada, T., and Mukai, T. (1995). *Jpn. J. Appl. Phys.* **34**, L1332–L1335.

Naritsuka, S., Nishikawa, Y., Sugawara, H., Ishikawa, M., and Kokubun, Y. (1991). *J. Electron. Mater.* **20**, 687–690.

Nelson, R. J., and Holonyak, N., Jr. (1976). *J. Appl. Phys.* **47**, 1704–1707.

Nishikawa, Y., Suzuki, M., and Okajima, M. (1993). *Jpn. J. Appl. Phys.* **32**, 498–501.

Nishizawa, J., Suto, K., and Teshima, T. (1977). *J. Appl. Phys.* **48**, 3484–3495.

Nishizawa, J., Koike, M., and Jin, C. C. (1983). *J. Appl. Phys.* **54**, 2807–2812.

Nitta, K., Okajima, M., Nishikawa, Y., Itaya, K., and Hatakoshi, G. (1992). *Electron. Lett.* **28**, 1069–1070.

Nojima, S., Tanaka, H., and Asahi, H. (1986). *J. Appl. Phys.* **59**, 3489–3494.

Nozaki, C., and Ohba, Y. (1989). *J. Appl. Phys.* **66**, 5394–5397.

Nozaki, C., Ohba, Y., Sugawara, H., Yasuami, S., and Nakanisi, T. (1988). *J. Crys. Growth* **93**, 406–411.

Nuese, C. J., Richman, D., and Clough, R. B. (1971). *Metall. Trans.* **2**, 789–793.

Nuese, C. J., Sigai, A. G., and Gannon, J. J. (1972). *Appl. Phys. Lett.* **20**, 431–434.

Nuese, C. J., Sigai, A. G., Abrahams, M. S., and Gannon, J. J. (1973). *J. Electrochem. Soc.* **120**, 956–965.

Ohba, Y., Ishikawa, M., Sugawara, H., Yamamoto, M., and Nakanisi, T. (1986). *J. Cryst. Growth* **77**, 374–379.

Okuda, H., Ishikawa, M., Shiozawa, H., Watanabe, Y., Itaya, K., Nitta, K., Hatakoshi, G., Kokubun, Y., and Uematsu, Y. (1989). *IEEE J. Quantum Electron.* **25**, 1477–1482.
Okuno, Y., Uomi, K., Aoki, M., Taniwatari, T., Suzuki, M., and Kondow, M. (1995). *Appl. Phys. Lett.* **66**, 451–453.
Onton, A., and Chicotka, R. J. (1970). *J. Appl. Phys.* **41**, 4205–4207.
Patel, D., Hafich, M. J., Robinson, G. Y., and Menoi, C. S. (1993). *Phys. Rev. B* **48**, 18031–18036.
Prins, A. D., Sly, J. L., Meney, A. T., Dunstan, D. J., O'Reilly, E. P., Adams, A. R., and Valster, A. (1995a). *J. Phys. Chem. Solids* **56**, 349–352.
Prins, A. D., Sly, J. L., Meney, A. T., Dunstan, D. J., O'Reilly, E. P., Adams, A. R., and Valster, A. (1995b). *J. Phys. Chem. Solids* **56**, 423–427.
Rennie, J., Okajima, M., Watanabe, M., and Hatakoshi, G. (1993). *IEEE J. Quantum Electron.* **29**, 1857–1862.
Richard, T. A., Holonyak, N., Jr., Kish, F. A., Keever, M. R., and Lei, C. (1995). *Appl. Phys. Lett.* **66**, 2972–2974.
Rupprecht, H., Woodall, J. M., and Petit, G. D. (1967). *Appl. Phys. Lett.* **11**, 81–83.
Schneider, J. R. P., Jones, E. D., Lott, J. A., and Bryan, R. P. (1992). *J. Appl. Phys.* **72**, 5397–5400.
Schnitzer, I., Yablonovitch, E., Caneau, C., and Gmitter, T. J. (1993). *Appl. Phys. Lett.* **62**, 131–133.
Scifres, D. R., Macksey, H. M., Holonyak, N., Jr., and Dupuis, R. D. (1972). *J. Appl. Phys.* **43**, 1019–1022.
Shimbo, M., Furukawa, K., Fukuda, K., and Tanzawa, K. (1986). *J. Appl. Phys.* **60**, 2987–2989.
Stinson, L. J., Yu, J. G., Lester, S. D., Peanasky, M. J., and Park, K. (1991). *Appl. Phys. Lett.* **58**, 2012–2014.
Sugawara, H., Ishikawa, M., and Hatakoshi, G. (1991a). *Appl. Phys. Lett.* **58**, 1010–1012.
Sugawara, H., Ishikawa, M., Kokubun, Y., Nishikawa, Y., and Naritsuka, S. (1991b). U.S. Pat. 5,048,035.
Sugawara, H., Itaya, K., Ishikawa, M., and Hatakoshi, G. (1992a). *Jpn. J. Appl. Phys.* **31**, 2446–2451.
Sugawara, H., Itaya, K., Nozaki, H., and Hatakoshi, G. (1992b). *Appl. Phys. Lett.* **61**, 1775–1777.
Sugawara, H., Ishikawa, M., Kokubun, Y., Nishikawa, Y., Naritsuka, S., Itaya, K., Hatakoshi, G., and Suzuki, M. (1992c). U.S. Pat. 5,153,889.
Sugawara, H., Itaya, K., and Hatakoshi, G. (1993). *J. Appl. Phys.* **74**, 3189–3193.
Sugawara, H., Kazuhiko, I., and Hatakoshi, G. (1994a). *Jpn. J. Appl. Phys.* **33**, 5784–5787.
Sugawara, H., Itaya, K., and Hatakoshi, G. (1994b). *Jpn. J. Appl. Phys* **33**, 6195–6198.
Sugiura, K., Domen, K., Sugawara, M., Anayama, C., Kondo, M., Tanahashi, T., and Nakajima, K. (1991). *J. Appl. Phys.* **70**, 4946–4949.
Suzuki, M., Ishikawa, M., Itaya, K., Nishikawa, Y., Hatakoshi, G., and Kokubun, Y. (1991). *J. Cryst. Growth* **115**, 498–503.
Suzuki, M., Itaya, K., Nishikawa, Y., Sugawara, H., and Okajima, M. (1993). *J. Cryst. Growth* **133**, 303–308.
Suzuki, M., Itaya, K., and Okajima, M. (1994). *Jpn. J. Appl. Phys.* **33**, 749–753.
Suzuki, T., Gomyo, A., Iijima, S., Kobayashi, K., Kawata, S., Hino, I., and Yuasa, T. (1988). *Jpn. J. Appl. Phys.* **27**, 2098–2106.
Sze, S. M. (1981). "Physics of Semiconductor Devices," 2nd ed. Wiley, New York.
Tan, I.-H., Dudley, J. J., Babic, D. I., Cohen, D. A., Young, B. D., Hu, E. L., Bowers, J. E., Miller, B. I., Koren, U., and Young, M. G. (1994). *IEEE Photon. Technol. Lett.* **6**, 811–813.

Tanaka, H., Kawamura, Y., and Asahi, H. (1986). *Electron. Lett.* **22**, 707–708.
Tanaka, H., Kawamura, Y., and Asahi, H. (1987a). *Electron. Lett.* **23**, 166–168.
Tanaka, H., Kawamura, Y., Nojima, S., Wakita, K., and Asahi, H. (1987b). *J. Appl. Phys.* **61**, 1713–1719.
Thompson, G. H. B. (1980). "Physics of Semiconductor Laser Devices." Wiley, New York.
Thornton, R. L., Burnham, R. D., and Streifer, W. (1984). *Appl. Phys. Lett.* **45**, 1028–1030.
Tsang, W. T. (1978). *Appl. Phys. Lett.* **33**, 245–248.
Tsang, W. T. (1980). *Appl. Phys. Lett.* **36**, 11–14.
Wada, H., Ogawa, Y., and Kamijoh, T. (1993). *Appl. Phys. Lett.* **62**, 738–740.
Watanabe, M. O., and Ohba, Y. (1986). *J. Appl. Phys.* **60**, 1032–1037.
Watanabe, M. O., and Ohba, Y. (1987). *Appl. Phys. Lett.* **50**, 906–908.
Yablonovitch, E., Hwang, D. M., Gmitter, T. J., Florez, L. T., and Harbison, J. P. (1990). *Appl. Phys. Lett.* **56**, 2419–2421.
Young, D. B., Babic, D. I., DenBaars, S. P., and Coldren, L. A. (1992). *Electon. Lett.* **28**, 1873–1874.
Yow, H. K., Houston, P. A., and Hopkinson, M. (1995). *Appl. Phys. Lett.* **66**, 2852–2854.

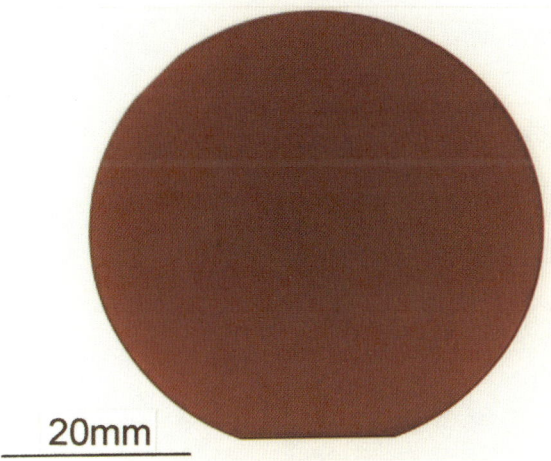

FIG. 25. (Chapter 5) Transmitted light image of a 50-mm-diameter wafer-bonded GaP-AlGaInP–GaP light-emitting diode (LED) wafer (Hofler et al., 1996). The uniform red color across the wafer surface is indicative of the absence of nonbonded areas, large voids, and other defects associated with the wafer-bonding process.

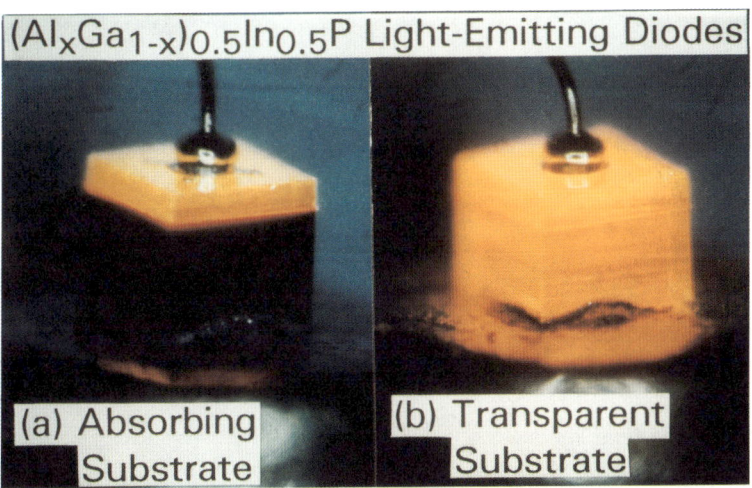

FIG. 26. (Chapter 5) Photomicrographs of (a) GaAs absorbing substrate (AS) and (b) wafer-bonded GaP transparent substrate (TS) $(Al_x Ga_{1-x})_{0.5} In_{0.5}P$ light-emitting diode (LED) chips (8.5 × 8.5 mil^2) operating at 50 mA (dc) at $\lambda \sim 600$ nm (Kish et al., 1994a). Light radiates evenly from all surfaces of the wafer-bonded TS LED, which is indicative of uniform curent flow throughout the device.

FIG. 32. (Chapter 8) The first actual light-emitting diode (LED) traffic light installed in Japan in 1994 using InGaN–AlGaN blue-green, AlInGaP yellow and GaAlAs red LEDs.

FIG. 33. (Chapter 8) The actual light-emitting diode (LED) full-color display installed in Japan in 1994 for the first time. The blue InGaN–AlGaN LEDs, green gallium phosphide (GaP) LEDs and red GaAlAs LEDs are used as three primary color LEDs.

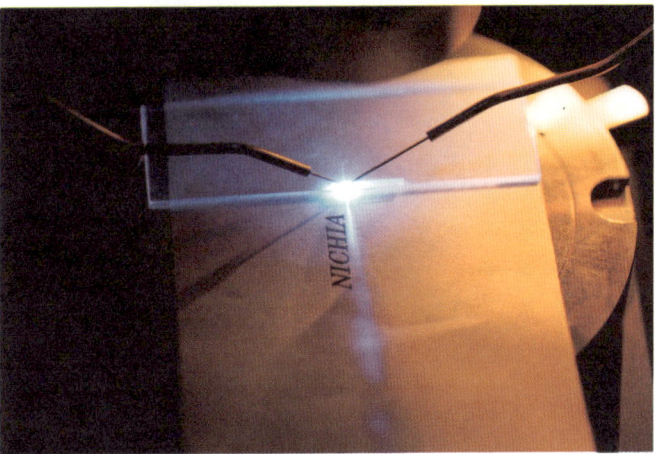

Fig. 50. (Chapter 8) The streak line of the laser emission from a light-emitting diode chip that was operated under pulsed current injection of 660 mA at room temperature.

CHAPTER 6

Applications for High-Brightness Light-Emitting Diodes

Mark W. Hodapp

OPTOELECTRONICS DIVISION
HEWLETT-PACKARD COMPANY
SAN JOSE, CALIFORNIA

I. INTRODUCTION	228
II. PHOTOMETRY AND COLOR MEASUREMENT PRINCIPLES	230
1. *Photometry and Radiometry*	230
2. *Converting from Luminous Intensity to Luminous Flux*	233
3. *Color Measurement*	247
III. AUTOMOTIVE SIGNAL LIGHTING	251
1. *Characteristics of the Automotive and Truck Market*	251
2. *Automotive Luminous Intensity and Color Requirements*	252
3. *Benefits of Light-Emitting Diode Technology*	252
4. *Luminous Flux Requirements*	259
5. *Optics Design Considerations*	263
6. *Electrical Design Considerations*	268
7. *Thermal Design Considerations*	273
IV. AUTOMOTIVE INTERIOR LIGHTING	277
1. *Characteristics of the Automotive and Truck Market*	277
2. *Benefits of Light-Emitting Diode Technology*	279
3. *Lighting Considerations for an Instrument Cluster Warning Light*	284
4. *Cavity Design for an Instrument Cluster Warning Light*	289
5. *Legend Optimization of an Instrument Cluster Warning Light*	294
V. TRAFFIC SIGNAL LIGHTS	296
1. *Characteristics of the Traffic Signal Market*	296
2. *Construction of a Traffic Signal Head*	298
3. *Comparison Between Incandescent and Light-Emitting Diode Traffic Signals*	298
4. *Luminous Flux Requirements*	302
5. *Electrical Design Considerations*	307
VI. LARGE-AREA DISPLAYS	319
1. *Characteristics of the Large-Area Display Market*	319
2. *Alternative Large-Area Display Technologies*	320
3. *Benefits of Light-Emitting Diode Technology*	324
4. *Principles of Color Mixing*	326
5. *Character Height Considerations*	329
6. *Drive Circuit Considerations*	330

VII. LIQUID CRYSTAL DISPLAY BACKLIGHTING 337
   1. *Characteristics of the Liquid Crystal Display Backlighting Market* 337
   2. *Liquid Crystal Display Operation* 338
   3. *Alternative Liquid Crystal Display Backlighting Technologies* 340
   4. *Benefits of Light-Emitting Diode Technology* 342
   5. *Direct-View Liquid Crystal Display Backlighting Techniques* 345
   6. *Edgelighting Liquid Crystal Display Backlighting Techniques* 347
   7. *Fiber-Optic Light-Emitting Diode Backlighting Techniques* 351
     References 355

I. Introduction

Light-emitting diode (LED) technology has flourished for over 25 years. The initial GaAsP–gallium arsenide (GaAs) (standard-red) technology and the GaAsP–GaP (high efficiency red, yellow, and green) technology introduced in the early 1970s has become a mainstay for indicators, numeric displays, and alphanumeric displays. In 1996, these technologies represent several billion dollars of worldwide sales. These products have found their way into a diverse range of markets including aerospace and avionics, motor vehicles, industrial applications, telecommunications, and consumer products. Despite some limitations in colors and efficiencies, the GaAsP–GaAs and GaAsP–gallium phosphide (GaP) LED materials technologies provide a number of benefits including long operating lifetimes, ruggedness, low current operation, and compatibility with transistor–transistor logic (TTL) and complementary metal-oxide semiconductor (CMOS) digital circuitry.

Over the past few years, the field of LED materials technologies has been rapidly expanding with new materials that have higher efficiencies and new colors. AlGaAs materials technology offers a higher efficiency red light-emitting material as compared with GAsP–GaP. Transparent substrate (TS) AlGaAs (637 nm λ_d) has a luminous efficiency approaching 10 lm/W, which is 10 times more efficient than is the GaAsP–GaP red LED materials technology. AlInGaP materials technology offers the same color range as does GaAsP–GaP materials technology: yellow-green (560 nm λ_d), greenish-yellow (570 nm λ_d), yellow (585 nm λ_d), amber (590 nm λ_d), orange (600 nm, 605 nm λ_d), and reddish orange (615, 622, and 626 nm λ_d). AlInGaP materials technology has a luminous efficiency of 10 lm/W for absorbing substrate (AS) and 20 lm/W for TS in the amber, orange, and reddish-orange color ranges, which is 10 to 20 times more efficient than the GaAsP–GaP LED materials technologies. AlInGaP green (560 nm, 570 nm λ_d) materials technology has a luminous efficiency of 5 lm/W, which is 2 to

4 times higher than GaP–GaP LED materials technology. The newest LED material technology, InGaN–sapphire, offers both new colors and high efficiency. InGaN LED materials technology is available in several blue colors (450 to 480 nm λ_d), as well as a bluish-green color (500 nm λ_d) and a green color (520 nm λ_d), with luminous efficiencies in the range of 3 to 15 lm/W. Although not as efficient as the AlGaAs and AlInGaP LED materials technologies, InGaN is much more efficient than the earlier silicon carbide (SiC) LED blue technology, which had a luminous efficiency of about 0.02 lm/W.

These new LED materials technologies are already opening up a number of new applications, in which LED technology provides benefits compared with other lighting technologies. Just as with the older GaAsP–GaP LED materials technology, these new LED material technologies offer the same benefits as the older GaP based LED material. However, now the luminous efficiency exceeds that of a filtered incandescent bulb. These new applications can replace an incandescent bulb in the 1-W power range with a single LED lamp, and an incandescent bulb in the 150-W power range with an array of LED lamps. For most of these applications, the incandescent bulb is filtered to a red, amber, green, or blue color. These new LED materials technologies offer the same color ranges. Future applications may include unfiltered incandescent bulbs because light from multiple LED die (i.e., blue and amber) can roughly approximate the color obtained from the unfiltered incandescent bulb.

In this chapter, five new applications have been selected in which the highly efficient AlGaAs, AlInGaP, and InGaN LED materials technologies can be utilized. Each section features a different set of applications and discusses the traditional methods of lighting, the benefits that LED technology provides, and some of the design issues of the LED approach. These applications include automotive lighting—both signal lighting on the exterior of the vehicle as well as indicator lighting within the vehicle. Another application that is discussed is traffic signal lighting—both red/yellow/bluegreen traffic signals and pedestrian cross-walk traffic signals. Large-area displays are discussed. These displays range in size from airport status boards to stadium scoreboards and can be designed for monochrome, bicolor, or full-color operation with full-motion video capabilities. The final application that is discussed is LCD backlighting for transmissive and transflective liquid crystal displays (LCDs). Again, monochrome, bicolor, and full-color applications are discussed. These new LED materials technologies are already commercially available in a number of new products in these five featured applications. Together, these five new applications are the newest frontiers for LED technology and can be expected to be responsible for considerable growth in the LED semiconductor industry.

II. Photometry and Color Measurement Principles

1. PHOTOMETRY AND RADIOMETRY

Photometry and radiometry deal with measurements of electromagnetic radiation. Both measurement systems define the measurement of flux and intensity. The basic concepts of flux and intensity are shown in Fig. 1. Electromagnetic flux (ϕ) is the total electromagnetic energy per unit time emitted by the radiation source. Generally, the convention is that the electromagnetic flux is measured in a sphere surrounding the radiation source.

Radiometry deals with measurements of electromagnetic radiation at all wavelengths. Unit symbols are subscripted with an e and unit names are prefaced with the term *radiant*. The unit of radiant flux (ϕ_e) is called a Watt (W). By contrast, photometry deals with measurements of electromagnetic radiation at wavelengths that are visible to the human eye. Unit symbols are subscripted with a v to denote visible and unit names are prefixed with the term *luminous*. The unit of luminous flux (ϕ_v) is called a lumen (lm). Photometry differs from radiometry in that all measurements are multiplied by a factor for human eye sensitivity.

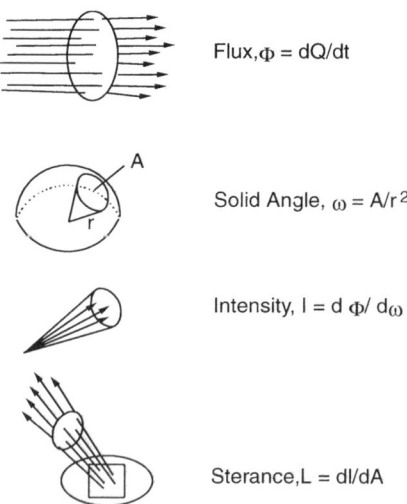

FIG. 1. Sketches showing basic concepts of optical flux, intensity, and sterance measurements. (From Gage *et al.* 1977. Reprinted with the permission of The McGraw-Hill Companies.)

In 1931, the CIE (Commission Internationale de L'Eclairage) adopted a human eye weighting factor, based on extensive testing on human subjects. This luminosity factor $V'(\lambda)$, also called the $\bar{y}$ *color matching function*, is shown in Fig. 2. The highest human eye sensitivity occurs at 555 nm and has a radiant flux to luminous flux conversion factor of 683 lm/W. The luminous flux at any wavelength can be calculated by the following equation:

$$\phi_v = 683 \text{ lm/W} \int \phi_e(\lambda) V'(\lambda) \, d\lambda$$

where ϕ_v is the luminous flux, $\phi_e(\lambda)$ is the radiant spectrum of the radiation source (W/nm units), and $V'(\lambda)$ is the 1931 CIE luminosity function.

One thing to watch out for is that the radiometric measurement system uses the same unit of Watt as is used for electrical measurements. In both measurement systems the term *Watt* refers to the energy per unit time generated by the radiation source. Thus, it is possible to define two totally different measurements with the same units. The term *radiant efficiency*, with units W/W, refers to the energy conversion efficiency of the radiation source in converting electrical power into radiant flux. The term *luminous efficiency*, with units lm/W, refers to the energy conversion efficiency of the light source

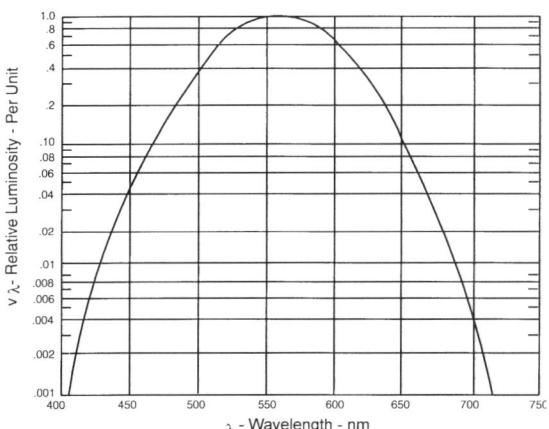

Note: Peak luminous efficacy of 683 lm/W occurs at a peak wavelength.

FIG. 2. 1931 CIE (Commission Internationale de L'Eclairage) luminosity curve $V'(\lambda)$. Note. Peak luminous efficacy of 683 lm/w occurs at a peak wavelength of 555 nm. (From Gage et al., 1977. Reprinted with the permission of The McGraw-Hill Companies.)

in converting electrical power into luminous flux. The term *luminous efficacy*, with units lm/W, refers to the ratio of luminous flux to radiant flux generated by the light source. For example, a TS AlInGaP amber LED lamp has a forward voltage of 2.4 V at 50 mA. Thus, the electrical power consumption is equal to (2.4 V) (50 mA), or 120 mW. It generates about 5 mW of radiant flux and 2400 mlm of luminous flux. Thus, the radiant efficiency is equal to 5 mW/120 mW, or 4.17%; the luminous efficiency is equal to 2400 mlm/120 mW, or 20 lm/W; and the luminous efficacy is equal to 2400 mlm/5 mW, or 480 lm/W.

Unless otherwise specified, the convention used in this chapter is that the unit lm/W, refers to the luminous efficiency of the light source (luminous flux output/electrical power input).

As shown in Fig. 1, the intensity of a radiation source is defined as the divergence of electomagnetic flux from the radiation source at a defined angular orientation. The intensity I is equal to the incremental amount of electromagnetic flux $d\phi$ emitted into an incremental solid angle ($d\omega$). The solid angle ω in steradians (sr) is equal to the area on the surface of a sphere subtended by the solid angle, divided by the square of the radius of the sphere. Note that there are 4π steradians of solid angle captured in a sphere surrounding the radiation source. Generally, the intensity of the radiation source is measured with a very small solid angle of about 10^{-3} steradians. For many radiation sources, the intensity varies substantially at different angular positions. Thus, the specific angular position for the intensity measurement must be defined. The unit of radiant intensity I_e is a W/sr. The unit of luminous intensity I_v is called a candela (cd) or lm/sr.

The convention used in this chapter is that the intensity is measured at the angular position giving the highest reading, unless otherwise specified. This angular position is called the optical axis.

The measurement of luminous intensity described earlier should not be confused with the measurement of mean spherical candelas (mscd) or mean spherical candle power (mscp) commonly used for small incandescent bulbs. The mean spherical candela or mean spherical candle power is defined as the average luminous intensity emitted by the light source in a sphere surrounding the light source. Because there are 4π steradians of solid angle captured in a sphere, the mean spherical candela or mean spherical candle power, in candelas, is equal to the luminous flux, in lumens, emitted by the light source divided by 4π:

$$I_v \text{ (mscd)} = I_v \text{ (mscp)} = \phi_v/\omega_{\text{sphere}} = \phi_v/4\pi$$

Luminous sterance (L_v) is another useful photometric measurement. Luminous sterance, as shown in Fig. 1, is defined as the luminous intensity of the light source per unit light emitting area. The metric units for luminous sterance are cd/m². Luminous sterance measurements correlate to the qualitative term of *brightness*. Sources with a higher luminous sterance appear to be much brighter than those with a lower luminous sterance.

$$L_v = I_v/A$$

where L_v is the luminous sterance of defined area A, I_v is the luminous intensity of defined area A, A is the defined area of optical measurement and is less than or equal to the total emitting area.

Assuming that the defined area is equal to the total emitting area, then the luminous sterance is equal to the luminous intensity divided by the emitting area. For two light sources of different light-emitting areas with the same luminous sterance, the light source with the largest area would have the highest luminous intensity. Both light sources would appear to have the same brightness. On the other hand, for two light sources of different light-emitting areas with the same luminous intensity, then the light source with the smallest light-emitting area, would have the highest luminous sterance and the highest brightness. Luminous sterance can vary significantly over the light-emitting area so that the mechanical position and size of the area used for the measurement must be defined. Alternatively, multiple luminous sterance measurements can be made across the light-emitting surface and averaged together.

2. Converting from Luminous Intensity to Luminous Flux

It is often useful to convert the radiation pattern of a light source into a value of luminous flux. As illustrated in Fig. 3, the term *radiation pattern* refers to the values of luminous intensity at all angles in a sphere surrounding the light source. Consider a sphere drawn around the light source, with the light source at the center. If the surface of the sphere is equally divided into small segments such that the luminous intensity is constant over each angular segment, then the total luminous flux emitted by one segment is equal to

$$\phi_v = I_v \omega = I_v A/r^2$$

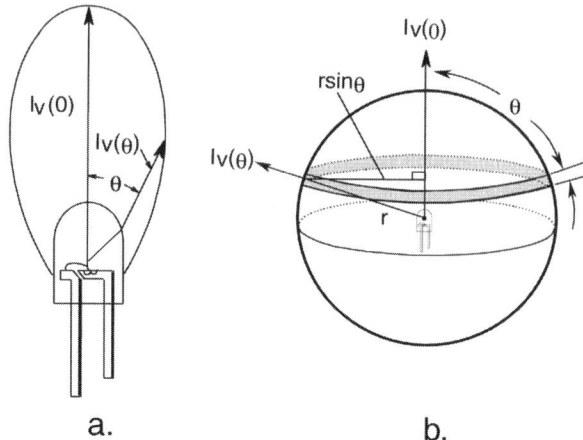

FIG. 3. (a) Typical radiation pattern for a light-emitting diode LED lamp. (From Hodapp, 1990. Reprinted with permission from SAE Paper 900474 ©1990 Society of Automotive Engineers, Inc.) (b) Horizontal band drawn around an LED lamp. (From Gage et al., 1977. Reprinted with the permission of The McGraw-Hill Companies.)

where

ϕ_v = total luminous flux emitted into the defined angular segment

I_v = luminous intensity for the defined angular segment

ω = solid angle

A = area of the angular segment on the surface of the sphere

r = radius of the sphere

Thus, the total luminous flux emitted by the light source is equal to the summation of all luminous flux values calculated previously for all segments on the sphere at which the luminous intensity is greater than zero.

In most cases, the radiation pattern of an LED lamp is rotationally symmetric about the optical axis, as shown in Fig. 3. The amount of incremental luminous flux emitted at an off-axis angle of θ degrees is equal to the luminous intensity emitted at an off-axis angle of θ degrees, $I_v(\theta)$, multiplied by the incremental solid angle of the horizontal band on the surface of the sphere. Figure 3 shows a hemisphere drawn around the LED lamp such that the optical axis is pointing toward the upper pole. The area of the horizontal band is equal to the circumference, $2\pi r \sin\theta$, times the width of the band, $rd\theta$. Thus, the incremental luminous flux is equal to

$$d\phi_v(\theta) = I_v(\theta)d\omega = I_v(\theta)dA/r^2 = I_v(\theta)(2\pi r \sin\theta)(rd\theta)/r^2 = 2\pi I_v(\theta)\sin\theta d\theta$$

Thus, the total luminous flux included between angles 0 and θ degrees emitted by a rotationally symmetric light source can be calculated by the following integral:

$$\phi_v(\theta) = 2\pi \int_0^\theta \sin(\theta) I_v(\theta)\,d\theta$$

where $\phi_v(\theta)$ is the total luminous flux emitted at angles from 0 to θ degrees off-axis and $I_v(\theta)$ is the off-axis luminous intensity at angle θ.

In some cases, it is more convenient to integrate the normalized radiation pattern (for example, most LED lamp data sheets provide a normalized radiation pattern and min/typ/max luminous intensity values) and multiply the result by the on-axis luminous intensity, as shown:

$$\phi_v(\theta) = 2\pi I_v(0) \int_0^\theta \sin(\theta) \hat{I}_v(\theta)\,d\theta$$

where $\hat{I}_v(\theta)$ is the normalized off-axis luminous intensity $I_v(\theta)/I_v(0)$ at θ degrees off-axis.

The *zonal constant integration technique* is one convenient method to numerically integrate the radiation pattern of a light source into a value of luminous flux. The zonal constant method calculates the luminous flux emitted into a narrow band on the surface of the sphere by summing the luminous intensities in the narrow band and multiplying the summation by a solid angle factor, called the zonal constant.

The value of the zonal constant C_z can be derived by first calculating the surface area A for one horizontal band on the surface of the sphere and then converting the surface area value into the subtended solid angle ω.

Figure 4(a) shows the surface area of a horizontal band on the sphere such that the lower circle is of height h_1 and the upper circle is of height h_2. If the width of the horizontal band is small, then the area of the horizontal band is approximately equal to the circumference of the midpoint of the horizontal band multiplied by the width of the band. If angle θ_{MID} is defined as the angle midway between θ_1 and θ_2, and angle θ_ω is defined as the angular width of the horizontal band $\theta_1 - \theta_2$ in radians, then the surface area of the horizontal band is approximately equal to

$$A \cong (2\pi r_\omega)(r\theta_\omega)$$

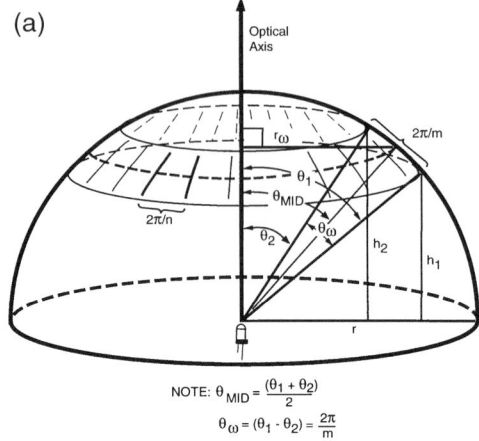

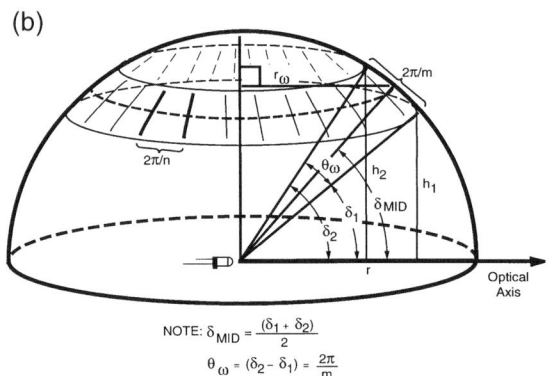

FIG. 4. The surface area of a horizontal band on the sphere such that the lower circle is of height h_1 and the upper circle is of height h_2. (a) Sphere with optical axis pointing toward upper pole. (b) Sphere with optical axis pointing to horizontal great circle.

where

$\theta_{\text{MID}} = (\theta_1 + \theta_2)/2$

r_ω = radius of the circle described by 360 degrees rotation of angle θ_{MID}

$2\pi r_\omega$ = circumference of the horizontal band at midpoint

$\theta_\omega = \theta_1 - \theta_2$, in radians

$r\theta_\omega$ = width of the horizontal band

However, r_ω is equal to ($r \sin\theta_{\text{MID}}$), and therefore the equation can be rewritten as:

$$A \cong (2\pi r \sin\theta_{MID})(r\theta_\omega)$$

$$A \cong 2\pi r^2 \theta_\omega \sin\theta_{MID}$$

where θ_ω is expressed in radians, or as

$$A \cong \frac{4\pi^2}{360°} r^2 \theta_\omega° \sin\theta_{MID}$$

where θ_ω is expressed in degrees.

If the sphere is equally divided into m vertical divisions, as shown in Fig. 5, then the vertical angular increment θ_ω is equal to $2\pi/m$, and the formula can be rewritten as:

$$A \cong \frac{4\pi^2 r^2}{m} \sin\theta_{MID}$$

Thus, the solid angle of the horizontal band ω and the zonal constant $C_z(\theta_{MID})$ are equal to:

$$\omega = \frac{A}{r^2} \cong \frac{4\pi^2}{m} \sin\theta_{MID}$$

$$C_Z(\theta_{MID}) = \frac{4\pi^2}{m} \sin\theta_{MID}$$

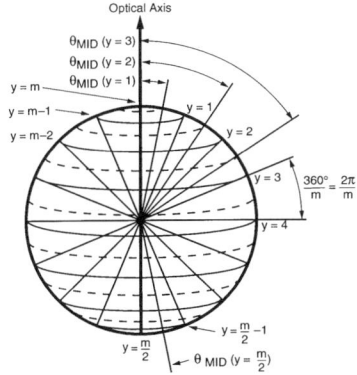

$\theta_{MID} = \left(\frac{360°}{m} y - \frac{360°}{2m}\right)$, for $y = 1, 2, \ldots \frac{m}{2}$

FIG. 5. Angular width of a horizontal band.

The exact solid angle ω of the horizontal band can be shown to be equal to

$$\omega = 4\pi \sin\left(\frac{\theta_\omega^\circ}{2}\right) \sin\theta_{\text{MID}}$$

Comparing the zonal constant solid angle approximation with the exact solid angle of the horizontal segment yields an error of

$$\% \text{ Error} = \frac{\dfrac{\pi}{360^\circ}\theta_\omega^\circ - \sin\left(\dfrac{\theta_\omega^\circ}{2}\right)}{\sin\left(\dfrac{\theta_\omega^\circ}{2}\right)} \times 100\%$$

Thus, the zonal constant approximation, $C_z(\theta_{\text{MID}})$ gives an error of 0.5% if the width of the horizontal band is equal to 20 degrees. Therefore, for small horizontal bands, the approximation error is negligible.

The optical axis of the light source can be oriented in any direction within the sphere. For the problem described earlier, the optical axis is pointed toward the upper pole so that the off-axis angles θ_1, θ_2, and θ_{MID} describe a circular band around the optical axis, where $I_v(\theta)$ is relatively constant. If I_v is constant over the range of subtended angles for the horizontal band at angle θ_{MID}, then the luminous flux emitted by the horizontal band is equal to:

$$\phi_v(\theta_{\text{MID}}) \cong \frac{4\pi^2}{m} \sin\theta_{\text{MID}} I_v(\theta_{\text{MID}}) \cong C_z(\theta_{\text{MID}}) I_v(\theta_{\text{MID}})$$

The total luminous flux emitted between angles 0 and θ degrees off-axis can be calculated by summing up the horizontal bands between 0 and θ degrees. If the sphere is divided into m vertical divisions, then angular width of each band is equal to $360^\circ/m$, as shown in Fig. 5. The midpoint θ_{MID} of each band is:

$$\theta_{\text{MID}} = \frac{360^\circ}{m} y - \frac{360^\circ}{2m}$$

where $y = 1, 2, \ldots m/2$.

If the radiation pattern is rotationally symmetric, then the total luminous flux emitted between angles 0 and θ degrees is equal to

$$\phi_v(\theta) \cong \sum_{y=1}^{m\theta/360^\circ} C_z(y) I_v\left(\frac{360^\circ}{m} y - \frac{360^\circ}{2m}\right)$$

where

$\phi_v(\theta)$ = total luminous flux emitted between 0 and θ degrees off-axis

m = number of vertical divisions of the sphere

$\dfrac{360°}{m}$ = angular width of the horizontal band

$\dfrac{m\theta}{360°}$ = number of horizontal bands between angles 0 and θ degrees

$$C_Z(y) = \frac{4\pi^2}{m} \sin\left(\frac{360°}{m} y - \frac{360°}{2m}\right).$$

Even if the radiation pattern is not rotationally symmetric, the luminous flux can still be calculated by subdividing the band into small enough horizontal segments such that the luminous intensity is constant for each horizontal segment. If the band is divided into n small horizontal segments, then the horizontal angular increment is $2\pi/n$ radians or $360°/n$ degrees. Then, the luminous flux emitted by one horizontal band can be calculated as

$$\phi_v(\theta_{\text{MID}}) \cong \frac{4\pi^2}{nm} \sin\theta_{\text{MID}} \sum_{x=1}^{n} I_v\left(\frac{2\pi}{n} x, \theta_{\text{MID}}\right)$$

$$\cong C_z(\theta_{\text{MID}}) \sum_{x=1}^{n} I_v\left(\frac{2\pi}{n} x, \theta_{\text{MID}}\right)$$

where angles are expressed in radians, or

$$\phi_v(\theta_{\text{MID}}) \cong \frac{4\pi^2}{nm} \sin\theta_{\text{MID}} \sum_{x=1}^{n} I_v\left(\frac{360°}{n} x, \theta°_{\text{MID}}\right)$$

$$\cong C_z(\theta_{\text{MID}}) \sum_{x=1}^{n} I_v\left(\frac{360°}{n} x, \theta°_{\text{MID}}\right)$$

where angles are expressed in degrees. Now, the zonal constant $C_z(\theta_{\text{MID}})$ is equal to:

$$C_Z(\theta_{\text{MID}}) = \frac{4\pi^2}{nm} \sin\theta_{\text{MID}}$$

Note that the only difference between the zonal constant for a rotationally symmetric light source and a nonsymmetric light source is that the zonal constant for the nonsymmetric light source is equal to the solid angle of the band divided by the number of segments n for which the intensity values were summed. In the case of the rotationally symmetric light source, $n = 1$.

Similarly, the total luminous flux emitted between angles 0 and θ degrees off-axis can be calculated even if the radiation pattern is not rotationally symmetric by subdividing the horizontal band into small enough horizontal segments such that the luminous intensity is constant for each segment. If the band is divided into n small horizontal segments, then the horizontal angular increment is $2\pi/n$ radians or $360°/n$ degrees. Then, the equation can be written as

$$\phi_v(\theta) \cong \sum_{y=1}^{m\theta/360°} C_z(y) \sum_{x=1}^{n} I_v\left(\frac{360°}{n} x, \frac{360°}{m} y - \frac{360°}{2m}\right)$$

where:

$$C_z(y) = \frac{4\pi^2}{nm} \sin\left(\frac{360°}{m} y - \frac{360°}{2m}\right)$$

In many cases, the numerical integration can be simplified by orienting the light source so that the optical axis is pointing from the center of the sphere toward a great circle of the sphere that is oriented horizontally. In this orientation, $I_v(H°, V°)$ refers to the luminous intensity at an angle of H degrees horizontal (left and right) and an angle of V degrees vertical (up and down) relative to the optical axis. With this orientation, angles δ_1, δ_2, and δ_{MID} are defined as the vertical angles from the plane of the horizontal great circle of the sphere to the horizontal band, as shown in Fig. 4(b). In this orientation, for a horizontal band with midpoint δ_{MID}, the luminous intensity values within the band would cover the following range: $I_v(H°, V°)$, where $0° \leq H \leq 360°$; $\delta_1 \leq V \leq \delta_2$. This orientation simplifies the numerical integration of many lighting standards because the tables are organized in rows and columns of luminous intensities at $H°$ horizontally left and right and $V°$ vertically up and down relative to the optical axis. Similarly to the approximation described earlier, the area of the horizontal band as shown in Fig. 4(b), is approximately equal to the circumference of the midpoint multiplied by the width of the horizontal band. If angle δ_{MID} is defined as the angle midway between δ_1 and δ_2, and angle θ_ω is defined as the angular width of the horizontal band $\delta_2 - \delta_1$, in radians, then the surface area of the horizontal band is approximately equal to:

$$A \cong 2\pi r^2 \theta_\omega \cos \delta_{MID}$$

where

$$\delta_{MID} = (\delta_1 + \delta_2)/2 \qquad \theta_\omega = \delta_2 - \delta_1 \quad \text{in radians}$$

If the sphere is equally divided into m vertical divisions, then the vertical angular increment θ_ω is equal to $2\pi/m$, and the formula can be rewritten as

$$A \cong \frac{4\pi^2 r^2}{m} \cos \delta_{\text{MID}}$$

Thus, the solid angle of the horizontal band ω and the zonal constant $C_z(\delta_{\text{MID}})$ are equal to

$$\omega = \frac{A}{r^2} \cong \frac{4\pi^2}{m} \cos \delta_{\text{MID}} \qquad C_Z(\delta_{\text{MID}}) = \frac{4\pi^2}{m} \cos \delta_{\text{MID}}$$

If the luminous intensity is constant within the horizontal band (note that this commonly occurs only if the light source has a spherical radiation pattern, i.e., I_v = constant), then the luminous flux emitted by the horizontal band is equal to

$$\phi_v(\delta_{\text{MID}}) \cong \frac{4\pi^2}{m} \cos \delta_{\text{MID}} I_v(\delta_{\text{MID}})$$

$$\cong C_z(\delta_{\text{MID}}) I_v(\delta_{\text{MID}})$$

Similarly to the previous example, if the radiation pattern is not rotationally symmetric, the total luminous flux in the horizontal band can be calculated by subdividing the band into small enough segments such that the luminous intensity is constant for each segment. If the band is divided into n small horizontal divisions, the equation can be written as

$$\phi_v(\delta_{\text{MID}}) \cong \frac{4\pi^2}{nm} \cos \delta_{\text{MID}} \sum_{x=1}^{n} I_v\left(\frac{2\pi}{n} x, \delta_{\text{MID}}\right)$$

$$\cong C_z(\delta_{\text{MID}}) \sum_{x=1}^{n} I_v\left(\frac{2\pi}{n} x, \delta_{\text{MID}}\right)$$

where angles are expressed in radians, or

$$\phi_v(\delta_{\text{MID}}) \cong \frac{4\pi^2}{nm} \cos \delta_{\text{MID}} \sum_{x=1}^{n} I_v\left(\frac{360°}{n} x, \delta_{\text{MID}}^°\right)$$

$$\cong C_z(\delta_{\text{MID}}) \sum_{x=1}^{n} I_v\left(\frac{360°}{n} x, \delta_{\text{MID}}^°\right)$$

where angles are expressed in degrees. Now, the zonal constant $C_z(\delta_{\text{MID}})$ is equal to

$$C_Z(\delta_{\text{MID}}) = \frac{4\pi^2}{nm} \cos \delta_{\text{MID}}$$

Note that the zonal constant can either be defined in terms of the angle θ relative to an optical axis pointing toward the upper pole, or in terms of the angle δ relative to an optical axis pointing toward the horizontal great circle.

An application of the second zonal constant example is to convert the desired luminous intensity radiation pattern of a government-regulated light source into an equivalent value of luminous flux. For example, an automotive manufacturer may want to calculate the total luminous flux needed to meet an automotive turn signal luminous intensity specification. Or a traffic signal manufacturer may want to calculate the total luminous flux needed to meet the Institute of Traffic Engineers (ITE) traffic signal luminous intensity specification. Because these specifications define the minimum luminous intensities over a certain range of degrees, left and right, and up and down, then the numerical integration is greatly simplified by making each integration zone equal to one vertical angle in the table of specified minimum luminous intensities.

Figure 6(a) shows a sphere equally subtended into m vertical divisions and n horizontal divisions. If the optical axis of the signal lamp is pointing toward the horizontal great circle of the sphere and $I_v(H = 0°, V = 0°)$ is facing the viewer, then the total luminous flux could be calculated as follows:

$$\phi_v = \sum_{y=1}^{m/4} C_z(y) \sum_{x=1}^{n} I_v\left(\frac{360°}{n} x, \frac{360°}{m} y - \frac{360°}{2m}\right)$$

$$+ \sum_{y=-1}^{-m/4} C_z(y) \sum_{x=1}^{n} I_v\left(\frac{360°}{n} x, \frac{360°}{m} y + \frac{360°}{2m}\right)$$

where

$m/4$ = number of horizontal bands in each hemisphere

$\dfrac{360°}{m} y - \dfrac{360°}{2m}$ = off-axis angle of the midpoint of each horizontal band in the upper hemisphere.

$\dfrac{360°}{m} y + \dfrac{360°}{2m}$ = off-axis angle of the midpoint of each horizontal band in the lower hemisphere.

$$C_z(y) = \frac{4\pi^2}{nm} \cos\left(\frac{360°}{m} |y| - \frac{360°}{2m}\right)$$

Note, that the left- and right-hand terms of the equation calculate the luminous flux in the upper and lower hemispheres respectively. For example,

6 APPLICATIONS FOR HIGH-BRIGHTNESS LIGHT-EMITTING DIODES

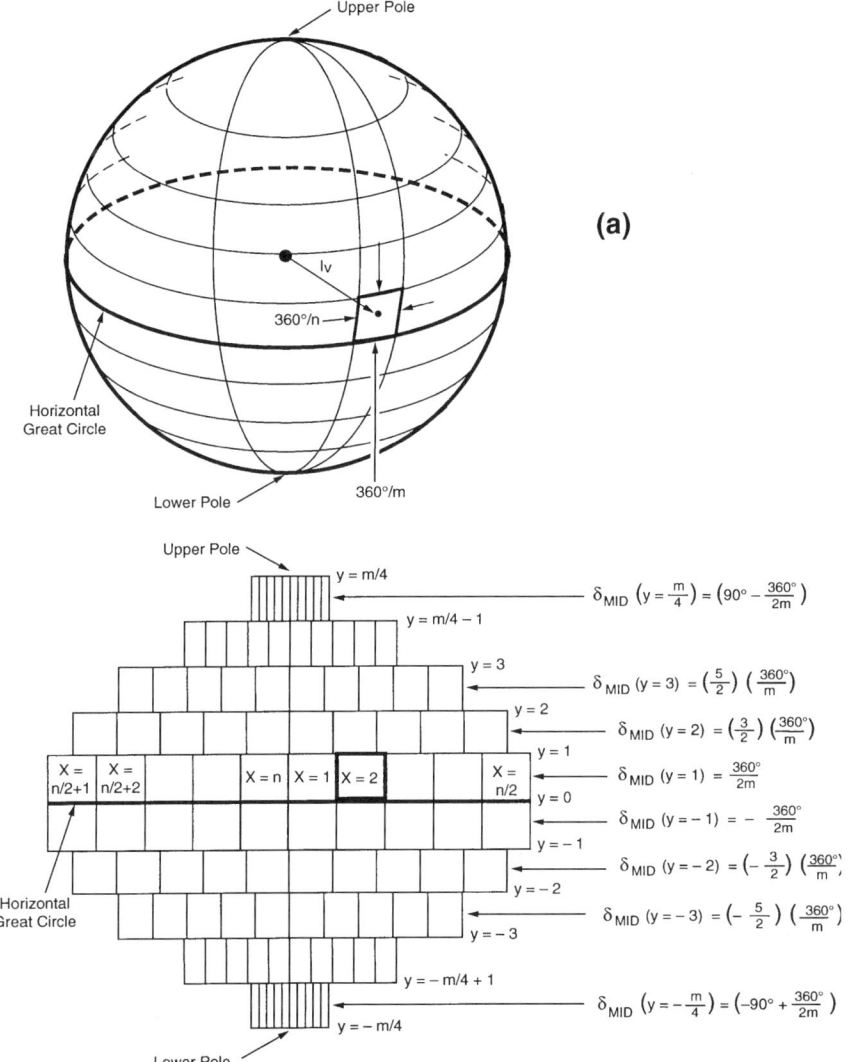

FIG. 6. Two approaches for zonal constant integration method. (a) A sphere equally subtended into m vertical divisions and n horizontal divisions. (b) A sphere equally subtended into m vertical divisions and n horizontal divisions such that the midpoint of the center band is equal to the horizontal great circle.

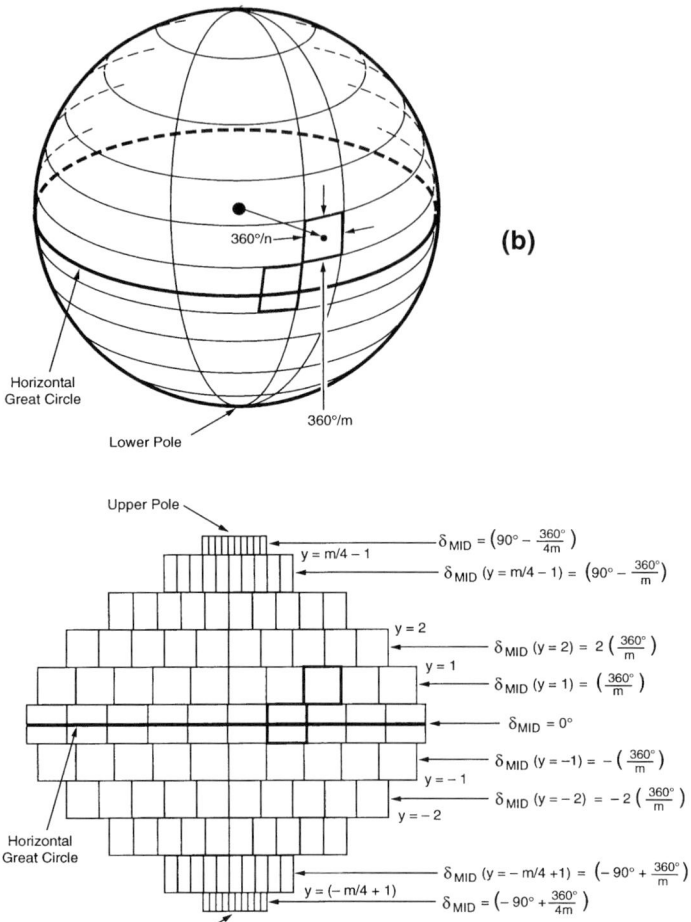

FIG. 6. Continued.

suppose that the sphere is divided into 10-degree segments. Thus, $m = 360°/10°$, or 36 and $n = 360°/10°$, or 36. Then, the double summation would become

$$\phi_v = \sum_{y=1}^{9} C_z(y) \sum_{x=1}^{36} I_v(10° x, 10° y - 5°)$$
$$+ \sum_{y=-1}^{-9} C_z(y) \sum_{x=1}^{36} I_v(10° x, 10° y + 5°)$$

where

$$C_z(y) = \frac{4\pi^2}{36^2} \cos(10°|y| - 5°)$$

Thus, for this example, the luminous intensity would need to be specified at vertical angles δ_{MID} of $\pm 5, 15, 25, 35, 45, 55, 65, 75,$ and 85 degrees.

One slight modification to this method would be to rotate the sphere vertically by half of a horizonal band. The effect would be to create a horizontal band with the midpoint equal to the horizontal great circle and to create horizontal bands at vertical angles of $\pm 360°(y - \tfrac{1}{2})/m$ for values of $y = 1, 2, \ldots (m/4 - 1)$. There would also be halfwidth horizontal bands at the north and south poles of the sphere. This approach is shown in Fig. 6(b). The benefit of this method is that the luminous intensities can be specified at more commonly used vertical angles. With this method, the total luminous flux could be calculated as follows:

$$\phi_v = \frac{4\pi^2}{nm} \sum_{x=1}^{n} I_v\left(\frac{360°}{n} x, 0°\right) + \sum_{y=1}^{(m/4-1)} C_z(y) \sum_{x=1}^{n} I_v\left(\frac{360°}{n} x, \frac{360°}{m} y\right)$$

$$+ \sum_{y=-1}^{-(m/4-1)} C_z(y) \sum_{x=1}^{n} I_v\left(\frac{360°}{n} x, \frac{360°}{m} y\right)$$

$$+ \frac{4\pi^2}{2nm} \cos\left(90° - \frac{360°}{4m}\right) \sum_{x=1}^{n} I_v\left(\frac{360°}{n} x, 90° - \frac{360°}{4m}\right)$$

$$+ \frac{4\pi^2}{2nm} \cos\left(-90° + \frac{360°}{4m}\right) \sum_{x=1}^{n} I_v\left(\frac{360°}{n} x, -90° + \frac{360°}{4m}\right)$$

where

$$C_z(y) = \frac{4\pi^2}{nm} \cos\left(\frac{360°}{m} |y|\right)$$

Note that the first term of the equation calculates the luminous flux in the band around the horizontal great circle; the second and third terms calculate the luminous flux in all of the full-width bands in the upper and lower hemispheres, respectively; and the fourth and fifth terms calculate the luminous flux in the half-width band next to the upper and lower poles, respectively. In many applications, the luminous intensity is nonzero for a small number of degrees up and down and left and right. Thus only the nonzero bands of the first, second, and third terms need to be summed, and the fourth and fifth terms of the equation equal zero.

Again, suppose that the sphere is divided into 10-degree segments. Then the double summation would become

$$\phi_v = \frac{4\pi^2}{36^2} \sum_{x=1}^{36} I_v(10° x, 0°) + \sum_{y=1}^{8} C_z(y) \sum_{x=1}^{36} I_v(10° x, 10° y)$$

$$+ \sum_{y=-1}^{-8} C_z(y) \sum_{x=1}^{36} I_v(10° x, 10° y)$$

$$+ \frac{2\pi^2}{(36)^2} \cos(87.5°) \sum_{x=1}^{36} I_v(10° x, 87.5°)$$

$$+ \frac{2\pi^2}{(36)^2} \cos(-87.5°) \sum_{x=1}^{36} I_v(10° x, -87.5°)$$

where

$$C_z(y) = \frac{4\pi^2}{36^2} \cos(10°|y|)$$

Thus, for this example, the luminous intensity would need to be specified at vertical angles δ_{MID} of 0 degrees and $\pm 10, 20, 30, 40, 50, 60, 70, 80$ and 87.5 degrees.

As an example of using this latter zonal constant method, the luminous flux for an automotive brake lamp can be calculated. The minimum luminous intensity for a US automotive brake light is given in Table I. These numbers are taken either from National Highway Traffic Safety Administration (NHTSA) regulation FMVSS108 or from the Society of Automotive

TABLE I

MINIMUM US AUTOMOTIVE STOP LAMP PHOTOMETRIC REQUIREMENTS (CANDLES)[a]

	20L	10L	5L	V	5R	10R	20R
10U				16		16	
5U	10	30		70		30	10
H		40	80	80	80	40	
5D	10	30		70		30	10
10D				16		16	

Data from National Highway Traffic Safety Administration (NHTSA) regulation FMVSS108 and from the Society of Automotive Engineers (SAE) specification J586.
[a] Maximum intensity is 300 cd. Minimum specification for one compartment lamp.

6 APPLICATIONS FOR HIGH-BRIGHTNESS LIGHT-EMITTING DIODES

TABLE II
ZONAL CONSTANT INTEGRATION OF US AUTOMOTIVE STOP LAMP

	20L	15L	10L	5L	V	5R	10R	15R	20R	Horizontal Sum	$C_z \cos V$	θ_v
10U	(5)	(10)	(14)	16	(26)	16	(14)	(10)	(5)	116	7.500×10^{-3}	0.87 lm
5U	10	(18)	30	(50)	70	(50)	30	(18)	10	286	7.586×10^{-3}	2.17 lm
H	(10)	(20)	40	80	80	80	40	(20)	(10)	380	7.615×10^{-3}	2.89 lm
5D	10	(18)	30	(50)	70	(50)	30	(18)	10	286	7.586×10^{-3}	2.17 lm
10D	(5)	(10)	(14)	16	(26)	16	(14)	(10)	(5)	116	7.500×10^{-3}	0.87 lm
											Total θ_v	8.97 lm

"Parentheses denote estimated minimum luminous intensity.

Engineers (SAE) specification J586. Note that minimum luminous intensities are specified over a range of 10 degrees up and down and 20 degrees left and right.

The first step is to estimate the minimum luminous intensities for the unspecified coordinates (e.g., 5U, 5L). A reasonable assumption could be that the intensity at each unspecified coordinate is equal to the average of the intensities of all four adjacent coordinates. Using this approach, the minimum luminous intensities of the unspecified coordinates could be calculated as shown in Table II.

Since this example assumes that the luminous intensity is constant over an area of 5 degrees horizontal by 5 degrees vertical, $n = m = 360°/5° = 72$. Thus, the integration would have a zonal constant equal to

$$C_Z(\delta) = \frac{4\pi^2}{72^2} \cos \delta = (7.615 \times 10^{-3}) \cos \delta$$

Then the summation would be calculated as shown in Table II. Thus, about 9.0 lumens of luminous flux would be required to meet the minimum luminous intensity specifications for a US automotive brake light.

3. COLOR MEASUREMENT

The color of a light source is another important visible property. Most lighting specifications refer to color in terms of the 1931 CIE (Commission Internationale de L'Eclairage) chromaticity color coordinates. Thus, the 1931 CIE color system should be reviewed.

Figure 7 shows the 1931 CIE chromaticity diagram. In general, the color of any light source can be represented as an (x, y) coordinate in this color space. The color coordinate (x, y) can be calculated by integrating the spectrum of the light source with three different functions called the 1931 CIE $\bar{x}, \bar{y}$, and $\bar{z}$ color matching functions. These 1931 CIE color matching functions are shown in Fig. 8. Thus, the color coordinate (x, y) can be calculated as follows:

$$x = \frac{X}{X + Y + Z}$$

$$y = \frac{Y}{X + Y + Z}$$

$$X = \int \bar{x}(\lambda)s(\lambda)\ d\lambda$$

$$Y = \int \bar{y}(\lambda)s(\lambda)\ d\lambda$$

$$Z = \int \bar{z}(\lambda)s(\lambda)\ d\lambda$$

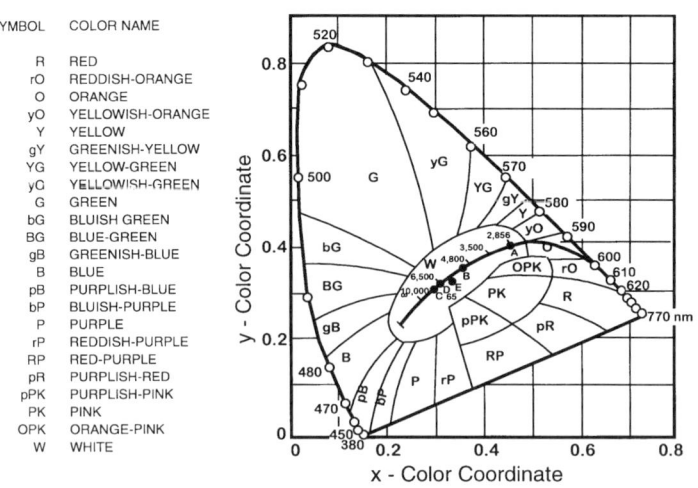

FIG. 7. 1931 CIE (Commission Internationale de L'Eclairage) color space. (From Gage *et al.*, 1977. Reprinted with the permission of The McGraw-Hill Companies.)

6 APPLICATIONS FOR HIGH-BRIGHTNESS LIGHT-EMITTING DIODES

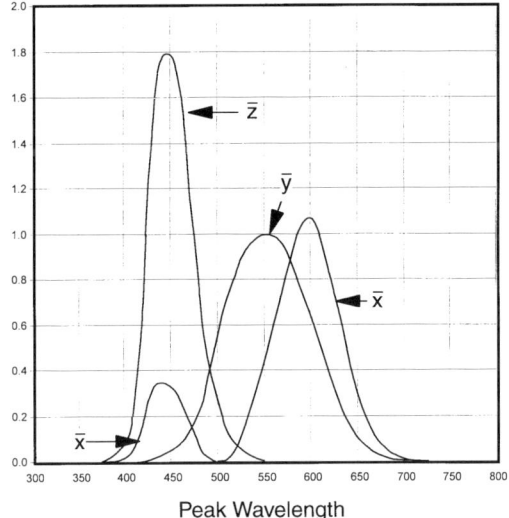

FIG. 8. 1931 CIE (Commission Internationale de L'Eclairage) color matching functions.

where:

(x, y) = color coordinate of the light source

$\bar{x}(\lambda)$ = 1931 CIE $\bar{x}$ color matching function

$\bar{y}(\lambda)$ = 1931 CIE $\bar{y}$ color matching function, also known as $V'(\lambda)$, the luminosity function

$\bar{z}(\lambda)$ = 1931 CIE $\bar{z}$ color matching function

$s(\lambda)$ = spectrum of light source (W/nm units)

Pure colors are located on the extreme outer edges of the diagram and white is located in the center. Also shown in Fig. 7 is the locus of points for blackbody radiators of different color temperatures. These color coordinates correspond to the nontechnical definition of "white" light. CIE has defined several white colors called *illuminants* that are the color coordinates for several commonly used illumination sources. The CIE illuminants and their definitions are as follows:

Illuminant A = (0.4476, 0.4074), 2856 K incandescent source
Illuminant B = (0.3484, 0.3516), direct sunlight, approximately 4870 K

Illuminant C = (0.3101, 0.3162), overcast sunlight, approximately 6770 K
Illuminant D_{65} = (0.3128, 0.3292), daylight, approximately 6504 K
Illuminant E = (0.3333, 0.3333), equal energy.

(Note. CIE illuminant E is an imaginary color that has equal amounts of X, Y, and Z energies.)

Color coordinates within the diagram can be considered to be a mixture of white light and monochromatic highly saturated light. As the point moves inward, the percentage of white light increases, whereas that of monochromatic highly saturated light decreases.

The dominant wavelength is defined as the single monochromatic wavelength that appears to be same color as the light source. The dominant wavelength can be determined by drawing a straight line from one of the CIE white illuminants, through the (x, y) coordinate to be measured, until the line intersects the outer locus of points along the spectral edge of the chart. The dominant wavelength is the wavelength of the intercept of the straight line and the outer edge of the space. This concept is shown in the graph of Fig. 9. The color purity is the weighted average of the (x, y) coordinate relative to the coordinate of the illuminant and the coordinate of the dominant wavelength:

$$\text{Color purity} = \frac{\sqrt{(x - x_i)^2 + (y - y_i)^2}}{\sqrt{(x_d - x_i)^2 + (y_d - y_i)^2}} \times 100\%$$

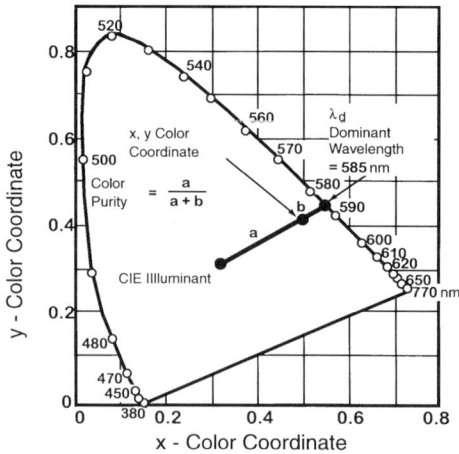

FIG. 9. Dominant wavelength calculation. CIE, Commission Internationale de L'Eclairage. (From Gage et al., 1977. Reprinted with the permission of The McGraw-Hill Companies.)

where (x, y) and (x_i, y_i) are the color coordinates of the light source and the CIE illuminant, respectively, and (x_d, y_d) is the color coordinate of the intercept of the straight line and the edge of the color space, also known as the dominant wavelength λ_d.

Unless otherwise specified, the convention used in this chapter is that the CIE illuminant C is used in all color calculations because most of the applications discussed are designed for outdoor viewing.

III. Automotive Signal Lighting

1. CHARACTERISTICS OF THE AUTOMOTIVE AND TRUCK MARKET

The worldwide automotive-truck market is a vast new market for LED technology for signal lighting. According to *Automotive News*, in 1994, 36 million cars and 15 million trucks were produced worldwide (*Automotive News*, 1995). Each vehicle requires red stop lamps, red-amber rear turn signals, and amber front turn signals. Many countries also require red center high mount stop lamps (CHMSL) for cars and light trucks. Red-amber clearance and side markers also are required for heavy trucks. Traditionally, incandescent bulbs have been used for these automotive signal lighting applications; however, the increase in performance of LED technology has made it a serious contender.

Automotive signal lighting is regulated by various agencies throughout the world. In the United States, NHTSA regulation FMVSS108 regulates the photometric intensity and acceptable color ranges of signal lighting, and detailed automotive lighting specifications are written by the Society of Automotive Engineers (SAE). In Europe the Economic Commission for Europe (ECE) and in Japan the Japanese Industrial Standard (JIS) define the automotive signal lighting requirements. For the rest of the world these requirements are defined individually by country or by the International Standards Organization (ISO). These various specifications cover all aspects of automotive signal lighting including mounting position, geometric visibility, illuminated area, environmental considerations, color, luminous intensity, and beam radiation pattern. Although these specifications were written around the use of incandescent bulbs, other lighting technologies such as LED technology are now being used.

Over the past several years, NHTSA, SAE, ECE, and JIS have made considerable progress in harmonizing their automotive signal lighting standards. This process requires the specification review committee of each country to review the lighting specifications of the other countries, and then to revise its standards to minimize differences. The closer that the lighting specifications agree in test conditions and limits, the easier that automotive manufacturers will be able to export their vehicles to the other countries, thus helping to reduce world trade barriers.

2. Automotive Luminous Intensity and Color Requirements

The previously mentioned automotive signal lighting specifications define the on-axis minimum and maximum luminous intensities ($I_v(H, V)$; where ($H = V = 0°$) as well as certain off-axis points. Table III gives the minimum on-axis luminous intensity standards for passenger cars in the United States, Europe, and Japan. Figure 10 shows the signal lighting colors for the United States, Europe, and Japan. In general, the specifications define two colors, red and amber, for most of the signaling and marking functions. The color ranges are defined as regions in the 1931 CIE chromaticity diagram that are bounded by a range of dominant wavelengths of monochromatic highly saturated colored light and a line parallel (or approximately parallel) to the edge of the color space. Thus, the color regions could be defined in terms of dominant wavelength and minimum color purity. Note, that the US, European, and Japanese specifications define the red color as a dominant wavelength (λ_d) of approximately $\lambda_d \geqslant 610$ nm and the amber color as a dominant wavelength range approximately 590 nm $\leqslant \lambda_d \leqslant$ 595 nm.

3. Benefits of Light-Emitting Diode Technology

Light-emitting diode (LED) technology provides a number of benefits for automotive signal lighting. Because incandescent bulb technology has tradi-

TABLE III

Minimum on-Axis Luminous Intensity Requirements for Automotive Signal Lamps

Function	Color	United States (cd)	Japan (cd)	Europe (cd)
CHMSL	Red	25	—	25
Stop		80	80	60
Tail		2.0	2.0	4.0
Rear turn		80	—	—
Rear turn	Amber	130	200	50
Front turn		200[a]		175[b]

[a] > 100 mm away from low beam headlamp.
[b] > 40 mm away from headlamp.
CHMSL, center high mount stop lights.

6 Applications for High-Brightness Light-Emitting Diodes

SAE J578 Color Ranges

Amber Color Ranges

COUNTRY	MIN	MAX
US	587nm	596nm
JAPAN	592nm	595nm
ECE	589nm	595nm

Red Color Ranges

COUNTRY	MIN	MAX
US	611 nm	-
JAPAN	610 nm	-
ECE	610 nm	-

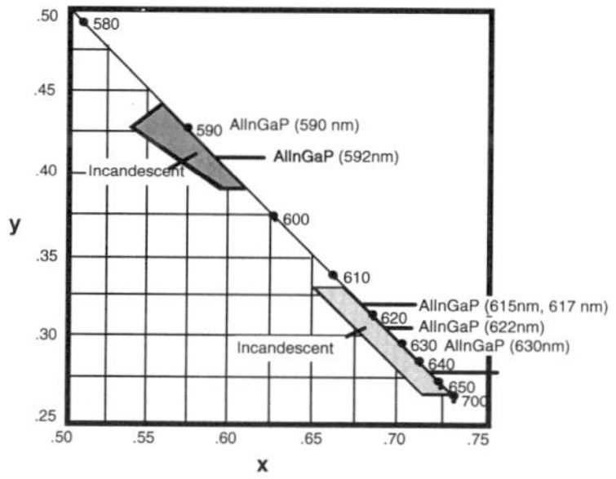

FIG. 10. Required SAE J578 color ranges for automotive signal lamps. SAE, Society for Automotive Engineers; AlInGaP, Aluminum Indium Gallium Phosphide LED technology; TS AlGaAs, Transparent Substrate Aluminum Gallium Arsenide LED technology.

tionally been used for this application, it would be natural to compare an LED signal lamp with its incandescent counterpart. However, with the advent of neon lighting technology, it is also appropriate to compare LED with neon as well.

To better understand the benefits of LED technology, the construction of an incandescent brake lamp can be compared with its LED counterpart. A sketch of a typical incandescent brake lamp is shown in Fig. 11(a). One or more socketed incandescent bulbs are mounted in an injection-molded plastic housing. Because the incandescent bulb acts as an optical point source, the bulb is normally mounted at the focal point of a parabolic reflector. The reflector collimates part of the luminous flux generated by the bulb. Generally, the reflector is formed by either painting or vacuum metalizing the inside of the plastic housing. A colored plastic lens is attached in the front of the plastic housing. The plastic lens is commonly constructed of polycarbonate or acrylic, which is also injection molded. The color of emitted light is determined by lens thickness and the amount of dye added to the plastic molding compound. Generally, small pillow lenses are molded into the inner surface of the plastic lens. These pillow lenses serve to break up the image of the incandescent bulb filament and may also spread the collimated luminous flux into the desired optical beam pattern. Because the same housing may also be used for turn signal and backup lamp functions, the colored plastic lens may be fabricated with a two-shot (red-clear or red-amber) or three-shot (red-amber-clear) injection molding process.

The construction of an LED brake lamp can now be compared. A sketch of this hypothetical brake lamp is shown in Fig. 11(b). An array of LED lamps is interconnected using either a printed circuit board or stamped metal lattice. Because each LED lamp has a narrow radiation pattern, an optical reflector is not required. However, for best luminous flux utilization,

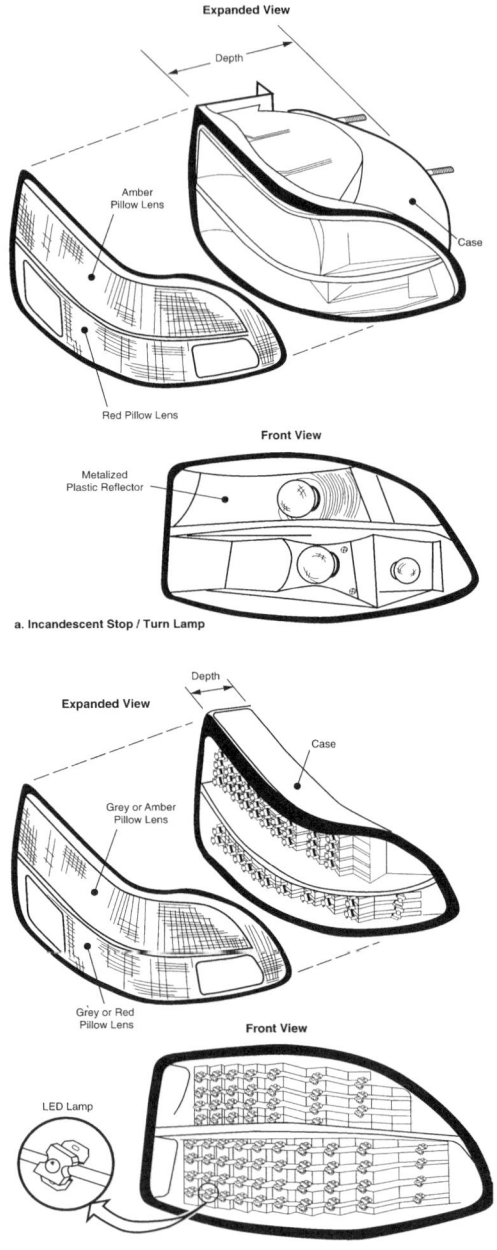

FIG. 11. Expanded view of (a) an incandescent and (b) a light-emitting diode (LED) stop-turn lamp.

each LED lamp should be mounted within a small metalized reflector. The LED array is attached to the inside of the plastic housing, and a plastic lens is attached to the front of the plastic housing. Because the LED lamp is a colored light source, the outer lens does not need to be colored. Similar to the incandescent design, pillow lenses are molded into the inner surface of the outer lens. These pillow lenses primarily serve to break up the images of the LED lamps. Most likely, the pillow lens design will be somewhat different from the incandescent pillow lens design owing to the different optical beam pattern of the LED array and the smaller size of the LED lamp images.

Table IV compares the incandescent bulb and LED lamp brake lights. The incandescent design has some advantages over its counterpart: it has the lowest cost source of luminous flux, is available in a wide range of luminous flux outputs, and is virtually independent of ambient temperature.

The incandescent design does have a number of disadvantages. The filament of the bulb can be broken by mechanical shock and vibration. Because the lifetime of the bulb is usually shorter than the normal lifetime of the vehicle, bulbs must be mounted in sockets for easy repair. The bulb also generates considerable heat, which puts a number of mechanical constraints on the size and volume of the housing, and the plastic materials used for the housing and lens. In addition, the large size of the bulb and limited number of bulbs used for the brake lamp tend to require that the outer case be fairly deep to accommodate the optical reflector. Because the

TABLE IV

COMPARISON BETWEEN LIGHT-EMITTING AUTOMOTIVE DIODES (LEDs) AND INCANDESCENT SIGNAL LAMPS

Lamp	Advantages	Disadvantages
Incandescent	Low cost luminous flux Socketed for easy repair Light output independent of temperature	Filament shock sensitive MTBF less than vehicle life Lifetime less than vehicle life Hight heat and power Slow response time Color determined by exterior lens
LED	Shock resistant MTBF much longer than vehicle life Low heat and power Fast response time Color detrmined by light source	Higher cost and luminous flux Limited junction temperature Not repairable Light output varies with temperature

MTBF, mean time between failure.

emitted color of the brake lamp is determined primarily by the plastic lens, the color range is dictated by process variations in the molding compound and injection molding process. Also, a broken or cracked lens no longer generates the desired color. Finally, the bulb takes 200 to 300 milliseconds to turn on. This time lag reduces the amount of time other drivers have to react.

The LED design has a number of advantages. It is very resistant to mechanical shock and vibration. The mean time between failure (MTBF) due to random catastrophic failures of LED lamps is several orders of magnitude higher than for incandescent bulbs. The lifetime of the LED lamp is determined by a phenomenon called light-output degradation in which the light output gradually diminishes as the LED is operated; however, light-output reduction is generally small over the normal lifetime of the vehicle. Therefore, the LED brake lamp can be designed as a sealed unit that is not repairable, eliminating a common type of field repair.

The TS AlGaAs and AlInGaP LED technologies have considerably higher luminous efficiencies than do red or amber filtered incandescent bulbs. This plus a more efficient optical radiation pattern allow the LED brake lamp to be operated at a substantially lower electrical current. Typical automotive applications reduce the drive current to approximately one quarter that of the incandescent brake lamp. The lower electrical consumption provides secondary benefits of potentially reducing the size and weight of the electrical wiring harness in the vehicle. The lower power consumption also reduces the internal temperature increase within the brake lamp assembly, which may allow the use of lower temperature plastics for the outer lens and housing.

The LED lamps are much smaller than are incandescent bulbs. An array of LED lamps is used to replace a single incandescent bulb so that the optical reflector (if used) is much smaller. Therefore, the LED brake lamp assembly can be designed to be much thinner than its incandescent counterpart, helping to reduce the depth of the outer housing. This, in turn, allows the use of simplified construction techniques for the body panel of the vehicle's outer skin that houses the brake light assembly. In addition, the LED lamps array easily can be formed into different shapes or designed to light around the corner of the vehicle, allowing for a great many styling options.

Because the LED lamp is a colored light source, the dye used in the molding compound of the outer lens has minimal impact on the color emitted by the brake lamp. A colorless or neutral density outer lens could even be used, which provides even more styling flexibility.

Finally, the turn-on response time of the LED lamp is on the order of

100 ns. This shorter time lag increases the amount of time other drivers have to react, providing a safety benefit.

The LED design has several disadvantages compared with the incandescent design. Considering the light source only as a source of luminous flux, LED lamps are more expensive. The cost per lumen has dropped significantly over the past few years, with the advent of TS AlGaAs and AlInGaP LED materials technologies. Higher power LED lamp packages also have been developed, further reducing the cost per lumen; however, it is still higher than for bulbs. When the complete system cost is considered, however, the cost of an LED automobile signal lamp is not significantly different from that of an incandescent automotive signal lamp.

Compared with an incandescent bulb, each LED lamp generates a much smaller amount of luminous flux. For example, an industry standard TS AlGaAs 5-mm (T-$1\frac{3}{4}$) lamp generates about 300 mlm of luminous flux at a 20-mA drive current. At 70 mA, a HPWT-DH00 TS AlInGaP high-flux lamp generates about 2500 mlm. A typical automotive incandescent bulb produces over 100 lm.

Unlike incandescent bulbs, LED lamp light output varies with temperature. For example at 55°C the luminous flux output of a TS AlGaAs or AlInGaP LED lamp decreases by about 30% compared with the luminous flux output at 25°C.

Finally, LED lamps have a limited maximum junction temperature. Those used for automobile signal lighting are constructed by casting an LED chip inside an optically clear epoxy package. The LED chip is electrically connected to the metal pins of the package with an electrically conductive epoxy on one side and a gold wire on the other. The maximum internal junction temperature of the lamp is limited by the thermal expansion properties of the epoxy package compared with the gold wire and lead frame. For many LED lamps, the maximum internal junction temperature is 110°C, although some are available with that of 125°C. If the LED lamp is temperature-cycled or power-temperature-cycled at temperatures higher than this maximum junction temperature, the wire bond can break prematurely, resulting in a catastrophic failure. Due to the limitations on maximum junction temperature, more care needs to be taken in the thermal design of an LED automotive signal lamp compared with an incandescent automotive signal lamp.

Neon technology is currently being considered as an alternative for automotive signal lighting. Table V compares neon and LED automotive signal lamps. With neon technology, light is created by generating a high-voltage electrical discharge in a glass tube filled with neon or other gases. The primary limitation of neon lighting is that a high-voltage

TABLE V

COMPARISON BETWEEN LIGHT-EMITTING DIODE (LED) AND NEON AUTOMOTIVE SIGNAL LAMPS

Lamp	Advantages	Disadvantages
Neon	Light output independent of voltage (9–16 V) due to ballast Uniform CHMSL lighting Light output independent of temperature	Tube shock sensitive Operates at lethal voltages Ballast adds complexity Heavy and bulky Ballast can generate EMI Expensive
LED	Shock resistant Low heat and power Lower cost Redundant light sources	Light output varies with temperature and applied voltage More difficult to get uniform appearance

CHMSL, center high mount stop lights; EMI, electromagnetic interference.

electrical potential must be generated to cause the neon gas to ionize, allowing electrical current to flow through the tube. Because most automotive electrical systems are 12 or 24 V dc, some type of step-up electrical circuit, called a *ballast*, must be used to create the approximately 1000-V electrical signal needed to ionize the gas.

Compared with LED technology, neon technology has several advantages for automotive signal lighting. The optical properties of the energized neon tube create a uniform strip of light ideal for CHMSL application. Where a uniform area of light is needed, such as in turn and brake signals, the neon tube can be shaped into a series of S-shaped curves. The light output of the neon tube also is independent of temperature, assuming that the electrical ballast output is independent of temperature. Also, the requirement of the electrical ballast allows a circuit to be designed that regulates the light output of the neon tube over a wide range of input voltages.

Compared with LED technology neon technology has several disadvantages for automotive signal lighting. Because the tube operates at lethal voltages, the signal light must be designed to be fail-safe in the event of damage caused by automobile accidents. Because the electrical ballast can be a source of electromagnetic interference (EMI), the outer case must be properly shielded to prevent EMI emission, and the electrical wires must be properly terminated to prevent conducted EMI transmission. The neon signal light tends to require higher electrical power consumption than does an LED signal light. The cost of the neon signal light also tends to be higher than an equivalent LED signal light.

4. LUMINOUS FLUX REQUIREMENTS

The first step in the design of an LED automotive signal light is to determine approximately how many LED lamps are needed to meet the minimum government-regulated photometric output. The easiest way to do this is to integrate the minimum luminous intensity values to calculate the minimum luminous flux required. Then, the number of LED lamps can be estimated by dividing the total luminous flux required by the luminous flux emitted by each LED lamp.

The zonal constant method can be used to convert the luminous intensity radiation pattern of the light source into the total luminous flux, as shown previously in Part II, Section 2. For example, about 9.0 lumens of luminous flux is needed to meet the minimum US photometrics for an automotive brake light according to NHTSA FMVSS108 or SAE specification J586.

A table of minimum on-axis luminous intensities for automotive signal lamps has been given previously (Table III). These specifications cover the NHTSA FMVSS, ECE, and JIS standards. Using the zonal constant integration method, the minimum luminous fluxes for the same automotive signal lamps are given in Table VI.

The values of luminous flux given in Table VI may be unrealistic. For example, the radiation pattern achieved may exceed the minimum luminous intensity values at some points. Or, perhaps, the luminous intensity is greater than zero at some points outside the specified range of angles. The calculation also does not account for transmission losses through a colored lens or a neutral density lens. For these reasons, the minimum luminous flux needed is somewhat higher than the values given in Table VI. The equation shown below can be used to estimate a more realistic minimum luminous

TABLE VI

MINIMUM PHOTOMETRIC FLUX REQUIREMENTS FOR AUTOMOTIVE SIGNAL LAMPS

Function	Color	United States (lm)	Japan (lm)	Europe (lm)
CHMSL	Red	2.9	—	2.9
Stop		9.0	9.2	5.6
Tail		0.23	0.23	0.37
Rear turn		9.0	—	—
Rear turn	Amber	15.1	22.6	4.6
Front turn		22.4		16.2

CHMSL, center high mount stop lights.

flux:

$$\phi_{min} = \phi_{spec} \left(\frac{1}{T_{filter}}\right)\left(\frac{1}{T_{glass}}\right)\left(\frac{1}{1-L_{pattern}}\right)(F_{guard})$$

where

ϕ_{spec} = minimum luminous flux needed to meet specifications

T_{filter} = transmission of the plastic outer lens

T_{glass} = transmission of the glass window (rear-window-mounted CHMSL only)

$L_{pattern}$ = luminous flux loss due to radiation pattern inaccuracy

F_{guard} = optional photometric guard band

Table VII shows more realistic minimum luminous flux requirements for the different automotive signal lamps given in Table III. Note that by using the assumptions given, almost twice as much red and 2.25 times as much amber luminous flux are required.

The next step is to divide the minimum luminous flux required by the signal lamp by the luminous flux emitted by each LED lamp. Several factors determine the luminous flux emitted by the LED lamp:

First, the LED lamp materials technology varies by a factor of approximately 4:1 in its luminous efficiency. Small packaging variations also can

TABLE VII

"Realistic" Minimum Photometric Flux Requirements for Automotive Signal Lamps

Function	Color	United States (lm)	Japan (lm)	Europe (lm)
CHMSL	Red	5.8	—	5.8
Stop		17.9	18.3	11.1
Tail		0.46	0.46	0.73
Rear turn		17.9	—	—
Rear turn	Amber	33.7	50.4	10.2
Front turn		50.0		36.2

Assumptions: T_{FILTER} = 0.90 (red), 0.80 (amber); T_{GLASS} = 1.00; $L_{PATTERN}$ = 0.30; F_{GUARD} = 1.25.
CHMSL, center high mount stop lights.

6 APPLICATIONS FOR HIGH-BRIGHTNESS LIGHT-EMITTING DIODES 261

effect the radiation pattern and optical beam. After manufacture, therefore, LED lamps are tested for either luminous intensity or luminous flux and sorted into bins. Each bin typically has a maximum to minimum ratio of about 2:1. Obviously, the particular bin used to assemble each signal lamp has a significant impact on the final light output.

Second, the luminous intensity and luminous flux of the LED lamp decreases with increasing temperature:

$$\phi_v(T) = \phi_v(25°C)\, e^{K(T-25°C)}$$

where

$$K = -0.013 \text{ for TS AlGaAs} \qquad K = -0.010 \text{ for AlInGaP}$$

Because government regulations were originally written around the use of incandescent bulbs and also due to the difficulty of making photometric measurements over temperature, all photometric tests normally are specified at 25°C. However, SAE J1889 recommends the use of a 30-minute warmup time prior to photometric testing to allow the LED signal lamp to reach thermal equilibrium. Thus, even if all photometric testing is done at 25°C, the internal heating within the LED lamp will have an effect on the final light output.

Third, the luminous intensity and luminous flux of the LED lamp is proportional to the drive current. This relationship generally is shown on a graph on the data sheet. Some LED lamp data sheets also show the effects of internal heating due to increased drive currents on the same graph. An example of this type of graph is shown in Fig. 12 for the Hewlett-Packard HPWA-M/DX00.

Fourth, only some of the luminous flux emitted by the LED lamp may be used by the secondary optics in the LED signal light. Without some type of secondary optics, light emitted at off-axis angles greater than 22 degrees (corresponding to 10 degrees up and 20 degrees right) does not contribute to the government photometric specification.

All of these factors may be combined into an equation that calculates the total flux output of the LED lamp as a function of drive current, temperature, and luminous flux utilization:

$$\phi_{\text{LED}} = (\phi_{\text{BIN}})(\Delta\phi/\Delta T_A)(\phi(I_F, \theta_{JA}))(T_{\text{optics}})$$

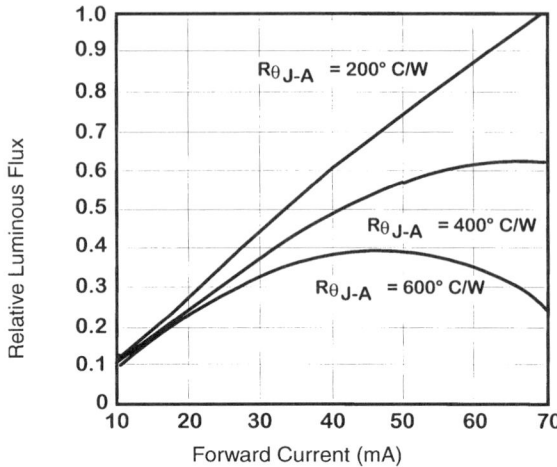

FIG. 12. Luminous flux versus forward current for HPWA-MH00 AlInGaP light-emitting diode lamp. (From Optoelectronics Designer's Catalog, Hewlett-Packard Company, 1996. Reprinted with permission from Hewlett-Packard Company.)

where

ϕ_{LED} = useful luminous flux emitted by the LED lamp

ϕ_{BIN} = minimum luminous flux for a specified luminous flux test bin

$\Delta\phi/\Delta T_A$ = luminous flux reduction if photometric specifications must be met at elevated ambient temperature.

$\phi(I_F, \theta_{JA})$ = normalized luminous flux output of the LED lamp based on the drive current and internal heating (increase in junction temperature over ambient)

T_{optics} = amount of luminous flux used by the secondary optics

Table VIII shows how this equation can be used to calculate the usable luminous flux generated by several Hewlett-Packard LED lamps. Note that the assumptions used in the calculation reduce the amount of luminous flux by a factor of about 60%.

The last step in determining the number of LED lamps needed for an automotive signal lamp is to divide the realistic luminous flux requirement (Table VII) by the useful luminous flux output of the LED lamp (Table VIII). These results are shown in Table IX.

Note that the number of LED lamps needed for an automotive signal lamp vary significantly based on the type of LED lamp used. For example,

TABLE VIII
EXAMPLES SHOWING CALCULATION OF LIGHT-EMITTING DIODE LAMP LUMINOUS FLUX OUTPUT

Device	ϕ_V TYP at 70 mA (mlm)	ϕ_{BIN} (mlm)	$\Delta\phi/\Delta T$	ϕ (60 mA, 400°C/W)	T_{OPTICS}	ϕ_{LED} (mlm)
HPWR-M300	800	D, 675	1.00	0.68	0.60	275
HPWA-M/DX00	1250	D, K 675	1.00	0.62	0.60	251
		E, L 990	1.00	0.62	0.60	368
HPWT-M/DX00	2500	F, M 1440	1.00	0.62	0.60	536
		G, N 2250	1.00	0.62	0.60	837

for a CHMSL, 48 Hewlett-Packard HLMP-C100 TS AlGaAs T-1¾ LED lamps, driven at 25 mA each, would be required to meet government specifications. At the other extreme, only 8 Hewlett-Packard HPWT-XX00 TS AlInGaP high-flux LED lamps, driven at 60 mA each, would be required to meet the same specification. These TS AlInGaP high-flux LED lamps allow automotive stop and turn signal lamps to be built using fewer LED lamps than a typical T-1¾ LED CHMSL.

5. OPTICS DESIGN CONSIDERATIONS

The optics design of an LED automotive signal lamp would be somewhat different from an incandescent automotive signal lamp due to the different

TABLE IX
EXAMPLE SHOWING THE MINIMUM NUMBER OF LIGHT-EMITTING DIODE LAMPS NEEDED TO MEET MINIMUM US PHOTOMETRIC REQUIREMENTS

Lamp	θ_{LED} (mlm)	CHMSL (5.8 lm)	Stop Lamp (17.9 lm)	Amber Rear Turn (33.7 lm)	Amber Front Turn (50 lm)
HLMP-C100	125	48	144	—	—
HLMA-DL00	140	—	—	240	360
HPWA-M/DXXX	D, 251	24	72	132	200
	E, 368	16	48	92	136
HPWT-M/DXXX	F, 536	12	32	64	92
	G, 837	8	20	40	60

CHMSL, Center high mount stop lights.

radiation patterns of the light sources. An incandescent bulb typically has a spherical radiation pattern. Thus, the bulb might be mounted at the focus of a paraboloid reflector in order to collimate part of the luminous flux. By contrast, an LED lamp has a relatively narrow beam of light. Thus, it is less critical that a reflector be used to collimate the luminous flux.

Figure 13 shows a graph of the luminous intensity radiation patterns of several LED lamps that might be used for automotive signal lighting. Figure 14 shows the luminous flux distribution for the same LED lamps. Note that in some cases less than half of the total luminous flux is emitted at angles less than 20 degrees. However, the use of either a reflector or convex fresnel lens mounted over the LED lamp could be used to collect and redirect this luminous flux. For example, the Hewlett-Packard HPWA-MH00 has about 18% of its total luminous flux emitted at angles between 0 and 20 degrees off-axis. However, an additional 46% of its total luminous flux is emitted at angles between 20 and 45 degrees off-axis.

Finally, a pillow lens might be used to break up the images of the LED lamps. A pillow lens consists of an array of small convex lenses molded into the outer plastic lens. Pillow lenses do not need to be aligned to the LED lamps provided that the pitch between the pillow lenses is much smaller than the spacing between the LED lamps. The curvature of the pillow lenses should be designed to minimize the additional divergence of the luminous flux emitted by the LED lamps and optional secondary optics.

The optics design of a LED CHMSL can be used as an example of the optical concepts discussed earlier. The minimum US photometric pattern is

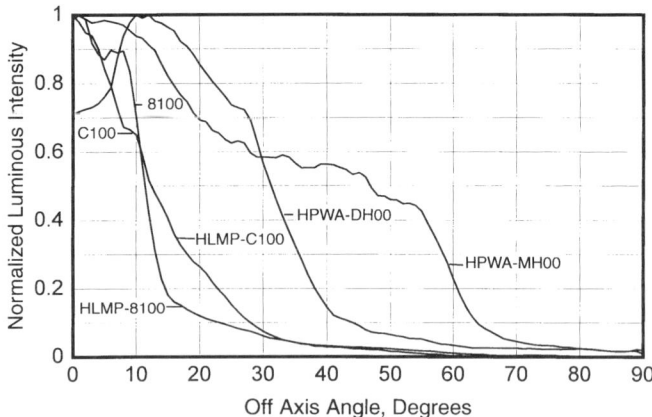

FIG. 13. Typical radiation patterns for several Hewlett-Packard light-emitting diode lamps used for automotive signal lamps.

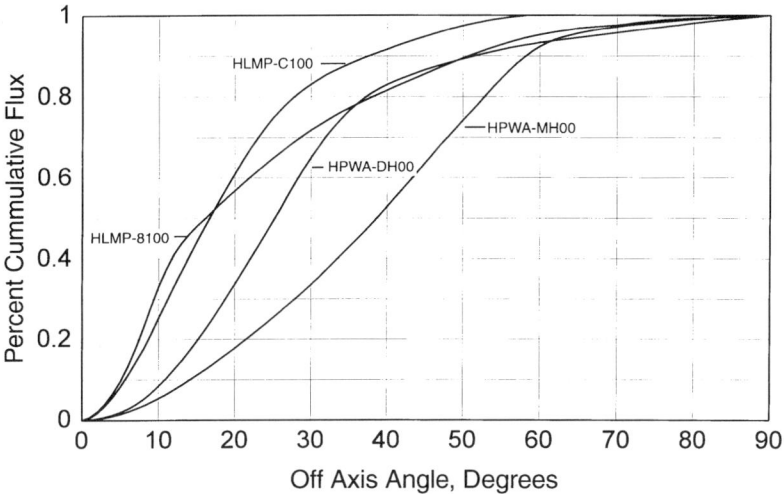

FIG. 14. Typical luminous flux versus off-axis angle for several Hewlett-Packard light-emitting diode lamps used for automotive signal lamps.

shown in Table X. As calculated previously in Table VII, about 5.8 lumens of luminous flux is needed, provided most of the luminous flux is contained within the desired range of viewing angles. Figure 15 shows several possible optics designs for an LED CHMSL. The first design (Fig. 15(a)), uses 19-degree viewing angle HLMP-8100 TS AlGaAs T-1$\frac{3}{4}$ LED lamps mounted behind a pillow lens array. Referring to Fig. 14, with this design, only about 45% of the total luminous flux generated by the LED lamps is emitted into a circular cone corresponding to the angles of 0 to 14 degrees off-axis (the

TABLE X

MINIMUM US AUTOMOTIVE CENTER HIGH MOUNT STOP LIGHT PHOTOMETRIC REQUIREMENTS (cd)[a]

	10L	5L	V	5R	10R
10U	8		16		8
5U	16	25	25	25	16
H	16	25	25	25	16
5D	16	25	25	25	16
10D	16	25	25	25	16

[a]Maximum intensity is 130 cd.

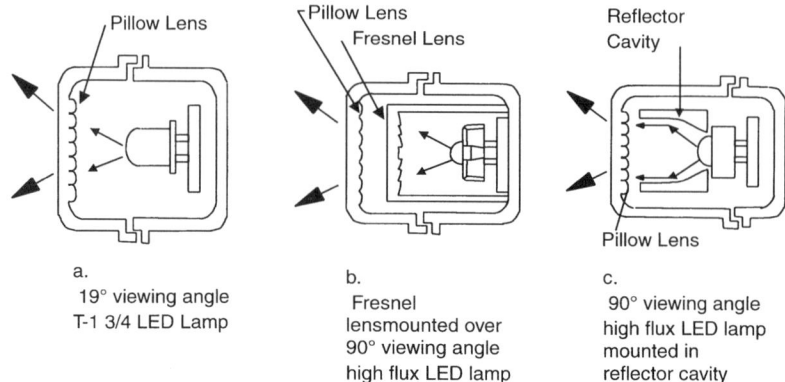

FIG. 15. Three approaches for the optics design of an automotive center high mount stop light. (a) A 19-degree viewing angle T-1¾ light-emitting diode (LED) lamp. (b) Fresnel lens-mounted over a 90-degree viewing angle high-flux LED lamp. (c) A 90-degree viewing angle high-flux LED lamp mounted in a reflector cavity.

angle formed by the photometric test point of 10 degrees up and 10 degrees left; i.e., $\sqrt{10^2 + 10^2} = 14°$). Additional luminous flux could be utilized by mounting either a convex lens over the lamp or a small reflector surrounding the lamp to capture more of this wasted luminous flux.

A second optical design (Fig. 15(b)) uses the Hewlett-Packard HPWR-M300 TS AlGaAs high-flux LED lamp mounted behind a convex fresnel lens. The fresnel lens is used to collimate the light generated by the lamp. For example, if the optical image of the LED lamp was mounted at the focal point of an $f1.0$ convex fresnel lens, then all of the luminous flux emitted at angles of 0 to 26.5 degrees off-axis will be collimated. Referring to Fig. 14, this corresponds to about 27% of the total luminous flux emitted by the HPWR-M300. Using conventional refractive optics, the smallest f number practical for a plastic convex fresnel lens is about $f0.65$. Thus, it is practical for a refractive plastic convex fresnel lens to capture the luminous flux for off-axis angles up to 37.4 degrees. This corresponds to about 47% of the luminous flux emitted by the HPWR-M300. Using a narrower angle LED lamp, such as the HPWA-DH00 AS AlInGaP high-flux lamp, allows a higher percentage of the luminous flux to be collected. Referring to Fig. 14, for a $f1.0$ convex fresnel lens, then about 52% of the total luminous flux of the HPWA-DH00 would be collimated (for angles of 0 to 26.5 degrees off-axis). For an $f0.65$ convex fresnel lens, about 80% of the luminous flux of the HPWA-DH00 would be collimated.

The primary disadvantage of the convex fresnel lens approach is the added expense of the lens and the need to provide precise alignment of the LED lamp to the lens.

Another optical design (Fig. 15(c)) uses the Hewlett-Packard HPWA-MH00 AS AlInGaP high-flux LED lamp mounted in a metalized plastic reflector. With this approach, it is possible to collect almost all of the luminous flux emitted by the LED lamp and redirect it within the desired 14 degree off-axis cone angle. This approach is the most efficient for utilizing the available luminous flux, and the most expensive.

The same optics designs in Figs. 15(a) to 15(c) could be used in the design of a LED brake lamp or turn signal. The minimum US photometric radiation pattern has been shown previously in Table I. This pattern is consistent for the brake lamp and red rear turn signals. The amber rear and front turn signals have a similar radiation pattern with higher values of luminous intensity. As calculated previously in Table VII, about 17.9 lumens of red luminous flux is needed for the stop or rear turn signals and 33.7 lumens or 50.0 lumens of amber luminous flux is needed for the rear or front turn signals.

If the signal lamp is designed to mount in the extreme left and right corners of the motor vehicle, then the signal lamp may need to wrap around the corner of the vehicle. The LED lamps can be mounted as shown in Fig. 16(a), in which all LED lamps are facing rearward such that each row of lamps is mounted a fixed distance behind the pillow lens. This approach

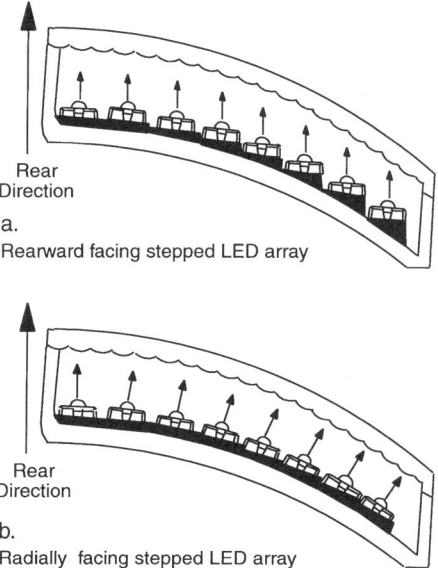

FIG. 16. Two approaches for the optics design of an automotive stop-turn lamp. Rearward (a) and radially (b) facing stepped light-emitting diode array.

provides both the most efficient optical radiation pattern and uniform illumination when viewed from the rear of the vehicle. When viewed from the sides of the vehicle, however, the signal lamp may exhibit vertical bright and dark bands due to the large steps between the adjacent rows of LED lamps. Another scheme is shown in Fig. 16(b), in which the lamps are angled perpendicular to the curvature of the outer pillow lens surface at a fixed distance behind the pillow lens. This approach does not provide quite as efficient an optical radiation pattern because the beam is spread outward. The advantage of this approach is that the signal lamp provides uniform illumination when viewed from all sides of the vehicle. (It is possible to get some vertical bright-dark bands at the outward edges of the lamp when viewed from the rear of the vehicle, if the step sizes are very large). The approach outlined in Fig. 16(b) also tends to make the signal lamp assembly somewhat thicker than that in Fig. 16(a).

6. Electrical Design Considerations

The electrical design of an LED automotive signal lamp is different from that of an incandescent one because LED lamps cannot be driven from a 12-V electrical system without some means of external current limiting. The typical electrical characteristics of LED lamps designed for automotive signal lighting are shown in Fig. 17. Note that at forward voltages less than the turn-on voltage, negligible electrical current flows through the lamp. Note also that at voltages higher than the turn-on voltage, the slope of the forward I-V curve is relatively constant. These characteristics suggest that the forward voltage of the LED lamp can be modeled as shown:

$$I_F = 0 \quad \text{for } V_F \leq V_{\text{turn-on}}$$

$$V_F = V_{\text{turn-on}} + R_S I_F \quad \text{for } V_F > V_{\text{turn-on}}$$

where

I_F = forward current through the LED lamp

V_F = forward voltage across the LED lamp

$V_{\text{turn-on}}$ = voltage where the straight line approximation of the I-V curve intersects the voltage axis at zero current

R_S = 1/slope of the straight line approximation of the I-V curve

= $\Delta V_F / \Delta I_F$.

6 APPLICATIONS FOR HIGH-BRIGHTNESS LIGHT-EMITTING DIODES 269

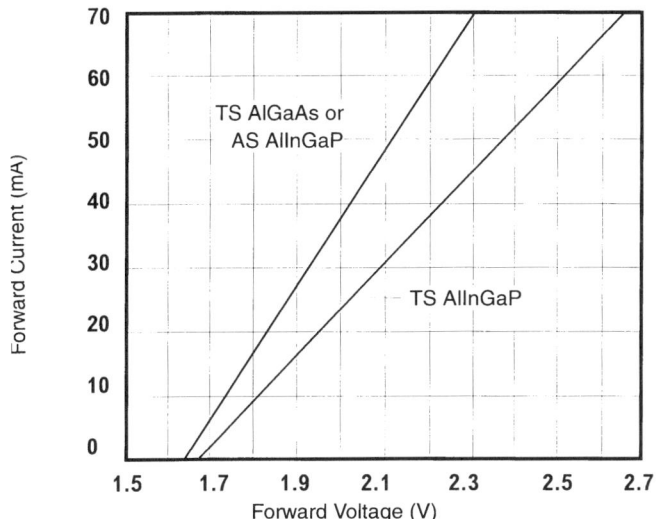

FIG. 17. Typical forward current versus forward voltage for absorbing substrate (AS) AlInGaP, transparent substrate (TS) AlInGaP, and TS AlGaAs light-emitting diode lamps. (From Optoelectronics Designer's Catalog, Hewlett-Packard Company, 1996. Reprinted with permission from Hewlett-Packard Company.)

In general, TS AlGaAs LED lamp technology has a slightly lower turn-on voltage than do AS AlInGaP and TS AlInGaP LED lamp technologies. The TS AlInGaP LED lamp technology also has a somewhat higher series resistance than either the TS AlGaAs or the AS AlInGaP lamp technologies.

A typical 12-V LED automotive signal lamp electrical drive circuit is shown in Fig. 18. Note that the lamps are driven in series strings, with an external current limiting resistor for each string. The signal lamp is constructed of multiple series strings, with up to five LED lamps connected in series in a single string. The value of R can be calculated as follows:

$$R = \frac{V_S - nV_F}{I_F}$$

where V_S is the design voltage, V_F is the forward voltage across LED lamp at the forward current through the circuit, I_F, and n is the number of LED lamps connected in series. Using the forward voltage model, the value of R can be calculated:

$$R = \frac{V_S - nV_{\text{turn-on}} - nR_S}{I_F}$$

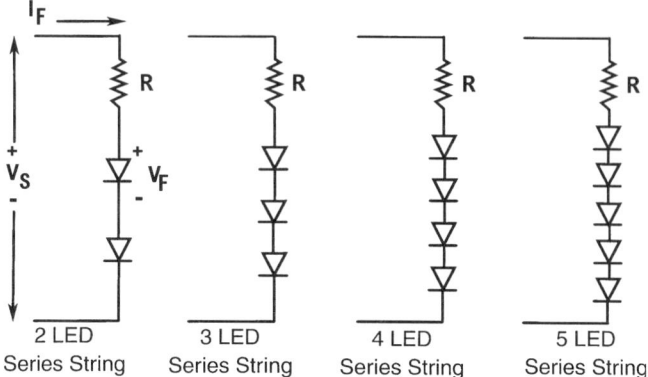

FIG. 18. Typical drive circuits for light-emitting diode (LED) automotive signal lamps. R, resistor; V_F, forward current; V_S, design voltage; V_F, forward voltage across the LED lamp at forward current I_F.

As V_S varies around the nominal design voltage, the current through each string varies, as shown in Fig. 19. Note, that the forward current variation due to a variation in applied voltage is higher with more LED lamps in series. On the other hand, fewer LED lamps per string means that more strings are required to drive the total number of LED lamps. Increasing the

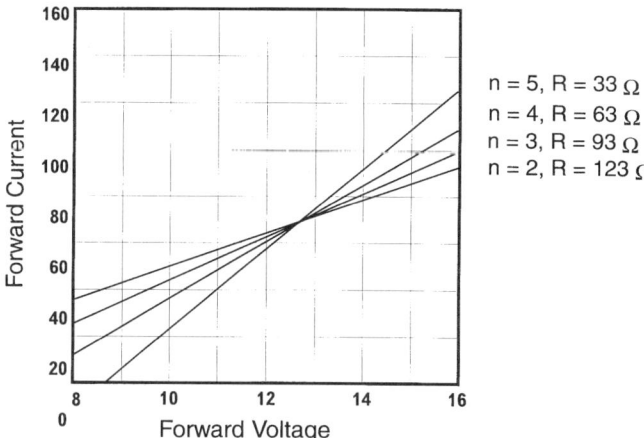

FIG. 19. Forward current versus applied voltage for light-emitting diode automotive signal lamp circuits. R, resistor; n, number of LED lamps in string.

number of strings increases the overall power dissipation of the signal light and the power rating for each current limiting resistor. For example, a 48-LED lamp CHMSL could be constructed using 12 strings of four lamps per string, or 16 strings of three lamps per string, or 24 strings of two lamps per string. Assuming a 25-mA drive current, the overall drive current would be 300 mA for the 12-string, 400 mA for the 16-string, and 600 mA for the 24-string design. Most LED CHMSL signal lamps currently in production use strings with 4 LED lamps per string, which is a good trade-off between good current regulation and low overall power consumption.

It is possible to connect LED lamps in parallel as a means to simplify the overall drive circuit for the automotive signal lamp. Two possible circuit configurations are shown in Fig. 20, in which it is important that the LED lamps have matched forward voltage characteristics. If one LED lamp has a lower forward voltage at the desired drive current than another LED lamp parallel to it, then the current through the first LED lamp will increase until both LED lamps have the same forward voltage. If the current variations are too large, then the LED lamps could have objectionable brightness variations or could result in driving some lamps at too high a drive current. It is possible to create matched sets of LED lamps by 100% electrical testing the LED lamps at the desired forward current and binning them into tightly grouped sets.

In some cases, it may be necessary to drive the automotive signal lamp at two different luminous intensity levels. One example might be a combined automotive tail light and brake light. The US automotive tail light specification requires a minimum luminous intensity of 2.0 candelas and a maximum luminous intensity of 18 candelas. One approach might be to drive the LED signal lamp with the circuit shown in Fig. 21. First, the resistor R is determined such that the LED lamps are driven at the desired

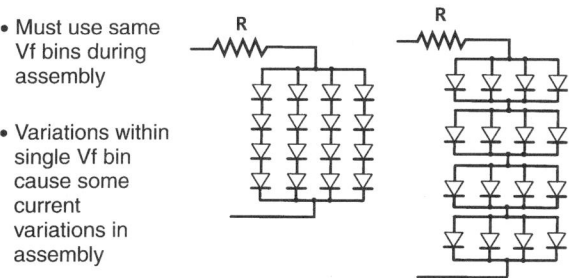

Fig. 20. Parallel drive circuits for light-emitting diode automotive signal lamps. R, resistor; V_f, forward current.

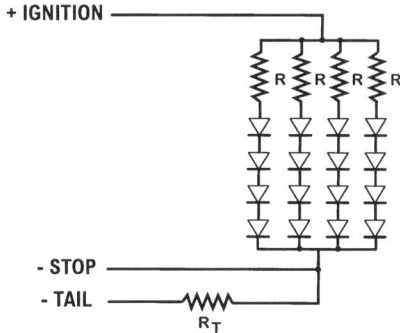

FIG. 21. Drive circuit designed for two-current operation. R, R_T, resistors.

current under the brake light condition. Then, the resistor R_T is determined such that the LED lamp array is driven at a much lower drive current for the tail lamp function. An alternative approach could be to pulse-width-modulate the current through the LED array by rapidly switching the current on and off. If this chopping is done at a frequency higher than about 100 Hz, the human eye perceives that the time averaged luminous intensity is equal to the peak luminous intensity multiplied by the duty cycle (ratio of on-time to total time).

Finally, the LED automotive signal lamp design needs to take into account the possibility of high-voltage electrical transients. This possibility in the automotive electrical system is due to the rapid interruption of inductive loads throughout the vehicle (e.g., ignition coil and alternator). The magnitude and pulse shape of these electrical transients is defined by several automotive electrical specifications, including ISO 7647-1 and SAE J1113. The worst-case positive electrical transients can have peak voltages up to 86.5 V, with a duration up to 400 ms (ISO 7647-1, pulse 5, level 4). The worst-case negative electrical transients can have peak voltages down to −300 V, with a duration up to 300 μs (ISO 7647-1, pulse 6, level 4). Electrical transients of these magnitudes can permanently damage LED lamps.

The TS AlGaAs, and AlInGaP materials technologies are capable of handling repetitive peak forward currents in the range of several hundred milliamps for pulse durations of several hundred microseconds. However, pulses of several hundred milliamps for several hundred milliseconds can cause significant thermal heating of the LED die, which could lead to failures of the LED wire bond or LED junction. The TS AlGaAs and AlInGaP materials technologies are more susceptible to damage due to negative transients, if the voltage is sufficient to cause reverse breakdown

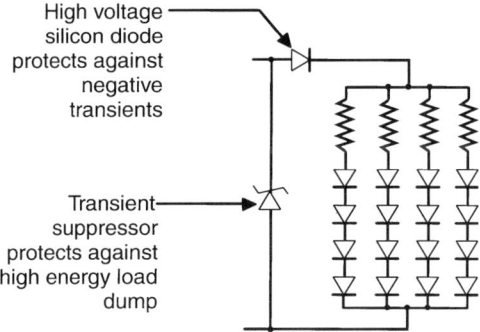

FIG. 22. Electrical transient protection circuitry.

and significant reverse currents, because the current does not tend to spread equally throughout the junction area of the LED chip.

Figure 22 shows a typical electrical transient protection circuit. The high-voltage silicon diode with a $V_{BR} \geqslant 300$ V protects against high-voltage negative transients by limiting the current through the LED array. The transient suppressor protects against high-energy positive transients by clamping the voltage across the LED array. The reverse protection silicon diode is more important than the transient suppressor because even the lower voltage levels of ISO 7647-1, pulse 6 (-120 V) are capable of damaging the LED array, whereas only the highest voltage levels of ISO 7647-1, pulse 5 (86.5 V) might damage the LED array.

7. THERMAL DESIGN CONSIDERATIONS

The thermal design of an LED automotive signal lamp should be different from that of an incandescent automotive signal lamp. Because an incandescent bulb generates a considerable amount of heat and the bulb itself is capable of withstanding high temperatures, the thermal design of an incandescent signal lamp is primarily concerned with the ability of the plastic case, lens, and electrical connections to withstand high temperatures. The LED lamp array typically generates only one fourth the heat, and therefore the demands on the plastic case, lens, and electrical connections are less. However, the maximum internal temperatures of the LED lamp must be kept below the recommended upper limits to ensure reliable operation. Exceeding the maximum junction temperature of the LED lamp can lead to premature failure of the wire bond, especially due to thermal cycling and powered thermal cycling.

The primary thermal path of the heat generated by the LED chip is through the metal pin to which the LED chip is attached. For TS AlGaAs LED material technology, this pin corresponds to the anode pin. For AlInGaP and GaP LED materials technology, this pin corresponds to the cathode pin. After traveling though the pin, the heat flows from the solder connection through the traces on the printed circuit board. Next, the heat is transferred to the outer case by a combination of conduction and convection. Finally, the heat flows through the outer case until it finally is dissipated. The system can be modeled as follows. (note that the units are °C for T, W for P, and °C/W for θ):

$$T_J = T_A + P_D(\theta_{J\text{-}PIN} + \theta_{PIN\text{-}PCB} + \theta_{PCB\text{-}CASE} + \theta_{CASE\text{-}AIR})$$

$$\theta_{PIN\text{-}AIR} = \theta_{PIN\text{-}PCB} + \theta_{PCB\text{-}CASE} + \theta_{CASE\text{-}AIR}$$

$$T_J = T_A + P_D(\theta_{J\text{-}PIN} + \theta_{PIN\text{-}AIR})$$

$$T_J = T_{PIN} + P_D(\theta_{J\text{-}PIN})$$

$$T_{PIN} = T_A + P_D(\theta_{PIN\text{-}AIR})$$

where

T_J = LED lamp junction temperature

T_A = ambient temperature

T_{PIN} = LED lamp pin temperature, as measured on the underside of the printed circuit board; cathode pin for AlInGaP and anode pin for TS AlGaAs

P_D = LED lamp power dissipation ($I_F V_F$)

and where the following are the resistances to heat flow:

$\theta_{J\text{-}PIN}$ = LED lamp junction to LED lamp pin

$\theta_{PIN\text{-}PCB}$ = LED lamp pin to printed circuit board traces

$\theta_{PCB\text{-}CASE}$ = printed circuit board to case

$\theta_{CASE\text{-}AIR}$ = case to air

$\theta_{PIN\text{-}AIR}$ = LED lamp pin to air

$\theta_{J\text{-}AIR}$ = LED lamp junction to air

6 APPLICATIONS FOR HIGH-BRIGHTNESS LIGHT-EMITTING DIODES

The various components of the thermal model may be measured by attaching small thermocouples to the complete automotive signal lamp and measuring the temperature increase of each thermocouple due a fixed amount of power being applied. The LED junction temperature can be measured indirectly by measuring the change in forward voltage of the LED lamp due to heating. For most LED materials technologies, the change in forward voltage is approximately -2 mV/°C. For most LED lamps, the thermal resistance $\theta_{J\text{-PIN}}$, is specified on the LED lamp data sheet.

With a little algebraic manipulation, the thermal resistance model can be solved for power dissipation as a function of junction temperature and thermal resistance as follows

$$T_J = T_A + P_D(\theta_{J\text{-AIR}})$$

$$P_D = (T_J - T_A)/\theta_{J\text{-AIR}}$$

The equation can be used to create the graph of maximum recommended power dissipation versus ambient temperature, by considering the limits of maximum power dissipation and maximum junction temperature as follows:

For

$$T_A \leq T_{J\text{-MAX}} - (P_{D\text{-MAX}})(\theta_{J\text{-AIR}}) \qquad \text{then } P_D = P_{D\text{-MAX}}$$

For

$$T_{J\text{-MAX}} - (P_{D\text{-MAX}})(\theta_{J\text{-AIR}}) < T_A \leq T_{J\text{-MAX}} \qquad \text{then } P_D = (T_{J\text{-MAX}} - T_A)/\theta_{J\text{-AIR}}$$

For

$$T_A > T_{J\text{-MAX}} \qquad \text{then } P_D = 0$$

Because P_D is equal to $I_F V_F$, the graph of maximum recommended power consumption versus ambient temperature can be converted to a graph of maximum dc forward current versus ambient temperature.

Let

$$P_{D\text{-MAX}} = (I_{\text{FMAX}})(V_{\text{FMAX}})$$

For

$$T_A \leq T_{J\text{-MAX}} - (I_{\text{FMAX}})(V_{\text{FMAX}})(\theta_{J\text{-AIR}}) \qquad \text{then } I_F = I_{\text{FMAX}}$$

For

$$T_{J\text{-MAX}} - (I_{\text{FMAX}})(V_{\text{FMAX}})(\theta_{J\text{-AIR}}) < T_A \leq T_{J\text{-MAX}}$$
$$\text{then } I_F = (T_{J\text{-MAX}} - T_A)/(V_F \theta_{J\text{-AIR}})$$

For

$$T_A > T_{J\text{-MAX}} \quad \text{then } I_F = 0$$

An example of this graph for the Hewlett-Packard HPWA-M/DXX-type LED lamp is shown in Fig. 23.

Reducing the thermal resistance $\theta_{J\text{-AIR}}$ of the LED lamp allows it to be driven at a higher forward current as shown by Fig. 23, and also provides a higher luminous flux output as shown previously in Fig. 12. Because the thermal resistance, $\theta_{J\text{-PIN}}$ is determined by the LED lamp package, good thermal design involves reducing the thermal resistance $\theta_{\text{PIN-AIR}}$. There are several techniques used to reduce thermal resistance in LED automotive signal lamps. Because the primary thermal path from the LED lamp package is through the soldered pin connections, using heavy traces on the printed circuit board lowers the overall thermal resistance by reducing the $\theta_{\text{PIN-PCB}}$. Ventilating the case or attaching the printed circuit directly to the case reduces the overall thermal resistance by reducing the $\theta_{\text{PCB-CASE}}$. Using thermally conductive plastic or metal for the case itself reduces the overall

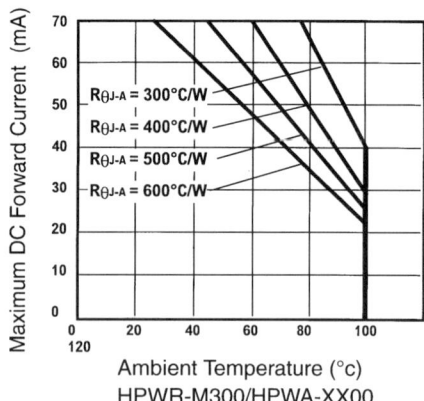

FIG. 23. Maximum dc forward current versus ambient temperature for HPWA-M300/HPWA-XX00 light-emitting diode lamp. (From Optoelectronics Designer's Catalog, Hewlett-Packard Company, 1996. Reprinted with permission from Hewlett-Packard Company.)

thermal resistance by reducing the $\theta_{CASE\text{-}AIR}$. Removing some of the power dissipation from within the case, such as by mounting the current limiting resistor outside the case, also reduces the overall thermal resistance. Finally, the packing density of the LED lamps on the printed circuit board has a direct effect on thermal resistance. A single in-line array of LED lamps a lower thermal resistance than does tightly packed X-by-Y array of LED lamps. As a general rule, a single in-line array of LED lamps used for an automotive CHMSL might have a thermal resistance in the range of 300 to 750 °C/W.

IV. Automotive Interior Lighting

1. CHARACTERISTICS OF THE AUTOMOTIVE AND TRUCK MARKET

The worldwide automotive-truck industry is another vast new market for LED technology for lighting within vehicles. A number of lighting applications currently use a mixture of LED lamps and incandescent bulbs, including instrument clusters, heater and air conditioning controls, radios, and convenience lighting on accessory and electric window switches. In fact, the potential for interior lighting far exceeds that for exterior signal lighting. Although, today, many interior applications use GaP LED technology, the AlGaAs, AlInGaP, and InGaN LED materials technologies are opening up new applications that are currently using incandescent bulbs.

Automotive interior lighting applications are not government regulated as are signal lighting applications. However, each car manufacturer has guidelines on the colors and luminous sterance levels required for each application. These applications generally have five distinct colors. Some lamps warn drivers of hazards, others alert drivers to certain activated functions. These lamps usually are red, amber, green, or blue, where the color indicates the severity of the problem and level of attention required. For example, the "Seat belt" and "Low oil" warning lights usually are red. The "Engine needs servicing" and "Low fuel" warning lights usually are amber. The "Left turn" and "Right turn" signal warning lights usually are green. The head-lamp "High beam" warning light usually is blue. Figure 24 shows the approximate color coordinates in the 1931 CIE chromaticity diagram for these signal colors. Each dot represents the color coordinate for a given signal color by each car manufacturer. A rectangular box has been drawn around each group of points representing each signal color.

The fifth color provides night-time general illumination to identify and backlight the analog gauges, push buttons, and controls. This general

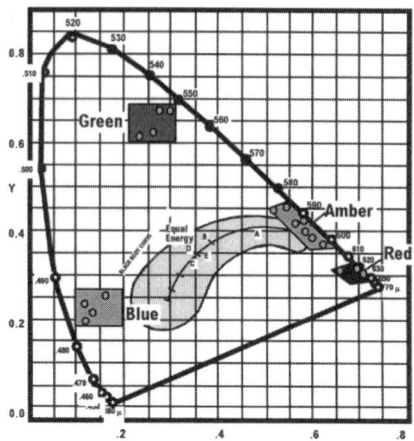

FIG. 24. Automotive instrument cluster warning light colors. Each coordinate represents one manufacturer's desired color.

illumination color has been selected by the automotive stylists to uniquely identify each car line. Figure 25 shows the approximate color coordinates in the 1931 CIE chromaticity diagram for a number of general illumination colors. These color coordinates correspond to colors ranging from mixtures of white and blue, green, or yellow, to orange and red pure. Regardless of

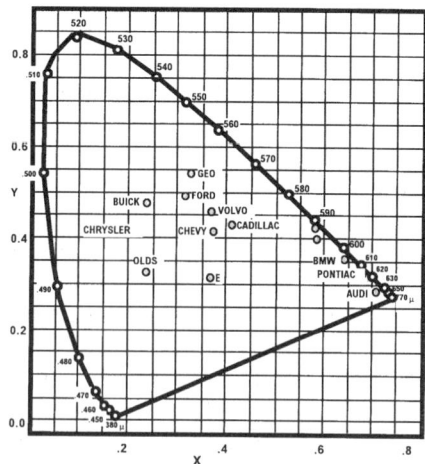

FIG. 25. Automotive instrument cluster night-time gauge illumination colors.

TABLE XI

LUMINOUS STERANCE RANGES FOR AUTOMOTIVE INTERIOR LIGHTING

Application	Color	Desired Luminous Sterance
Instrument cluster warning lights	Red	80–300 cd/m^2
	Amber	100–400 cd/m^2
	Green	
	Blue	10–30 cd/m^2
Instrument cluster gauge night illumination	All	5–20 cd/m^2
LCD backlighting-transflective display (night supplemental lighting)		10–30 cd/m^2 (measured through LCD)
LCD backlighting-transmissive display (daylight readable lighting)		100–400 cd/m^2 (measured through LCD display)
Switch day illumination		50–200 cd/m^2
Switch night illumination		1–10 cd/m^2

LCD, liquid crystal display.

the specific color, automotive stylists generally require that all applications of night-time illumination match the same targeted color coordinates for a given car line.

Warning lights intended to be visible in sunlight conditions typically have luminous sterance levels in the range of 80 to 400 cd/m^2. For warning lights intended to be visible primarily at night, such as the headlamp high beam, the luminous sterance levels are typically in the range of 10 to 30 cd/m^2. Other nighttime general illumination applications, such as analog gauge scales, push buttons, and controls typically have a luminous sterance level of 5 to 15 cd/m^2. Table XI summarizes the typical luminous sterance levels for automotive interior lighting applications.

2. BENEFITS OF LIGHT-EMITTING DIODE TECHNOLOGY

Light-emitting diode technology provides a number of benefits for automotive interior lighting some of which are the same as for automotive signal lighting. There also are several others. Because incandescent bulbs traditionally have been used for these applications, we compare LED lamps with their incandescent counterparts and restrict this discussion to high-performance applications for which the newer high-performance LED materials technologies are needed.

Figure 26(a) shows a sketch of a typical instrument cluster warning lamp utilizing an incandescent bulb. The bulb is typically attached to either a rigid or flexible printed circuit board utilizing a quarter-turn socket base. The printed circuit board is then mounted at the base of a white plastic cavity. The cavity serves to reflect the luminous flux outward as well as to separate the individual warning lights and prevent the light source from falsely illuminating adjacent warning lights. A piece of diffusing film material is placed on the outer surface of the white cavity. The desired legend symbol or text is typically silk screened on the rear surface of the diffusing film. The top surface of the diffusing film is typically silk screened with a neutral density gray "deadfront" ink to minimize stray sunlight from falsely illuminating the legend. A transparent colored ink is also silk screened on the back of the diffusing film. This colored layer creates the desired signal color or general illumination color.

An LED lamp can be used in the same instrument cluster warning lamp. Figure 26(b) shows an equivalent LED design. The primary difference is that because the LED lamp is a colored light source, the colored silk-screen ink on the diffusing film can be eliminated. Table XII compares the incandescent bulb and LED lamp instrument cluster warning lamps.

The incandescent design has a number of advantages. First, the incandescent bulb is the lowest cost source of luminous flux. For this application (in the United States), type #37 incandescent bulbs are commonly used. These bulbs operate at approximately 90 mA and generate a luminous intensity of 0.5 mscd and a luminous flux of 6.3 lm of white light. A second advantage is that the bulb is usually mounted in a quarter-turn base, allowing the bulb to be field repairable. Unfortunately, the only way to access the bulbs is by removing the dashboard and instrument cluster from the vehicle. A third advantage is that the color of the graphics is determined

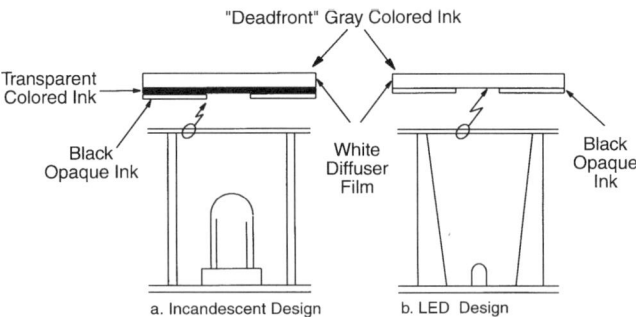

FIG. 26. Incandescent (a) and light-emitting diode (b) instrument cluster warning light construction.

TABLE XII

COMPARISON BETWEEN LIGHT-EMITTING DIODE (LED) AND INCANDESCENT INSTRUMENT CLUSTER WARNING LIGHTS

	Advantages	Disadvantages
Incandescent	Low cost luminous flux Socketed for field repair Color determined by graphics Directly compatible with flexible pcb	Filament shock sensitive MTBF and lifetime less than vehicle life High heat (approximately 1.2 W/bulb) High current (approximately 100 mA/bulb) High in-rush current More expensive graphics Large size
LED	Comparible price to incandescent Shock resistant MTBF much longer than vehicle life Low heat (approximately 0.25 W/LED plus resistor) Low current (approximately 20 mA/LED) No in-rush current Simplified graphics Compatible with automated pcb assembly Small size	Higher cost luminous flux Limited junction temperature Not field repairable Color selection limitations Light output varies with temperature Not optimized for flexible pcb

pcb, printed circuit board; MTBF, mean time between failure.

by the silk-screening colors. This allows the same bulb part number (possibly two or three if different amounts of luminous flux are needed) to be used for all warning light and general illumination applications.

The incandescent design has some disadvantages. The filament of the incandescent bulb is sensitive to shock and vibration. The MTBF and the lifetime of the bulb also are typically less than the life of the vehicle, which is a special concern for general illumination applications in which the bulbs are on continuously at night. Another disadvantage is that they operate at high electrical currents and dissipate considerable heat. A typical bulb operates at 90 mA at 12 V, which corresponds to a power dissipation of 1.2 W. This high power dissipation can cause thermal problems for the electronic circuitry within the instrument cluster. Furthermore, the bulb has a very high inrush current (approximately 1 A), when the bulb is initially energized (approximately for 1 ms). As the instrument cluster circuit design has evolved from mechanical sensors to electronic control, the high currents and even higher inrush currents add to the cost of the electronic driver circuitry. The higher internal temperature due to the high electrical power consumption also possibly requires the use of high-temperature plastics. In addition, while colored graphics add flexibility to the incandescent design, the multiple silk-screen colors add costs to the graphics. A final disadvantage is that the quarter-turn incandescent bulbs are physically quite large and limit the information density within the instrument cluster. Because the trend has been to provide more information to the driver, this size limitation tends to restrict the front panel layout.

The LED lamp instrument cluster warning lamp provides a number of advantages over the incandescent design. The LED lamp is very resistant to shock and vibration. The MTBF of the LED lamp is several orders of magnitude higher than that of the incandescent bulb. The lifetime of the LED lamp is longer than the expected lifetime of the vehicle. Together these advantages allow the LED lamp to be permanently soldered to the printed circuit board, just as with any other electronic component. Another advantage of the LED design is reduced electrical current operation and reduced power dissipation. Typical LED lamps used for instrument cluster warning lamps can generate adequate luminous flux at a drive current of 20 mA. Thus, the LED lamp operates at approximately one fifth of the operating current of the incandescent bulb, and the power dissipation within the 12-V instrument cluster is reduced to 0.25 W. If needed, up to four LED lamps (assuming a 2-V forward voltage) could be driven from the same 20-mA drive current. Also, LED lamps do not have a higher current drain when initially energized. The reduced drive current and lack of a high inrush current provide savings to the cost of the electronic driver circuits. Another cost savings results from elimination of the colored silk-screen layers on the diffusing graphics films. The LED lamps also are much smaller than are incandescent bulbs, allowing both more flexibility in the layout of the front panel graphics and more information to be displayed. Finally, the components of the LED lamp design are compatible with high-speed automated assembly processes. Depending on the package used, LED lamps are available on paper tape carriers compatible with high-speed radial automatic insertion equipment or on plastic film carriers compatible with high-speed surface-mount pick-and-place insertion equipment. In general, the LED lamps can be handled just like any other electronic component, which can reduce manufacturing assembly costs.

The LED lamps have some disadvantages compared with incandescent bulbs for instrument cluster warning lamps. First, LED lamps are more expensive than are incandescent bulbs as a source of luminous flux. However, most of the luminous flux can be utilized because all of the

6   APPLICATIONS FOR HIGH-BRIGHTNESS LIGHT-EMITTING DIODES

luminous flux generated by the LED lamp falls in a narrow spectrum. For an instrument cluster warning light, a total luminous flux output of 100 to 300 mlm of colored light is sufficient to generate the desired legend luminous sterance. In many applications, a single LED lamp can be used for each warning lamp cavity. In this case, the price of the LED lamp itself is comparable with or even lower than the incandescent bulb currently in use.

A second disadvantage of the LED design is that the LED lamp used in an instrument cluster warning lamp has a limited maximum junction temperature. The instrument cluster may have to operate at higher ambient temperatures than would an automotive signal lamp. The instrument cluster electronics also may generate considerable heat, which subjects all of the electronics to elevated ambient temperatures. For best reliability and lowest manufacturing costs, the LED lamps mounted within the instrument cluster generally are not field repairable. Thus, it is important to design the instrument cluster to operate the LED lamps at temperatures lower than the recommended maximum junction temperature.

A third disadvantage of the LED design is a more limited color availability than that for a filtered incandescent bulb. Referring to Fig. 24, both AlGaAs and AlInGaP LED materials technologies are capable of generating the desired red signal color. The AlInGaP, and in some cases, GaP LED material technologies are capable of generating the desired amber signal color. In fact, two different amber-colored LED lamps are available with dominant wavelengths of 585 to 590 nm and in the 600 to 605 nm color ranges. The desired green signal color can be achieved with InGaN materials technology. Alternatively, GaP and AlInGaP materials technologies can be used if the dominant wavelength is shifted to the 560 to 570 nm color range. The desired blue signal color can be achieved with InGaN materials technology.

However, the general illumination colors shown in Fig. 25 present serious challenges to LED technology. In general, most of the colors within the center of the chromaticity diagram can be achieved by mixing the color of InGaN blue materials technology with either GaP or AlInGaP materials technologies. Or, in some cases, the spectrum of the InGaN blue LED can be selectively filtered in order to shift the color coordinate inward. However, both of these approaches are relatively expensive due to the high cost of InGaN materials technology. The remaining general illumination colors, such as the oranges and reds, can easily be achieved using AlInGaP materials technology.

The last disadvantage of LED lamps for instrument cluster warning lighting is that they are not optimized for use with inexpensive polyester flexible printed circuit board material. Adding a quarter-turn socket base with a discrete current limiting resistor to the LED lamp adds significant cost. However, the market trend is moving away from polyester flexible printed circuit board material due to needs of higher packing density that can be achieved with a double-sided printed circuit and the need to install other electronic components such as integrated circuits, resistors, and capacitors that are more optimized for a rigid printed circuit board.

3. LIGHTING CONSIDERATIONS FOR AN INSTRUMENT CLUSTER WARNING LIGHT

There are a number of applications for LED lamps in the interior of automobiles and trucks. In many of these applications, low-performance GaP LED materials technology provides adequate lighting. This section discusses the use of LED lamps for instrument cluster warning lights, because this application needs the higher performance of AlInGaP, AlGaAs, and InGaN LED materials technologies.

In Fig. 26(b), we have seen the basic concept of an instrument cluster warning light. The depth of the white cavity typically varies from 15 to 40 mm, as measured from the top surface of the printed circuit board to the rear surface of the diffusing film.

In 1994, experiments were made on a number of LED lamp packages for potential instrument cluster warning lights. These LED lamps were evaluated for their lighting performance and the results were published in a SAE paper (Hodapp, 1994a). The paper evaluated the luminous sterance and uniformity of a piece of backlit white diffusing film mounted over a production instrument cluster warning light. A range of instrument cluster cavity depths was evaluated by milling the rear of the case at several depths without modifying the top features of the case. A sketch of the basic experiment is shown in Fig. 27. Four commonly available LED lamp packages (Fig. 28) were mounted at the bottom of the white plastic cavity and evaluated for lighting. Where available, LED lamps with different viewing angles were evaluated, to determine the effect of lamp viewing angle, if any, on the lighting. A piece of Makrofol BL 6-2 white diffuser film (Bayer Plastics) was mounted over the cavity. For each lamp, the average luminous sterance was measured through the white diffuser film by averaging the luminous sterance of a 2.5-mm-diameter spot in the center of the legend and a 2.5-mm-diameter spot in the corner of the legend. The center-to-corner ratio of the same luminous sterance readings also was calculated. All tests were done with the LED lamp or lamps driven at a current of 20 mA each (Fig. 29). In general, the figure of merit for the performance of a given LED lamp was that the average luminous sterance should be as high as possible, while still obtaining a ratio of luminous sterances (center–corner) less than 2:1.

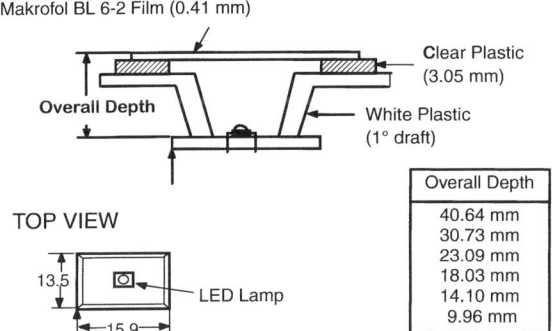

FIG. 27. (a) Top view and (b) side view of an instrument cluster warning light experiment. LED, light-emitting diode (Makrofol BL 6-2 film; Baya Plastics). (After Hodapp, 1994a, reprinted with permission from SAE paper 941046, ©1994 Society of Automotive Engineers, Inc., and Hodapp, 1993, reprinted with permission from ISATA, 42 Lloyd Park Avenue, Croydon CR0 5SB, England.)

The results of these tests are a series of graphs showing how each lamp family performs at different instrument cluster depths. For example, two of these graphs are shown in Figs. 30 and 31. Figure 30 shows the ratio of luminous sterances versus cavity depth for the HSMX-TXXX TS AlGaAs surface-mount LED lamp, whose package is shown in Fig. 28(d). Note that the lamp provides uniform lighting (a ratio of 2.0 or less) over a range of 18- to 40-mm cavity depths. At shallower cavity depths, the ratio increases quickly. The lamp may still be usable at cavity depths down to 14 mm, if multiple lamps are used in the same cavity or if the legend area is reduced.

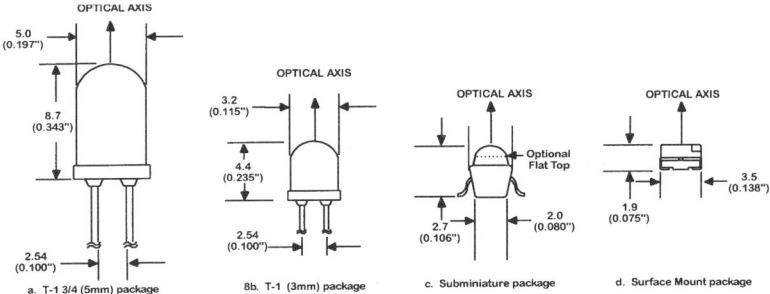

FIG. 28. Typical light-emitting diode lamp packages. (a) T-1$\frac{3}{4}$ (5-mm) (b) T-1 (3-mm), (c) subminiature, and (d) surface-mount packages.

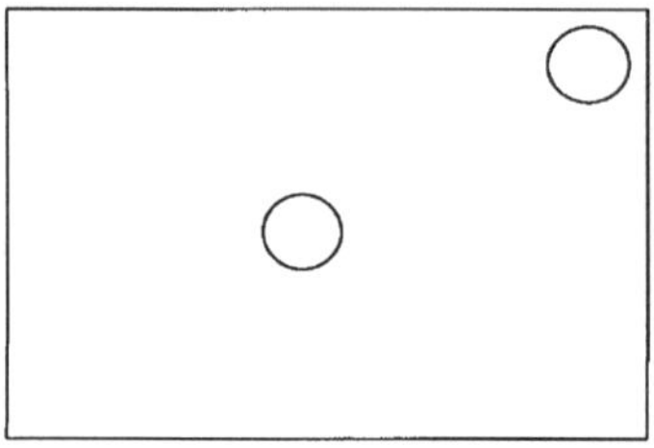

$$Lv \text{ (average)} = \frac{Lv \text{ (center)} + Lv \text{ (corner)}}{2}$$

$$Lv \text{ (ratio)} = \frac{Lv \text{ (center)}}{Lv \text{ (corner)}}$$

FIG. 29. Luminous sterance measurements (measured with 2.5-mm-diameter spot) used for warning light experiment. All experiments were done at 20-mA forward current.

Figure 31 shows the average luminous sterance versus cavity depth for the same surface-mount lamp. Note that the luminous sterance is roughly proportional to the number of LED lamps used in the cavity. Also note that the luminous sterance is relatively independent of cavity depth; that is, the luminous sterance increases by a factor of about 1.6 over a range of cavity depths from 18 to 40 mm.

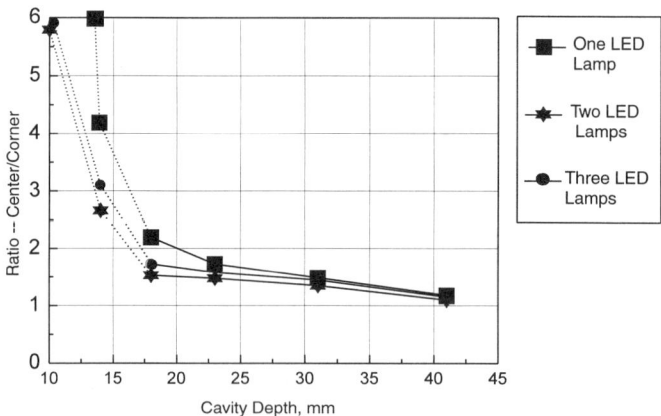

FIG. 30. Ratio of luminous sterances versus warning light depth for transparent substrate (TS) AlGaAs HSMX-TXXX light-emitting diode (LED) lamps. Dashed lines indicate cavity depths where luminous sterance ratio is greater than 2.0. (After Hodapp, 1994a. Reprinted with permission from SAE paper 941046, ©1994 Society of Automotive Engineers, Inc.)

6 APPLICATIONS FOR HIGH-BRIGHTNESS LIGHT-EMITTING DIODES 287

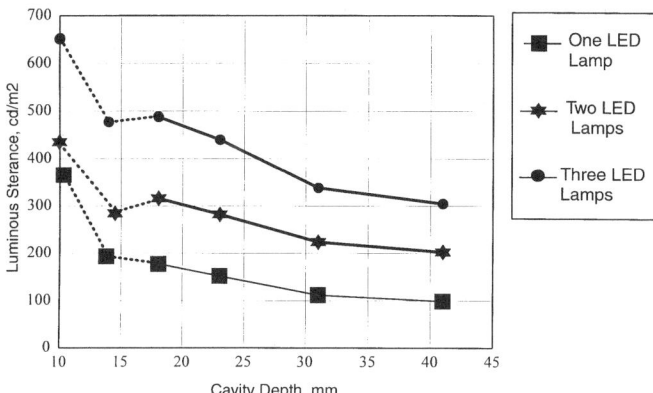

FIG. 31. Average luminous sterance versus warning light depth for transparent substrate (TS) AlGaAs HSMX-TXXX light-emitting diode (LED) lamps. (After Hodapp, 1994a. Reprinted with permission from SAE paper 941046, ©1994 Society of Automotive Engineers, Inc.)

Figures 32 and 33 show the performance of several other TS AlGaAs LED lamp packages shown in Figs. 28(a) to 28(c). Figure 32 shows the center–corner ratio of luminous sterances at different cavity depths. A 19-degree viewing angle T-1$\frac{3}{4}$ LED lamp is uniform only at the 41mm cavity depth. A 30-degree viewing angle T-1$\frac{3}{4}$ LED lamp is uniform at 31-

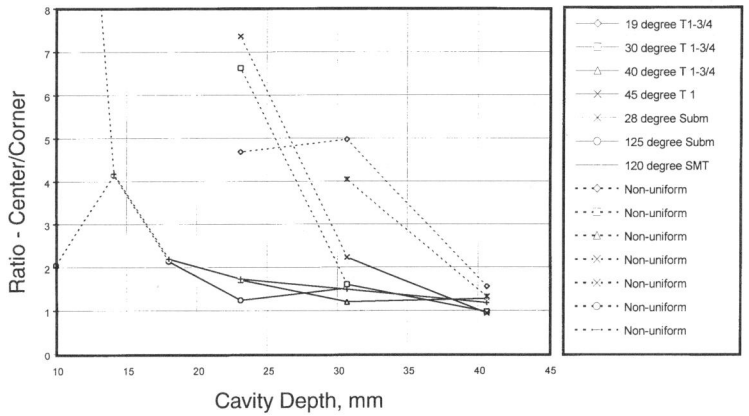

FIG. 32. Ratio of luminous sterances versus warning light depth for several transparent substrate (TS) AlGaAs light-emitting diode lamp packages. SMT, surface mount LED lamp (Fig. 28d); Subm, subminiature LED lamp (Fig. 28c); T1-$\frac{3}{4}$ LED lamp (Fig. 28a); T1 LED lamp (Fig. 28b). (After Hodapp, 1994a. Reprinted with permission from SAE paper 941046, ©1994 Society of Automotive Engineers, Inc.)

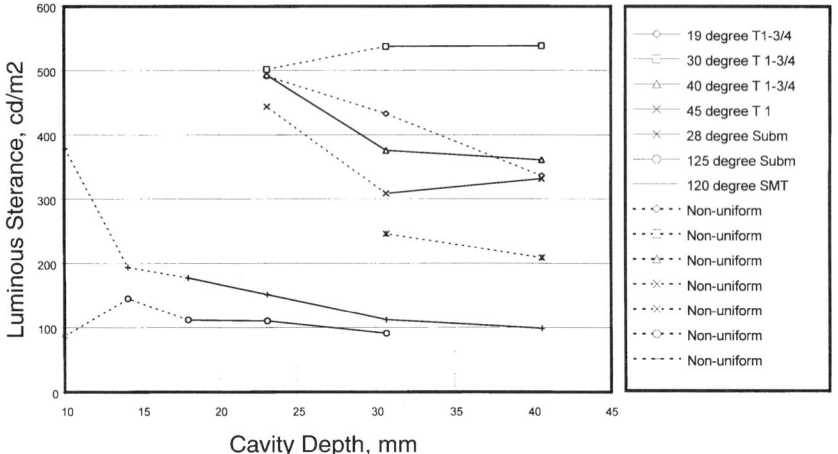

FIG. 33. Average luminous sterances versus warning light depth for several transparent substrate (TS) AlGaAs light-emitting diode lamp packages. (After Hodapp, 1994a. Reprinted with permission from SAE paper 941046, ©1994 Society of Automotive Engineers, Inc.)

and 41-mm cavity depths. A 40-degree viewing angle T-1$\frac{3}{4}$ LED lamp is uniform at cavity depths of 23, 31, and 41mm. A 45-degree viewing angle T-1 LED lamp is uniform at cavity depths of 31 and 41 mm. Figure 33 shows the average luminous sterance of the same four lamps. Note that all four lamps generate a luminous sterance in excess of 300 cd/m² at cavity depths at which they provide uniform lighting.

Figures 32 and 33 also show the performance of several TS AlGaAs SMT LED lamps. A 15-degree viewing angle subminiature LED lamp is uniform only at the 41-mm cavity depth. A 75-degree viewing angle subminiature LED lamp is uniform at cavity depths down to 18 mm. A 120-degree viewing angle SMT LED lamp is uniform at cavity depths down at 18 mm. Figure 33 shows the average luminous sterance of the same four lamps. The 15-degree subminiature LED lamp generates a luminous sterance of 200 cd/m² at the 41-m cavity depth. The other three lamps generate a luminous sterance of 100 to 175 cd/m² at cavity depths at which they provide uniform lighting.

One conclusion reached in reviewing these data is that the viewing angle of the LED lamp should be selected based on the intended instrument cluster cavity depth. For a deep instrument cluster, the optimum LED lamp would have a narrow viewing angle. For a shallow instrument cluster, the optimum LED lamp would have a wider viewing angle. With a wide viewing angle LED lamp in a deep instrument cluster, the luminous sterance of the

6 APPLICATIONS FOR HIGH-BRIGHTNESS LIGHT-EMITTING DIODES   289

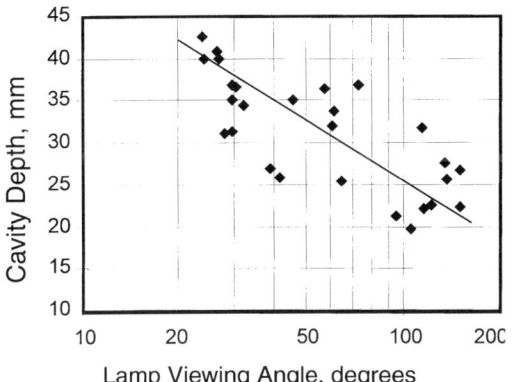

FIG. 34. Minimum viewing angle versus warning light cavity depth. Cavity depth where the L_v ratio is two to one. The viewing angle is $2\theta_{\frac{1}{2}}$.

legend is not as bright as is the optimum narrow angle LED lamp. Conversely, using a narrow angle LED lamp in a shallow instrument cluster may not properly illuminate the corners of the legend. Obviously, the size of the legend affects the minimum viewing angle versus cavity depth. A graph showing minimum viewing angle versus cavity depth for the 14- by 16-mm legend area used in the paper (Hodapp, 1994a) is shown in Fig. 34.

A second conclusion reached by viewing these data is that the undiffused T-1$\frac{3}{4}$ and T-1 LED lamp packages (which have narrower viewing angles) out-perform the SMT LED lamp packages (which tend to have wider viewing angles). Although the results are based on limited test data, the through-hole LED lamps out-perform the SMT LED lamps by a factor greater than 2 to 1. This difference in performance is large enough to allow sizable reductions in lamp count (i.e., from two lamps to one lamp per warning light) or to allow the use of less efficient and cheaper LED materials technology (i.e., from TS AlGaAs to AS AlGaAs, or possibly from AS AlInGaP to GaP).

4. CAVITY DESIGN OF AN INSTRUMENT CLUSTER WARNING LIGHT

The small size and low power dissipation of LED lamps allow the warning light cavity to be optimized for higher luminous flux output. Sloping the reflector cavity walls helps to reflect the light rays emitted by the LED lamp toward the diffuser film. The basic concept is shown in Fig. 35. For an incandescent design, θ_w is generally very small due to the large

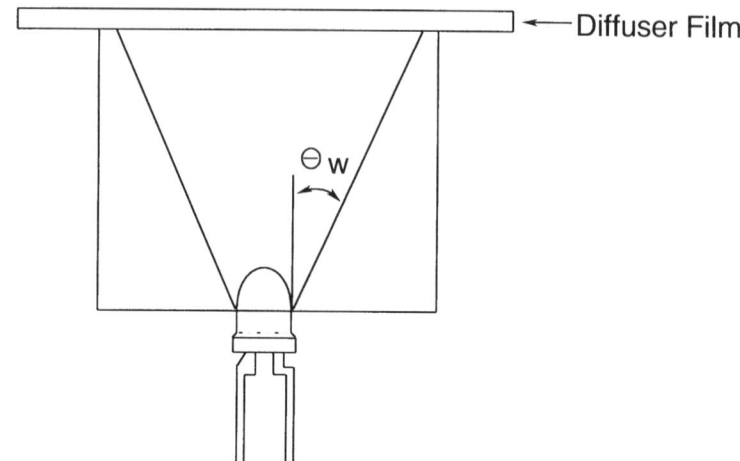

FIG. 35. Straight wall reflector cavity.

size and high power dissipation of the bulb. This allows maximum air volume in the cavity. In an LED design, the base of the cavity can be almost touching the LED package. This allows θ_w to be increased. The benefit of sloping the cavity walls is that when light rays spectrally reflect off the cavity walls, the light will be reflected more normal to the diffuser film surface. This

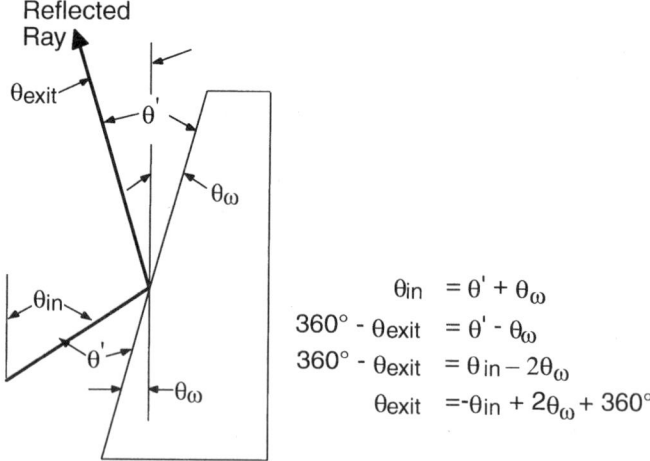

FIG. 36. Principle of specular reflections. (After Hodapp, 1990. Reprinted with permission from SAE paper 900474, ©1994 Society of Automotive Engineers, Inc.)

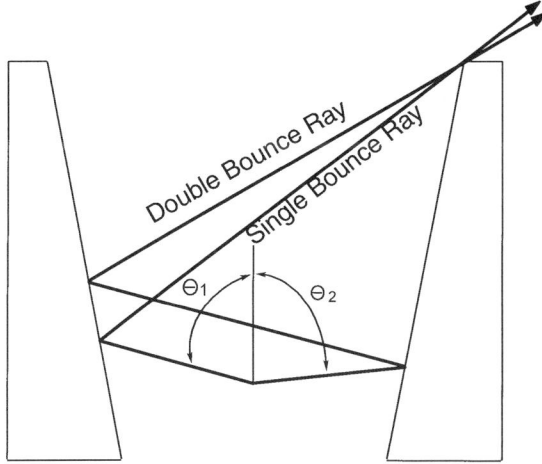

FIG. 37. Single- and double-bounce specular reflections.

allows the luminous flux generated by the light source to be emitted from the reflector cavity with less loss. The angle of the reflected ray of light is related to the emitted ray of light by the formula shown below, and drawn in Fig. 36:

$$\theta_{exit} = 360° - \theta_{in} + 2\theta_w$$

Based on this formula, and on the mechanical dimensions of the cavity, it is possible to calculate which range of angles of emitted light rays are emitted directly from the cavity. It is also possible to calculate the angles of emitted light rays that make a single bounce off the cavity walls, that make two bounces off the cavity walls, and so on. Figure 37 shows this concept. Note that the ray of light denoted as θ_1 makes a single reflection off one cavity wall before being emitted from the cavity. Light ray θ_2 reflects off one cavity wall, then off the adjacent cavity wall, before being emitted from the cavity. Then, by knowing the angular distribution of luminous flux generated by the light source, the reflectance of the cavity walls, and assuming only specular reflections, it is possible to estimate the total luminous flux emitted by the reflector cavity as follows:

$$\phi_T = \phi_{\theta d} + R(\phi_{\theta 1} - \phi_{\theta d}) + \sum_{n=2}^{\infty} R^n(\phi_{\theta n} - \phi_{\theta n-1})$$

where

ϕ_T = percent luminous flux emitted by the cavity

$\phi_{\theta d}$ = percent luminous flux emitted by the light source from angle 0 to θ_d

$\phi_{\theta 1}$ = percent luminous flux emitted by the light source from angle 0 to θ_1

$\phi_{\theta n}$ = percent luminous flux emitted by the light source angle 0 to θ_n

R = cavity wall reflectance

Figure 38 shows an example of a simple straight-walled cavity design. For this design, the maximum cavity wall angle is equal to

$$\theta_w = \tan^{-1}\left(\frac{\text{Exit} - \text{Entrance}}{2\,\text{Depth}}\right)$$

Two cavity designs were compared. The first design uses a straight-walled cavity with a 1-degree draft to facilitate demolding of the instrument cluster case. The second design has sloped cavity walls where the cavity wall angle is as large as possible, rounded down to the closest integer value. For both designs, the reflection angles θ_n, were calculated for up to four reflections. The direct angle θ_d also was calculated where light is emitted from the cavity without interacting with the cavity walls. The results of these calculations are shown in Table XIII.

A suitable lamp was then chosen. Based on the results shown earlier in Fig. 33, the HSMX-TXXX SMT LED lamp-type package was selected,

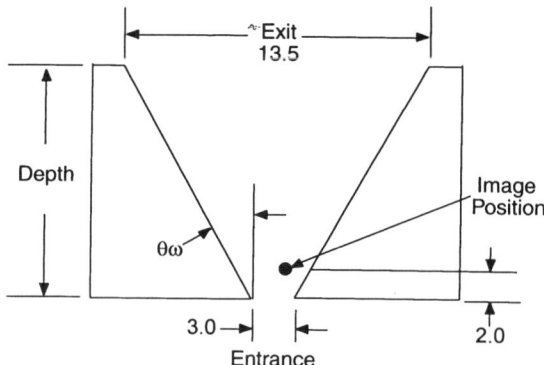

FIG. 38. An example of straight wall cavity design.

6 APPLICATIONS FOR HIGH-BRIGHTNESS LIGHT-EMITTING DIODES

TABLE XIII
REFLECTION ANGLES VERSUS CAVITY DEPTH

Depth (mm)	θ_ω (degrees)	θ_d (degrees)	θ_1 (degrees)	θ_2 (degrees)	θ_3 (degrees)	θ_4 (degrees)
40.6	1	9.9	28.0	42.2	55.9	64.9
	7		29.3	47.6	67.9	85.2
30.7	1	13.2	35.6	50.0	63.4	71.0
	9		38.6	61.5	84.4	103.4
23.1	1	17.7	44.4	59.7	70.2	76.4
	12		50.7	78.7	104.6	126.1
18.0	1	22.9	52.5	66.6	75.3	80.4
	16		64.0	97.7	127.5	>135
14.1	1	29.1	60.1	72.4	79.4	83.6
	20		78.6	116.9	>135	>135
10.0	1	40.2	69.6	78.9	84.0	87.2
	27		101.2	>135	>135	>135

because it can be used over a wide range of cavity depths. The lamp has a very wide viewing angle. Next, the cumulative luminous flux versus off-axis angle was measured, as shown in Fig. 39.

Finally, the total luminous flux emitted from the white reflector cavity was calculated based on the reflection angles (θ_d, θ_1, θ_2, θ_3, θ_4) shown in Table XIII, the cavity reflectance R ($0.20 \leqslant R \leqslant 0.80$), and the cumulative lumi-

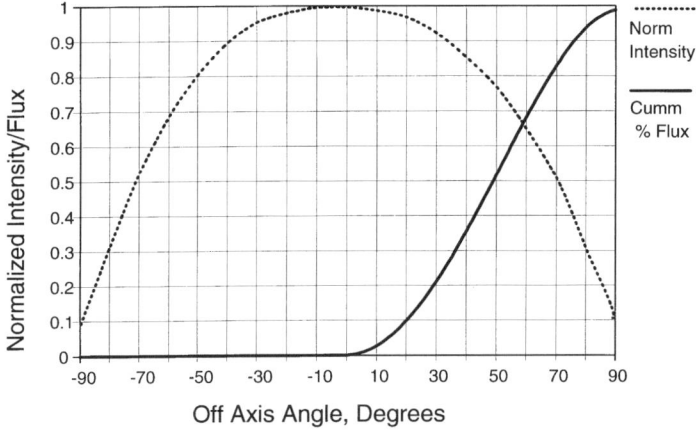

FIG. 39. Radiation pattern and cumulative optical flux versus off-axis angle for HSMX-TXXX light-emitting diode lamp.

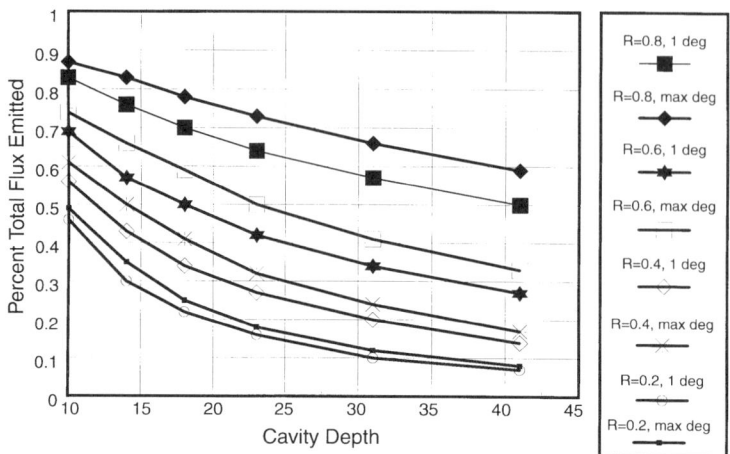

Fig. 40. Percentage of optical flux emitted from the cavity design example.

nous flux versus off-axis angle of the LED lamp shown in Fig. 39. The results of these calculations are shown in Fig. 40.

Note that sloping the cavity wall increases the total luminous flux emitted from the cavity by a factor of 10% to 20%, depending on wall reflectance and cavity depth. However, an even stronger relationship is observed in the total luminous flux emitted, depending on the cavity wall reflectance. In some cases, increasing the wall reflectance by 20% increases the total luminous flux emitted by the cavity by a factor of 50%! Note, that reducing cavity depth does tend to increase the amount of total luminous flux emitted. This increased cavity efficiency may account for the improvement in luminous sterance at shallow cavity depths as has been shown in Fig. 31.

5. Legend Optimization of an Instrument Cluster Warning Light

A third consideration in the design of the instrument cluster warning light is the proper optimization of the diffuser film graphics. Historically, the graphics film has been created by silk screening different ink layers on top and on the back of a piece of diffuser film. The top layer might consist of a neutral density gray "deadfront" surface that is designed to minimize front surface reflections and match the reflected light from the warning light cavities with the other parts of the instrument cluster. On the back of the diffuser film are several layers of ink. One layer is used to opaque portions of the graphics. Several semi-transparent colored layers are used for analog gauge scales and lettering. Additionally, transparent colors are silk-screened on the back of the film to generate the signal lighting colors and general illumination (night-time) colors. A typical graphics film might have 10 to 12 different silk-screen layers. The cross-section of a typical warning light cavity has been shown previously Fig. 26. With the LED lamp, the transparent colored ink layers can be removed from the graphics, not only reducing the overall silk-screen costs of the graphics film but increasing the luminous sterance of the warning light legend. Depending on the ink color, eliminating the transparent ink can increase the luminous sterance by a factor of two times or more!

However, it is also important to consider the diffusing properties of the graphics film. If a film has good diffusing properties, then light rays passing through the film are randomly scattered. The effect causes the image of the illuminated legend to appear to be emitted at the plane of the diffusing film. Without sufficient diffusing properties, the light appears to be emitted by a light source located behind the film. In this case the legend may not be easily readable, or may be readable only over a very restricted range of viewing angles. The differences between these optical properties are shown in Fig. 41. The diffusing properties of the legend graphics are determined not only by the properties of the base diffusing film but also by the diffusing properties

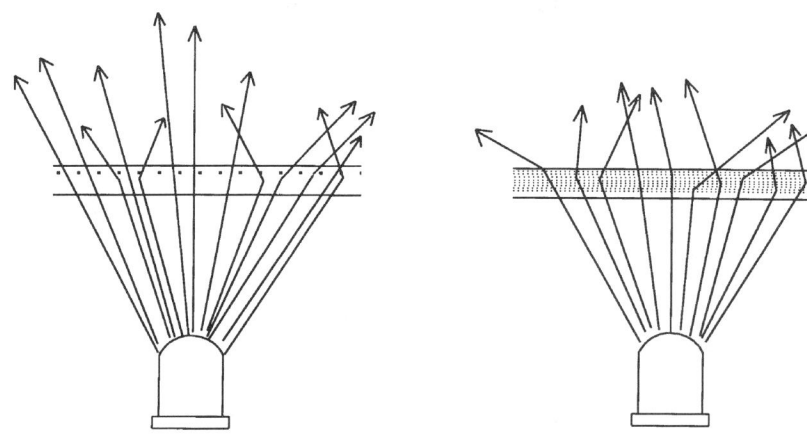

a. Poor diffusing properties   b. Good diffusing properties

FIG. 41. Diffusing properties of legend graphics. (a) Poor diffusing properties. Note that the image of the light-emitting diode (LED) lamp is visible through the film. (b) Good diffusing properties. Note that the light appears to be emitted from the film instead of the LED lamp.

of all the ink layers screened on the film. Removing the transparent ink layer on the film tends to reduce the overall diffusing properties of the graphics. Coupling this with the more directional beam of light generated by the LED lamp may cause the viewing angle of the legend to be restricted and in worst-case situations may prevent the legend from being fully illuminated.

The solution to this problem can be as simple as choosing a different base film material with higher diffusing properties than the base film used for the incandescent design. The overall result may be to sacrifice some of the luminous sterance improvement obtained by removing the transparent ink layer, while significantly increasing the readability of the legend graphics on-axis as well as at off-axis angles. Thus, the cost savings realized by eliminating several silk-screen layers can be achieved without reducing the viewing angle of the warning light legend.

V. Traffic Signal Lights

1. Characteristics of the Traffic Signal Market

The worldwide traffic signal market is a vast new market for LED technology for signal heads and pedestrian traffic control. According to E Source, Inc., the US Department of Transportation estimates that there are 260,000 signalized intersections in the United States alone (Houghton, 1994). Each intersection includes a minimum of 12 red, yellow, and blue-green traffic signal lights (four tricolor heads). Many intersections include a redundant set of red, yellow, and blue-green left turn signals, as well as orange and white pedestrian crossing signals. E Source estimates that an average intersection may have 20 traffic signal lights illuminated simultaneously, out of a total of 52 traffic signals lights. Thus, the United States, alone, represents a total market of 13.5 million traffic signal lights. Traditionally, incandescent lamps have been used to illuminate these signals. The desire to reduce electrical power consumption and related energy costs is driving the conversion to nonincandescent traffic signals. The increase in performance of LED technology and the advent of InGaN LED materials technology makes it a serious contender for traffic signal lighting.

Traffic signal lighting is regulated by each country. In the United States, the Institute of Transportation Engineers (ITE) (1991) has written overall specifications for traffic signal lights. Additionally, state and local governments create their own final specifications that define the signal colors, minimum luminous intensities, beam radiation pattern, mounting, and

6 APPLICATIONS FOR HIGH-BRIGHTNESS LIGHT-EMITTING DIODES 297

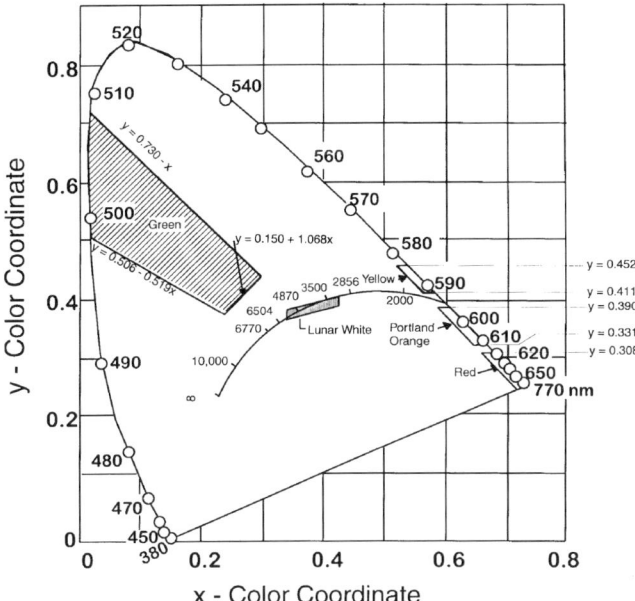

FIG. 42. Institute of Transportation Engineers (ITE) traffic signal colors. Width of yellow, portland orange, and red acceptable color ranges are exaggerated for better clarity. (After ITE. 1991. Reprinted with the permission of the Institute of Transportation Engineers.)

environmental requirements. Although these specifications were originally written around incandescent lamps, other lighting technologies such as LEDs are now being used.

The ITE has defined five distinct colors for traffic signal lighting, as shown in Fig. 42. The colors red, yellow and blue-green are used to represent "Stop", "Caution", and "Go" for the standard tricolor traffic signal head. The colors portland orange and lunar white are used to represent the "Don't walk" and "Walk" signals for pedestrian traffic control.

Local, county, and state governments are responsible for proper maintenance of traffic signals, depending on the location of the intersection. Due to the safety hazards resulting from burned-out signal lights (and potential for high liability costs due to traffic accidents), municipalities normally re-lamp traffic intersections on a proactive basis in order to minimize the occurrence of signal burnouts. In addition, provisions must still be made for emergency repair in the advent of a signal burnout. In some cases, redundant signal lights are used at major intersections to minimize the probability of complete signal burnouts.

2. Construction of a Traffic Signal Head

Figure 43 shows the construction of a traffic signal head. A sketch of a typical incandescent traffic signal head is shown in Fig. 43(a). A socketed incandescent lamp is mounted at the focus of a parabolic reflector. The reflector collimates part of the luminous flux generated by the lamp. A colored glass or uv stabilized plastic lens is mounted over the reflector. The lens is designed to filter the white incandescent light to achieve the desired signal color. Finally, a black visor may be mounted over the lens to enhance visibility. The visor is designed to minimize stray sunlight from falsely illuminating the signal indication and may also restrict viewing of the traffic signal for certain off-axis angles.

A sketch of a generic LED traffic signal module is shown in Fig. 43(b). An array of LED lamps is soldered on a printed circuit board, and the board is mounted behind a glass or plastic lens. The lens is only lightly colored because LED lamps directly generate the desired signal color. Similar to the incandescent signal head, a black visor may be mounted over the lens.

3. Comparison Between Incandescent and Light-Emitting Diode Traffic Signals

Table XIV compares the incandescent lamp and LED lamp traffic signal modules. The incandescent lamp has the advantage of being the lowest cost source of luminous flux. Twelve-inch-diameter traffic signal heads

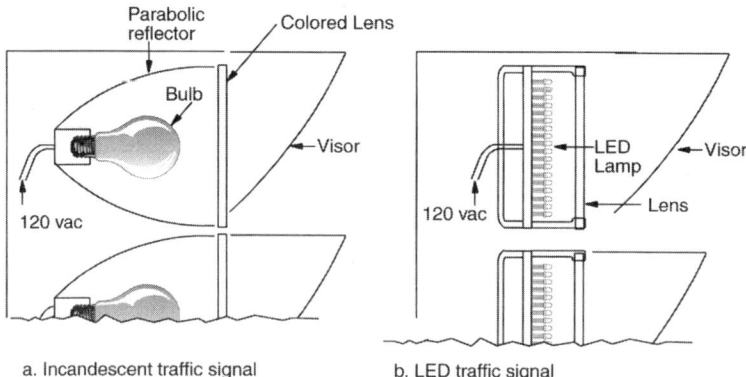

FIG. 43. (a) Incandescent traffic signal head construction and (b) light-emitting diode (LED) traffic signal module construction.

TABLE XIV
Comparison Between Incandescent and Light-Emitting Diode (LED) Traffic Signal Designs

Lamp	Advantages	Disadvantages
Incandescent	Low cost luminous flux Socketed for field repair Light output independence of temperature Operates directly from 120 Vac	High power consumption High maintenance cost Color determined by lens
LED	Much higher MTBF Much longer lifetime Built-in redundancy Low power consumption Lower maintenance cost Color determined by LED lamp More vandal resistant	Higher cost luminous flux Requires external drive circuit Light output varies with temperature

Vac, ac line voltage; MTBF, meantime between failure.

commonly use single 150-W incandescent lamps. Eight-inch-diameter traffic signal heads and pedestrian crossing signals commonly use single 67-W incandescent lamps. These incandescent lamps are designed to operate directly off 120-V ac current, through some type of traffic signal controller.

The primary disadvantages of the incandescent lamp traffic signal heads are the high power consumption of the lamps and the high maintenance costs due to frequent re-lamping. Figure 44 shows a sketch of a typical intersection. In this example, 20 traffic signal heads are illuminated simultaneously. Based on the mixture of 8- and 12-inch traffic signal heads, pedestrian crosswalk signals, and lamp duty cycles specified, the average electrical power consumption of the traffic signal lights of the signalized intersection is 2.070 kW or 18,133 kWh per year. At an energy cost of 8 cents/kWh, this corresponds to an electricity cost of $1450 per year, for signal operation at this intersection.

The cost of properly maintaining the intersection is also substantial. According to E Source, most municipalities re-lamp all red and blue-green traffic signals annually and yellow traffic signals every two years (Houghton, 1994). This replacement requires a two-man crew with a bucket truck, and typically costs about $250 per intersection. In addition, many transit agencies require the immediate replacement of failed red traffic signals and replacement of failed yellow and blue-green traffic signals the next business day in order to minimize the municipality's liability.

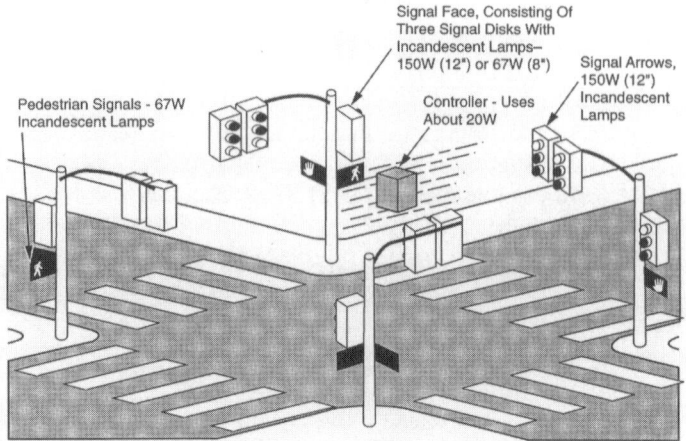

FIG. 44. Typical signalized intersection. (After Houghton, 1994. Reprinted with the permission of E Source, Inc.)

Another disadvantage of incandescent traffic signal heads is that the traffic signal color of the traffic signal unit is determined by the color of the outer lens. Thus, vandalism that can destroy the glass lens (and possibly the incandescent lamp) results in an ambiguous white light signal condition, and its many consequences.

The LED traffic signal modules have a number of advantages over incandescent signal heads. The major benefits of LED technology are primarily lower electrical power consumption, which reduces electricity costs, and secondarily longer life, which reduces maintenance costs.

The primary benefit of LED traffic signal modules is the reduced electrical power consumption. Typical 8- and 12-inch red LED traffic signal modules operate at about 15 and 20 W, respectively. Since left turn signals only use an array of LED lamps in the shape of an arrow, their power consumption is about 9 W. Thus, in the intersection shown in Fig. 44, replacing just the eight 12-inch red traffic signal heads and the four 12-inch red arrows with LED traffic signal modules represents an average electrical power savings of 1.132 kW (over half of the total of the 2.070 kW for the complete incandescent traffic signal intersection). This represents a savings of 9916 kWh per year. At an energy cost of 8 cents/kWh, this corresponds to an electricity cost savings of $793 per year. With an average cost of $200 per module, the payback on the use of red LED traffic signal modules is about 3 years, based only on the electricity cost savings.

With the exception of the portland orange "Don't walk" pedestrian cross-walk symbols, the other colors represent less of an energy cost savings, because the duty cycle is relatively low. However, the portland orange "Don't walk" pedestrian cross-walk symbol has a typical duty cycle of 75%, and therefore it also represents a significant energy savings. At 12 W each, the eight LED "Don't walk" cross-walk symbols represent an average electrical power savings of 330 W, or a savings of 2891 kWh per year ($231/year at 8 cents/kWh).

When LED traffic signal modules were first introduced, manufacturers claimed that the lifetime of the LED signal module was 10 years, or 43,800 hours at a 50% duty cycle. After this time, the LED signal modules could be expected to be fully functional, with less than 50% reduction in luminous intensity. After considerable field trials with TS AlGaAs red LED lamps, the expected lifetime was reduced to the range of 5 to 6 years. Meanwhile, in 1996, AlInGaP red, yellow, and portland orange LED technologies have become available that have the potential for meeting the 10-year-lifetime goal. In addition, InGaN blue-green LED technology has been introduced, which is currently undergoing field trials in Japan and in New Hampshire.

Even if only the red incandescent traffic signal heads are replaced with red LED traffic signal modules, the primary safety concerns of red lamp burnout are minimized. This may allow the remaining incandescent yellow and blue-green traffic signal heads to be replaced on a less frequent maintenance schedule, perhaps every 2 or 3 years. Naturally, the largest maintenance savings will occur if all five traffic signal colors are replaced by LED lamp technology.

The LED traffic signal modules typically provide a built-in redundancy that tends to make them more dependable than are incandescent signals.

Most LED traffic signal modules use several series strings of LED lamps. For example, a 12-inch red LED traffic signal module might have 3 to 9 series strings with 70 to 75 LED lamps per string (210 to 675 lamps total). The failure of one LED lamp in an LED traffic signal module would only affect a single series string, which would still allow the traffic signal to operate at reduced [$\frac{2}{3}$ (67%) to 8/9 (89%)] light output.

Failure of an LED traffic signal module due to vandalism is a reduced concern. E Source cites a study done by the state of Oregon (Houghton, 1994). A "shoot-out" was conducted in December 1993 to see how resistant an LED signal module is to destructive vandalism. After sustaining several shots fired from a 9-mm weapon at a distance of 25 feet, only 112 of the 600 LEDs in the 12-inch traffic signal module under test had failed. Because the color of the signal head is determined by the LED lamps and not the lens, there was no effect on the color of the signal. The damaged LED signal module was still a viable red signal at a distance of 50 to 75 feet.

The primary disadvantage of LED traffic signal modules is high cost. When 12-inch TS AlGaAs red LED traffic signal modules were first introduced in 1992, their price was in the range of $350. In 1996, the price of a 12-inch AlInGaP red LED traffic signal module is in the range of $200. By the year 2000, InGaN blue-green traffic signal modules are expected to be priced competitively with AlInGaP LED signal modules. As shown earlier, with tricolor LED signal modules installed at an intersection the energy cost savings and maintenance costs savings become significantly large.

One disadvantage of LED traffic signals is the need for a more complicated external drive circuit, which is included in the price of the LED traffic signal modules. Another disadvantage of LED traffic signals is that the light output of the signal varies with ambient temperature. However, drive circuits can be designed to minimize the change in light output due to variations in ac line voltage, and temperature. Examples of these circuits will be discussed in Part V, Section 5.

4. Luminous Flux Requirements

The first step in the design of an LED traffic signal module is to determine approximately how many LED lamps are needed to meet the minimum photometric luminous intensity output. This can be done by integrating the minimum luminous intensity values over the defined spatial radiation pattern to calculate the required minimum luminous flux. Then the number of LED lamps can be estimated by dividing the total luminous flux required by the luminous flux emitted by each LED lamp.

TABLE XV
Minimum Luminous Intensities and Zonal Constant Integrations for 8- and 12-inch Red Traffic Signals per Institute of Transportation Engineers (ITE) Specification ST-017

8-inch traffic signal minimum photometrics (cd)	27.5L	22.5L	17.5L	12.5L	7.5L	2.5L	V	Total	$C_z \cos V$	ϕ_v
2.5D			29	67	114	157	157	734	7.608×10^{-3}	5.58
7.5D	12	21	48	76	105	119	119	762	7.550×10^{-3}	5.73
12.5D	10	14	24	33	38	43	43	324	7.435×10^{-3}	2.41
17.5D	5	7	10	12	17	19	19	140	7.263×10^{-3}	1.02
									Total ϕ_v	14.76 lm

12-inch traffic signal minimum photometrics (cd)	27.5L	22.5L	17.5L	12.5L	7.5L	2.5L	V	Total	$C_z \cos V$	ϕ_v
2.5D			90	166	295	399	399	1900	7.608×10^{-3}	14.46
7.5D	19	45	105	171	238	266	266	1688	7.550×10^{-3}	12.74
12.5D	19	26	40	52	57	59	59	506	7.435×10^{-3}	3.76
17.5D	19	24	26	26	26	26	26	294	7.263×10^{-3}	2.14
									Total ϕ_v	33.10 lm

Total is equal to horizontal sum, less V position, times two.

The zonal constant method outlined in Part II, Section 2, can be used to calculate the total luminous flux. Table XV shows the minimum luminous intensity values for 8-inch and 12-inch red traffic signal heads as outlined by ITE specification ST-017. Please note that Table XV shows only half of the minimum luminous intensity values, because the values for 2.5° R, 7.5° R, 12.5° R, 17.5° R, 22.5° R, and 27.5° R are the same as the corresponding left angles. The spatial radiation pattern can be integrated by assuming that the luminous intensity is constant over an area of 5 degrees horizontal and 5 degrees vertical, with $n = m = 350°/5° = 72$, and vertical angles of 2.5, 7.5, 7.5, 12.5, and 17.5 degrees. Thus, the integration would be calculated as shown in Table XV. Therefore, a minimum of 14.8 or 33.1 lumens of luminous flux would be required to meet the 8- or 12-inch red traffic signal specification, respectively.

TABLE XVI

MINIMUM LUMINOUS INTENSITIES AND ZONAL CONSTANT INTEGRATIONS OF 8- AND 12-INCH TRAFFIC SIGNALS PER INSTITUTE OF TRANSPORTATION ENGINEERS (ITE) SPECIFICATION ST-017

Traffic Signal	Minimum Luminous Intensity (2.5 L, 2.5 D) (cd)	Minimum Luminous Flux (lm)
8-inch red	157	14.8
8-inch yellow	726	68.1
8-inch green	314	29.5
12-inch red	399	33.1
12-inch yellow	1848	153.3
12-inch green	798	66.2

Similarly, the minimum luminous intensity values for the 8- and 12-inch yellow and blue-green traffic signal heads given in ITE specification ST-017 can be integrated to calculate the total luminous flux. Table XVI shows the minimum luminous intensity values for these other signals and their corresponding minimum luminous flux values. Please note that the radiation patterns for the yellow and blue-green traffic signal heads are the same as for the red traffic signal heads of the same diameter. Thus, the approximate luminous intensity values for all angular positions can be calculated by dividing the (2.5L, 2.5D) point shown in Table XVI by the corresponding (2.5L, 2.5D) point shown in Table XV and multiplying the radiation patterns in Table XV by this factor.

The next step would be to estimate the luminous flux generated by each LED lamp. As outlined in Part III, Section 4, a number of factors determine the luminous flux emitted by each LED lamp. The primary factors are the LED lamp drive current, operating temperature, and LED lamp luminous intensity or luminous flux test bin. Table XVII shows typical specifications for several T-$1\frac{3}{4}$ LED lamps that are suitable for traffic signal heads and pedestrian "Don't walk" cross-walk signals. Again, using a similar equation as shown in Part III, Section 4, it would be possible to estimate the luminous flux losses due to internal heating as well as the luminous flux that is not utilized by the outer lens (and possibly by any secondary reflector optics), to arrive at a more accurate value of usable luminous flux generated by each LED lamp. However, for the sake of this analysis, we assume that the typical luminous flux values given in Table XVII should be halved to account for these losses.

The last step in determining the approximate number of LED lamps needed to achieve the minimum ITE luminous intensity specification for

TABLE XVII

Typical Luminous Flux and Color of T-1¾ Lamps for Light-Emitting Diode Traffic Signal Modules

Lamp	Color	Material	Dominant Wavelength (nm)	Viewing Angle (degrees)	Typical V at 20 mA	Typical Luminous Flux at 20 mA (mlm)
HLMP-DG30	Red	AS AlInGaP	626	30	1.9V	380
HLMP-DJ15	Portland orange		605	15		570
HLMP-DL30	Yellow		590	30		
Nichia NSPE-510S	Blue-green	InGaN	505	30	3.5V	700

AS AlInGaP, absorbing substrate Aluminum Indium Gallium Phosphide LED technology; InGaN, Indium Gallium Nitride LED technology.

traffic signals, is to divide the luminous flux requirement (Table XVI) by the luminous flux generated by each LED lamp (Table XVII). The values for luminous flux given in Table XVI may be unrealistic, because the calculations did not take into account lens transmission losses and inaccuracies in the actual generated radiation pattern. Using an equation similar to that shown in Part III, Section 4, it would be possible to estimate these various luminous flux losses to arrive at a more realistic minimum luminous flux requirement. However, for the sake of this analysis, let us assume that the minimum luminous flux values given in Table XVI should be doubled to account for these losses. These calculations are shown in Table XVIII. Using a factor of 4 times (twice as much luminous flux required divided by half the luminous flux per LED lamp) in doing the calculations, roughly 350 red T-1¾ LED lamps, 1080 yellow T-1¾ LED lamps, or 380 T-1¾ blue-green LED lamps driven at 20 mA each would be needed to meet the ITE 12-inch traffic signal specification.

Using design techniques similar to those shown in Part III, Section 5, it is possible to substantially reduce the number of LED lamps needed to meet the ITE traffic signal specification. The luminous flux generated by each LED lamp can be used much more efficiently by mounting either a convex fresnel lens or metalized reflector over each LED lamp. A pillow lens with rectangular-shaped pillows also could be used to break up the individual lamp images and to spread the light into the desired radiation pattern. Furthermore, high-flux LED lamps, such as the Hewlett-Packard HPWT-XX00 TS AlInGaP, can be driven at a higher forward current to generate a larger amount of luminous flux per LED lamp.

TABLE XVIII

Number of Light-Emitting Diode (LED) Lamps Needed to Meet Institute of Transportation Engineers (ITE) Traffic Signal Specification

	Minimum Luminous Flux (lm)	Realistic Luminous Flux (lm)	LED Lamp	Typical Luminous Flux at 20 mA (mlm)	Half Typical Luminous Flux at 20 mA (mlm)	Number LED Lamps
8-inch red	14.8	29.5	HLMP-DG30	380	190	155
8-inch yellow	68.1	136.2	HLMP-DL30	570	285	478
8-inch green	29.5	58.9	NSPE-510S	700	350	168
12-inch red	33.1	66.2	HLMP-DG30	380	190	348
12-inch yellow	153.3	306.7	HLMP-DL30	570	285	1076
12-inch green	66.2	132.5	NSPE-510S	700	350	379

One caution in these calculations is that the ITE specification calls for an oval-shaped radiation pattern, approximately 10 degrees vertical by 30 degrees horizontal viewing angle ($2\theta_{1/2}$). The viewing angle $2\theta_{1/2}$ is equal to two times the off-axis angle $\theta_{1/2}$ which is the off-axis angle at which the luminous intensity is half the on-axis value. Without secondary optics mounted over the LED lamps or incorporated in the outer lens, the LED lamps specified in the example create a round 30-degree viewing angle ($2\theta_{1/2}$) radiation pattern, which wastes a great amount of luminous flux. LED lamps with various viewing angles are available and the selection of the optimum LED lamp depends on the optical properties of the outer lens and secondary optics.

The ITE specification was written around the use of 665 and 1950 lm incandescent lamps for the 8- and 12-inch traffic signal heads, operating at a color temperature between 2600 and 2856 K with commonly available reflectors and glass and plastic colored lenses. The specification was written long before LED lamps were available and does not address some of the issues peculiar to LED technology, such as light-output variation due to ambient temperature. The specification also seems to have been written around the radiation patterns obtained with a particular set of reflectors and colored lenses without considering the human factors such as the minimum luminous intensity needed for each color based on human perception experiments on the different colors and radiation patterns needed for readability at specified distances. Until these experiments are conducted and the results approved, the existing higher luminous intensity requirements for yellow and blue-green traffic signals (as compared with red ones) could slow the adoption of yellow and blue-green LED traffic signals. However, the ITE specification as it is currently written shows that LED technology is technically and economically feasible for red traffic signals.

5. ELECTRICAL DESIGN CONSIDERATIONS

The electrical design of an LED traffic signal module would be completely different from an incandescent traffic signal head, but would use the same traffic signal controller. The incandescent design would use a 120-V ac-compatible lamp driven by a mechanical or a solid state relay. The LED design would probably incorporate an ac-to-dc circuit, because LED lamps are designed to operate from dc voltages with some type of external current limiting circuit.

A block diagram of a potential LED traffic signal circuit is shown in Fig. 45, which designed to replace the incandescent lamp using the existing traffic signal controller. First, the 120-V ac is full-wave rectified and filtered. The rectifier circuit produces an unregulated dc voltage of approximately 150 V. The traffic signal module consists of an array of LED lamps mounted on a printed circuit board. Although it may be possible to drive the LED lamp array directly from the ac line, such an approach leads to severe forward current variations and has the potential to subject the LED lamps to damaging electrical transients.

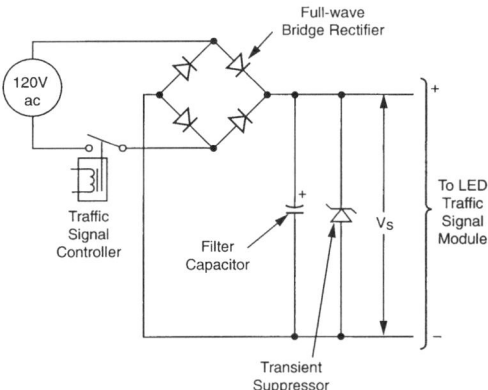

FIG. 45. Block diagram of light-emitting diode (LED) traffic signal module circuit.

The design of the traffic signal module LED lamp array should consider several factors.

First, the forward current through the LED lamps should be regulated so as not to cause noticeable luminous intensity variations within the traffic signal module due to the effects of variations in ac line voltage, ambient temperature, and forward voltage of the LED lamps themselves. In addition, operation of the LED lamps at excessive forward current for a long period of time can cause unacceptable luminous intensity degradation and even catastrophic failure.

Second, the circuit should be designed to minimize luminous intensity variations of the traffic signal module due to the effects of ac line voltage, ambient temperature, and LED electrical characteristic parametric variations.

Third, the design should minimize electrical power consumption to maximize energy savings.

Fourth, the design should provide for some redundancy so that the random loss of one LED lamp does not cause the entire LED traffic signal module to fail. With sufficient redundancy, a significant portion of the LED lamps would be fully functional so as not to noticeably impair the operation of the traffic signal.

Fifth, the drive circuit should be as inexpensive as possible, while meeting the other four design goals.

Because LED lamps have a forward voltage of about 2 V, they are normally driven as series connected strings in order to bring the combined forward voltage closer to the value of the unregulated dc voltage generated by the circuit in Fig. 45. The use of series strings, compared with driving each LED separately, substantially improves the electrical power efficiency of the traffic signal module. Figure 46 shows several possible circuit designs for the LED lamp array.

In Fig. 46(a), each LED string is driven by its own constant current source. The number of LED lamps per string should be chosen such that the current source operates properly over the worst-case circuit variations:

$$nV_{FMAX} \leqslant V_{SMIN} - V_{DROP\text{-}OUT}$$

where

n = number of LED lamps per string

V_{FMAX} = maximum forward voltage at the design current

V_{SMIN} = minimum dc voltage that occurs at the minimum ac line voltage

$V_{DROP\text{-}OUT}$ = lowest input-output voltage so that the current source can properly regulate the electrical current

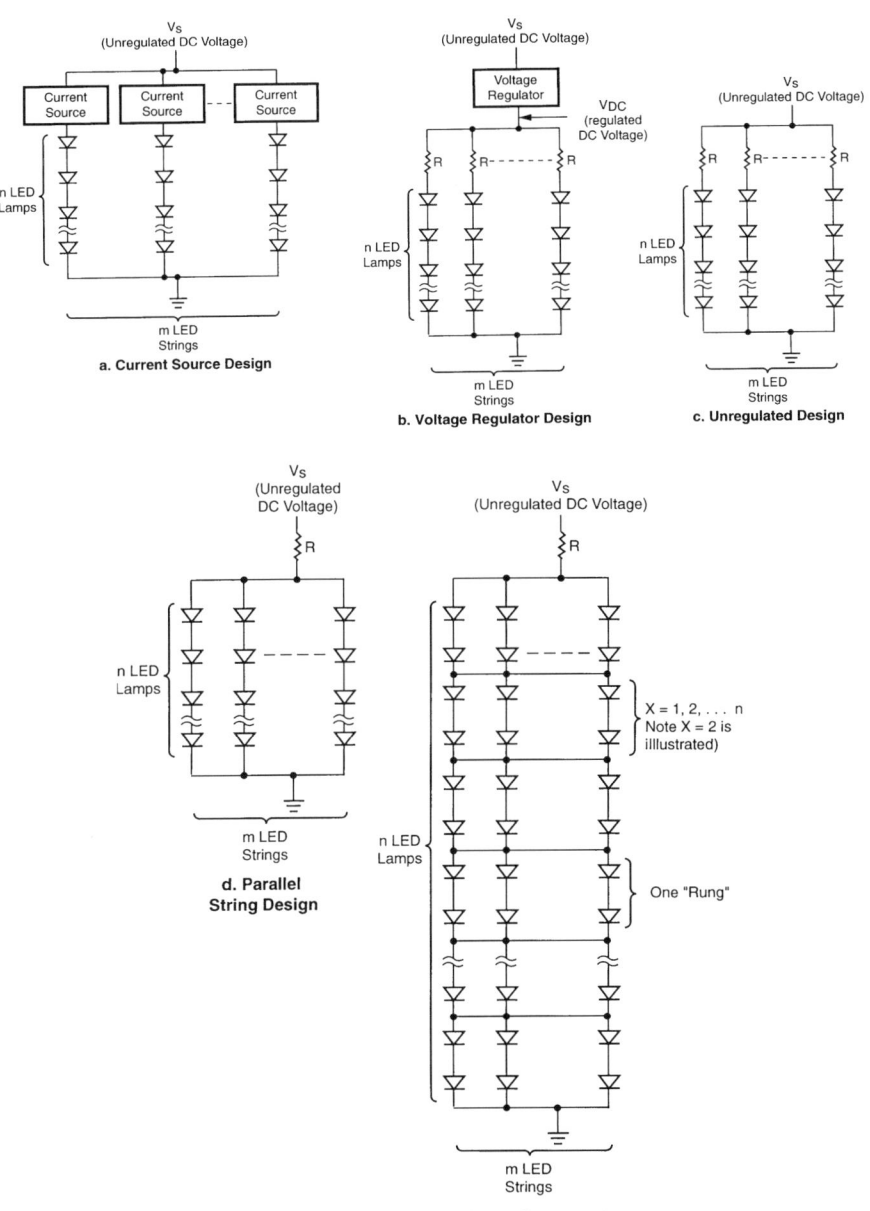

FIG. 46. Drive circuits designs for light-emitting diode (LED) traffic signal module. (a) Current source, (b) voltage regulator, (c) unregulated, (d) parallel string, and (e) cross-connected parallel string designs. R, resistor.

Thus, the forward current through the LED lamps will be constant regardless of ac line voltage variations, ambient temperature, and electrical characteristic variations of the LED lamps.

In Fig. 46(b), a voltage regulator generates a constant dc voltage. Then, a current limiting resistor regulates the current through each LED string. The output voltage V_{DC} of the voltage regulator should be chosen so that

$$V_{DC} \leqslant V_{S\,MIN} - V_{DROP\text{-}OUT}$$

The current limiting resistor acts as a current source, providing that the voltage dropped across the resistor is much larger than the sum of the forward voltage variations of the LED lamps. Thus the forward current through the LED lamps will be independent of ac line voltage variations, although it will vary somewhat owing to LED lamp forward voltage.

As described in Part III, Section 6, the LED lamp can be electrically modeled as a voltage source $V_{\text{turn-on}}$ in series with a resistor R_S. Thus, the forward voltage is equal to $V_{\text{turn-on}}$ plus the voltage drop across R_S. This equation can be used over the voltage range of 5 mA $\leqslant I_F \leqslant$ 50 mA (for AlInGaP LED lamps).

$$V_F = V_{\text{turn-on}} + R_S I_F + (\Delta V_F / \Delta T)(T - 25°C)$$

where

$V_{\text{turn-on}}$ = voltage where the straight line approximation of the I-V curve intersects the voltage axis at zero current

R_S = 1/slope of the straight line approximation of the I-V curve, $\Delta V_F / \Delta I_F$

$\Delta V_F / \Delta T$ = thermal coefficient of the forward voltage, V/°C

For the linearized model, the model parameter R_S varies much more than the model parameter $V_{\text{turn-on}}$. For most LED lamps $\Delta V_F / \Delta T$ varies by approximately -2 mV/°C.

For the circuit shown in Figure 46(b), the basic equation to calculate the resistor value R is shown below. Note that this circuit is identical to the automotive signal lamp circuit described in Part III, Section 6.

$$R = \frac{V_{DC} - nV_F}{I_F}$$

where

V_{DC} = regulated dc voltage

n = number of LED lamps per string

V_F = forward voltage of each LED lamp at the forward current I_F

I_F = desired forward drive current

Using the linearized forward voltage model, the current through the LED string can be written as

$$I_F = \frac{V_{\text{DC}} - nV_{\text{turn-on}} - n\Delta V_F/\Delta T}{R + nR_S}$$

In order to provide good current regulation, good design practice would be to choose the string length n such that R is greater than nR_S. This causes the forward current to be controlled by the external resistor R rather than the less well controlled series resistance R_S of the LED lamps. Over temperature, the variation $\Delta V_F/\Delta T$ can have a large effect on the forward current through the LED string. The forward current variation becomes more pronounced for longer string lengths.

In Fig. 46(c), the current limiting resistors are connected directly to the unregulated dc voltage. The forward current through the LED lamps will vary somewhat due to variations in the line voltage, ambient temperature, and electrical characteristics of the LED lamps.

The same equation described previously for Fig. 46(b) can be used to calculate the current through each LED string, except that the term V_{DC} in the equation would be replaced by V_S, the unregulated dc voltage. Due to the additional variation in V_S for a given LED string length, the overall current variation through the LED strings will be higher than the circuit that is shown in Fig. 46(b).

In Fig. 46(d), all LED strings are connected in parallel and all LED lamps are driven from a common resistor. For this circuit to work properly, the number of lamps per string must be the same. Even so, forward voltage variations of the LED lamps will cause the current to split unevenly across the different strings. In addition, the forward current through the LED lamps will vary somewhat due to the variations in ac line voltage, temperature, and LED lamp forward voltage.

Assuming that all LED lamps have the same electrical characteristics, we give the equation to calculate the resistor value R:

$$R = \frac{V_S - nV_F}{mI_F}$$

where

V_S = unregulated dc voltage

n = number of LED lamps per string

V_F = forward current of each LED lamp at the forward current I_F

m = number of parallel strings

The variations in forward current through each LED string can be estimated by using the linearized forward current model. Because the variation in the series resistance R_S is much larger than the variation in turn-on voltage $V_{\text{turn-on}}$, the equations can be simplified by assuming that $V_{\text{turn-on}}$ is the same for all LED lamps and the forward current variation is simply due to variations in R_S. (More accurate modeling could be done using the ideal diode model, with computer simulation of the random placement of different forward voltage LED lamps in the array.) Then, a worst-case scenario might be that all LED lamps in a single string have minimum series resistance R_L. A second string might have all LED lamps with a maximum series resistance R_H. All other strings might have typical LED lamps with series resistance R_T. This worst-case scenario is shown in Fig. 47. The forward currents I_L, I_H, and I_T can be calculated as shown below:

$$R = \frac{V_S - nV_{\text{turn-on}} - nR_T I_F}{mI_F}$$

$$I_L = \frac{V_S - nV_{\text{turn-on}}}{R\left[1 + \frac{R_L}{R_H} + (m-2)\frac{R_L}{R_T}\right] + nR_L}$$

$$I_H = \frac{V_S - nV_{\text{turn-on}}}{R\left[1 + \frac{R_H}{R_L} + (m-2)\frac{R_H}{R_T}\right] + nR_H}$$

$$I_T = \frac{V_S - nV_{\text{turn-on}}}{R\left[(m-2) + \frac{R_T}{R_L} + \frac{R_T}{R_H}\right] + nR_T}$$

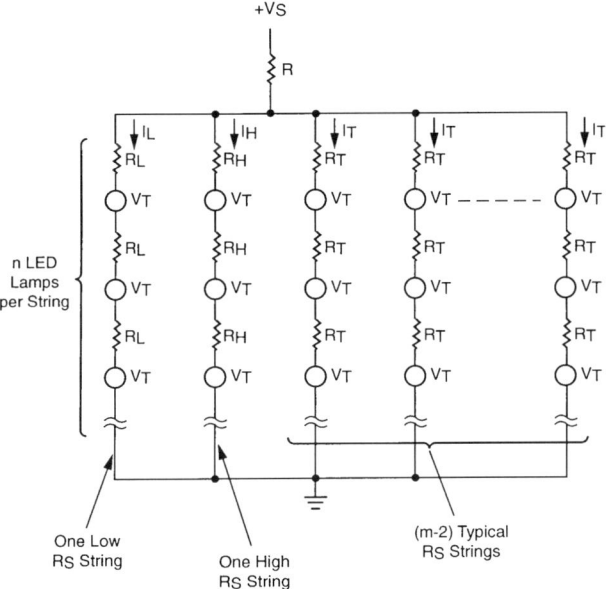

FIG. 47. Worst-case light-emitting diode (LED) forward current modeling for a parallel drive circuit.

where

R = external series resistor

V_S = unregulated dc voltage

I_F = design forward current per string

n = number of LED lamps per string

m = number of paralleled strings

I_L, I_H = actual forward current for worst-case minimum and worst-case maximum R_S strings, respectively

I_T = actual forward current for a typical R_S string

$V_{\text{turn-on}}$ = turn-on voltage for the linearized forward voltage model

R_L, R_H = worst-case minimum and worst-case maximum series resistance for the linearized forward voltage model, respectively

R_T = typical series resistance for linearized forward voltage model

Compared with the unregulated circuit shown in Fig. 46(c), the worst-case LED string currents for the paralleled string circuit will have larger forward current variations. Note that for the same linearized model parameters, the external resistor for the circuit in Fig. 46(c), R', is equal to mR, where R is the external resistor calculated previously. Again, assuming that all LED lamps are at their worst-case extremes, the forward currents I_L and I_H for the circuit in Fig. 46(c) are calculated as

$$R' = mR = \frac{V_S - nV_{\text{turn-on}} nR_T I_F}{I_F}$$

$$I_L = \frac{V_S - V_{\text{turn-on}}}{R' + nR_L} = \frac{V_S - nV_{\text{turn-on}}}{mR + nR_L}$$

$$I_H = \frac{V_S - nV_{\text{turn-on}}}{R' + nR_H} = \frac{V_S - nV_{\text{turn-on}}}{mR + nR_H}$$

where R' is the external series resistor (Fig. 46(c)): Comparing these formulas, note that the maximum string current I_L for the circuit in Fig. 46(d) is much higher than is I_L for the circuit in Fig. 46(c):

$$\frac{I_L(\text{Fig. 46(d)})}{I_L(\text{Fig. 46(c)})} = \frac{mR + nR_L}{R\left[1 + \frac{R_L}{R_H} + (m-2)\frac{R_L}{R_T}\right] + nR_L}$$

since

$$\left[1 + \frac{R_L}{R_H} + (m-2)\frac{R_L}{R_T}\right] \ll m$$

Conversely, note that the minimum string current, I_H for the circuit in Fig. 46(d) is much lower than is I_H for the circuit in Fig. 46(c):

$$\frac{I_H(\text{Fig. 46(d)})}{I_H(\text{Fig. 46(c)})} = \frac{mR + nR_H}{R\left[1 + \frac{R_H}{R_L} + (m-2)\frac{R_H}{R_T}\right] + nR_H}$$

6 APPLICATIONS FOR HIGH-BRIGHTNESS LIGHT-EMITTING DIODES 315

since

$$\left[1 + \frac{R_H}{R_L} + (m-2)\frac{R_H}{R_T}\right] \gg m$$

In conclusion, the primary benefit for the paralleled series string circuit in Fig. 46(d) is lower cost due to eliminating all but one of the current limiting resistors. However, the cost savings comes at the expense of current regulation. Paralleling the series strings requires that the LED lamps have tightly controlled forward voltages, which could be achieved by tightly binning for forward voltage at the desired drive current.

In Fig. 46(e), all LED strings are connected in parallel and all LED lamps are driven from a common resistor. This circuit differs from the circuit shown in Fig. 46(d) in that each LED lamp or LED lamp substring is connected in parallel, where x is the number of LED lamps per substring and each group of LED substrings is then connected in series. With $x = 1$, all LED lamps are connected in parallel, like the rungs of a ladder. With $x = 2$, two LED lamps are connected in series and then paralleled with the other substrings. Note that if x equals the number of LED lamps per string, then this circuit is identical to the circuit shown in Fig. 46(d). For this circuit to work properly, the number of LED lamps per substring in a given "rung" must be the same. Forward current variations of the LED lamps will cause the current to split unevenly across the different substrings in a given rung. Also, the forward current through the LED substrings will vary somewhat due to the variations in ac line voltage, temperature, and LED lamp forward voltage. Unlike all of the other circuits shown in Fig. 46, for the circuit in Fig. 46(e) the failure of a single LED lamp does not cause an entire string of LED lamps to fail, only the lamps in a given substring.

The equations described previously for the paralleled series string circuit in Fig. 46(d) also apply to this design. The primary benefit for this circuit is increased redundancy. On the other hand, the nominal forward current variations are likely to be higher for the cross-connected parallel string circuit than for the parallel string circuit. Considering the statistical distribution of LED lamps, the likelihood of all lamps in a substring (since $x < n$) being at their worst-case parametric extremes is much higher than all lamps in a given series string being at their worst-case extremes. For example, for a 60 LED lamp series string design with $x = 2$, the probability of two lamps in a given substring for one of the rungs being at their worse-case extremes is much higher than all 60 lamps in a non-cross-connected series string being at their worst-case extremes.

Earlier, it was stated that the five design goals for a LED traffic signal module were good forward current regulation through the LED lamps, minimal changes in luminous intensity, low overall power consumption, redundancy in case of a random LED lamp failure, and low circuitry cost. Table XIX compares the five circuit designs for these design goals.

The current source circuit offers the best forward current regulation because the forward current through the LED strings is independent of line voltage, temperature, and LED parametric variations. The voltage regulator circuit provides an LED string current independent of line voltage. However, temperature and LED parametric variations affect the LED string current. The magnitude of the current variation depends on the LED string length. For the unregulated circuit, the LED string current varies due to line voltage, temperature, and LED parametric characteristics. The unregulated circuit has less current variation than does the voltage-regulated circuit due to temperature and LED parametric characteristics because a higher voltage is dropped across the resistor. For the parallel and cross-connected parallel string circuits, the LED string current varies due to line voltage, temperature, and LED parametric characteristics. The magnitude of the current

TABLE XIX

COMPARISON BETWEEN FIVE LIGHT-EMITTING DIODE (LED) TRAFFIC SIGNAL MODULE CIRCUITS

	Current Source Circuit	Voltage Regulator Circuit	Unregulated Circuit	Parallel String Circuit	Cross-Connected Parallel String Circuit
String current regulation					
ac line voltage	①	①	③	③	③
Ambient temperature	①	③	②	②	②
LED forward voltage	①	③	②	⑤	⑤
Combined	①	③	③	⑤	⑤
Luminous intensity					
ac line voltage	①	①	②	②	②
Ambient temperature	③	②	②	②	②
LED forward voltage	①	②	②	⑤	⑤
Combined	③	③	④	⑤	⑤
Power consumption					
ac line voltage	②	②	③	③	③
Ambient temperature	①	②	②	②	②
LED forward voltage	①	①	①	①	①
Combined	②	②	③	③	③
Redundancy	③	③	③	③	①
Circuit cost	⑤	③	①	①	①

Figure of merit: best ① ② ③ ④ ⑤ worst.

variations due to line voltage and temperature are the same as the unregulated circuit. The LED parametric variations can cause unacceptable forward current variations in both paralleled circuits unless the forward voltage characteristics of the LED lamps are well controlled. In practice, the parallel string circuit provides somewhat better current regulation than does the cross connected circuit because, with longer strings, the probability of all LED lamps in a given LED string being at their parametric extremes is much lower.

The second design goal is to provide a constant luminous intensity independent of line voltage, temperature, and LED electrical characteristic variations. For all LED materials technologies, the luminous intensity varies with temperature. Assuming that the luminous intensity is directly proportional to forward current, the luminous intensity over temperature can be calculated as:

$$I_v(I_F, T) = I_v(I_{test}, 25°C) \left(\frac{I_F}{I_{test}}\right) (e^{K(T - 25°C)})$$

where $I_v(I_F, T)$ is the luminous intensity at the forward current I_F and ambient temperature T, $I_v(I_{test}, 25°C)$ is the data sheet luminous intensity at the forward current I_{test} and 25°C, and K is -0.010 for AlInGaP.

Note, that for all but the constant current circuit, the LED forward voltage varies with temperature such that colder temperatures result in a lower string current, which tends to partially offset the increase in luminous intensity of the LED material.

Referring to Table XIX, the constant current circuit generates a luminous intensity that is independent of line voltage and LED electrical parameter variations. However, it provides the largest change in luminous intensity due to temperature. For AlInGaP LED technology, the luminous intensity increases by 92% at $-40°C$ and is reduced by 25% at 55°C, as compared with 25°C. The voltage regulator circuit generates a luminous intensity independent of line voltage. However, ambient temperature and LED parametric variations affect the luminous intensity, depending on the LED string length. The unregulated, parallel, and cross-connected circuits have the same luminous intenstity variations due to line voltage and ambient temperature for a given string length. For the same string length, these luminous intensity variations are somewhat higher than for the voltage regulated circuit. For the parallel and cross-connected circuits, the LED electrical characteristic parametric variations can cause unacceptable luminous intensity variations, unless the LED lamps are well matched.

The third design goal is to operate at the lowest electrical power consumption possible. Under nominal conditions for a given number of LED strings, all five circuits would have the same power consumption:

$$P_D = (I_F)(V_S)(m)$$

where P_D is the overall power consumption, I_F is the nominal string current, V_S is the nominal unregulated dc line voltage, and m is the number of strings.

Referring back to Table XIX, the current source circuit would provide the lowest worst-case power consumption at the highest line voltage conditions. The voltage regulator circuit would come in a close second due to the additional power consumption caused by LED electrical parametric variations, and to the reduced LED forward voltage at higher temperatures. The three unregulated circuits would have significantly higher power consumption at the highest line voltage as compared with the two regulated circuits.

The fourth design goal is to provide as much redundancy as possible in case LED lamps fail due to random acts of vandalism. For all of the circuits, except the cross-connected circuit, the failure of one LED lamp would cause an entire LED string to fail. For the current source, voltage regulator, and unregulated circuits, the overall luminous intensity of the LED traffic signal module would be reduced by $(m - 1)/m$, where m is the number of strings. For the parallel circuit, the failure of one string would result in the remaining $(m - 1)$ strings being driven at a higher forward current so that the overall luminous intensity would change by a smaller amount. The cross-connected circuit would have the best redundancy because the failure of one LED lamp would cause only one LED substring to fail, and the remaining substrings in the same rung would be driven at a somewhat higher forward current.

The last design goal is to have as low a circuit cost as possible. The current source circuit would have the highest circuit cost because it requires m independent current sources. The voltage regulator circuit would have the second highest cost because it requires one voltage regulator in addition to m current limiting resistors. In most cases, the voltage regulator will also need a heatsink, which tends to further increase the cost. The unregulated circuit requires m resistors, so its cost will be less than the voltage regulated circuit. Assuming that 60 LED lamps are used per string, and assuming a drive current of 20 mA, a nominal forward voltage of the LED lamp of 1.9 V, and a worst-case dc voltage of 165 V, each resistor needs to be capable of handling 1.5 W. The parallel and cross-connected circuits only require a single resistor for the entire LED traffic signal module, so that they would have the lowest overall cost. Using the same assumptions described earlier (60 LED/string, 20 mA, 1.9 V, 165 V max), with m equal to 10 strings, the resistor needs to be capable of handling 15 W. Because higher power resistors tend to be more expensive than are lower power resistors, the overall circuit cost savings for the parallel and cross-connected circuits might not be as large as anticipated. The LED lamps also might be more expensive due to the need for tighter electrical characteristics.

In conclusion, the best overall drive circuit for the LED traffic signal module is probably either the voltage regulator or unregulated circuit. If the ac line voltage is expected to vary significantly, then the voltage regulated circuit has the edge. On the contrary, if the ac line voltage is tightly controlled, then the unregulated circuit would be the most cost-effective design. In addition, the LED string length is a key design parameter, because it affects the variation in LED current due to LED electrical characteristics, ambient temperature for both circuits, and line voltage for the unregulated circuit.

VI. Large-Area Displays

1. CHARACTERISTICS OF THE LARGE-AREA DISPLAY MARKET

The worldwide large-area display market is another vast new market for LED technology. There are a wide range of applications that require monochrome, bicolor, and full-color graphics and alphanumeric displays. A list of some of these applications is given in Table XX. Traditionally, those displays have been created using a number of active display technologies such as direct-view incandescent bulbs, fiber optics, and cathode ray tubes (CRTs). Passive display technologies such a flip-disk also can be used. Until recently, the use of LED technology for these applications has been limited by the low performance and limited colors available. Today, AlInGaP, TS AlGaAs, and InGaN LED materials technologies provide bright red, amber, green, and blue colors. Mixtures of these colors can be used to create full-color large-area displays.

The large-area display market is very diverse. For retail applications, signs are purchased and leased by individual businesses, for example, to sell advertising or to provide information to customers in order to increase sales. Highway signs are purchased by local, state, and federal highway commissions. These signs are used to provide roadway control and to alert drivers to hazardous conditions. Portable highway signs are purchased by local roadway construction companies to alert drivers to hazardous conditions and to reduce their liability costs. Large-area displays are used in sports

TABLE XX

LARGE-AREA DISPLAY APPLICATIONS

	Monochrome	Bi-Color	Full-Color
Outdoor applications			
Highway message signs	◆	◆	
Portable highway message signs	◆		
City billboards	◆	◆	◆
Retail window advertising signs	◆	◆	◆
Outdoor large advertising signs		◆	◆
Stadium signs	◆	◆	◆
Stadium scoreboards	◆	◆	
Bus and train signs	◆	◆	
Public information signs	◆	◆	◆
Indoor applications			
One line advertising signs	◆	◆	
Airport terminal signs	◆	◆	
Bus and train station signs	◆	◆	
Arena signs	◆	◆	◆
Arena scoreboards	◆	◆	
Hotel information signs	◆	◆	
Exit signs	◆		
Stock market signs		◆	
Wagering and gaming signs	◆	◆	
Fast food menu signs	◆	◆	

arenas, bus and railway stations, and airports to provide information and for advertising.

2. ALTERNATIVE LARGE-AREA DISPLAY TECHNOLOGIES

There are a number of lighting technologies that could be used to fabricate a large area display. Figure 48 shows the construction of one pixel of a large full-color display. The first approach might be to use three light sources with three colored filters. Any color including white then can be generated by separately varying the luminous intensities of the three light sources. To generate white light, all three sources would be illuminated at approximately the same luminous intensity. A second approach would be to use four light sources with three colored filters and one unfiltered (white) light source. With this approach, generating white light is greatly simplified because only one source would be illuminated. Again, by separately modu-

6 APPLICATIONS FOR HIGH-BRIGHTNESS LIGHT-EMITTING DIODES   321

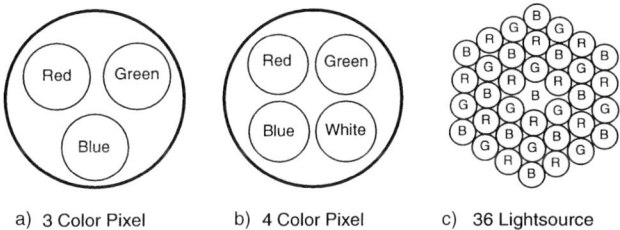

FIG. 48. Pixel construction of a large full-color display.

lating the intensity of the three colored light sources, any color could be generated. The third approach would use very small light sources arranged as triads of red, green, and blue in a closely packed array. Again, by separately varying the luminous intensity of each color, any color including white could be generated. The primary optical difference between these three techniques is the differing ability to mix the three primary colors in order to create a uniformly colored pixel. The approach using a number of small light sources arranged as a closely packed array provides more uniform color mixing as well as a redundancy such that the failure of a single source would have negligible effect on the operation of the pixel.

The diameter of the pixel shown in Fig. 48 can vary over a large range. Small displays might have a 5-mm pixel diameter, whereas displays designed for the outside of large buildings could give an overall pixel diameter of 150 mm. In general, the luminous intensity of each colored light source would need to scale with the pixel area in order to maintain a constant luminous sterance. The actual realization of each pixel would depend at least to some extent on the size of the pixel.

The pixel shown in Fig. 48 could be realized with a number of lighting technologies including CRTs, filtered incandescent bulbs, fiber-optic bundles, LED diodes, and flip disks. Figure 49 shows several cross sections of a single large-area display pixel using these different lighting technologies.

Figure 49(a) shows the construction of a CRT technology pixel. The CRT technology large-area display uses individual CRTs for each pixel of the display. Large displays use three or four individual CRTs (one per primary color) for a given pixel.

Figure 49(b) shows the construction of a filtered incandescent technology pixel. Filtered incandescent technology displays use individual bulbs for each primary color for a given pixel. The filtering can be accomplished either with a plastic filter mounted over each bulb or by using a silicone rubber boot mounted over the bulb. Each incandescent bulb is generally socketed to allow for easy replacement.

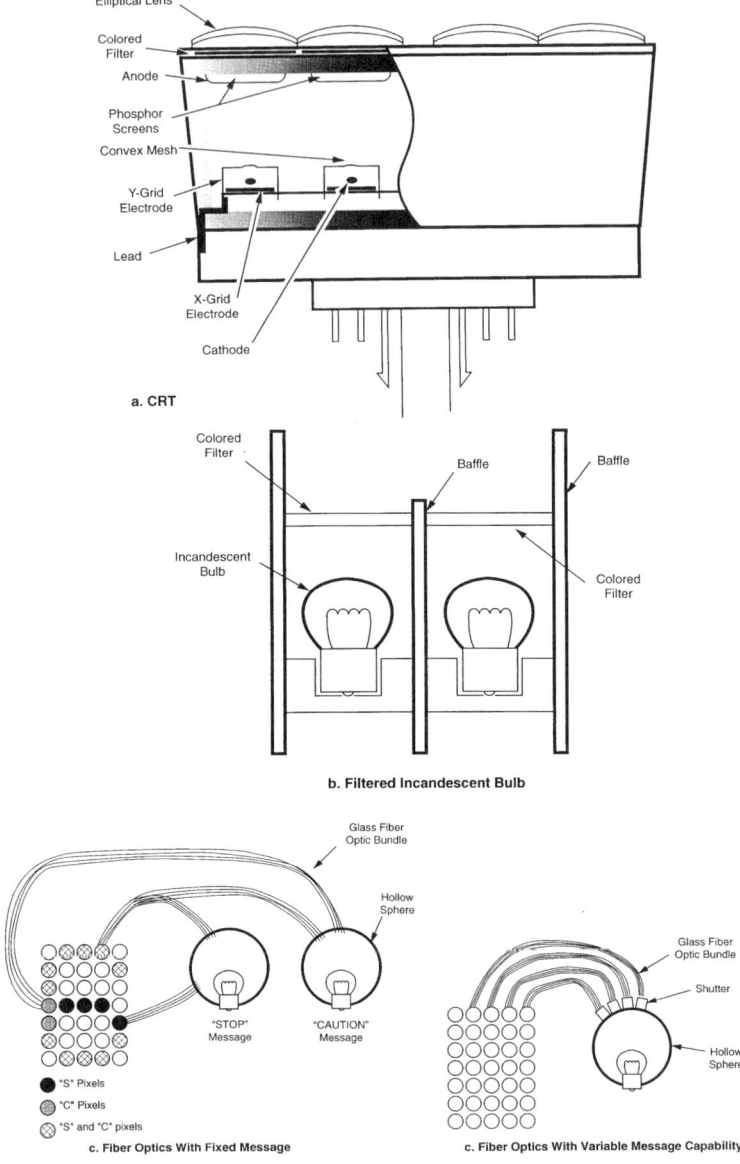

FIG. 49. Light source technologies used for large area displays. (a) Cathode-ray tube (CRT). (b) Filtered incandescent bulb. (c) Fiber optics with fixed and variable message capabilities. (d) LED lamp motherboard and LED lamp clusters. (e) Flip disk. (Fig. 49(a) after Mitsubishi Electronics America Inc., 1992. Reprinted with the permission of Mitsubishi Electronics America, Inc.)

6 Applications for High-Brightness Light-Emitting Diodes

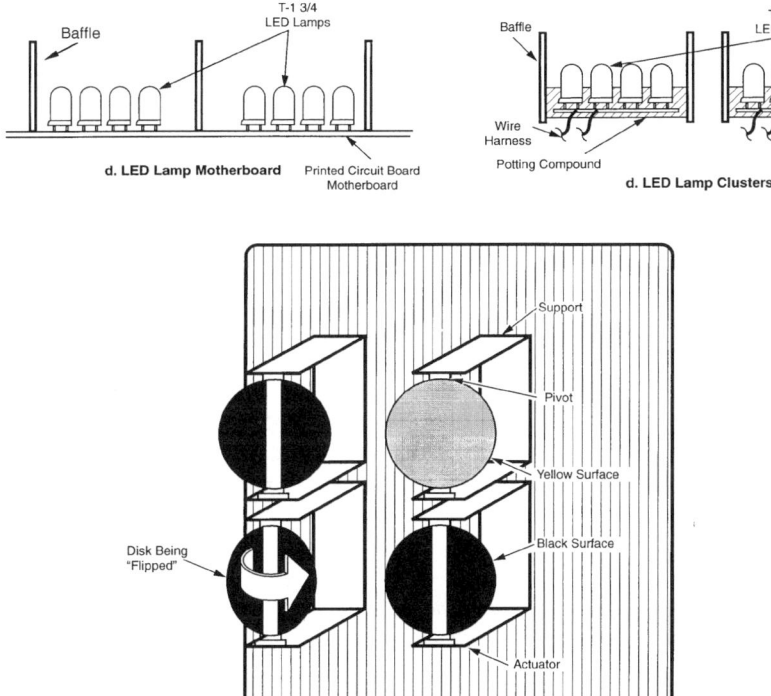

Fig. 49. *Continued.*

Figure 49(c) shows the construction of a fiber-optic technology pixel. Fiber-optic technology generally uses a central light source illuminating a large array of fiber-optic bundles. The fiber-optic bundles are mechanically routed to each pixel in the display. For a general-purpose display, each fiber-optic bundle would be switched on or off through a mechanical or optical shutter. For a display designed to display a limited number of fixed messages, a central light source might be used for each message. Then the fiber-optic bundles would be routed to the pixels that are illuminated by the particular message.

Figure 49(d) shows the construction of an LED technology pixel. The LED technology display is assembled into a variety of configurations, depending on the pixel size. Small pixels (5-mm diameter or less) might use a single T-$1\frac{3}{4}$ lamp per pixel for a monochrome display or a bicolor T-$1\frac{3}{4}$ LED lamp per pixel for a bicolor display. A full-color display would require three discrete T-1 red, green, and blue lamps or possibly a single multichip

T-1¾ LED lamp per pixel. Larger pixels would be created by clustering a number of T-1¾ red, green, and blue LED lamps together, as shown in Fig. 48(c).

Figure 49(e) shows the construction of a flip-disk technology pixel. Flip-disk technology uses a flat plastic disc that is highly reflective or colored on one side and black on the other side. The disc is rotated 180 degrees by an electrical pulse or by a small servo-motor. Flip-disk technology uses sunlight for daytime illumination and supplemental front lighting for night illumination.

3. BENEFITS OF LIGHT-EMITTING DIODE TECHNOLOGY

Light-emitting diode technology provides a number of benefits for large-area displays. In general, the characteristics of long lifetime and high efficiency for LED technology lead to the benefits of low maintenance and energy costs. Several competing technologies require mechanical shuttering or moving parts, and the advantage of an electronically controlled light source provides obvious reliability benefits. However, LED technology provides other benefits as well.

Previously, five technologies—CRT, filtered incandescent, fiber-optic, LED, and flip-disk—have been discussed for large-area displays. In certain applications, each technology provides some benefits over the others. Table XXI compares of these five lighting technologies.

Referring to Table XXI, flip-disk technology offers the lowest cost technology for a large-area display, and it is especially attractive for monochrome displays. Filtered incandescent technology is the second least

TABLE XXI

COMPARISON BETWEEN CATHODE-RAY TUBE (CRT), FILTERED INCANDESCENT, FIBER-OPTIC, FLIP-DISK, AND LIGHT-EMITTING DIODE (LED) LARGE-AREA DISPLAY TECHNOLOGIES

	CRT	Filtered Incandescent	Fiber Optics	LED	Flip Disk
Cost	H	M	M	H	L
Power and consumption	H	H	H	M	L
Reliability	H	L	L	H	M
Brightness	H	H	H	M	L
Weight	H	M	H	L	L
Speed of response	H	M	L	H	L
Off axis viewing angle	H	H	H	M	H

L, low; M, medium; H, high.

expensive technology. When used as a display with a limited number of fixed messages, the custom fiber-optic display is relatively inexpensive. However, the cost of individually switching each fiber-optic bundle for a general-purpose display is relatively expensive. For LED technology, the primary cost factor is the need for true full-color capability versus a monochrome (red-yellow-green) or bicolor (red-green) configuration, because the price for blue LED technology is currently much more expensive than are the other colors. Cost also scales directly with pixel size. Generally, the most expensive technology is CRT technology due to the high cost of the tube and the need for expensive drive circuitry.

For power consumption, flip-disk technology is the clear winner, especially for daylight applications. The lowest power consumption for active display technologies is provided by LED technology. Portable LED road signs have found a niche because they can be operated by solar power and a small battery, whereas incandescent signs require a small motor-generator to provide the necessary electrical power.

Light-emitting diode and CRT technologies displays offer the highest reliability of the different technologies available. In both cases, the display lifetimes range from 20 K to 100 K hours, and the possibility of an individual light source catastrophic failure is very low. Filtered incandescent and fiber-optic display technologies are limited by the lifetime of the incandescent bulb, which usually range from 5 K to 20 K hours. Because the lifetime of flip-disk technology signs is limited by the wear out of the flip-disk "flipper" mechanism, the best applications for these displays are those in which the message is changed infrequently.

Cathode-ray tube displays offer the highest brightness display technology, followed by filtered incandescent and fiber-optic technology. Today's perception is that LED technology is not as bright as filtered incandescent technology. However, the advent of InGaN and TS AlInGaP LED materials technologies will certainly change this perception. In general, flip-disk technology is a passive display technology is readable in direct sunlight but requires illumination at night.

Light-emitting diode and flip-disk technologies offer the lowest weight large-area displays, which provides some benefits for portable message signs. Fiber-optic and CRT technologies offer relatively heavy displays owing to the weight of the CRT display and fiber-optic cabling.

Cathode-ray tube and LED display technologies offer the fastest switching speeds, which makes them the obvious choices for full-motion video displays. Filtered incandescent displays are limited by the response time of the bulb filament, which is generally in the 10 to 100 millisecond range. Fiber-optic and flip-disk displays are limited by the speed of the electromechanical shutter or flipper mechanism.

Cathode-ray tube, filtered incandescent, fiber-optic, and flip-disk technologies displays generally have a wide Lambertian $(I_V(\theta) = I_V(0)\cos\theta)$ viewing angle unless it is limited by the outer housing. With LED technology, the viewing angle is limited by the radiation pattern of the LED lamp package. Depending on the desired application, LED displays can be designed with viewing angles ranging from under 10 degrees to Lambertian by using appropriate lamps. Similar to the other display technologies, the viewing angle may be limited by the outer housing.

4. Principles of Color Mixing

The basic principles of color measurement have been discussed previously in Part II, Section 3. In general, the color of any light source can be represented as an (x, y) coordinate in the 1931 CIE chromaticity diagram. Pure colors are located on the extreme outer edges of the diagram and white is located in the center. Color coordinates within the diagram can be considered to be a mixture of white light and monochromatic highly saturated light. As the point moves inward, the percentage of white light increases while the percentage of monochromatic highly saturated light is reduced.

The dominant wavelength is defined as the single monochromatic wavelength that appears to be same color as the light source. The dominant wavelength can be determined by drawing a straight line from one of the CIE white illuminants through the (x, y) coordinate to be measured, until the line intersects the outer locus of points along the spectral edge of the chart. Then the dominant wavelength is the wavelength of the intercept of the straight line and the outer edge of the space.

For two light sources of different colors with the same light-emitting areas, the color coordinate of the color mixture will be located on a straight line between the (x, y) coordinates of the two light sources, and the position of the color coordinate of the color mixture will be proportional to the relative luminous intensities of the two light sources. This concept is shown as a graph in Fig. 50. If the color coordinates and luminous intensities of the two light sources are denoted as (x_1, y_1) and I_{v1}, and (x_2, y_2), and I_{v2}, then the color coordinate of the color mixture can be calculated as shown below. Similarly, for three light sources of different colors, the color coordinate of the color mixture will be located within a triangle where the apexes of the triangle correspond to the three color coordinates. The color coordinate of the color mixture can be calculated as follows:

$$(x_c, y_c) = \text{color coordinate of color mixture}$$

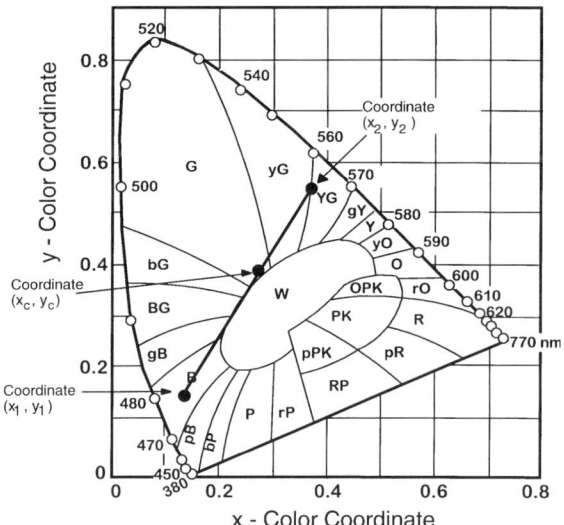

FIG. 50. Principles of mixing two colors.

$$x_c = \frac{x_1 I_{v1} + x_2 I_{v2} + x_3 I_{v3}}{I_{v1} + I_{v2} + I_{v3}}$$

$$y_c = \frac{y_1 I_{v1} + y_2 I_{v2} + y_3 I_{v3}}{I_{v1} + I_{v2} + I_{v3}}$$

where I_{vi} is the luminous intensity of light source i, and (x_i, y_i) is the color coordinate of light source i. Note that for two light sources, $I_{v3} = 0$. This concept is shown as a graph in Fig. 51. Note, that the choice of colors for the three light sources determines the available color range and color balance of the full-color display.

Figure 52 shows the color range for a bicolor LED display using GaP (572 nm) green and AS AlGaAs (637 nm) red lamps. This display would provide a series of saturated colors in the ranges greenish-yellow, yellow, yellowish-orange, orange, reddish-orange, and red. Although this display may be called a "multicolor" display, its color range is fairly limited.

Figure 53 shows the color range for a tricolor LED display in which an InGaN (480 nm) blue LED lamp has been added to the GaP green and AS AlGaAs red LED lamps. This display provides true full-color capability with the addition of white and mixtures of blue with green and blue with red. While this display is considerably better than the one shown in Fig. 52, it still has limitations, in displaying deep-green colors.

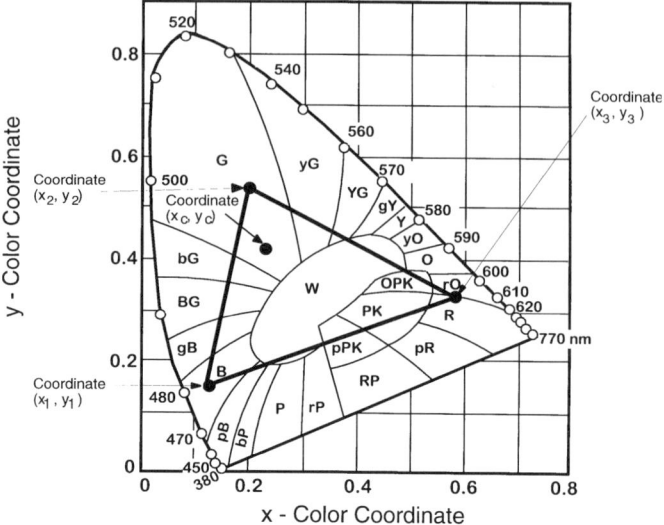

FIG. 51. Principles of mixing three colors.

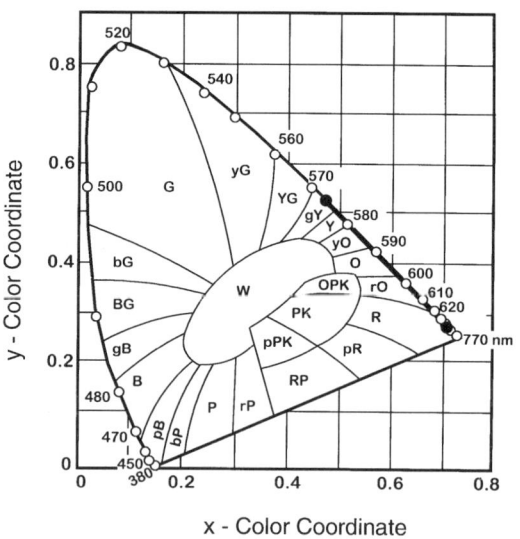

FIG. 52. Color range of a red-green bicolor light-emitting diode display.

6 APPLICATIONS FOR HIGH-BRIGHTNESS LIGHT-EMITTING DIODES   329

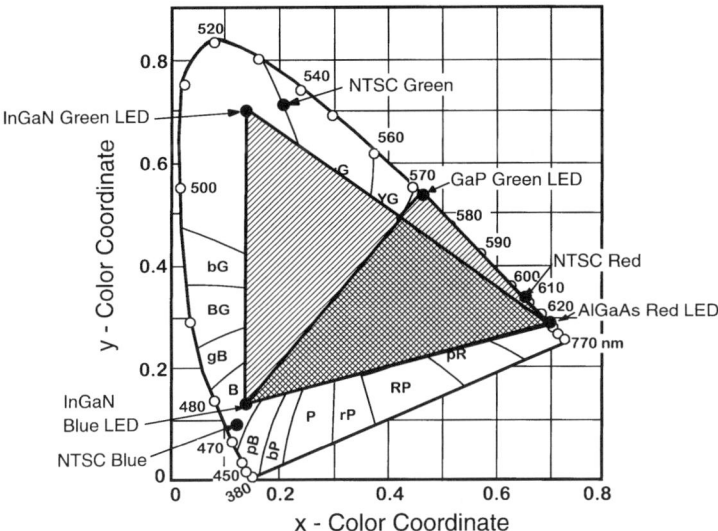

FIG. 53. Color range of a red-green-blue full-color light-emitting diode (LED) display. GaP, gallium phosphide; InGaN, Indium Gallium Nitride LED technology; NTSC, National Television System Committee; AlGaAs, Aluminum Gallium Arsenide LED technology. (After Hodapp, 1994. Reprinted with the permission of the Society of Photo-Optical Instrumentation Engineers.)

Figure 53 also shows the color range for a tricolor LED display using InGaN (480 nm) blue, InGaN (515 nm) green, and AS AlGaAs (637 nm) red LED lamps. This display provides true full-color capability and provides a larger color range in the blue-green and green-red color regions. Altogether, the color range is not too different from the National Television System Committee (NTSC) television phosphors. Compared with the NTSC television, the LED lamps provide deeper green and deeper red colors.

5. CHARACTER HEIGHT CONSIDERATIONS

The size of a large-area display needs to be selected based on an understanding of the intended application. One guideline to use is to choose a pixel size such that the smallest alphanumeric or numeric font is readable at the expected maximum viewing distance of the display. The visual acuity of the human eye refers to the ability to properly see objects at different viewing distances. A human eye with 20/20 vision is able to resolve 5 minutes of arc. Thus, the minimum character height of a single line of numeric-alphanumeric text needs is be

$$H = 2D \tan\left(\frac{\theta}{2}\right)$$

where H is the character height, D is the viewing distance, and θ is the minimum angle subtended by the human eye, based on visual acuity. For 20/20 vision, $\theta = 0.0833°$. For example, at a viewing distance of 15 m, the character height of the display would need to be 21.8 mm to be readable by someone with 20/20 vision. The character height would need to be proportionally larger to accommodate someone with less than perfect vision. In the same example, the character height would need to be 54.5 mm to be readable by someone with 20/50 vision at a viewing distance of 15 m.

Then the minimum pixel diameter can be calculated based on the minimum-size x-by-y matrix of a single alphanumeric character, the relative size of the pixel to the pixel pitch, and the maximum viewing distance. For example, if the minimum size of the alphanumeric display is 5 by 7 (35 dots per character), and the pixel diameter is half of the pixel pitch, then the pixel diameter is calculated as follows:

$$13P = H = 2D \tan\left(\frac{\theta}{2}\right)$$

where P is the pixel diameter, H is the character height, D is the viewing distance, and θ is the minimum angle subtended by the human eye, based on visual acuity. Note that the value of 13 is determined by the number of vertical rows in the display matrix (y) plus (y-1) spaces between pixels, assuming that the space between pixels is of the same dimension as the pixel diameter. Using these assumptions, the minimum pixel diameter versus viewing distance is shown in Table XXII for several different viewing distances and display fonts. The recommended pixel diameter should be at least twice these minimum values in order to accommodate persons with less than perfect visual acuity.

6. Drive Circuit Considerations

These are a number of possibilities to drive a LED large-area display. This section will shed some light on the types of drive circuits that are possible. No single drive circuit is optimal for all applications. In general, the complexity of the drive circuitry is proportional to the number of the pixels in the display (more on this later), whether the display is monochrome, bicolor, or tricolor, and whether each pixel or pixel component can

TABLE XXII
MINIMUM PIXEL DIAMETER VERSUS VIEWING DISTANCE[a]

Viewing Distance (m)	Minimum Character Height (mm)	Minimum Pixel Diameter[b]		
		5 × 7 Matrix (mm)	7 × 9 Matrix (mm)	10 × 15 Matrix (mm)
10	14.5	1.1	0.9	0.5
20	29.1	2.2	1.7	1.0
50	72.7	5.6	4.3	2.5
100	145	11.2	8.6	5.0
200	291	22.4	17.1	10.0
500	727	55.9	42.8	25.1
1000	1454	112	85.6	50.2

[a] Based on human visual acuity of 20/20 vision, where the eye can resolve 5 minutes of arc.
[b] Pixel diameter is equal to one-half the pixel pitch.

be independently dimmed in order to accommodate a variable luminous sterance for gray-scale effects. In this section, we look at all of these options at the circuit block diagram level. The implementation of the overall drive circuitry of the entire LED large-area display is largely dependent on the overall capabilities of the sign (i.e. text, graphics, or full-motion video).

Figure 54 shows the two basic methods used to drive an LED lamp. In Fig. 54(a), each LED pixel is dc driven. This means that each pixel requires its own drive circuit. The drive circuit could be as simple as a one-bit memory, an open collector drive transistor, or a logic gate and a resistor to limit the current through the LED lamp(s). For this drive scheme, the luminous intensity can be calculated as follows:

$$I_v \propto (I_F)^n$$

$$I_v(I_F) = I_v(I_{\text{TEST}}) \left(\frac{I_F}{I_{\text{TEST}}}\right)^n$$

where

$I_v(I_F)$ = luminous intensity at the forward current I_F

$I_v(I_{\text{TEST}})$ = luminous intensity at the test current I_{TEST}

I_F = desired drive current

I_{TEST} = current at which the LED lamp is tested

n = linearity factor ≈ 1.00.

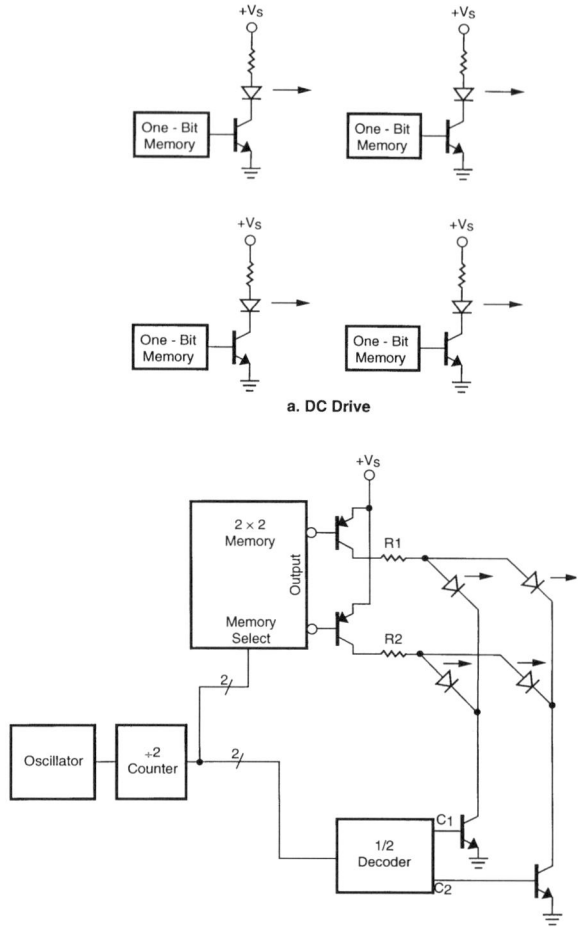

FIG. 54. Basic drive current approaches. (a) dc drive. (b) Multiplexed array.

In Fig. 54(b), each LED pixel is driven as part of a multiplexed array. In a multiplexed x, y array, each LED pixel shares an array of drivers. The overall operation of the circuit is as follows: An x word by y output memory is used to store pixel data for the display. The divide by x counter is used to sequentially select each of the x words. The y outputs of the memory are connected to y individual output drivers. Simultaneously, the output of the counter is decoded by a one of x decoder, such that only one of the x outputs is enabled at any time. Each of the x outputs of the decoder are

6 APPLICATIONS FOR HIGH-BRIGHTNESS LIGHT-EMITTING DIODES

connected to an output driver. Thus, the rows (r_1, r_2) are enabled based on the digital word data stored in the memory at the same time that the appropriate column (c_1, c_2) is enabled. If the counter cycles at an overall refresh rate greater than 100 Hz, then the human observer perceives that the entire display is uniformly illuminated. For this drive scheme, the luminous intensity can be calculated as follows:

$$I_{v\,\text{AVERAGE}} \propto (I_{F\,\text{AVERAGE}})^n$$

$$I_v(I_{F\,\text{AVERAGE}}) = I_F(I_{\text{TEST}}) \left(\frac{I_{F\,\text{AVERAGE}}}{I_{\text{TEST}}}\right) \frac{\eta(I_{F\,\text{AVERAGE}})}{\eta(I_{\text{TEST}})}$$

$$I_{F\,\text{AVERAGE}} = (I_{F\,\text{PEAK}})(D)$$

where:
  $I_{v\,\text{AVERAGE}}$ = time-averaged luminous intensity
  $I_{F\,\text{AVERAGE}}$ = time-averaged forward current
  n = linearity factor ≈ 1.00
  $I_v(I_{F\,\text{AVERAGE}})$ = time-averaged luminous intensity at the time-averaged forward current $I_{F\,\text{AVERAGE}}$
  $I_v(I_{\text{TEST}})$ = luminous intensity at the test current I_{TEST}
  I_{TEST} = current at which the LED lamp is tested
  $\eta(I_{F\,\text{PEAK}})$ = relative efficiency of the LED at current $I_{F\,\text{PEAK}}$
  $\eta(I_{\text{TEST}})$ = relative efficiency of the LED at current I_{TEST}
  $I_{F\,\text{PEAK}}$ = peak current at which the LED is pulsed
  D = duty cycle, the time LED is pulsed on divided by the refresh time.

In comparing the cost of the dc drive scheme versus the multiplexed drive scheme, note that in the dc drive scheme, the number of drivers scales directly with the number of pixels being driven. In the multiplexed drive scheme, the number of drivers is roughly proportional to the square root of the number of pixels being driven. In both drive schemes the memory size scales directly with the number of pixels. In addition, the multiplexed drive scheme requires some additional scanning and decoding circuitry. For large displays, because the reduction in output drivers far outweighs the additional scanning and decoding circuitry, the multiplexed drive scheme is generally more cost-effective than is the dc drive scheme.

The maximum peak current of the LED lamp limits the maximum size of one dimension of the multiplexed array. The maximum peak current for a given LED lamp is determined by an acceptable rate of light-output degradation over the anticipated operating life of the display panel. The time-averaged luminous intensity is primarily determined by the average forward current through the LED lamp (the relative efficiency η is relatively constant at higher peak currents). Thus, the maximum size of one dimension of the x, y array is approximately equal to the ratio of I_{FPEAK} divided by $I_{Faverage}$. For example, suppose the maximum peak current for a given LED lamp is 200 mA and that a time-averaged forward current of 20 mA is needed for good visibility. Then the maximum size of the array is limited to x by 10.

The time-averaged luminous intensity of a pixel designed for outdoor viewing would be much higher than the time-averaged luminous intensity of a pixel designed for indoor applications. Thus, the outdoor sign would need to be driven with a higher duty cycle than would the indoor sign. This means that the outdoor sign would probably be driven as several smaller x, y array modules. A benefit of the TS AlGaAs, AlInGaP, and InGaN LED materials technologies, is that the higher luminous efficiencies allow the sign to be driven at lower averaged currents, which in turn, allows the drive circuitry to multiplex a larger matrix and reduces circuitry costs.

The drive circuits shown in Fig. 54 are designed to drive each pixel at a fixed current. The circuitry can be modified to vary the luminous intensity or time-averaged luminous intensity of the entire array. First, for a limited range of luminous intensity variations, the supply voltage V_S could be varied. Most LED lamps will match fairly well over a 10:1 forward current range (e.g., example, over a range of 2 to 20 mA). At low currents, LED lamps tend to be less well matched. For a larger dynamic range, a PWM scheme is recommended. With this approach the peak or dc forward current through the LED lamp is fixed, and the current is rapidly switched on and off. The time-averaged luminous intensity then is equal to the initial luminous intensity multiplied by the duty cycle of the PWM waveform.

Figure 55 shows how the circuits shown in Fig. 54 would be adapted for PWM dimming control. For best results the PWM circuitry used for the multiplexed drive circuit in Fig. 55(b) should be synchronous to the refresh circuit. With PWM dimming, the dynamic range of luminous intensity is determined by the switching speeds of the LED lamps and their drivers and the multiplex rate (for the multiplexed drive circuit). For example, for a multiplex rate of 100 Hz and a 1 of 10 duty cycle, each column would be energized for 1 ms. If the switching speed of the drive circuitry is 1 μs (LED lamps typically have a switching speed less than 100 ns), then the dynamic range could be as large a 1000:1.

These basic circuits can be modified to allow the intensity of each pixel to be varied independently. For example, the luminous intensity could be

6 APPLICATIONS FOR HIGH-BRIGHTNESS LIGHT-EMITTING DIODES 335

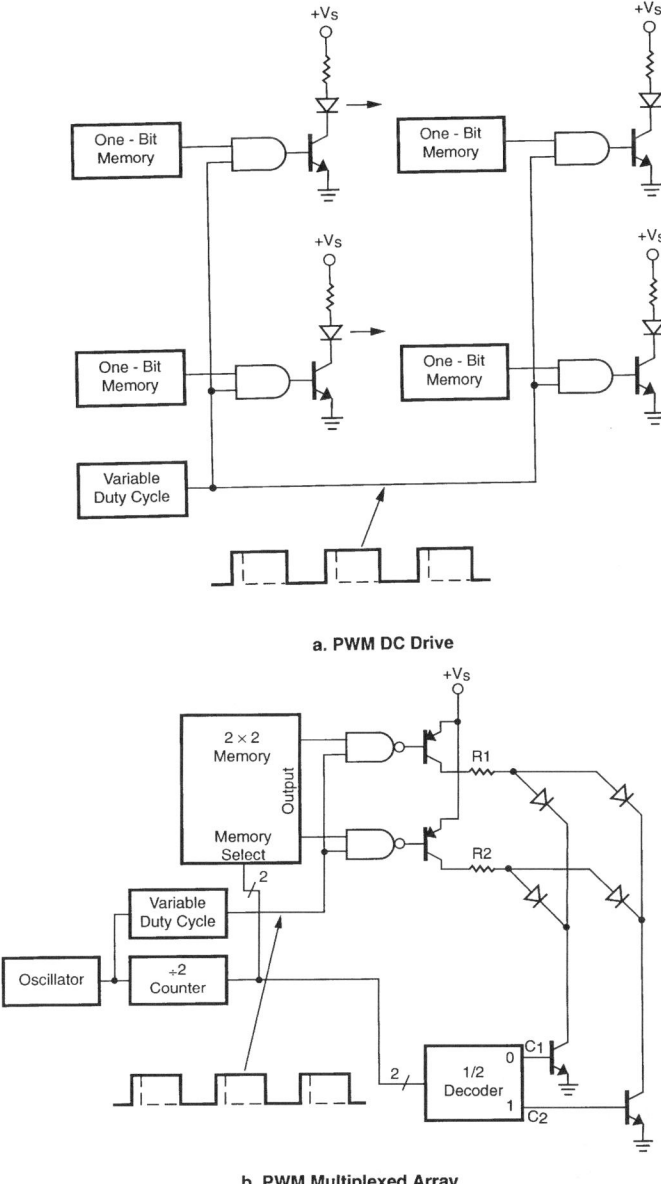

FIG. 55. PWM dimming scheme. (a) PWM dc drive. (b) PWM multiplexed array.

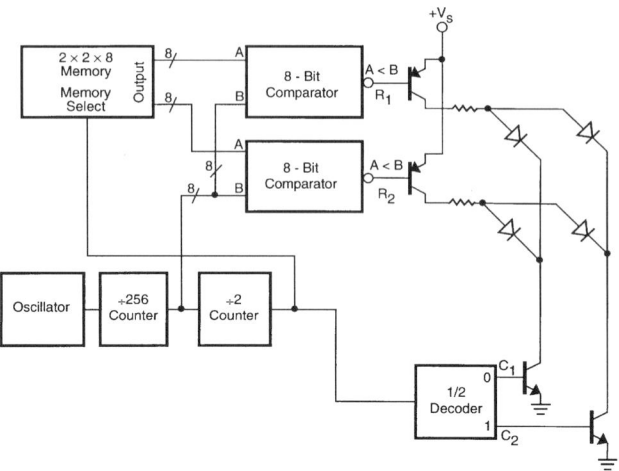

FIG. 56. Multiplexed drive eight-bit dimming level circuit.

varied over an eight-bit dynamic range. A dynamic range of 256:1 is larger than that which can be accommodated by modulating the dc or peak forward current while maintaining good matching within the LED array. However, a PWM scheme could be used. Instead of using a single memory bit per pixel, an 8-bit memory per pixel is needed.

Figure 56 shows a multiplexed driver circuit with an eight-bit dimming range. The 2×2 memory has been replaced with a $2 \times 2 \times 8$ memory (2×16 memory) in order to store 8 bits per pixel. Two 8-bit words are simultaneously read from the memory and digitally compared with the output of a divide by 256 counter. If the outputs of the memory are less than the output of the 8-bit counter, then the corresponding row driver is enabled. Thus, the dimming level 00_{16} corresponds to OFF and dimming level FF_{16} corresponds to 100% ON (ON for 255 of 256 counts). Simultaneously, as the two words are read from the memory, the corresponding column driver is turned on. In order to multiplex the display at a rate greater than 100 Hz, the oscillator would need to have a frequency of at least 51.2 KHz.

The basic circuits outlined in Figs. 54, 55, and 56 easily can be modified to accommodate bicolor and tricolor operation. Unfortunately, the dc drive schemes scale proportionately to the number of LED colors. For the multiplexed drive schemes, the additional colors can be added by increasing the size of the memory, adding additional row drivers, and using the existing column decoding and drive circuitry. (The extra row drivers may require the use of higher current column driver transistors).

VII. Liquid Crystal Display Backlighting

1. CHARACTERISTICS OF THE LIQUID CRYSTAL DISPLAY BACKLIGHTING MARKET

The worldwide LCD display market is another vast new market for LED technology for LCD backlighting. According to Stanford Resources, Inc., the worldwide LCD market had 1.1 trillion units in sales in 1995 (Castellano *et al.*, 1996). Markets for LCD displays include consumer products, industrial and medical applications, gas pumps, computers (laptops), video displays, and motor vehicles. A list of some of these applications is given in Fig. 57. Consider that at least 10% of these LCD displays are backlit with an active light source. The light source allows the display to be readable in darkened ambient lighting. The light source is also needed for full-color LCD displays, which tend to have low light transmission. The LED technology is a viable contender for LCD backlighting. The new AlInGaP, AlGaAs, and InGaN LED materials technologies offer high lighting efficiency and a broad color range.

The LCD displays are generally available in two forms. First, the LCD display manufacturer sells a complete display subsystem. The display

Market	Applications	Monochrome	Full-color	Graphics	Backlit
Computer	• Personal organizer	x		x	x
	• Calculator	x			
	• Notebook computer	x	x	x	x
	• Desktop computer		x	x	x
	• Printer	x			
Consumer	• Cellular telephone	x			x
	• Electronic game/toy	x	x	x	x
	• Watch	x			x
	• Audio/video equipment	x			x
	• Portable television		x	x	x
	• Appliance	x			
	• Camcorder viewfinder	x	x	x	x
Business/ Commercial	• Cash register/Bank terminal	x	x		x
	• Ticketing machine	x			x
	• Copier	x			x
	• Portable word processor	x			x
	• Fax	x			
	• Projection TV		x	x	x
	• Arcade games		x	x	x
Transportation	• Auto instrument cluster	x			x
	• Auto GPS		x	x	x
	• Marine instrumentation	x	x		x
	• Aircraft/train instrumentation		x	x	x
	• Aircraft/train entertainment				
Industrial	• Test equipment	x	x	x	x
	• Medical equipment	x	x	x	x
	• Data collection terminal	x	x	x	x
	• Process controller		x	x	x

FIG. 57. Liquid crystal display (LCD) backlighting applications.

subsystem includes an alphanumeric or graphics LCD display, all drive electronics mounted on a printed circuit board, all connectoring of the LCD display terminals to the drive electronics printed circuit board, and an optional LCD backlighting module. All the user need do is connect the display subsystem to a microprocessor controller, apply power, and download alphanumeric or graphics information. This approach is ideal for low to moderate volume applications that can use one of several standard display sizes and have sufficient physical space to accommodate the entire package. The user can also purchase the LCD display, integrated circuit drivers, and connectors separately to create a custom or semicustom display subsystem. This approach makes sense for high-volume applications in which the user needs custom features or, perhaps, has very tight physical constraints that do not allow the use of a standard module. Both of these approaches can utilize LCD backlighting. However, the second approach has the most diverse lighting requirements (brightness, color, dimmability, and full-color capabilities).

2. LIQUID CRYSTAL DISPLAY OPERATION

First, the operation of an LCD display should be reviewed. Figure 58 shows a cross-section view of three types of LCD displays. The LCD cell consists of two glass plates with indium tin oxide (ITO) electrodes that define the location, size, and shape of the pixels in the display. Between the glass plates is a thin layer of liquid crystal material. The liquid crystal material consists of long chains of electrically polarized organic molecules that are suspended in a liquid. The organic molecules are randomly aligned when no electrical potential is applied between the electrodes. However, when an electrical potential is applied, the molecules are all aligned in the same orientation. Light is polarized by a thin sheet of polarizer material before passing through the liquid crystal cell. After passing through the cell, the light passes through a second polarizer film. When the liquid crystal molecules are randomly aligned, the light passes through the cell without any change in polarization. However, when the liquid crystal molecules are aligned by the electric field, polarized light passing through the cell is re-oriented. The pixels are switched on and off because the second polarizer either transmits or absorbs the light, depending on whether the light, was re-oriented by the liquid crystal material.

For a reflective LCD display, as shown in Fig. 58(a), ambient light is used to illuminate the display. After passing through the display, a reflector reflects the light back through the display to the viewer. This display is highly visible in bright ambient lighting, is difficult to read in dim ambient

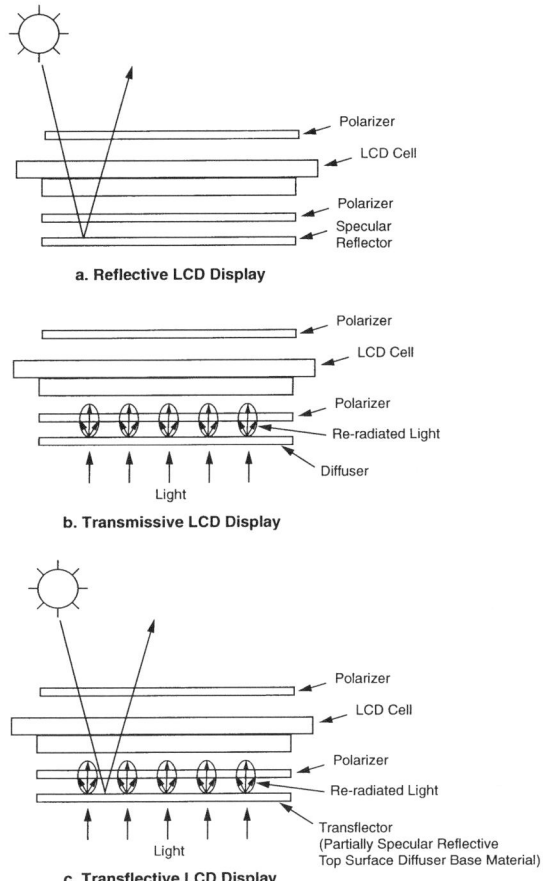

FIG. 58. (a) Reflective liquid crystal display (LCD). (b) Transmissive LCD. (c) Transflective LCD.

lighting, and is unreadable in the dark. A handheld calculator is an example of a reflective LCD display.

Figure 58(b) shows a transmissive LCD display. For this type of display, a light source is used to illuminate the back of the LCD display through a diffusive film. Then the light passes through the polarizers and liquid crystal cell to the viewer. Note, that without the light source, the LCD display would not be viewable (or very marginally viewable). This display is highly viewable in low ambient lighting conditions, although it may tend to wash out in direct sunlight. A laptop computer is an example of a transmissive LCD display.

TABLE XXIII
Luminous Sterance Ranges for Backlit LCD Displays

Application	LCD Type	Desired Luminous Sterance
Auto clock display	Transflective	2–10 cd/m^2
Auto GPS Display	Transmissive	60–400 cd/m^2

Figure 80(c) shows the cross-section of a transflective LCD display. For this type of display, a partially transmissive–reflective (transflective) film is mounted behind the liquid crystal cell and polarizers. Ambient light is partially reflected by this layer, similar to a reflective LCD display. In addition, a light source can be turned on behind the transflective film to backlight the LCD display in dim or dark ambient lighting, similar to the transmissive LCD display. An automotive clock is an example of a transflective LCD display, where the light source is energized when the head lamps are energized. Note that for both the transmissive and transflective LCD displays, the light source needs to project a uniform field of light such that the luminous sterance and color of light emitted through the diffuser or transflector are uniform. A transflective LCD display provides a higher degree of flexibility in different ambient lighting conditions, although it provides somewhat lower performance than does a reflective LCD display in direct sunlight and a transmissive LCD display in darkened ambient lighting.

Approximate luminous sterance ranges for LCD displays for different ambient lighting conditions are shown in Table XXIII. In all cases, the luminous sterance ranges refer to the luminous sterance as measured through the LCD display. Thus, the luminous sterance of the backlighting module, as measured through the diffuser or transflector film, would be equal to the value shown divided by the light transmission of the liquid crystal cell and polarizers.

3. Alternative Liquid Crystal Display Backlighting Technologies

There are a number of lighting technologies that can be used to backlight LCD displays. These technologies include incandescent bulbs, cold-cathode fluorescent tubes, electroluminescent panels, and LED lamps. Each technology provides some benefits over the others in certain applications. Figure 59 shows several cross sections of an LCD display backlighting module using these different lighting technologies.

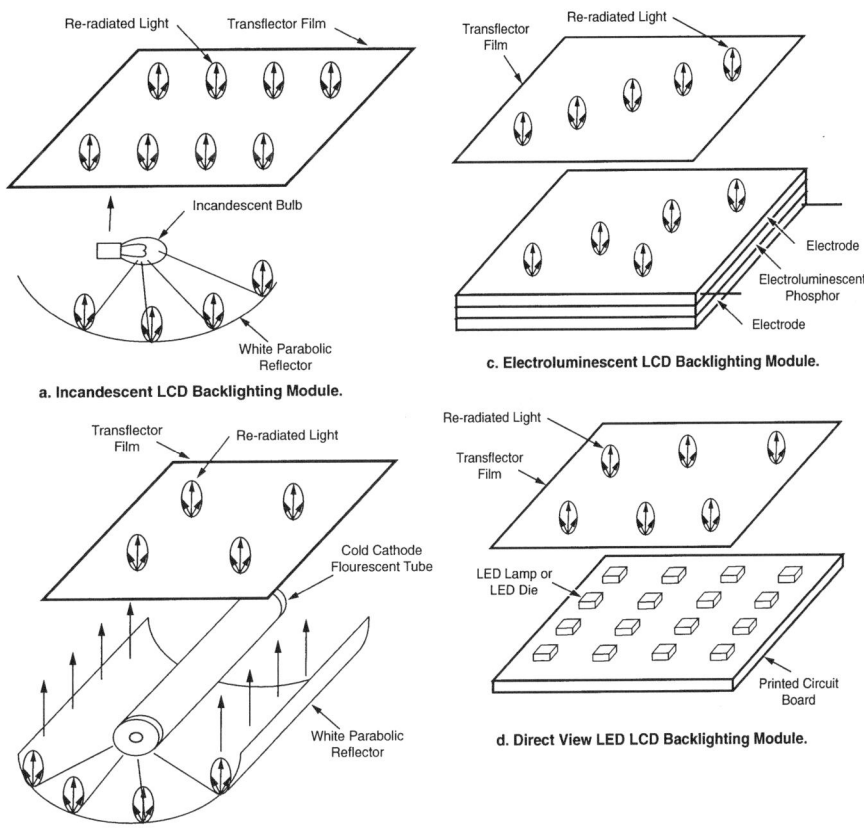

FIG. 59. Light source technologies used for liquid crystal display (LCD) backlighting modules. (a) Incadescent, (b) cold cathode fluorescent, (c) electroluminescent, and (d) direct-view light-emitting diode (LED) LCD backlighting modules.

Figure 59(a) shows the cross section of an incandescent LCD backlighting module. One or more incandescent bulbs are mounted behind the LCD diffuser–transflector film. A white reflector further diffuses the luminous flux from the filament into a uniform light field behind the film. Colored light can be created either by using a silicone rubber boot mounted over the bulb or by using a piece of colored diffuser–transflector film.

Figure 59(b) shows the cross section of a cold cathode fluorescent LCD backlighting module. The fluorescent tube is mounted behind the LCD diffuser–transflector film. These tubes are miniature versions of household fluorescent tubes, and are available in a number of sizes and shapes, ranging from linear tubes to serpentine shapes (to illuminate a large area). A white reflector is mounted behind the tube to further diffuse the light. An alternative approach is to mount the cold cathode tube behind one or more edges of the LCD display and to use a light pipe to spread the light evenly across the underside of the diffuser–transflector film. The fluorescent tubes are available in colors other than white. Colored light can be generated either by using a colored tube, or by using a colored diffuser–transflector film.

Figure 59(c) shows the cross section of an electroluminescent LCD backlighting module. A phosphor material is placed between two electrodes. Light is generated by applying a high-voltage ac signal between the two electrodes. Electroluminescent panels are available in a number of colors.

Figure 59(d) shows the cross section of a direct-view LED technology LCD backlighting module. Light-emitting diode die or packaged LED lamps are mounted on a white printed circuit board. The spacing between the LEDs is chosen to provide a uniform light field at the plane of the diffuser–transflector film. Alternative LCD backlighting techniques include edgelighting, where light is emitted into the edge of a light pipe mounted behind the diffuser–transflector film, and woven fiber optics, where light is emitted into a fiber-optic bundle and then extracted behind the LCD display.

4. Benefits of Light-Emitting Diode Technology

LED technology provides a number of benefits for LCD backlighting. In general, the characteristics of long lifetime and high efficiency of LED technology lead to the benefits of low repair costs, and low heat generation. LED lamps are very small and allow the creation of a thin LCD backlighting module, which reduces the cost of connectoring the LCD display to the drive electronics. The LED lamps operate at low voltages and can be driven directly from 5- or 12-V dc. These lamps have very fast switching times, which allow for PWM dimming, and potentially could allow the removal of the red-green-blue filters behind the LCD panel of a full-color LCD display by separately pulsing red, green, and blue LED lamps. Until the advent of InGaN blue LED technology, the primary limitation for LED technology was the limited color range (red, orange, amber, and yellow-green). However, using InGaN LED technology that can produce blue and deep-green colored light, the full color range, including white, is now possible.

Table XXIV compares these different lighting technologies for LCD backlighting. Each technology provides some benefits over the others in certain applications.

6 Applications for High-Brightness Light-Emitting Diodes

TABLE XXIV

Comparison Between Incandescent, Cold Cathode Fluorescent, Electroluminescent (EL), and Light-Emitting Diode (LED) Light Source Technologies Used for Liquid Crystal Display Backlighting

	Incandescent	Cold Cathode Fluorescent	EL	LED
Cost	L	H	H	M
Power and consumption	M	H	M	M
Brightness	H	H	L	M
Reliability	L	H	M	H
Thickness	H	H	L	M
Low voltage compatibility	H	L	L	H
Color capability	H	H	M	M
Response time	L	L	M	H

L, low; M, medium; H, high.

Incandescent technology offers the lowest cost LCD backlighting solution using a single bulb for a moderately sized display. Cold cathode fluorescent and electroluminescent technologies are the most expensive backlighting technologies because both require high voltages for proper operation. Thus, either a separate power supply or dc–dc or dc–ac conversion is required. Electroluminescent panels require 400- to 1000-Hz ac for proper operation, which entails some type of dc-to-ac chopping circuit. The cost of LED backlighting modules generally scales with the area and brightness of the LCD displays making these modules very cost-effective for small displays but less attractive for large displays.

Incandescent and cold cathode technologies provide the brightest LCD backlighting modules. Both technologies can provide a luminous sterance (for white light) in excess of 1000 cd/m^2, as measured through the diffuser–transflector film. In both cases, wavelength filtering to get a colored light reduces the luminous sterance. Electroluminescent technology is limited to a luminous sterance of less than 50 cd/m^2 for most applications. Electroluminescent panels can be driven harder but this reduces the lifetime of the panel. The LED technology is capable of producing a luminous sterance in excess of 500 cd/m^2 when assembled in a direct-view configuration. If less performance is required, LED lamps can be used in either an edgelighting configuration or in conjunction with a fiber-optic LCD backlighting module.

Cold cathode and LED technologies offer the highest reliability for LCD backlighting modules. In both cases, the lifetime of the backlighting module, defined as the operating time at which the light degrades to 50% of its initial volume, is in the range of 10 K to 100 K hours. In addition, the probability of catastrophic failure of the light source is very low. Electroluminescent LCD backlighting modules traditionally have had a limited lifetime of less than 5 K hours owing to light-output degradation. Furthermore, the rate of light-output degradation is higher in moist environments. Incandescent bulb technology generally has a catastrophic failure rate with a mean time between failure rate of 1 K to 10 K hours. However, long-life incandescent bulbs are available with MTBFs in excess of 20 K hours. The lifetime of the incandescent bulb is reduced by the presence of mechanical shock and vibration.

The thickness of the LCD backlighting module is another factor in selection, because the LCD is normally connected to the LCD drive circuitry by elastomeric connectors or by pins that are soldered directly to the LCD display. If the LCD backlighting module is too thick, then the cost of the elastomeric connectors are increased or a cutout must be routed in the printed circuit board directly behind the LCD display in order to accommodate the backlighting module. This cutout adds cost to the printed circuit board due to the wasted space, and may also entail additional connectoring. Incandescent and cold cathode fluorescent technology backlighting modules typically have a total thickness between 12 and 25 mm (as measured from the back of the diffuser film to the top of the printed circuit card mounted underneath the LCD panel) owing to the size of the light sources and the necessary distance between the light source and the white reflector. However, it is possible to construct an LCD backlighting module using small T-1 incandescent bulbs and reduce the overall depth to under 10 mm. It is also possible to use incandescent and cold cathode technology in an edgelighting design and to reduce the overall depth to under 8 mm. The electroluminescent panel is by far the thinnest LCD solution, with a panel thickness of under 2 mm. The LED backlighting modules can be developed with a thickness of about 7 to 10 mm for a direct-view design, between 2 and 5 mm for an edgelighting design, and under 1.2 mm for a fiber-optic design.

Incandescent and LED backlighting modules can be driven directly from the same supply used for the 5-V LCD driver circuitry. Both cold cathode fluorescent and electroluminescent backlighting modules can be driven from the 5-V dc supply using a step-up dc–dc or dc–ac converter. In the case of the cold cathode fluorescent backlighting module, if the product is connected to the ac line, then the high voltages might be generated by an additional filament winding on the power supply transformer. In the case of the electroluminescent backlighting module, some type of chopping circuit is required to generate the proper ac frequency.

As mentioned previously, both incandescent and cold cathode LCD backlighting modules generally produce white light. Almost any color can be generated through the proper wavelength filtering by the diffuser–transflector film with some loss in lighting efficiency. Electroluminescent backlighting modules were initially available with a blue-green zinc oxide phosphor. Today, a number of other phosphor colors are available that span the color space. Similarly, LED technology was initially limited to the dominant wavelength range of 560 to 660 nm. Today, with the advent of InGaN blue technology, the complete color range of 450 to 660 nm is available.

Last, the response time of the LCD backlighting module may be a factor in the selection of the module. A fast response time allows the backlighting module to be dimmed by pulse width modulation. Granted, the backlighting module may also be dimmed by varying the voltage or current applied to the unit. However, with some lighting technologies, the color of the light source varies with different voltage or circuit levels. Another application for a fast response time LCD backlighting module is to separately illuminate the LCD pixels with red, green, and blue light in order to generate a full-color LCD display without using a red-green-blue filter layer behind the liquid crystal cell. This application also requires that the LCD display have a fast response time in order to multiplex the colors at a 60-Hz, or faster, rate. Small incandescent bulbs generally have a switching speed in the 10- to 100-ms range. LED technology has, by far, the fastest response time with switching speeds under 100 ns for most LED materials technologies.

5. Direct-View Liquid Crystal Display Backlighting Techniques

There are several ways to use LED lamps (or die) to backlight LCD displays. The simplest approach is to mount the LED lamps directly behind the LCD diffuser–transflector. Figure 59 has shown the side view of a direct-view LCD backlighting module. With this scheme, the LED lamps generate a uniform field of light on the rear surface of the diffuser–transflector and the light is then re-radiated by the diffuser. The spacing between the top of the LED lamp printed circuit board is determined by the spacing between LED lamps, the arrangement of the lamps, the LED lamp radiation pattern, and the scattering properties of the diffuser–transflector film.

Figure 60 shows two different arrangements for LED lamps. Assuming that the lamps are mounted on the same pitch, the equilateral triangle layout allows a slightly smaller spacing between the back of the diffuser film and the top of the LED printed circuit board. For best results, the top of

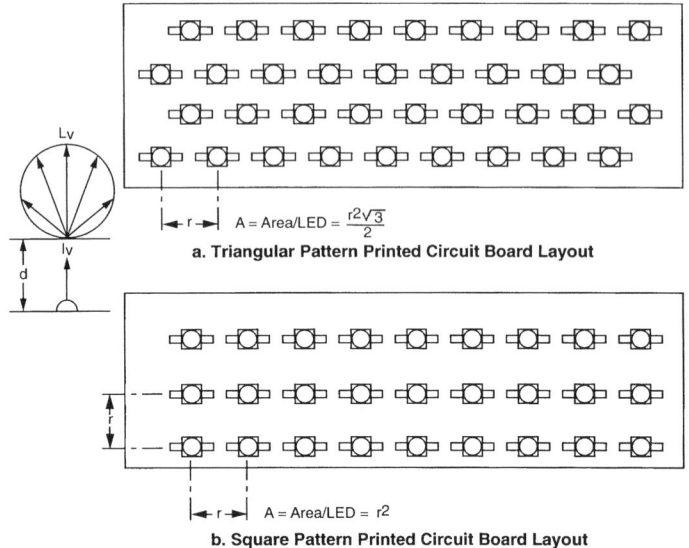

FIG. 60. Printed circuit board layouts for a direct-view liquid crystal display (LCD) backlighting module. (a) Triangular pattern. (b) Square pattern. LED, light-emitting diode. (Makrofol BL 6-2 diffuser film, Bayer Plastic.)

the printed circuit board can be coated with a white solder mask, which helps to distribute light uniformly within the backlit area.

A number of LED lamp packages can be used. Generally, the best LED lamps generate a very wide radiation pattern with good on-axis uniformity. A low profile package also helps to reduce the overall thickness of the module. With this direct-view approach, the GaP technology LED lamps generate sufficient luminous flux for most LCD applications. They can produce a luminous sterance as measured through the diffuser–transflector film in the range of 100 to 300 cd/m². The AlInGaP and TS AlGaAs LED materials technologies provide higher performance that can be utilized to reduce overall power consumption of the module. The InGaN materials technology provides additional colors in the blue and green region for added color capability.

Unpackaged LED die could also be used. They generate a very wide radiation pattern and have a very low profile. The lack of a reflector dish surrounding the die tends to reduce their overall lighting efficiency. Because the index of refraction of the LED die is on the order of 3.4 and air equals 1.0, some luminous flux will be lost at the chip-to-air interface. For best results, a transparent epoxy or junction coating material should be applied over the LED die in order to minimize the index of refraction differences between the two surfaces.

Using the triangular printed circuit configuration shown in Fig. 60(a), several wide-angle LED lamps were assembled into direct-view LCD backlighting modules. A sample of 0.20-mm-thick Bayer Plastic white Makrofol BL 6-2 diffuser film was mounted over the LED array. Green GaP HLMP-P505 and AS AlGaAs HLMP-P105 LED lamps generated a luminous sterance of 200 cd/m² when mounted with a spacing r of 7.62 mm and driven at 10 mA. Good uniformity was achieved with a spacing d of 13 mm between the back of the diffuser film and the top of the printed circuit board. With a lamp spacing r of 10.16 mm, the same lamps generated a luminous sterance of 120 cd/m² with a 10-mA drive current and a spacing d of 18 mm. By using a more highly diffusing material, the LED-to-diffuser spacing can be reduced. For example, a 3.0-mm-thick piece of white acrylic material, such as Rohm & Haas #7138 sheet material, allows the LED-to-diffuser spacing to be reduced by about 30%, with a 40% reduction in luminous sterance.

With this direct-view backlighting approach, the number of LED lamps needed for a given LCD display scales directly with the display area. For example, for the LED lamps mounted in a triangular layout pattern, the area illuminated by a single LED lamp is equal to

$$A = \frac{r^2 \sqrt{3}}{2}$$

where A is the backlit area and r is the pitch between LED packages. For example, it would take 50 LED lamps, mounted on a 7.62-mm triangular pitch or 30 LED lamps mounted on a 10.16-mm triangular pitch, to illuminate a 25- by-100-mm LCD display.

Bicolor and tricolor direct-view LCD backlighting modules could be designed using the same approach outlined in Fig. 60. Sketches of these printed circuit designs are shown in Fig. 61. To obtain adequate color mixing, the minimum spacing between the back of the diffuser film and the top of the LED printed circuit board needs to be increased, because luminous sterance variations across the surface of the diffuser–transflector film will cause color variations across the surface.

6. EDGELIGHTING LIQUID CRYSTAL DISPLAY BACKLIGHTING TECHNIQUES

A second approach to use LED lamps (or die) to backlight an LCD display is to use a light pipe. Figure 62 shows the side view of an edgelighting LCD backlighting module. A light pipe is a rectangular block

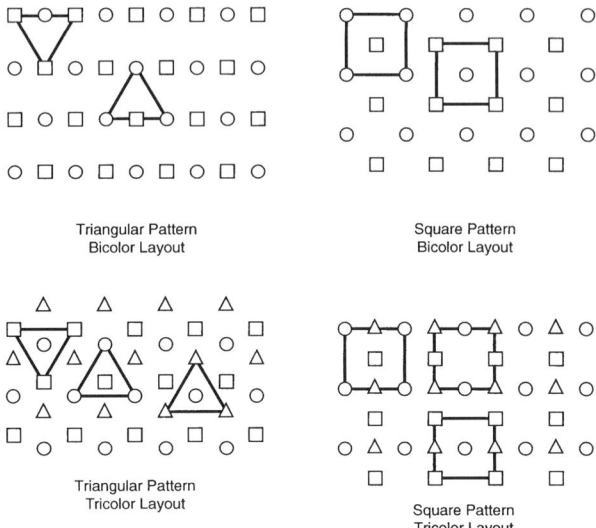

FIG. 61. Printed circuit board layouts. (a) Triangular pattern, bicolor layout. (b) Square pattern, bicolor layout. (c) Triangular pattern, tricolor layout. (d) Square pattern, tricolor layout.

of transparent or translucent plastic that is mounted directly behind the diffuser–transflector film. The rear surface of the plastic block is coated with a white reflective material. Light is then emitted into the block from one or more of the edges of the block. For best results the other edges of the plastic block should be coated with a white reflective material, or covered with a specular reflecting metal tape. This edgelighting approach can produce a luminous sterance as measured through the diffuser–transflector film in the range of 20 to 200 cd/m², depending on how many LED lamps are used and the size of the LCD display.

Light pipes use the principle of total internal reflection to efficiently transmit light. The principle of total internal reflection is defined by Snell's law, as shown:

$$n_1 \sin \theta_1 = n_2 \sin \theta_2$$

where

n_1 = index of refraction of the incidence medium

θ_1 = angle of light in the incidence medium, as measured from a line perpendicular to the surface

n_2 = index of refraction of the second medium

6 APPLICATIONS FOR HIGH-BRIGHTNESS LIGHT-EMITTING DIODES 349

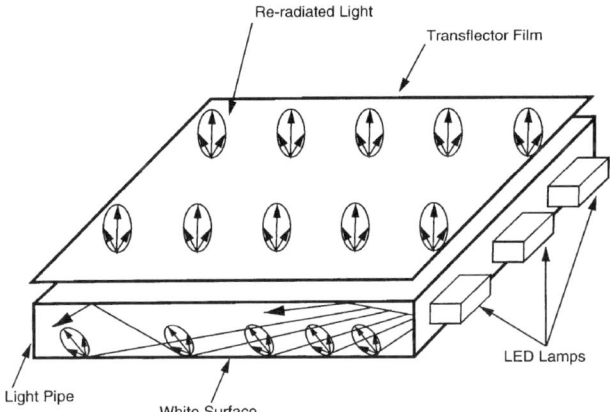

FIG. 62. Cross-section view of an edge-lighting liquid crystal display (LCD) backlighting module. Note. Metalized reflector tape covers the edge surfaces.

θ_2 = angle of light in the second medium, as measured from a line perpendicular to the surface.

Note, that when light passes from a medium with a low index of refraction into one with a higher index of refraction (i.e., air into plastic), the refracted light ray is more normal to the surface that is the incident light ray. Likewise, as light passes from a medium with a high index of refraction to one with a lower index of refraction (i.e., plastic into air), the refracted light ray is emitted with a more oblique angle than is the incident light ray.

When light passes from a medium with a high index of refraction into a medium with a lower index of refraction (i.e., plastic to air) it is possible for the refracted light ray to be emitted at an angle of 90 degrees relative to the normal. This angle is called the critical angle and is equal to

$$\theta_c = \mathrm{Sin}^{-1}\left(\frac{n_1}{n_2}\mathrm{Sin}\,90°\right) = \mathrm{Sin}^{-1}\left(\frac{n_1}{n_2}\right)$$

At incident angles greater than the critical angle, the light ray will be reflected back into the incident medium instead of being refracted into the second medium. This light ray is called a totally internally reflected light ray. Figure 63 shows how light is refracted from a medium with a high index of refraction to one with a lower index of refraction for different incident angles.

The white paint on the bottom of the light pipe helps to extract light from the pipe by scattering part of the light striking the white surface. The scattered light rays strike the opposite wall of the light pipe at angles less

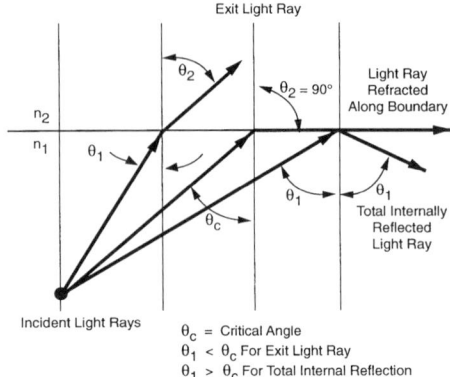

FIG. 63. Principle of total internal reflection. (After Hewlett-Packard Company. 1993. Reprinted with permission from Hewlett-Packard Company.)

than the critical angle and are emitted. A portion of the light rays striking the white reflective surface are total internally reflected back into the light pipe. These light rays are total internally reflected off the top surface and then strike the white surface again. This process is repeated over and over until the light is either emitted from the top surface, absorbed by the light pipe material, or lost at one of the interfaces.

For best luminous flux coupling, the entrance to the light pipe can be a concave surface, as shown in Fig. 64. For most LED lamps, the light rays emitted from the top surface of the lamp are diverging. The concave surface allows the light rays to enter the light pipe with minimal refraction. Then a curved surface on the sides of the light pipe reflects the light into a narrow beam. This, in turn, allows more of the luminous flux to be total internally

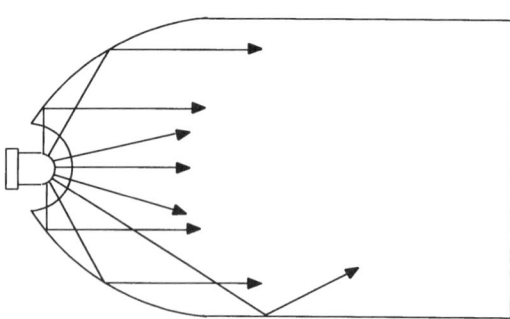

FIG. 64. Concave entrance port.

6 APPLICATIONS FOR HIGH-BRIGHTNESS LIGHT-EMITTING DIODES 351

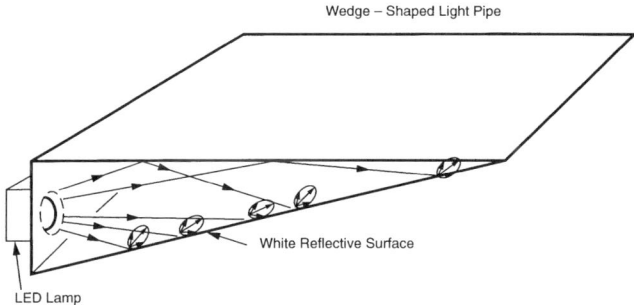

FIG. 65. Wedge-shaped light pipe. LED, light-emitting diode. (After Hewlett-Packard Company. 1993. Reprinted with the permission of Hewlett-Packard Company.)

reflected back into the light pipe, rather being refracted at the plastic-to-air interface.

For small LED displays with dimensions less than 50 by 100 mm, a flat planar light pipe, such as shown in Fig. 62, can be used. For larger LCD displays, a wedge-shaped light guide, as shown in Fig. 65, can be used. The light guides shown in Figs. 62 and 65 can either be molded out of transparent plastic or cut from plastic sheet stock. When using plastic sheet stock, the edges of the block should be polished or fire-polished to provide a smooth surface. One vendor of plastic sheet light pipe material is listed at the end of this chapter (Nitto Jushi Kogyo Co.).

It is also possible to reflect the light 90 degrees after it enters the light pipe. This allows the use of a LED lamp package that is mounted on a printed circuit directly below the LCD display rather than at 90 degrees. This approach is shown in Fig. 66. This approach would have more light losses than the designs shown in Figs. 62 and 65, but allows a wider selection of LED lamps. Also, this design tends to increase the thickness of the LCD backlighting module.

7. FIBER-OPTIC LIGHT-EMITTING DIODE BACKLIGHTING TECHNIQUES

A third way to use LED lamps to backlight an LCD display is to pipe light into a bundle of plastic fiber optics. The fiber-optic bundle is formed into a flat sheet behind the diffuser–transflector film. Then using one of several methods, the luminous flux is extracted from the fiber-optic sheet to evenly backlight the LCD display. A sketch of this method is shown in Fig. 67. This fiber-optic approach can produce a luminous sterance, as measured through the diffuser–transflector film, in the range of 1 to 20 cd/m^2 depending on the luminous flux output of the LED lamp and the size of the LCD display.

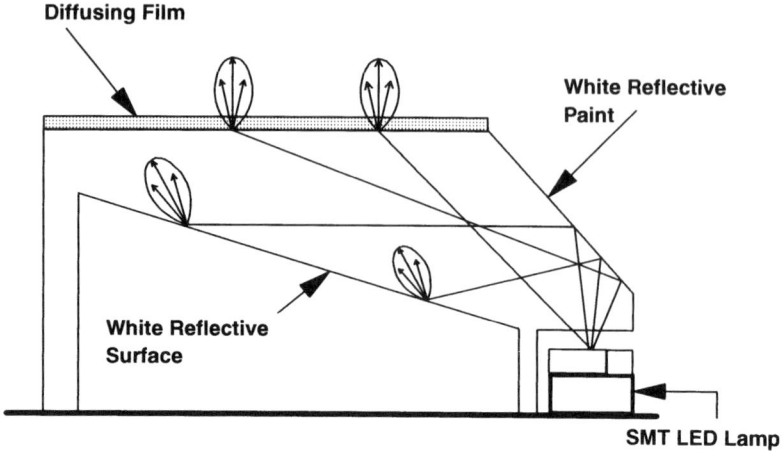

FIG. 66. Right angle light pipe.

Optical fiber consists of a transparent inner core and a thin outer cladding. The cladding material has lower index of refraction than does the core. Thus, light rays within the core can be totally internally reflected at the core-cladding interface, using the same optical principles discussed earlier. For total internal reflection to occur within the fiber, light must enter the fiber within a specified cone acceptance angle. Light rays entering the fiber at larger off-axis angles do not propagate down the fiber. The cone acceptance angle is equal to

$$C_A = 2\sin^{-1}(\sqrt{n_1^2 - n^22})$$

where C_A is the acceptance cone angle of the fiber, n_1 is the index of refraction of the core, and n_2 is the index of refraction of the cladding.

A sketch of the cone acceptance angle of the fiber is shown in Figure 68. The cone acceptance angle C_A is about 56 degrees for plastic fiber. Note that the cone acceptance angle is the total included angle. Thus, the light emitted

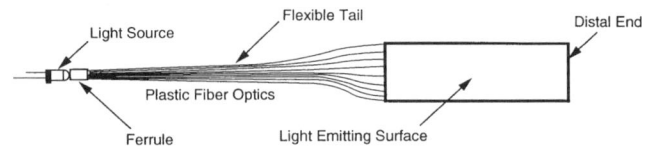

FIG. 67. Fiber-optic liquid crystal display backlighting module. (After Poly-Optical® Products, Inc. 1996. Reprinted with the permission of Poly-Optical Products, Inc.)

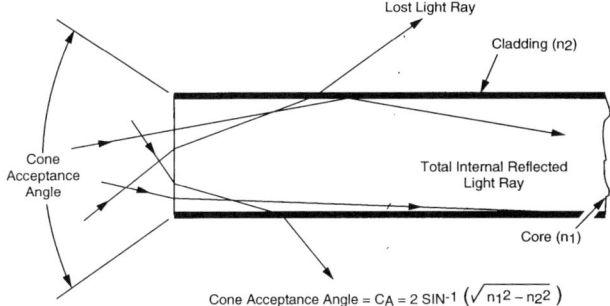

FIG. 68. Acceptance angle of fiber. (After Poly-Optical® Products, Inc. 1996. Reprinted with the permission of Poly-Optical Products, Inc.)

by the light source must strike the end of the fiber at an off-axis angle less than or equal to $C_A/2$.

The stranded fiber optics, or fiber-optic ribbon, is terminated in a sleeve called a ferrule. The ferrule is designed to align the LED light source to the end of the fiber-optic bundle. For best optical coupling, the ends of the fibers should be polished.

There are several techniques to extract light from the fiber-optic bundle. Lumitex, Inc. (1993) uses a technique of weaving the fiber, as shown in Fig. 69(a). With this approach, the light is emitted due to the sharp bends of the fiber. Uniformity across the backlit area is achieved with a variable fill thread pitch. Poly-Optical® Products, Inc. (1996) uses an abrasive process called Uniglow®

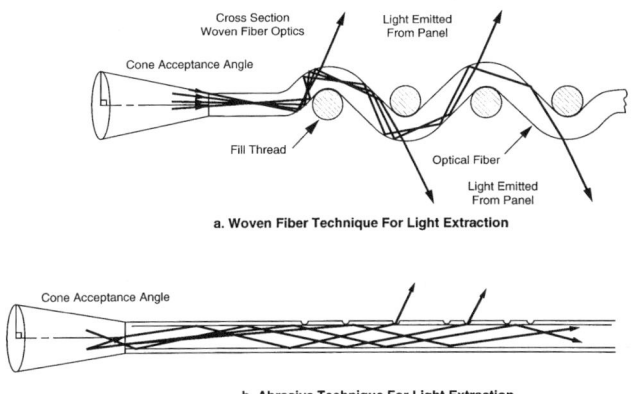

FIG. 69. Woven fiber (a) and abrasive (b) techniques for light extraction from fiber. (Fig. 69(a) from Lumitex, Inc., 1993. Reprinted with the permission of Lumitex, Inc.)

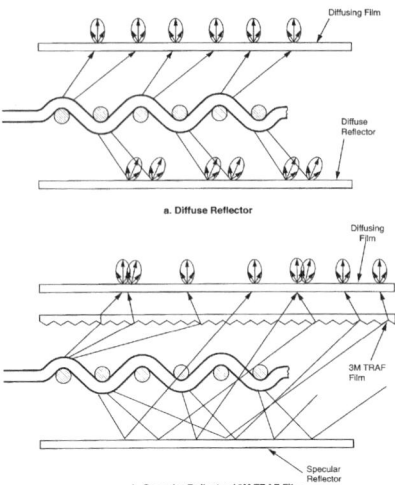

FIG. 70. Methods to increase light extraction efficiency. (a) Diffuse reflector. (b) Specular reflector/3M TRAF film.

where the cladding layer is selectively abraded across the light-emitting area. This process is illustrated in Fig. 69(b). Uniformity is achieved across the backlit area by varying the amount of abrasion across the surface. Serigraph, Inc. (1992) offers both monofilament fiber-optic bundles as well as a flat plastic fiber-optic ribbon. Serigraph also uses a precision mechanical abrasion process to selectively remove the cladding from the fiber. These vendors of fiber-optic LCD backlighting modules are listed at the end of this chapter.

Several options exist to increase the light extraction efficiency of the light-emitting surface. These techniques are shown in Fig. 70. Mounted underneath the fiber surface is a highly reflective material, which can be either a specular reflector or a white diffuse reflector. In either case, this reflective surface helps to redirect the light rays toward the underside of the transflector film. Please note that the radiation patterns of these two reflectors are significantly different. The white diffuse reflector tends to create a wide radiation pattern normal to the transflector film. The metalized reflector creates a more directional off-axis beam. A diffusing film can be mounted over the top surface of the fibers. This also tends to create a wide radiation pattern normal to the transflector film. Alternatively, a layer of 3M Transmissive Right Angle Film (TRAF) (3M Center, St. Paul, MN) can be mounted over the top surface of the fibers behind the diffuser film (3M Center, 1996). For light rays striking the TRAF in the range of 69 to 89 degrees off-axis, this material bends the light rays 90 degrees, so that they

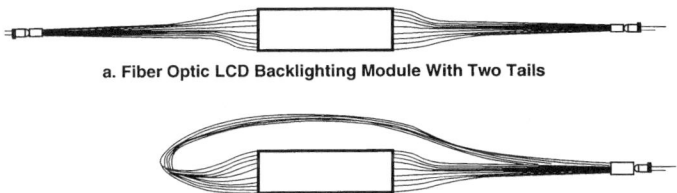

a. Fiber Optic LCD Backlighting Module With Two Tails

b. Fiber Optic LCD Backlighting Module With Two Tails In Common Ferrule

FIG. 71. Two tail techniques to double optical performance. (a) Fiber-optic liquid crystal display (LCD) backlighting module with two tails. (b) Fiber-optic LCD backlighting module with two tails in a common ferrule. (After Poly-Optical® Products, Inc. 1996. Reprinted with the permission of Poly-Optical Products, Inc.)

are more normal to the surface of the transflector film. The TRAF works especially well using a metalized reflector on the underside of the fibers.

Fiber-optic modules also are available in several other configurations besides the single-layer single-tail version shown in Fig. 67. One option is to create a module using two fiber-optic tails, as shown in Fig. 71. Using this approach, the luminous sterance of the light-emitting surface can be doubled by using two LED lamps. Alternatively, the two tails could be doubled back together in a common ferrule.

It is also possible to create multiple-layer fiber-optic modules, as shown in Fig. 72. This approach also can be used to increase the luminous sterance of the light-emitting surface. This approach also can be used to create a bicolor or full-color LCD backlighting module. A bicolor module would use two fiber-optic layers, with each layer connected to a different colored LED lamp. By modulating the luminous intensity of the LED lamps, a range of colors could be created. Using three separate fiber-optic layers and red,

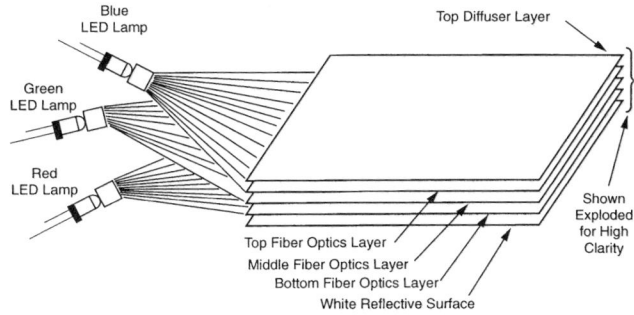

FIG. 72. Multiple-layer fiber-optic liquid crystal display (LCD) backlighting module for full-color applications. LED, light-emitting diode.

green, and blue LED lamps, any color within the color space defined by the lamps can be generated. For best results in color mixing, the fiber-optic layers need to enter the light-emitting surface from the same end. The color coordinate of the color mixture is determined by the relative luminous sterances of the two or three colors at each position across the light-emitting surface. Thus, even if the luminous sterance varies across the surface, a consistent color will be achieved as long as all of the luminous sterances of the layers vary proportionately.

REFERENCES

Automotive News (1995). "Market Data Book." Automotive News, Detroit, MI.
Castellano, J., Mentley, D., and Weichert, A. (1996). "Flat Information Displays," 7th ed., p. 131. Stanford Resources, Inc., San Jose, CA.
Gage, S., Evans, D., Hodapp, M., and Sorensen, H. (1977). "Optoelectronics Applications Manual," 1st ed., pp. 6.3, 6.5, 7.1, 7.10. McGraw-Hill, New York.
Hewlett-Packard Company (1993). "Light Guide Techniques Using LED Lamps, Optoelectron. Div. Appl. Brief I-003. Hewlett-Packard Company, Palo Alto, CA.
Hewlett-Packard Company (1996). "Optoelectronics Designer's Catalog," pp. 1–27, 1–28. Hewlett-Packard Company, Palo Alto, CA.
Hodapp, M. (1990). "Backlighting Automotive Telltales with LED Indicators," Pap. No. 900474. Society of Automotive Engineers, SAE Publications, Warrendale, PA.
Hodapp, M. (1993). "LED Instrument Cluster Design," 26th ISATA Proc., Pap. 93ME024, pp. 377–384. Automotive Automation Limited, Croydon, England.
Hodapp, M. (1994a). "Update: Using LEDs for Instrument Cluster Warning Lights," Pap. No. 941046. Society of Automotive Engineers, SAE Publications, Warrendale, PA.
Hodapp, M. (1994b). "LEDs as Indicators, Illuminators, and Full Color Displays," Photon. Consumer Electron., SPIE TTS, Vol. 1, pp. 66–75. SPIE, Bellingham, WA.
Houghton, D. (1994). "LED Traffic Lights, New Technology Signals Major Energy Savings," Publ. TU-94-1. E Source, Inc., Boulder, CO.
Institute of Transportation Engineers (ITE) (1991). "Lane-Use Traffic Control Heads," Publ. ST-017. ITE, Washington, DC.
Lumitex, Inc. (1993). "Creators of Woven Light," Rev. 1–94. Lumitex®, Inc., Strongsville, OH.
Mitsubishi Electronics America, Inc. (1992). "Mark III HB." Mitsubishi Electronics America, Inc., Diamond Vision Systems Divison, Norcross, GA.
Poly-Optical® Products, Inc. (1996). "Technical Reference Manual," pp. 3, 7, 15, 16. Poly-Optical Products, Inc., Irvine, CA.
Serigraph, Inc. (1992). "Fiber Optics." Serigraph®, Inc., West Bend, WI.
3M Center (1996). "3M Transmissive Right Angle Film (TRAF) II." 3M Electronic Display Lighting. 3M Center, St. Paul, MN.

CHAPTER 7

Organometallic Vapor-Phase Epitaxy of Gallium Nitride for High-Brightness Blue Light-Emitting Diodes

Isamu Akasaki and Hiroshi Amano

DEPARTMENT OF ELECTRICAL AND ELECTRONIC ENGINEERING
MEIJO UNIVERSITY
NAGOYA, JAPAN

I. INTRODUCTION	357
II. HISTORICAL OVERVIEW	359
1. *Hydride Vapor-Phase Epitaxy*	360
2. *Molecular-Beam Epitaxy*	361
3. *Organometallic Vapor-Phase Epitaxy*	362
III. DESIGN AND STRUCTURE OF NITRIDE-BASED SUPERBRIGHT LIGHT-EMITTING DIODES	379
1. *Layered Structure*	379
2. *Electrode*	383
IV. EFFICIENCY, WAVELENGTH, AND LIFETIME	384
V. LASER DIODES	385
VI. SUMMARY	387
References	388

I. Introduction

Wurtzite polytypes of group III nitrides, except for boron nitride (BN), have direct transition-type bandgap structures with bandgap energies ranging from 1.9 eV for indium nitride (InN) (Tansley and Foley, 1986) to 3.4 eV for gallium nitride (GaN) (Maruska and Tietjen, 1969), and to 6.2 eV for aluminum nitride (AlN) (Yim *et al.*, 1973) at room temperature. Figure 1 shows lattice constants and wavelengths corresponding to the bandgap energies of group III nitrides, including BN, which is known as an indirect material. The energy gap bowing parameters in AlGaN and GaInN alloys were reported to be 1 (Baranov *et al.*, 1982, Koide *et al.*, 1987) or 0

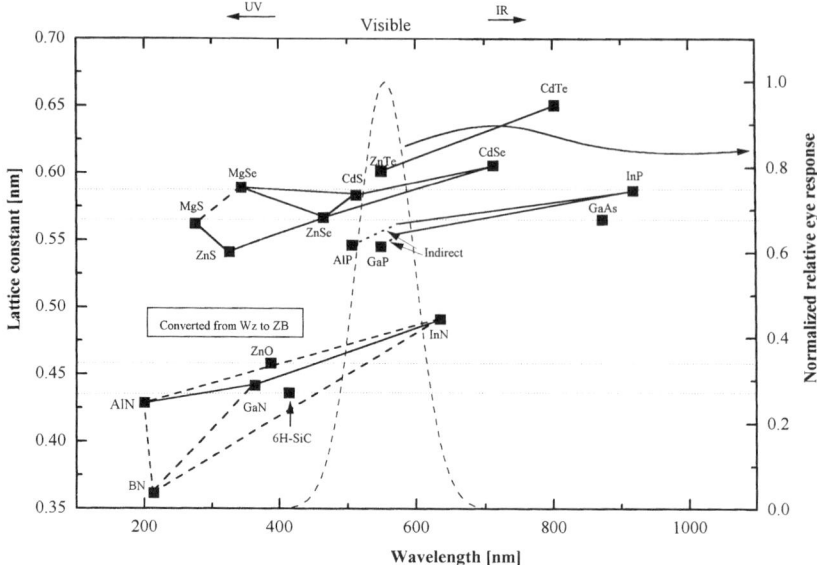

FIG. 1. Lattice constants and wavelengths corresponding to the bandgaps of III-V and II-VI compounds. UV, ultraviolet; IR, infrared; MgSe, magnesium selenide; MgS, magnesium sulfide; ZnS, zinc sulfide; ZnSe, zinc selenide; CdS, cadmium sulfide; ZnTe, zinc telluride; AlP, aluminum phosphide; GaP, gallium phosphide; CdSe, cadmium selenide; CdTe, cadmium telluride; GaAs, gallium arsenide; InP, indium phosphide; InN, indium nitride; ZnO, zinc oxide; AlN, aluminum nitride; BN, boron nitride; GaN, gallium nitride; SiC, silicon carbide.

(Wickenden et al., 1994) and 1 eV (Osamura et al., 1972), respectively. Those of AlInN, BAlN, BGaN, and BInN alloys are not determined at present. The lattice constants and the wavelengths corresponding to the bandgaps of other III-V compounds and wide bandgap II-VI compounds for the visible light emitters are also shown for comparison. As shown in Fig. 1, almost all the visible and near-ultraviolet (uv) regions can be covered by direct bandgap AlN, GaN, InN, and their alloys. Moreover, due to the strong bond between nitrogen and each group III atom, they are chemically and thermally stable. Therefore, group III nitrides have been one of the most promising candidates as materials for application to light emitters such as light-emitting diodes (LEDs) and laser diodes (LDs) from the uv to the visible region of the spectrum.

To realize such light emitters, it is essential to grow high-quality nitride single crystals and to control their electrical conductivity. During the last two decades before 1985, much pioneering work was done for GaN from the fundamental point of view and for practical use by many researchers. Unfortunately, it is fairly difficult to grow large-size bulk GaN crystals because the equilibrium vapor pressure of nitrogen at the growth temperature is extremely high (Karpinski *et al.*, 1984) compared with those of phosphorus (P), arsenic (As), and antimony (Sb) in other III-V compounds. Besides, it had been quite difficult to grow high-quality epitaxial nitride films with a flat surface free of cracks. This is caused by the lack of substrate materials with lattice constants and thermal expansion coefficients close to those of GaN and the nitride alloys. Moreover, it has been well known that undoped GaN is strongly n-type with high residual electron concentrations, and p-type GaN and GaN p-n junctions had not been realized until recently. These problems had prevented the fabrication of GaN-based light emitters for a long time.

Recent developments in growth technology and in the basic understanding of the growth mechanism for the heteroepitaxial growth of GaN on highly mismatched substrates (e.g., sapphire) have enabled us to grow very-high-quality thin GaN, AlGaN and GaInN films and their heterostructures with specular surfaces free of cracks. In addition, the mechanism of impurity incorporation and its activation in nitrides has been developed step by step. Conductivity control of both n-type and p-type nitrides has also been achieved. Zinc- (Zn-) or magnesium-(Mg-)related blue emission from the nitrides can be remarkably enhanced by low-energy electron beam irradiation (LEEBI) or thermal annealing. Moreover, optical and carrier confinement have been achieved by using heterostructures, and the emission wavelength has been controlled by the use of alloys as the active layer. These achievements as well as the earlier discovery of impurities, which form very efficient blue luminescence centers in the nitrides, have led to the fabrication of high-brightness uv–blue, bluish-green, green, and yellow LEDs with efficiencies in excess of 1%. Today, the GaN-based blue LED is commercially available. The brightness and the efficiency of these devices are comparable to the GaAlAs-based red LEDs and the AlGaInP-based orange to yellow LEDs.

In this chapter, advances in crystal growth and conductivity control are reviewed as the groundwork for a discussion of the recently developed high-brightness LEDs and LDs based on the group III nitrides.

II. Historical Overview

Up to now, three major methods have been developed for the epitaxial growth of GaN and the nitride alloys: hydride vapor-phase epitaxy (HVPE), molecular-beam epitaxy (MBE), and organometallic vapor-phase epitaxy (OMVPE). We briefly review the HVPE and MBE growth techniques in relation to the development of blue LEDs. Then we focus on OMVPE, the technique used for the fabrication of high-performance blue LEDs from the nitrides.

1. Hydride Vapor-Phase Epitaxy

Gallium nitride was first synthesized by Johnson et al. (1932), who used the direct reaction between metallic Ga and ammonia (NH_3). Lorenz and Binkowski (1962) reported photoluminescence (PL) in the uv region from synthesized GaN that was polycrystalline. The first single-crystalline GaN was grown by Maruska and Tietjen (1969) using HVPE on sapphire substrates. Gallium chlorides and NH_3 were used as source gases, which were reacted at about 1000°C. Through optical absorption experiments these authors found that GaN has a direct transition–type band structure with a bandgap energy of about 3.39 eV at room temperature. From their findings, GaN attracted much attention by many researchers for use in light-emitting devices, just as for GaAs, but for the short-wavelength region of the spectrum. Dingle et al. (1971) showed the first laser action using such GaN single crystals by optical pumping at 2 K. Following this success, Pankove et al. (1971) and Maruska et al. (1972) fabricated the first greenish-blue and violet metal–insulator–semiconductor (MIS) type of LEDs based on GaN using Zn- and Mg-related luminescence, respectively. Pankove (1973) fabricated a full-color LED based on GaN, the external quantum efficiency of which was about 0.1%, maximum. He used various Zn-associated luminescence centers. The mechanism of light emission and current-voltage characteristics are somewhat different (Pankove and Lamprey, 1974) from those of conventional p-n junction–type LEDs based on gallium arsenide (GaAs) or gallium phosphide (GaP). Monemar et al. (1980) proposed the origin of four luminescence centers; violet luminescence was attributed to Zn_{Ga}, and blue, yellow, and red luminescence was attributed to Zn_N complex. Jacob et al. (1978) optimized the growth conditions for the light-emitting insulating layer in the MIS-type structure and fabricated more efficient blue, green, and red LEDs, the structure of which is basically the same as that discussed previously. Ohki et al. (1981) developed a flip chip–type electrode, enabling the first practical GaN- based blue LEDs. The external quantum efficiency was 0.12%, maximum.

In the 1970s and early 1980s, HVPE was commonly used for the fabrication of MIS-type GaN LEDs. The major problems of fabricating MIS-type GaN LEDs by HVPE are (1) generation of many cracks and pits, and (2) poor controllability of the thickness of the insulating (i) layer that governs the operating voltage. One of the most knotty problems was the residual donor in undoped GaN. Usually, HVPE-grown GaN contains a high density of residual donors. In order to reduce the concentration of residual donors, thick films have been indispensable (Ilegems and Montgomery, 1973). Due to the large differences of the lattice constants and thermal expansion coefficients between GaN and sapphire, however, many cracks are generated at the interface between the GaN and the sapphire when the thickness of the GaN film exceeds several tens of μms (Itoh and Rhee, 1985). Another issue is thickness control. In MIS-type structures, the operating voltage strongly depends on the thickness of the acceptor-doped i-layer. Due to the poor controllability of thickness by HVPE, the operating voltage usually was approximately 10 V.

2. MOLECULAR-BEAM EPITAXY

Molecular-beam epitaxy is another important technique for the fabrication of LEDs because of the more precise controllability of thickness and alloy composition and the more abrupt heterojunction as compared with OMVPE. Moreover, it is easy to observe the growth process *in situ*. Gas source MBE (GSMBE) of GaN single crystals was proposed in 1974 (Akasaki and Hayashi, 1976). Metallic Ga and ammonia (NH_3) were used as the sources. Growth of AlN by MBE was reported by Yoshida *et al.* (1975) using metallic Al and ammonia as sources. In the late 1970s and the early 1980s, the residual donor concentration in undoped GaN was more than 10^{19} cm^{-3}. Yoshida *et al.* (1983) first tried to control the interface between GaN and sapphire. They used single-crystalline AlN (sc-AlN) coated sapphire as the substrate and found that the cathodoluminescence was enhanced. However, the residual donor was still more than 10^{19} cm^{-3} and the mobility less than 40 cm^2/V·sec. Insufficiency of active nitrogen, leading to N vacancies was believed to lead to such high residual donor densities.

The use of a nitrogen plasma decomposition technique for the growth of group III nitrides was demonstrated by Wauk and Winslow (1968). In this way, at present, the residual donor concentration is reduced to 10^{17} cm^{-3}. Silicon was used to control the conductivity of the n-type GaN. In 1993, p-type GaN was achieved by doping with Mg (Moustakas, 1993). No special treatment was necessary to activate Mg, because GaN was grown in a hydrogen-free atmosphere. Molnar *et al.* (1995) showed the first p-n–junction type GaN LED grown by MBE. Luminescence originating from the Mg-related luminescence centers in the GaN:Mg layer was observed. As concerns the heterojunction, Sitar et al. (1990) reported a GaN–AlN short-period superlattice. Satellite peaks from the GaN and AlN sublattices were clearly observed. Fujita et al. (1995) used AlGaN–GaN double heterostructure (DH) LEDs with Si and Mg co-doped GaN as the active layer, while Sakai et al. (1995) fabricated GaN–GaInN DH blue LEDs with power efficiencies of about 0.3%. These are the most efficient blue LEDs grown by MBE.

Bright blue LEDs were first fabricated by OMVPE. In the next section, we focus on the growth of the group III nitrides by OMVPE.

3. Organometallic Vapor-Phase Epitaxy

a. *Brief History of the Growth and Basic Properties of the Nitrides*

Manasevit et al. (1971) synthesized GaN and AlN on 6H-SiC and sapphire substrates by OMVPE for the first time, using triethylgallium (TEGa) and ammonia as source gases. These authors obtained highly c-axis–oriented films composed of grains or islands with sizes of about 10 to 20 μm. The first blue LED based on GaN grown by OMVPE was fabricated by Kawabata et al. (1984). The morphology of their films was similar to that of Manasevit et al. (1971). The residual donor concentration was still higher than 10^{18} cm^{-3}. The performance of their MIS-type LEDs was similar or inferior to those fabricated by HVPE in the early 1970s.

A great advance in crystal growth by OMVPE was achieved by Amano et al. (1986), who proposed the use of a low-temperature-deposited (LT) buffer layer. A typical OMVPE reactor for the growth of group III nitrides is depicted schematically in Fig. 2 (Hirosawa et al., 1993). A group V hydride such as ammonia and the group III organometallics were separately introduced into the reactor to prevent parasitic reactions between them. Figure 3 illustrates the growth process with (a) and without (b) the LT-buffer layer. Figure 4 shows a comparison of the surface morphologies of GaN grown on sapphire with (a) and without (b) the LT-buffer layer. As shown in Fig. 4, the flatness is much improved by using the LT-buffer layer. Figure 5 shows a cross-sectional transmission electron micrograph (TEM) of GaN grown with the LT-buffer layer. As seen in Fig. 5, the GaN film with the LT-buffer layer has a three-zone texture with regard to the density of faults: (a) a faulted zone with thickness of about 50 nm, which is just adjacent to the AlN buffer layer, (b) a semisound zone with thickness of about 100 to 200 nm, located nearest to the faulted zone, and (c) the uppermost layer with a thickness of the order of microns. It is found that, although defects such as dislocations are crowded in the former two zones,

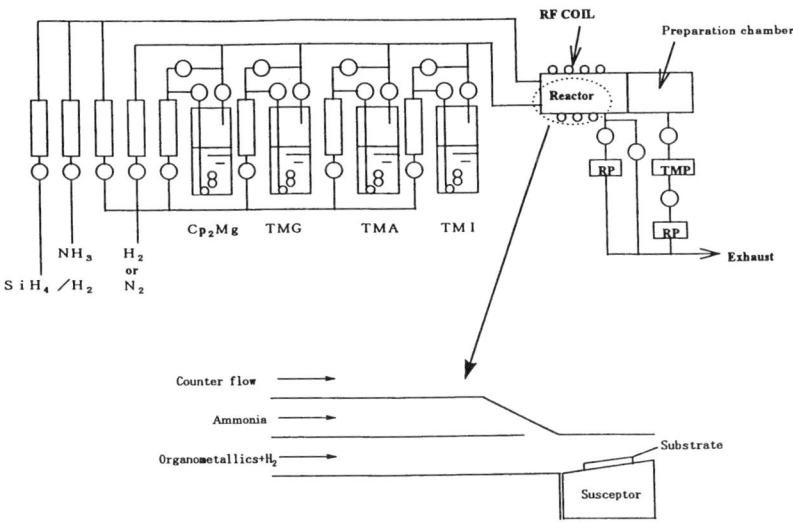

FIG. 2. Typical apparatus for the OMVPE growth of nitrides. Cp$_2$Mg, bis(cyclopentadienyl)magnesium; TMGa, trimethylgallium; TMAl, trimethyl aluminum; TMI, trimethylindium RF, radio frequency turbomolecular pump; TMP, trimethylphosphorus; RP, rotary pump; NH$_3$, ammonia; H$_2$, hydrogen; N$_2$, nitrogen; SiH$_4$, silane.

mostly in zone (a), they decreased abruptly for layers thicker than about 200 to 300 nm. So-called ω-mode X-ray rocking curves and $2\theta/\theta$-mode profiles from GaN grown on sapphire with and without the LT-buffer layer are shown in Fig. 6 (Koide et al., 1988). This shows that mosaicity is remarkably reduced using the LT-buffer layer. Figure 7 shows the PL spectra of the samples shown in Figs. 4(a) and 4(b). Near band-edge uv emission is dominant from the undoped GaN grown with the LT-buffer layer, whereas deep-level–related emission is dominant from the GaN grown without the buffer layer. Figure 8 shows the near band-edge PL of undoped GaN (solid line) and Si-doped GaN (dotted line) grown with the LT-buffer layer. (The sample is different from that used in Fig. 7(a).) In the spectrum from the undoped GaN, three strong free exciton lines, corresponding to the valence band splitting, can be clearly observed in addition to the donor and acceptor bound exciton lines. The intensity of the main exciton peak A is about two orders of magnitude higher than the impurity-bound exciton peaks. The full-width at half-maximum for the emission A is about 3 meV. The free exciton emission dominates the PL even at room temperature. Such a sharp and strong free exciton emission indicates the much improved crystalline quality. Figure 9 shows the temperature dependence of the free electron concentration and mobility of nominally undoped n-type GaN grown with

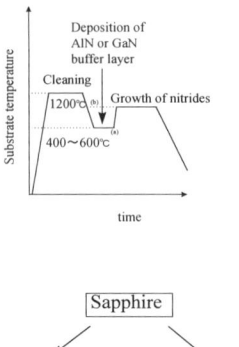
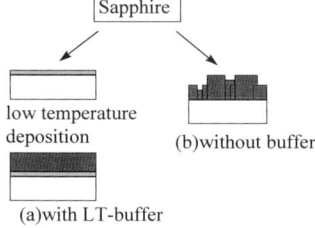

FIG. 3. Timing chart and the model for the growth of gallium nitride (GaN) by organometallic vapor-phase epitaxy with (a) and without (b) the LT-buffer layer. LT, low-temperature-deposited; AlN, aluminum nitride.

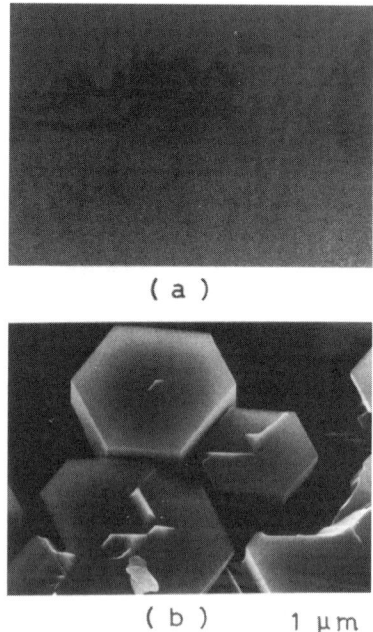

FIG. 4. Scanning electron micrograph of gallium nitride grown by organometallic vapor-phase epitaxy with (a) and without (b) the LT-buffer layer. LT, low-temperature-deposited.

FIG. 5. Cross-sectional transmission electron micrograph of gallium nitride (GaN) grown with the LT-buffer layers. (a) Faulted zone, about 50-nm thick, adjacent to the aluminum nitride buffer layer. (b) Semisound zone, about 100 to 200 nm thick, nearest the faulted zone. (c) Uppermost layer, with thickness on the order of microns. LT, low-temperature-deposited; AlN, aluminum nitride; α-Al_2O_3, sapphire.

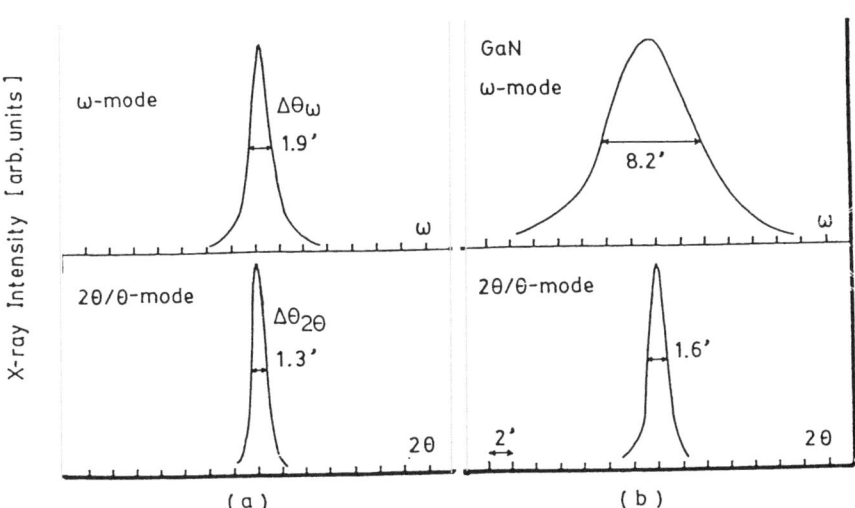

FIG. 6. X-ray rocking curve (ω-mode and $2\theta/\theta$-mode) of gallium nitride (GaN) grown with (a) and without (b) the LT-buffer layer. LT, low-temperature-deposited; AlN, aluminum nitride; α-Al_2O_3, sapphire.

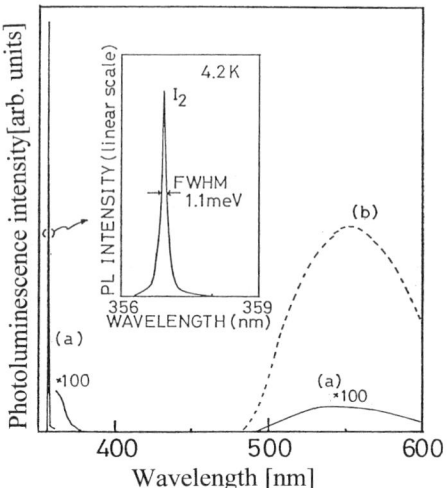

FIG. 7. Photoluminescence (PL) spectra of nominally undoped gallium nitride with (a) and without (b) the LT-buffer layer. LT, low-temperature-deposited; FWHM, full-width at half-maximum.

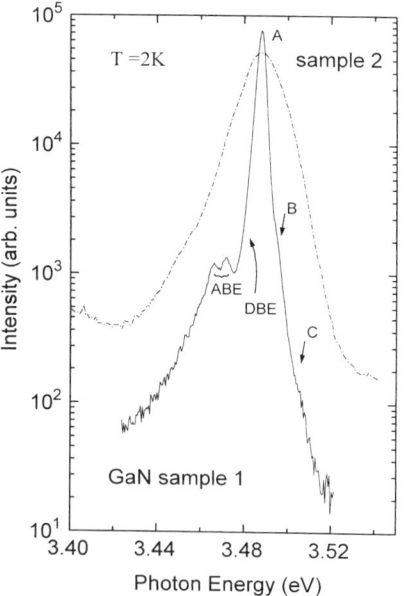

FIG. 8. Near band-edge photoluminescence from undoped (solid line) and silicon-doped (broken dotted line) gallium nitride (GaN) grown with the LT-buffer layer. LT, low-temperature-deposited.

7 ORGANOMETALLIC VAPOR-PHASE EPITAXY OF GALLIUM NITRIDE 367

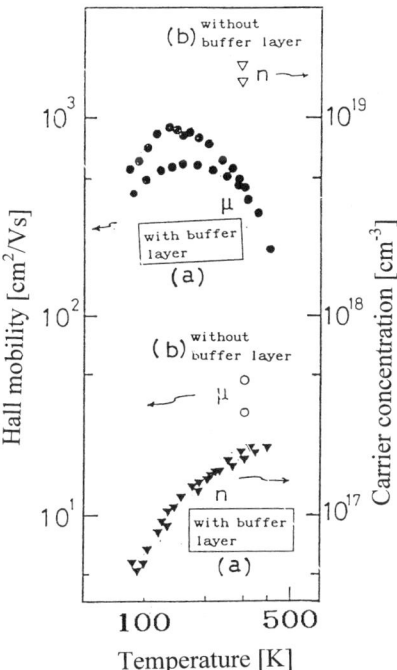

FIG. 9. Temperature dependence of free electron concentration and Hall mobility of undoped gallium nitride with (a) and without (b) the LT-buffer layer. LT, low-temperature-deposited.

(Fig. 9(a)) and without (Fig. 9(b)) the LT-buffer layer. By using the LT-buffer layer, the residual donor concentration is reduced by more than two orders of magnitude and the mobility is increased by an order of magnitude at room temperature. This also indicates that the electron mobility in GaN grown using the LT-buffer layer is governed by polar optical phonon scattering at high temperatures, whereas it is limited by ionized impurity scattering at low temperatures.

These results show that not only the crystalline quality but also the luminescence and the electrical properties have been dramatically improved simultaneously using the LT-buffer layer method. It should be stressed that such high-quality GaN layers can be realized that are thin enough to avoid generating cracks. Akasaki *et al.* (1989a) and Hiramatsu *et al.* (1991) elucidated the growth mode and the role of the buffer layer in detail, as shown in Figs. 10 and 11 (Amano *et al.*, 1988). Three-dimensional growth dominates when the GaN is grown directly on the sapphire substrate

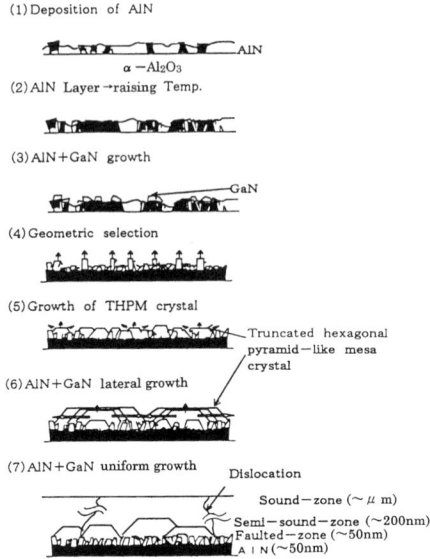

FIG. 10. Model for the growth of gallium nitride (GaN) grown by organometallic vapor-phase epitaxy using the LT-buffer layer. LT, low-temperature-deposited; AlN, aluminum nitride; α-Al_2O_3, sapphire; THPM, truncated hexagonal pyramid-like mesa.

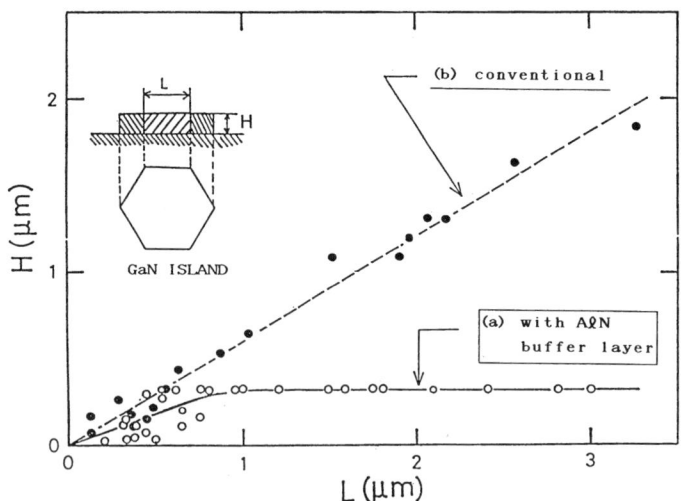

FIG. 11. Height (H) and width (L) of the initial stage of gallium nitride (GaN) islands on sapphire by organometallic vapor-phase epitaxy with (a) and without (b) the LT-buffer layer. LT, low-temperature-deposited; AlN, aluminum nitride.

(Fig. 11(b)), whereas quasi-two-dimensional growth dominates when GaN is grown with the LT-buffer layer (Fig. 11(a)). The essential role of the LT-buffer layer was found to be related to both the supply of high-density nucleation centers having the same orientation as the substrate and the promotion of lateral growth of the epitaxial film due to decrease in interfacial free energy between the epitaxial film and substrate. The effectiveness of the LT-buffer layer method in the improvement of the crystalline quality of the nitride alloys such as AlGaN (Koide et al., 1988) and GaInN was also proved. This buffer layer method was followed by Nakamura (1991), who used an LT-GaN buffer layer. Today, these low-temperature AlN or GaN buffer layers are indispensable and are standard for the growth of GaN and nitride alloys on sapphire substrates by OMVPE.

The resistivity of nominally undoped GaN grown by the LT-buffer layer method is much higher than that of HVPE grown GaN or OMVPE grown GaN without the LT-buffer layer due to the decrease in residual donor concentration. In order to obtain more conductive n-type GaN and to control the conductivity of n-type GaN, doping with a donor impurity is indispensable. Amano and Akasaki (1990) used Si. Details of the control of the n-type conductivity of the nitrides are discussed in the next section.

It was Amano et al. (1989) who achieved single-crystalline low-resistivity p-type GaN by doping with Mg, followed by an LEEBI treatment, for the first time. Later, p-type GaN was also obtained by thermal annealing under a hydrogen-free atmosphere (Nakamura et al., 1991). Details of the achievement and control of the conductivity of the p-type nitrides and the production of the GaN p-n junction are discussed in the next section.

Soon after the success of p-type GaN, Amano et al. (1989) and Akasaki et al. (1992a, b) fabricated GaN p-n junction LEDs and AlGaN–GaN DH LEDs with efficiencies of about 1.5%. The electron blocking layer used was p-type AlGaN.

In order to achieve visible emission, the following are essential: doping with impurities that form efficient blue luminescence centers, or using smaller bandgap materials such as GaInN, GaNP, or GaNAs. The incorporation of acceptor impurities in GaN strongly depends on the species and the growth conditions, especially the growth temperature. Figure 12 shows the temperature dependence of the ratio between normalized concentration of the impurity in GaN and flow rate of the precursor. Magnesium shows little dependence on growth temperature, whereas Zn shows a strong dependence (Akasaki et al., 1989b; Amano et al., 1990b). Sticking and evaporation govern the incorporation of Zn in GaN.

Polycrystalline GaInN was first synthesized by Osamura et al. (1972) using sputtering. Nagatomo et al. (1989) and Yoshimoto et al. (1991) grew

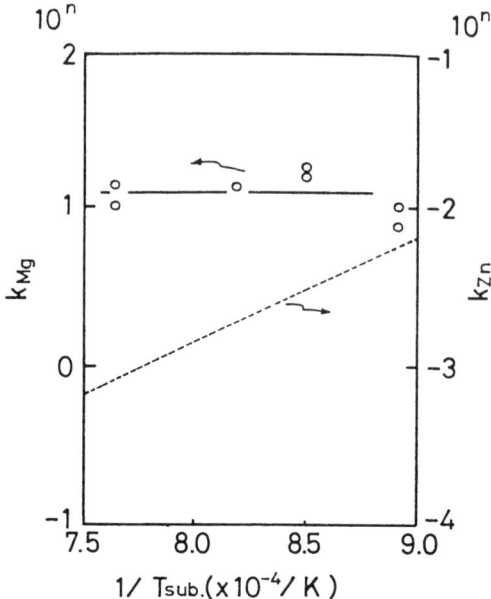

FIG. 12. Sticking coefficients of zinc and magnesium in gallium nitride as a function of substrate temperature.

sc-GaInN by OMVPE. Compared with AlGaN, the composition of the GaInN is strongly dependent on the growth temperature. Figure 13 shows the dependence of the InN molar fraction of GaInN on the growth temperature for a constant vapor composition. The GaInN growth process is affected by the sublimation of GaInN, or a change of the sticking coefficient of the In precursor, or both (Akasaki and Amano, 1995a). Therefore, precise control of growth temperature is essential for the growth of GaInN having the desired composition.

The wavelength dispersion of the refractive index of GaN at room temperature was measured in transmission and reflection by Ejder (1975). The refractive index of the $E \perp c$ component near the bandgap was found to be about 2.65. Later, Amano et al. (1993) measured the dispersion of the refractive index of the $E \perp c$ component using spectroscopic ellipsometry, and found a refractive index about 0.1 times larger than that reported by Ejder, possibly because of the difference in the effect of strain owing to heteroepitaxy. Akasaki and Amano (1994) measured the refractive indices of AlGaN and GaInN. The M_0 specific point shifts toward higher energy in proportion to the bandgap. Therefore, it is possible to fabricate waveguide structures using AlGaN and GaInN heterostructures.

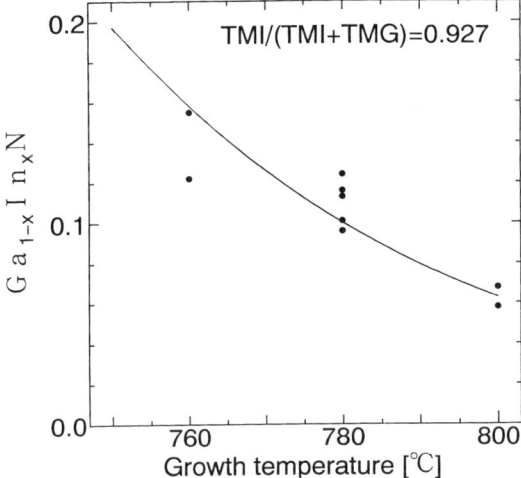

FIG. 13. Solid–vapor ratio of gallium (Ga) and indium (In) in GaInN alloys. Solid line is that estimated considering the vapor pressure of In. TMIn, trimethylindium; TMGa, trimethylgallium.

Nakamura et al. (1993a) first used GaInN as an active layer for violet LEDs. Nominally undoped GaInN or GaInN doped with Si was used. In the first report, p-type GaN was used as the electron blocking layer. Later, p-type AlGaN was employed (Nakamura et al., 1994a). Zinc- and Cd-related luminescence centers in the GaInN active layer were employed. These authors later reported that Si was co-doped in the active layer to enhance the blue emission. The fact that the existence of Si enhances the Zn-related violet emission from GaN was revealed by the group at Nagoya University (Khan et al., 1986). Figures 14 and 15 show the dependence of PL intensity of Zn-doped GaN (Khan et al., 1986) and GaInN (Nakamura et al., 1994b), respectively, on the concentration of co-doped Si. The PL intensity of the GaInN reaches a maximum at an Si concentration of about 1×10^{19} cm^{-3}.

Control of thickness and alloy composition are important issues for the fabrication of efficient nitride LEDs, because ternary compounds such as AlGaN and GaInN are lattice-mismatched systems, as shown in Fig. 1. Figure 16 shows the critical layer thicknesses of AlGaN and GaInN on GaN as a function of alloy composition as calculated from the theory of Matthews and Blakeslee (1974; Akasaki and Amano, 1994). It should be noted that the critical layer thickness for GaInN on GaN is much lower than that for AlGaN on GaN. Quantum well structures (QWSs) are useful

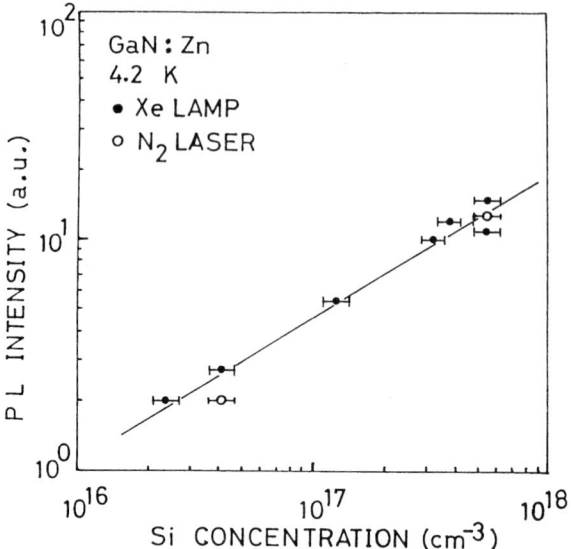

FIG. 14. Violet photoluminescence (PL) intensity of zinc-doped gallium nitride (GaN) as a function of silicon concentration. Zn, zinc; Xe, xenon; N_2, nitrogen.

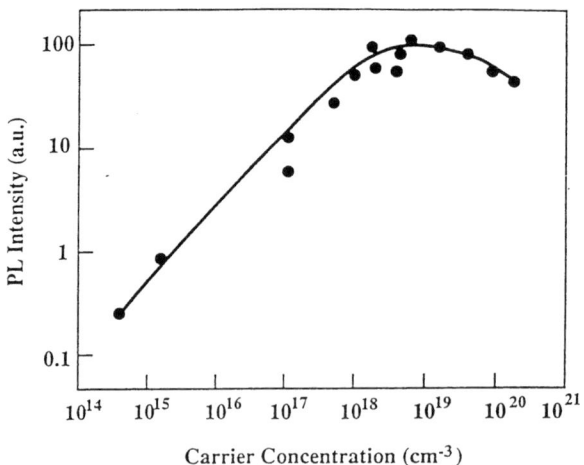

FIG. 15. Blue-green photoluminescence (PL) intensity of zinc-doped $Ga_{0.77}In_{0.23}N$ as a function of carrier concentration. (Reprinted from Nakamura et al., 1994, J. Appl. Phys. **76**, 8189–8191, with the permission of the American Institute of Physics.)

to check the controllability of thickness. Khan et al. (1990) and Itoh et al. (1991) first fabricated AlGaN–GaN multiple quantum well (MQW) structures. AlGaN–GaN MQW structures with GaN well thicknesses as small as 5 nm were fabricated. Layered structures of GaN–GaInN were fabricated by Nakamura et al. (1993b). Akasaki et al. (1995b) and Koike et al. (1996) characterized the GaInN–GaN MQW in detail by TEM, secondary ion mass spectroscopy (SIMS), and high-resolution X-ray diffraction (Akasaki and Amano, 1995b; Akasaki et al. 1995a; Koike et al., 1996). Figure 17 shows a cross-sectional TEM of a five-well GaInN–GaN MQW structure sandwiched between $Al_{0.05}Ga_{0.95}N$ layers. The dislocation density threading the MQW structure was found to be on the order of 10^9 cm^{-2} (Akasaki and Amano, 1995b). Figure 18 shows the $2\theta/\theta$ mode profile from a GaInN–GaN multilayered structure having six pairs of GaInN and GaN layers sandwiched between AlGaN layers. The experimental result agrees with the calculated result obtained using the kinematic theory. Four fringes can be clearly observed between the 0 and -1 order peaks of the MQW, which indicates that the interface is smooth and that the thickness and alloy

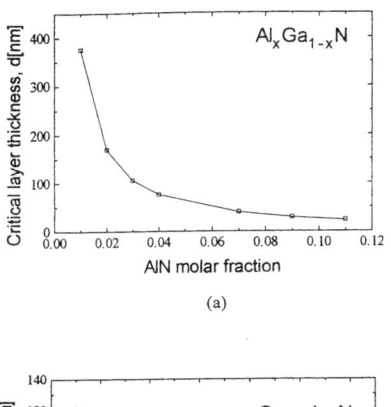

(a)

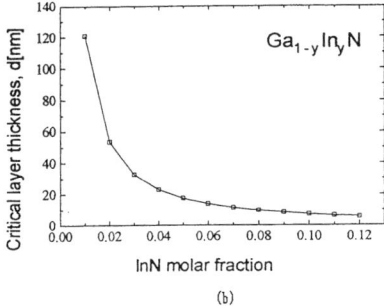

(b)

FIG. 16. Critical layer thicknesses of $Al_xGa_{1-x}N$ (a) and $Ga_{1-y}In_yN$ (b) on gallium nitride. AlN, aluminum nitride; InN, indium nitride.

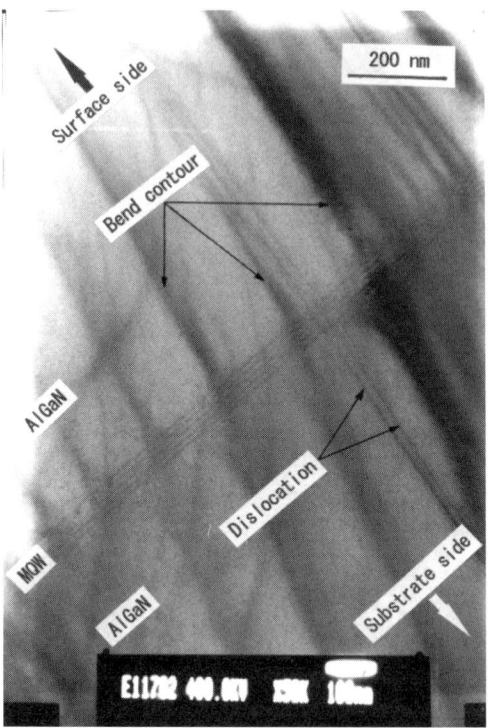

FIG. 17. Cross-sectional transmission electron micrograph of AlGaN–(GaInN–GaN)$_5$–AlGaN–LT-buffer–sapphire. MQW, multiquantum well; GaN, gallium nitride; LT, low-temperature-deposited.

composition are well controlled. Amano and Akasaki (1995) and Akasaki et al. (1995a) showed that the band-edge PL intensity increases with decreasing well thickness. The intensity of the MQW emission with thicknesses smaller than 2.5 nm is about 1.5 (at room temperature) or 3 (at 77 K; Fig. 19) orders of magnitude higher than those of bulk GaInN. Nakamura et al. (1995a, b) showed efficient blue, green, and yellow LEDs fabricated using very thin GaInN quantum well active layers. In these LEDs, the GaInN well layers were nominally undoped. Details of the performance of these LEDs are discussed in the following section.

b. *Conductivity Control*

n-Type. Sayyah et al. (1986) studied Si doping of GaN and AlGaN. They reported that silane (SiH$_4$) was harmful to the growth of GaN and

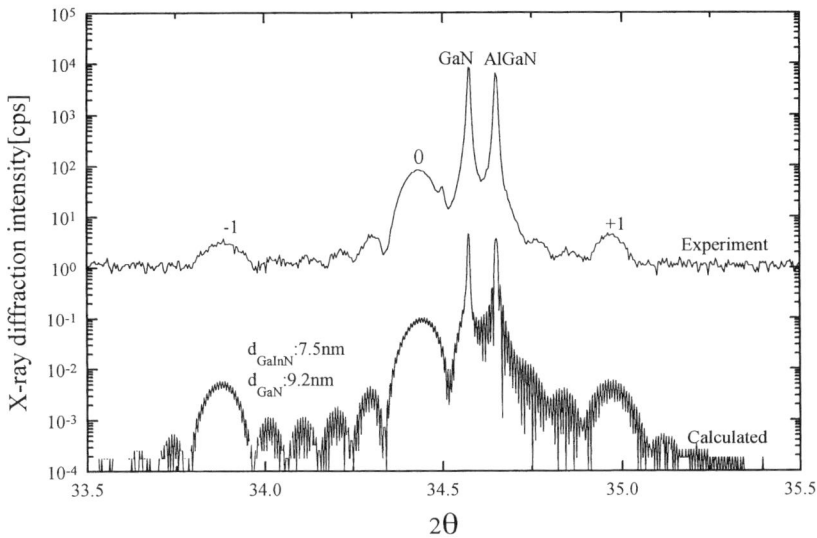

FIG. 18. High resolution X-ray diffraction ($2\theta/\theta$ scan profile) of $(GaInN-GaN)_6$–AlGaN–LT-buffer–sapphire MQW structure. GaN, gallium nitride; LT, low-temperature-deposited; MQW, multiquantum well.

AlGaN but did not affect the conductivity of the film. Nominally undoped GaN grown by OMVPE using the LT-buffer method is nominally highly resistive, owing to the reduced concentration of residual donors, as discussed above. Therefore, in order to obtain low-resistive n-type GaN, doping with a donor impurity is necessary. Amano and Akasaki (1990) and Murakami et al. (1991) showed that SiH_4 is suitable for Si doping during OMVPE growth of GaN and AlGaN using the LT-buffer method and that Si behaves as a donor. They also showed that the cathodoluminescence intensity increases with increasing Si concentration in n-type GaN and AlGaN. The electron concentration at room temperature can be controlled from the undoped level to levels of more than $10^{19}\,cm^{-3}$. Figure 20 shows the electron concentration and resistivity of GaN and $Al_{0.1}Ga_{0.9}N$ as a function of the SiH_4 flow rate. Si_2H_6 is also commonly used as an Si precursor (Rowland et al., 1995). The donor activation energy of Si in GaN and AlGaN, with AlN molar fractions less than 0.15, is around 33 meV. Nakamura et al. (1993b) demonstrated an enhancement of near band-edge emission of GaInN obtained by doping with Si. The incorporation rate of Si in GaN is not strongly dependent on the substrate temperature. Therefore, it is possible to control the concentration of Si in GaN by changing the flow rate of the precursor, either SiH_4 or Si_2H_6. Today, Si doping in

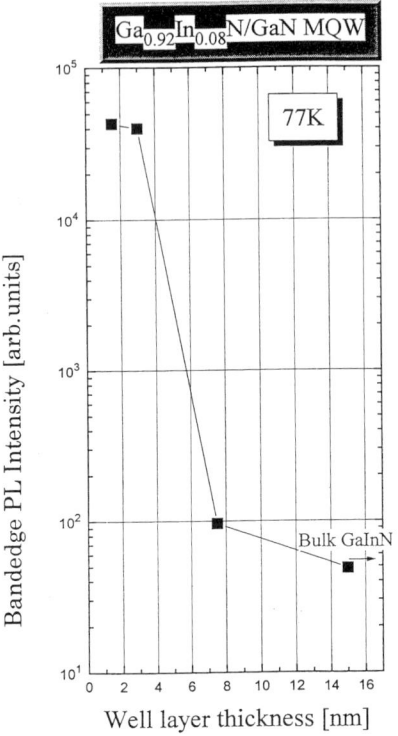

FIG. 19. Dependence of band-edge photoluminescence (PL) intensity on GaInN well layer thickness. GaN, gallium nitride; MQW, multiquantum well.

OMVPE growth with the LT-buffer layer is widely adopted for the conductivity control of the n-type nitrides GaN, AlGaN, and GaInN.

p-Type. Although through the years many groups attempted to produce p-type GaN, none succeeded until 1989. Good controllability of Mg concentration during OMVPE growth of GaN was reported (Akasaki *et al.*, 1989b, Amano *et al.*, 1990b) by using bis(cyclopentadienyl)magnesium (Cp_2Mg) as the Mg precursor. The added Mg was mostly inactive as-grown. An LEEBI treatment was used to activate the Mg to yield p-type GaN and p-n junction–type LEDs (Amano *et al.*, 1989). Figure 21 shows the first results of the free hole concentration of Mg-doped GaN after the LEEBI treatment as a function of the Mg concentration and the LEEBI treatment current. The solid line is the estimated hole concentration, assuming that all the added Mg is activated. In this calculation, the activation energy of the

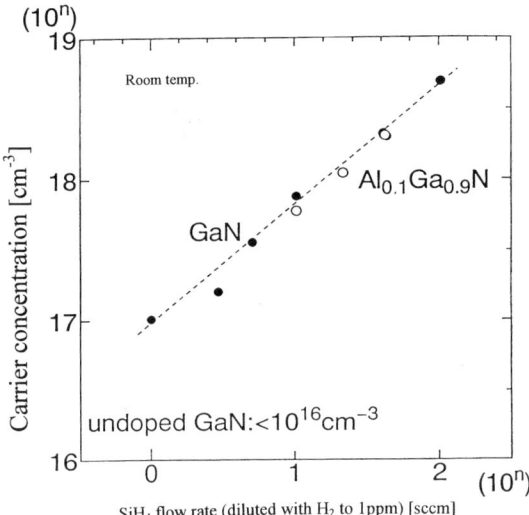

FIG. 20. Electron concentration of gallium nitride (GaN) and $Al_{0.1}Ga_{0.9}N$ as a function of silane (SiH_4) flow rate. H_2, hydrogen.

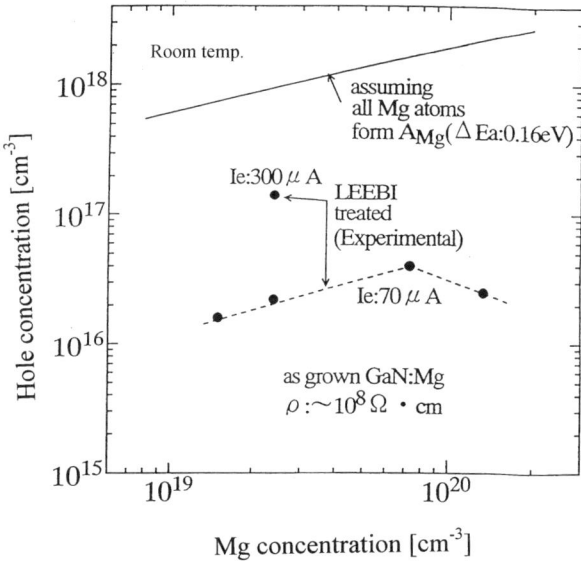

FIG. 21. Hole concentration of gallium nitride (GaN) as a function of magnesium (Mg) concentration. Solid line is the estimated hole concentration assuming that all of the added Mg is activated. LEEBI, low-energy electron beam irradiation.

Mg acceptor is assumed to be 160 meV, which was obtained from the temperature dependence of the hole concentration. A free hole concentration at room temperature of about 2×10^{17} cm^{-3} was achieved at that time. Today, free hole concentrations at room temperature of more than 10^{18} cm^{-3} are achieved. In 1991, p-type AlGaN was obtained in the same manner (Akasaki and Amano, 1991, 1992b). Later, Nakamura *et al.* (1991) achieved p-type GaN by thermal annealing. They showed that Mg is passivated by hydrogen in the as-grown state (Nakamura *et al.*, 1992). The reduction of hydrogen passivation is the dominant process required to activate the Mg when ammonia is used as the nitrogen source (Van Vechten *et al.*, 1992). Tanaka *et al.* (1994) and Yamasaki *et al.* (1995) determined activation energies of Mg in AlGaN and GaInN, respectively. Figure 22 shows the temperature dependence of the hole concentration of Mg-doped $Al_{0.08}Ga_{0.92}N$, GaN, and $Ga_{0.91}In_{0.09}N$. Contrary to the case for Si, the activation energy of Mg was found to depend on alloy composition. At present, in OMVPE, all the p-type GaN and group III nitride alloys are prepared by Mg doping using Mg precursors such as Cp_2Mg, followed by

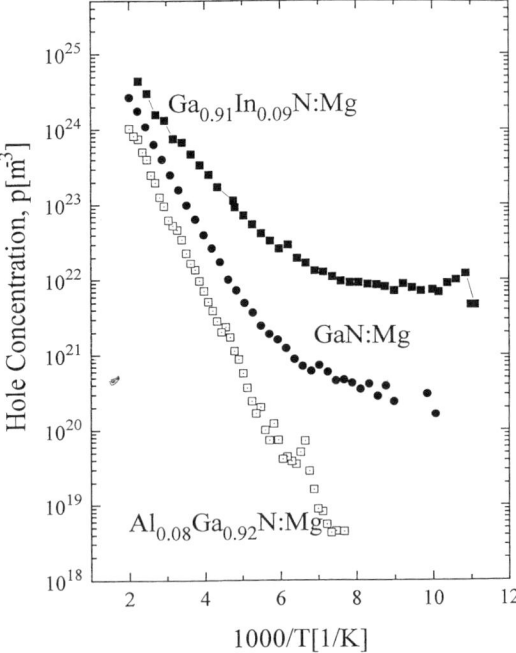

FIG. 22. Temperature dependence of the hole concentration of magnesium-doped $Al_{0.08}Ga_{0.92}N$, gallium nitride and $Ga_{0.91}In_{0.09}N$.

the LEEBI treatment or thermal annealing in a hydrogen-free atmosphere.

The production of high-quality nitrides by using the LT-buffer layer, conductivity control, and in particular, the realization of p-type GaN mentioned above, have led to high-performance blue LEDs and short-wavelength LDs.

III. Design and Structure of Nitride-Based Superbright Light-Emitting Diodes

1. LAYERED STRUCTURE

a. Gallium Nitride p-n Homojunction-Type Light-Emitting Diodes

Figures 23 and 24 show the electroluminescence (EL) spectrum and the band diagram of the GaN p-n junction LED under a forward biased condition. Two peaks are clearly observed. The shorter one originates from the near band-edge emission in nominally undoped n-type GaN; the longer

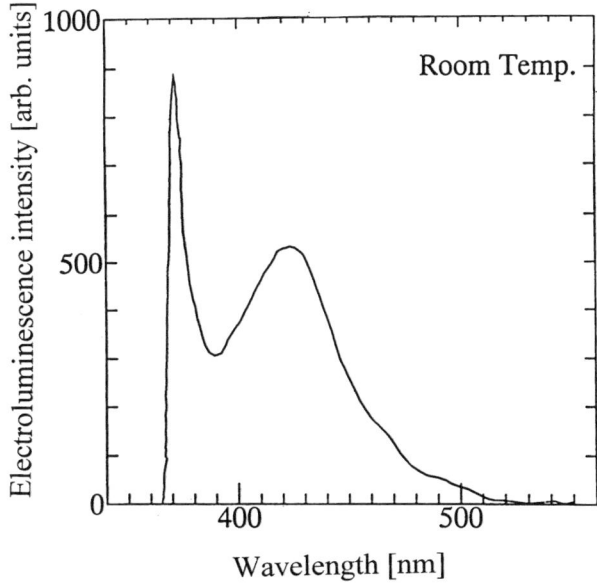

FIG. 23. Electroluminescence (EL) spectrum of gallium nitride p-n homojunction.

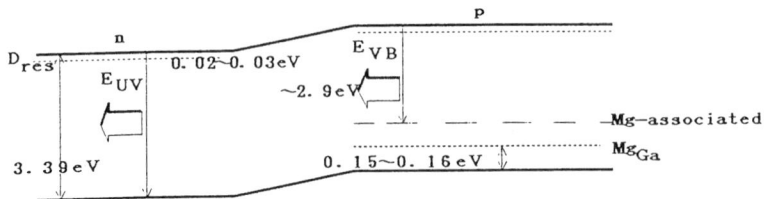

FIG. 24. Model for the electroluminescence from gallium nitride (GaN) p-n homojunction. Mg, magnesium.

one originates from the Mg-related emission center in p-type Mg-doped GaN. Emission from both n-type and p-type GaN can be clearly observed. This indicates that both electrons and holes are injected efficiently into the p-type and n-type layers, respectively.

b. AlGaN–GaInN DH–Type (Nominally Undoped Active Layer) Light-Emitting Diodes

In order to achieve carrier confinement for efficient recombination of electrons and holes, a heterostructure is essential. In order to achieve efficient electron confinement, a p-type cladding layer having a high hole concentration or a large conduction band discontinuity, or both, is necessary. Compared with the GaAs–AlAs system, the GaN–AlN and GaN–InN systems are more lattice-mismatched. Therefore, to reduce the lattice defects in the nitride DH structure, special attention should be paid to thickness and alloy composition. Figure 25 shows the band diagram of an $Al_{0.15}Ga_{0.85}N$–$Ga_{0.85}In_{0.15}N$ DH at a forward bias of 3.5 V. This was derived from the modified Harrison model (Gonda, 1986). The band discontinuities are estimated to be 0.4 eV for the conduction band and 0.24 eV for the valence band. In the EL spectrum of the DH AlGaN–GaN, near band-edge emission from the GaN active layer can be clearly observed (Akasaki and Amano, 1992b).

c. AlGaN–GaInN Asymmetric DH (Acceptor and Donor Co-Doped Active Layer) Light-Emitting Diodes

Figure 26 shows the schematic structure of an asymmetric DH LED with GaInN active layer. The cladding layers are AlGaN in order to reduce

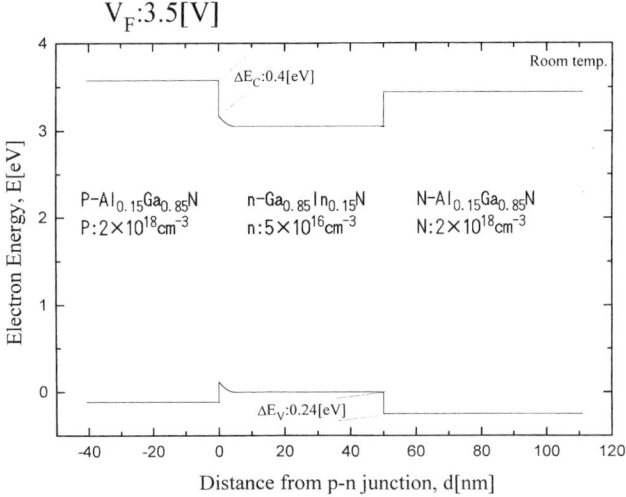

FIG. 25. Energy band diagram of $Al_{0.15}Ga_{0.85}N$–$Ga_{0.85}In_{0.15}N$ double heterostructure under forward bias.

electron leakage from the active layer to the p-type cladding layer. The active layer contains an acceptor, such as Zn, and a donor, such as Si, together (Figs. 14 and 15). As concerns the Zn-doped GaInN, the origin of the luminescence is basically the same as the violet luminescence center in Zn-doped GaN. Silicon was found to enhance this violet luminescence. Figure 27 shows a typical EL spectrum for this type of LED. The efficiency

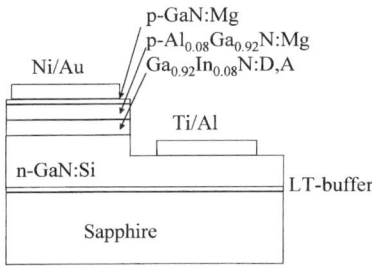

FIG. 26. Structure of asymmetric double heterostructure GaN-based blue light-emitting diode. Ni, nickel; Au, gold; Mg, magnesium; Al, aluminum; LT, low-temperature-deposited.

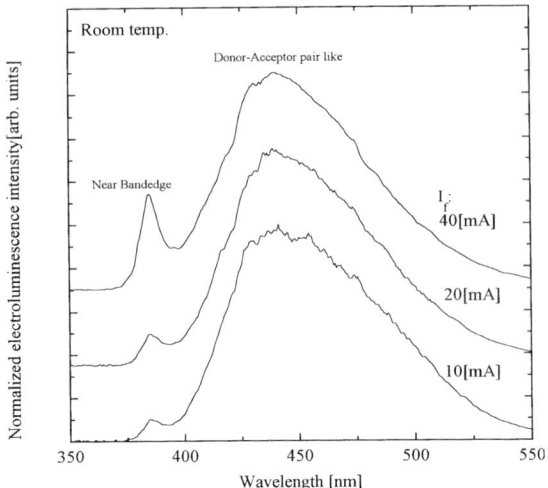

FIG. 27. Electroluminescence (EL) spectrum of asymmetric double heterostructure with GaInN active layer co-doped with donor, such as silicon, and acceptor, such as zinc. I_f, forward current.

and on-axis luminous intensity at a beam divergence of 16° are 3% and 2.5 cd, respectively. By changing the InN molar fraction in the GaInN from 0 to 0.23, very efficient violet, blue (Nakamura et al., 1994a; Koike et al., 1995), and green (Nakamura et al., 1994b) LEDs have been realized.

c. *Light-Emitting Diodes with Very Thin Active Layers*

In order to use band-to-band or near band-edge emission, a high-quality active layer is essential. However, if the lattice mismatch between the cladding layer and the active layer is large, the thickness of the active layer should be thinner than the critical layer thickness. Nakamura et al. (1995a,b), Amano and Akasaki (1995), and Akasaki et al. (1995a,b) fabricated quantum wells with active layer thicknesses as small as 2.0 nm (Nakamura et al., 1995a,b; Amano and Akasaki, 1995; Akasaki et al., 1995a). By controlling the InN molar fraction, very efficient LEDs that emit from the violet to the orange were realized in the nitrides (Nakamura et al., 1995b). The structure is basically the same as the previously mentioned DH, except for the very thin active layer. Figure 28 shows EL spectra. The InN molar fraction of the active layer was changed from 0.15 for the violet to 0.7 for the yellow LED.

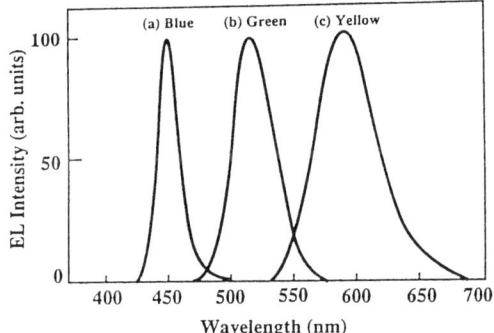

FIG. 28. Electroluminescence (EL) spectrum of a light-emitting diode having a very thin $Ga_{1-x}In_xN$ active layer $0.12 < x < 0.7$). (Reprinted from Nakamura et al., 1995b, Jpn. J. Appl. Phys. **34**, L797, with the permission of the publisher.)

2. ELECTRODE

Sapphire had been the most widely used substrate for epitaxial growth. This has required a special design for the electrical contacts to both the n-type and acceptor-doped layers. The nitride researchers in the 1970s made an n-type electrode on the side of the n-type layer. The electrode for the acceptor doped layer was formed on top of the film. Indium and Al were used as the electrodes for the n-type and acceptor-doped layers, respectively.

Ohki et al. (1981) used highly conducting polycrystalline n-type GaN grown by HVPE for the n-type electrode. Patterned silicon dioxide (SiO_2) was used to grow polycrystalline GaN. By using this technique, the electrodes for the n-type GaN can be formed more easily on the top surface. This has enabled fabrication of a flip-chip–type LED.

Koide et al. (1991) first used the reactive ion etching technique to etch the p-type layer, thus allowing formation of the n-type electrode on top of the etched surface. A chlorine-containing compound, such as boron chloride (BCl_3), is commonly used for the etchant gas.

Edmond et al. (1996) fabricated bright blue LEDs consisting of a Zn-doped GaInN active layer on a 6H-SiC substrate. The 6H-SiC is itself conducting, therefore, the n-type electrode can be formed on the rear surface of the 6H-SiC substrate. A shorting ring is used to reduce the effect of the nitride–SiC energy barrier.

Today, Al or Ti–Al is mostly used to form ohmic contact to n-type GaN (Foresi and Moustakas, 1993). Specific resistivities as low as $10^{-7}\,\Omega \cdot cm^2$ have been achieved. Titanium forms nitride compounds such as titanium nitride (TiN), which are metallic. This enables the formation of even more superior low-resistance contacts.

Compared with the n-type electrodes, optimization of the p-type electrode is still in progress. Metals having large work functions, such as gold (Au), nickel (Ni), and platinum (Pt), are commonly used to form these ohmic contacts. However, the specific resistivity is still higher than $10^{-3}\,\Omega\cdot cm^2$.

IV. Efficiency, Wavelength, and Lifetime

Figure 29 summarizes the change in the efficiency of nitride-based LEDs over the past two decades. As shown in Fig. 29, the efficiency began to increase greatly soon after the successes in producing high-quality crystals using the LT-buffer method, which resulted in p-n junction LEDs.

As concerns the emission wavelength, Pankove and Lamprey (1974) and Jacob *et al.* (1978) fabricated full-color GaN LEDs using various kinds of Zn-related luminescence centers. Nakamura *et al.* (1995b) fabricated LEDs

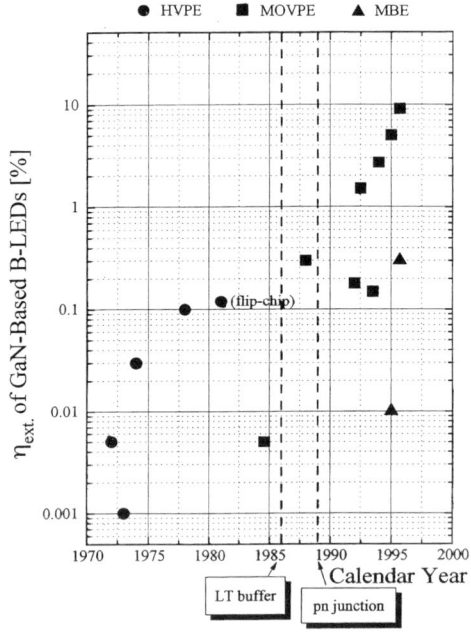

FIG. 29. Chronological table of the change of the external quantum efficiency ($\eta_{ext.}$) of GaN-based short-wavelength light-emitting diodes (LEDs). LT, low-temperature-deposited; HVPE, hydride vapor-phase epitaxy; MOVPE, metal organic vapor-phase epitaxy; MBE, molecular-beam epitaxy.

with emission wavelengths from 400 to 650 nm using GaInN active layers. The on-axis luminous intensity reaches 12 cd for green LEDs. The luminous efficiency is as high as 18 lm/W, which is close to half the maximum of the most efficient LED ever reported based on AlGaInP (Kish et al., 1994).

Ponce et al. (1995) found an extremely high density of dislocations, up to 10^{10} cm^{-2}, in the commercialized high-brightness blue LEDs. It is suggested that these dislocations are not electrically active in the nitrides. The rapid degradation of nitride-based blue LEDs observed early in development came from extrinsic origins. The intrinsic lifetime of nitride devices today is longer than 10,000 hours. Koike et al. (1996) reported that the dislocation density decreased to the order of 10^8 cm^{-2} in GaN–GaInN MQW structures grown on sapphire by OMVPE by using the LT-buffer layer.

V. Laser Diodes

The first stimulated emission and laser action by optical pumping was reported by Dingle et al. (1971) at a temperature of 2 K. Cingolani et al. (1986) reported high-temperature stimulated emission of up to 120 K and also reported an electron-hole plasma model as the mechanism of the stimulated emission in GaN. Amano et al. (1990b) reported the first stimulated emission by optical pumping at room temperature using high-quality GaN epitaxial layers grown with the LT-buffer layer. They also pointed out that, by the use of heterostructures, reduction of threshold was possible. This is due to the well-known effect of carrier and optical confinement by the heterostructure. Khan et al. (1991) reported the first surface mode stimulated emission by optical pumping. Kim et al. (1994) observed optical gain from AlGaN–GaN DH structures by optical pumping with low threshold at room temperature. In 1995, a threshold power for stimulated emission by optical pumping of 27 KW/cm^2 was reported, which corresponds to 7.4 KA/cm^2 for current injection (Akasaki and Amano, 1995b). Figure 30 shows the change in the threshold power (P_{th}) for stimulated emission by optical pumping from group III nitrides. The achievement of stimulated emission by optical pumping with low P_{th} and high-performance LEDs both strongly suggest the possibility of the realization of nitride-based LDs. In late 1995, Akasaki et al. (1995b) observed the onset of stimulated emission by current injection for the first time. Many nitride researchers believed that a nitride-based laser would be realized. In early 1996, Nakamura et al. (1996a, b, c) fabricated the first nitride-based LD. The structure is basically a separate confinement heterostructure. The

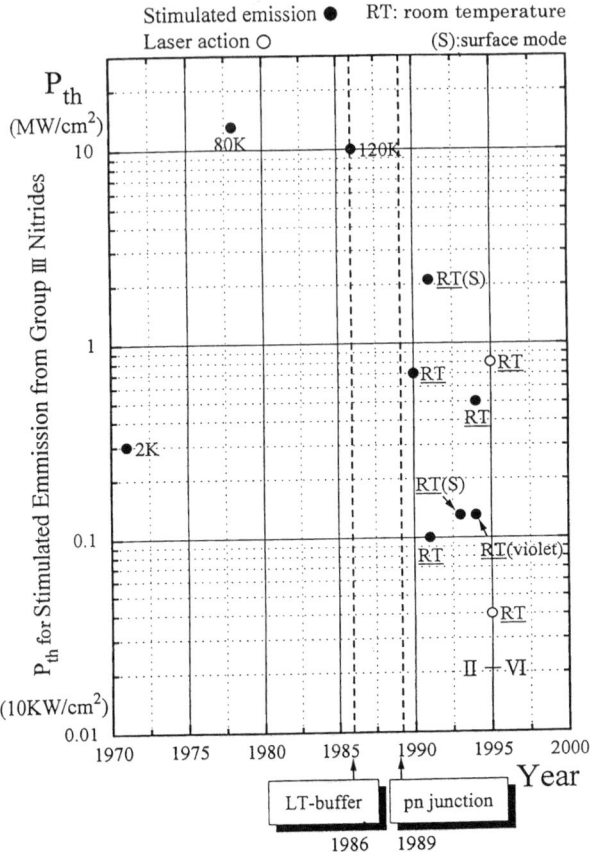

FIG. 30. Chronological table of the change of the threshold power (P_{th}) for stimulated emission from nitrides by optical pumping. LT, low-temperature-deposited.

active layer is an MQW having a number of well layers. The waveguide and cladding layers were GaN and AlGaN, respectively. Figure 31 shows the emission spectrum for operation under pulsed current injection. The duty cycle is 0.1%. The threshold current density is 4.0 KA/cm². The operating voltage at the threshold current is as high as 34 V, which is mainly caused by the high resistivity of the p-type electrode.

Akasaki et al. (1996) used a GaInN SQW structure as the active layer and succeeded in fabricating the shortest-wavelength semiconductor LD to date with a wavelength of 376.0 nm at room temperature. The threshold current density is 3.0 KA/cm², and the operating voltage at the threshold current density is 16 V.

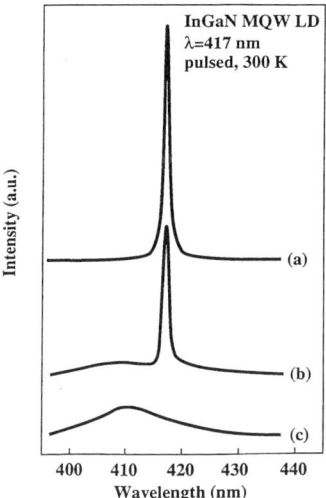

FIG. 31. Emission spectra from a GaN-based laser diode (LD) having a large number of GaInN wells. MQW, multiquantum well. (Reprinted from Nakamura *et al.*, 1996a, *Jpn. J. Appl. Phys.* **35**, L74, with the permission of the publisher.)

VI. Summary

This chapter gives a historical review of the fundamental research into GaN and related nitrides. As shown, an accumulation of a large quantity of outstanding work, in particular in the area of OMVPE growth, has led to the commercialization of bright blue LEDs and the realization of LDs based on the nitrides. The performance of these nitride-based light emitters is still progressing at present. However, there are still a variety of problems to be solved in the areas of materials science and device fabrication. Further progress in these areas will surely open new areas in optoelectronics and continue to impact the compound semiconductor world.

Acknowledgments

The authors are grateful to G. B. Stringfellow for his critical reading of our manuscript. We also are greatly indebted to the following collaborators who have made major contributions to this work: N. Sawaki, K. Hiramatsu, Y. Koide, T. Tanaka, B. Monemar, I. Suemune, M. Hashimoto, N. Koide, M. Koike, and Y. Toyoda.

References

Akasaki, I., and Amano, H. (1991). *Mater. Res. Soc. Fall Meet.* Abstr. 241.
Akasaki, I., and Amano, H. (1992a). *Mater. Res. Soc. Symp. Proc.* **242**, 383.
Akasaki, I. and Amano, H. (1992b). *Ext. Abstr., Int. Conf. Solid State Devices Mater., 1992,* Tsukuba, S-I-1.
Akasaki, I., and Amano, H. (1993). *Inst. Phys. Conf. Ser.* **136**, 249.
Akasaki, I., and Amano, H. (1994). *Mater. Res. Soc. Symp. Proc.* **339**, 443.
Akasaki, I., and Amano, H. (1995a). *J. Cryst. Growth* **146**, 455-461.
Akasaki, I., and Amano, H. (1995b). *Inst. Phys. Conf. Ser.* **145**, 19.
Akasaki, I., and Hayashi, I. (1976). *Ind. Sci. Technol.* **17**, 48 (in Japanese).
Akasaki, I., Amano, H., Koide, Y., Hiramatsu, K., and Sawaki, N. (1989a). *J. Cryst. Growth* **98**, 209.
Akasaki, I., Amano, H., Kitoh, M., and Hiramatsu, K., (1989b). *Electrochem. Soc. Spring Meet.*, Los Angeles.
Akasaki, I., Amano, H., Itoh, K., Koide, N., and Manabe, K., (1992a). *Symp. GaAs Relat. Comd., 16th. Conf. Ser.—Inst. Phys.* **129**, 851.
Akasaki, I., Amano, H., Itoh, K., Koide, N., and Manabe, K. (1992b).
Akasaki, I., Amano, H., and Suemune, I. (1995a). *Proc. Int. Conf. Silicon Carbide Relat. Mater.*, Kyoto, p. 7.
Akasaki, I., Amano, H., Sota, S., Sakai, H., Tanaka, T., and Koike, M. (1995b). *Jpn. J. Appl. Phys.* **34**, L1517.
Akasaki, I., Sota, S., Sakai, H., Tanaka, T., Koike, M., and Amano, H. (1996). *Electron Lett.* (to be published).
Amano, H., and Akasaki, I. (1990). *Mater. Res. Soc. Ext. Abstr.* **EA-21**, 165.
Amano, H., and Akasaki, I. (1995). *Ext. Abstr. Int. Conf. Solid State Devices Mater.*, Osaka, V-7, p. 683.
Amano, H., Sawaki, N., Akasaki, I., and Toyoda, Y. (1986). *Appl. Phys. Lett.* **48**, 353–355.
Amano, H., Akasaki, I., Hiramatsu, K., Koide, N., and Sawaki, N. (1988). *Thin Solid Films* **163**, 415.
Amano, H., Kito, M., Hiramatsu, K., Sawaki, N., and Akasaki, I. (1989). *Jpn. J. Appl. Phys.* **28**, L2112.
Amano, H., Asahi, T., and Akasaki, I. (1990a). *Jpn. J. Appl. Phys.* **29**, L205.
Amano, H., Kitoh, M., Hiramatsu, K., and Akasaki, I. (1990b). *J. Electrochem Soc.* **137**, 1639–1641.
Amano, H., Watanabe, N., Koide, N., and Akasaki, I. (1993). *Jpn. J. Appl. Phys.* **32**, L1000–L1002.
Baranov, B. V., Gutan, V. B., and Zhumakulev, U. (1982). *Sov. Phys.—Semicond.* (*Engl. Transl.*) **16**, 819.
Cingolani, R., Ferrara, M., and Lugara, M. (1986). *Solid State Commun.* **60**, 705.
Dingle, R., Shaklee, K. L., Leheny, R. F., and Zetterström, R. B. (1971). *Appl. Phys. Lett.* **19**, 5.
Edmond, J., Kong, H. S., Leonard, M., Dmtriev, V., Irvine, K., and Bulman, G. (1996). *Top. Workshop III-V Nitrides*, Nagoya, Abstr. A-3.
Ejder, E, (1975). *Phys. Status. Solidi.* **A6**, 445–448.
Foresi, J. S., and Moustakas, T. D. (1993). *Appl. Phys. Lett.* **62**, 2859–2861.
Fujita, Sz., Johnson, M. A. L., Rowland W. H., Jr., Hughes, W. C., He, Y. W., El Masry, N. A., Cook, J. W., Jr., and Schetzina, J. F., (1995). *Ext. Abstr. Int. Conf, Solid State Devices and Mater.*, Osaka, S-V-10, pp. 692–694.
Gonda, S. (1986). *Solid State Commun.* **60**, 249.

Hiramatsu, K., Itoh, S., Amano, H., Akasaki, I., Kuwano, N., Shiraishi, T. and Oki, K. (1991). *J. Cryst. Growth* **115**, 628.
Hirosawa, K., Hiramatsu, K., Sawaki, N., and Akasaki, I. (1993). *Jpn. J. Appl. Phys.* **32**, L1039.
Ilegems, M., and Montgomery, H. C. (1973). *J. Phys. Chem. Solids* **34**, 885.
Itoh, K., Kawamoto, T., Amano, H., Hiramatsu, K., and Akasaki, I. (1991). *Jpn. J. Appl. Phys.*, **30**, 1924.
Itoh, N., and Rhee, J. C. (1985). *J. Appl. Phys.* **58**, 1828.
Jacob, G., Boulou, M., and Bois, D. (1978). *J. Lumin.* **17**, 263.
Johnson, W. C., Parson, J. C., and Crew, M. C. (1932). *J. Phys. Chem.* **36**, 2561.
Karpinski, J., Jun, J., and Porowski, S. (1984). *J. Cryst. Growth* **66**, 1.
Kawabata, T., Matsuda, T., and Koike, S. (1984). *J. Appl. Phys.* **56**, 2367.
Khan, M. A., Skogman, R. A., Van Hove, J. M., Krishnankutty, S., and Kolbas, R. M. (1990). *Appl. Phys. Lett.* **56**, 1257.
Khan, M. A., Olson, D. T., Van Hove, J. M. and Kuznia, J. N. (1991). *Appl. Phys. Lett.* **58**, 1515.
Khan, M. R. H., Ohshita, Y., Sawaki, N., and Akasaki, I. (1986). *Solid State Commun.* **57**, 405.
Kim, S. T., Amano, H., Akasaki, I., and Koide, N. (1994). *Appl. Phys. Lett.* **64**, 1535–1536.
Kish, F. A., Steranka, F. M., DeFevere, D. C., Vanderwater, D. A., Park, K. G., Cuo, C. P., Osentowski, T. D., Peanasky, M. J., Yu, J. G., Flecher, R. M., Steigerwald, D. A., Craford, M. G., and Robbins, R. M. (1994). *Appl. Phys. Lett.* **64**, 2839.
Koide, N., Kato, H., Sassa, M., Yamasaki, S., Manabe, K., Hashimoto, M., Amano, H., Hiramatsu, K., and Akasaki, I., (1991). *J. Cryst. Growth* **115**, 639.
Koide, Y., Itoh, H., Khan, M. R. H., Hiramatsu, K., Sawaki, N., and Akasaki, I. (1987). *J. Appl. Phys.* **61**, 4540.
Koide, Y., Itoh, N., Itoh, K., Sawaki, N., and Akasaki, I. (1988). *Jpn. J. Appl. Phys.* **27**, 1156–1161.
Koike, M., Shibata, N., Kato, H., Yamasaki, S., Koide, N., Amano, H., and Akasaki, I., (1995). *Top. Workshop III-V Nitrides*, Nagoya, Abstr. C-2.
Koike, M., Yamasaki, S., Nagai, S., Koide, N., Asami, S., Amano, H., and Akasaki, I. (1996). *Appl. Phys. Lett.* **68**, 1403.
Lorenz, M. R. and Binkowski, B. B. (1962). *J. Electrochem. Soc.* **109**, 24.
Manasevit, H. M., Erdman, F. M., and Simpson, W. I., (1971). *J. Electrochem. Soc.* **118**, 1864.
Maruska, H. P., and Tietjen, J. J. (1969). *Appl. Phys. Lett.* **15**, 327.
Maruska, H. P., Rhines, W. C., and Stevenson, D. A. (1972). *Mater. Res. Bull.* **7**, 777.
Matthews, J. W., and Blakeslee, A. E. (1974). *J. Cryst. Growth* **27**, 118.
Molnar, R. J., Singh, R., and Moustakas, T. D. (1995). *Appl. Phys. Lett.* **66**, 268–270.
Monemar, B., Lagerstedt, O., and Gislason, H. P. (1980). *J. Appl. Phys.* **51**, 625.
Moustakas, T. D. (1993). *Electrochem. Soc. Meet.*, Honolulu, Ext. Abstr. 93-1, p. 958.
Murakami, H., Asahi, T., Amano, H., Hirtamatsu, K., Sawaki, N., and Akasaki, I. (1991). *J. Cryst. Growth* **115**, 648.
Nagatomo, T., Kuboyama, T., Minamino, H., and Omoto O. (1989). *Jpn. J. Appl. Phys.* **28**, L1334.
Nakamura, S. (1991). *Jpn. J. Appl. Phys.* **30**, L1705.
Nakamura, S., Senoh, M., and Mukai, T. (1991). *Jpn. J. Appl. Phys.* **30**, L1708.
Nakamura, S., Iwasa, N., Senoh, M., and Mukai, T. (1992). *Jpn. J. Appl. Phys.* **31**, 1258.
Nakamura, S., Mukai, T., Senoh, M., Nagahama, S., and Iwasa, N. (1993a). *J. Appl. Phys.* **74**, 3911.
Nakamura, S., Senoh, M., and Mukai, T. (1993b). *Appl. Phys. Lett.* **62**, 2390.
Nakamura, S., Mukai, T., and Senoh, M. (1994a). *Appl. Phys. Lett.* **64**, 1687.
Nakamura, S., Mukai, T., and Senoh, M. (1994b). *J. Appl. Phys.* **76**, 8189.
Nakamura, S., Senoh, M., Iwasa, N., and Nagahama, S. (1995a). *Appl. Phys. Lett.* **67**, 1868.
Nakamura, S., Senoh, M., Iwasa, N., and Nagahama, S. (1995b). *Jpn. J. Appl. Phys.* **34**, L797.

Nakamura, S., Senoh, M., Nagahama, S., Iwasa, N., Yamada, T., Matsushita, T., Kiyoku, H., and Sugimoto Y. (1996a). *Jpn. J. Appl. Phys.* **35**, L74–L76.
Nakamura, S., Senoh, M., Nagahama, S., Iwasa, N., Yamada, T., Matsushita, T., Kiyoku, H., and Sugimoto, Y. (1996b). *Jpn. J. Appl. Phys.* **35**, L217–L219.
Nakamura, S., Senoh, M., Nagahama, S., Iwasa, N., Yamada, T., Matsushita, T., Kiyoku, H., and Sugimoto, Y. (1996c). *Appl. Phys. Lett.* **68**, 2105–2107.
Ohki, Y., Toyoda, Y., Kobayashi, H., and Akasaki, I. (1981). *Inst. Phys. Conf. Ser.* **63**, 479.
Osamura, K., Nakajima, N., and Murakami Y. (1972). *Solid State Commun.* **11**, 617.
Pankove, J. I. (1973). *RCA Rev.* **34**, 336.
Pankove, J. I., and Lamprey M.A. (1974). *Phys. Rev. Lett.* **33**, 361.
Pankove, J. I., Miller, E. A., Richman, D., and Berkeyheiser, J. E. (1971). *J. Lumin.* **4**, 63.
Ponce, F. A., Krusor, B. S., Major, J. S., Jr., Plano, W. E., and Welch, D. F. (1995). *Appl. Phys. Lett.*, **67**, 410.
Rowland, L. B., Doverspike, K., and Gaskill D. K. (1995). *Appl. Phys. Lett.* **66**, 1495.
Sakai, H., Koide, T., Suzuki, H., Yamaguchi, M., Yamasaki, S., Koike, M., Amano, H., and Akasaki I. (1995). *Jpn. J. Appl. Phys.* **34**, L1429–L1431.
Sayyah, K., Chung, B. C., and Gershenzon M. (1986). *J. Cryst. Growth* **77**, 424.
Sitar, Z., Paisley, M. J., Yan, B., Ruan, J., Choyke, W. J., and Davis, R. F. (1990). *J. Vac. Sci. Technol.*, *B* [2] **8**, 316–322.
Tanaka, T., Watanabe, A., Amano, H., Kobayashi, Y., Akasaki, I., Yamazaki, S., and Koike, M. (1994). *Appl. Phys. Lett.* **65**, 593.
Tansley, T. L., and Foley, C. P. (1986). *J. Appl. Phys.* **59**, 3241.
Van Vechten, J. A., Zook, J. D., Hornig, R. D., and Goldenberg, B. (1992). *Jpn. J. Appl. Phys.* **31**, 3662.
Wauk, M. T., and Winslow, D. K. (1968). *Appl. Phys. Lett.* **13**, 286.
Wickenden, D. K., Bargeron, C. B., Bryden, W. A., Miragliova, J., and Kistenmacher, T. J. (1994). *Appl. Phys. Lett.* **65**, 2024.
Yamasaki, S., Asami, S., Shibata, N., Koike, M., Tanaka, T., Amano, H., and Akasaki, I. (1995). *Appl. Phys. Lett.* **66**, 1112.
Yim, W. M., Stofko, E. J., Zanzucchi, P. J., Pankove, J. I., Ettenberg, M., and Gilbert S. L. (1973). *J. Appl. Phys.* **44**, 292.
Yoshida, S., Misawa, S., and Itoh, A. (1975). *Appl. Phys. Lett.* **26**, 461.
Yoshida, S., Misawa, S., and Gonda, S. (1983). *Appl. Phys. Lett.* **42**, 427.
Yoshimoto, N., Matsuoka, T., Sasaki, T., and Katsui, A. (1991). *Appl. Phys. Lett.* **59**, 2251.

CHAPTER 8

Group III-V Nitride-Based Ultraviolet Blue-Green-Yellow Light-Emitting Diodes and Laser Diodes

Shuji Nakamura

RESEARCH AND DEVELOPMENT DEPARTMENT
NICHIA CHEMICAL INDUSTRIES, LTD.
TOKUSHIMA, JAPAN

I. INTRODUCTION	391
II. GALLIUM NITRIDE GROWTH	394
1. *Undoped Gallium Nitride*	394
2. *n-Type Gallium Nitride*	398
3. *p-Type Gallium Nitride*	399
4. *Gallium Nitride p-n Junction Light-Emitting Diodes*	405
III. InGaN GROWTH	409
1. *Undoped InGaN*	409
2. *Impurity Doping of InGaN*	413
IV. LIGHT-EMITTING DIODES	416
1. *InGaN/Gallium Nitride Double Heterostructure Light-Emitting Diodes*	416
2. *InGaN/AlGaN Double Heterostructure Light-Emitting Diodes*	419
3. *InGaN Quantum Well Structures*	425
4. *InGaN Light-Emitting Diodes with Quantum Well Structures*	426
5. *High-Brightness Green and Blue Light-Emitting Diodes*	430
V. InGaN MULTIPLE QUANTUM WELL STRUCTURE LASER DIODES	436
VI. SUMMARY	440
References	441

I. Introduction

Much research has been done on high-brightness blue light-emitting diodes (LEDs) and laser diodes (LDs) for use in full-color displays, full-color indicators, and light sources for lamps with the characteristics of high efficiency, high reliability, and high speed. For these purposes, II-VI materials such as zinc selinide (ZnSe) (Xie *et al.*, 1992), silicon carbide (SiC)

(Edmond *et al.*, 1994), and III-V nitride semiconductors such as gallium nitride (GaN) (Pankove *et al.*, 1971) have been investigated intensively for a long time. However, it was impossible to obtain high-brightness blue LEDs with a brightness over 1 cd and reliable LDs. Figure 1 shows the bandgap energy of various materials as a function of the lattice constant. Conventionally, AlGaAs material has been used for high-brightness red LEDs and infrared (IR) LDs. The lattice constant of AlGaAs almost equals that of gallium arsenide (GaAs) for all compositions. Therefore, GaAs substrates are suitable for AlGaAs growth. AlInGaP has been used for red, yellow, and green LEDs and red LDs. This system is also lattice-matched with GaAs substrates when the composition of AlInGaP is adjusted properly.

For II-VI materials, the lattice constant of ZnMgSSe equals that of GaAs when the composition is adjusted properly (Okuyama and Ishibashi, 1994). Also, the bandgap energies of II-VI materials are large enough to emit luminescence from the green to the blue spectral regions. Therefore, ZnMgSSe-, ZnSSe- and ZnCdSe-based materials have been intensively studied for blue and green light emission devices because high-brightness blue LEDs and LDs have not been achieved using other materials, such as AlGaAs and AlInGaP. Much progress has been achieved recently on green LEDs and LDs using II-VI-based materials. The recent performance of

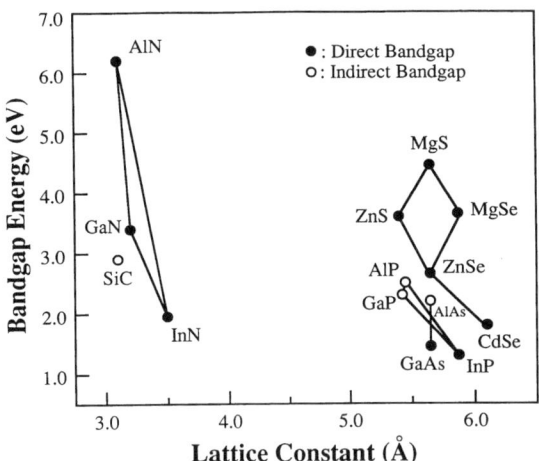

FIG. 1. Bandgap energy of various materials for visible emission devices as a function of their lattice constant. AlN, aluminum nitride; GaN, gallium nitride; InN, indium nitride; SiC, silicon carbide; MgS, magnesium sulfide; ZnS, zinc sulfide; MgSe, magnesium selenide; AlP, aluminum phosphide; ZnSe, zinc selenide; GaP, gallium phosphide; AlAs, aluminum arsenide; CdSe, cadmium selenide; GaAs, gallium arsenide; InP, indium phosphide. Reprinted from Nakamura (1995b) with the permission of the IEEE.

green LEDs is that the output power is 1.3 mW at 10 mA and that the peak wavelength is 512 nm (Eason et al., 1995). When the peak wavelength becomes shorter, into the blue spectral region, the output power decreases dramatically to about 0.3 mW at 489 nm. Also, the lifetime of these devices is about 1 hour for green LDs and 100 hours for green LEDs. These short lifetimes prevent II-VI-based devices from commercialization at present. It is considered that the short lifetime of these II-VI-based devices is caused by the crystal defects at a density of $10^5/cm^2$ because one crystal defect would cause the propagation of other defects, leading to failure of the devices. Another wide bandgap material for blue LEDs is SiC. The brightness of SiC blue LEDs is only between 10 and 20 mcd because of the indirect bandgap of this material. Despite this poor performance, 6H-SiC blue LEDs have been commercialized for a long time because there has been no competition for blue light-emitting devices (Koga and Yamaguchi, 1991).

For green devices, the external quantum efficiency of conventional green gallium phosphide (GaP) LEDs is only 0.1% owing to the indirect bandgap of this material, and the peak wavelength is 555 nm (yellowish green) (Craford, 1992). As another material for green emission devices, AlInGaP has been used. The present performance of green AlInGaP LEDs is an emission wavelength of 570 nm (yellowish green) and a maximum external quantum efficiency of 1% (Craford, 1992; Sugawara et al., 1994). When the emission wavelength is reduced to the green region, the external quantum efficiency drops sharply because the band structure of AlInGaP becomes nearly indirect. Therefore, high-brightness pure green LEDs that have a high efficiency of above 1% at the peak wavelength of between 510 and 530 nm with a narrow full-width at half-maximum (FWHM) have not been commercialized yet. Among II-VI materials, ZnSSe- and ZnCdSe-based materials have been intensively studied for use in green light-emitting devices, and much progress has been made recently. The recent performance of II-VI green LEDs is an output power of 1.3 mW, external quantum efficiency of 5.3% at 10 mA, and peak wavelength of 512 nm (Eason et al., 1995). However, the lifetime of II-VI-based devices is still short, which prevents commercialization of II-VI-based devices at present. Therefore, until 1994, there had been no high-brightness highly efficient high-power blue and green LEDs with a luminous intensity above 10 cd and an output power above 1 mW at 20 mA. In 1994, the first high-brightness InGaN/AlGaN double heterostructure (DH) blue LEDs with a luminous intensity of 1 cd were commercialized (Nakamura, 1994a,b,c; Nakamura et al., 1994a,b).

Gallium nitride and related materials such as AlGaInN are III-V nitride semiconductors with the wurtzite crystal structure and a direct-energy band structure suitable for light-emitting devices. The bandgap energy of AlGaInN varies between 6.2 and 1.95 eV, depending on its composition at room temperature (Fig. 1). Therefore, these III-V nitride semiconductors are useful for light-emitting devices, especially in the short-wavelength regions. With a bandgap energy of 3.4 eV at room temperature, GaN has been the most intensively studied. Recent research on III-V nitrides has paved the way for the realization of high-quality crystals of GaN, AlGaN, and GaInN, and of p-type conduction in GaN and AlGaN (Strite and Morkoç, 1992; Morkoç et al., 1994). The mechanism of acceptor-compensation, which prevents obtaining low-resistivity p-type GaN and AlGaN, has been elucidated (Nakamura et al., 1992a; Rubin et al., 1994; Brandt et al., 1994; Zavada et al., 1994). In magnesium-doped p-type GaN, Mg acceptors are deactivated by atomic hydrogen produced from ammonia (NH_3) gas used as the nitrogen source during GaN growth. High-brightness blue LEDs have been fabricated on the basis of these results, and luminous intensities over 1 cd have been achieved (Nakamura, 1994a; Nakamura et al., 1994a). These LEDs are now commercially available. Also, high-brightness single quantum well (SQW) structure blue, green, and yellow InGaN LEDs with a luminous intensity above 10 cd for green LEDs have been achieved and commercialized (Nakamura et al. 1995a, b, c). By combining these high-power high-brightness blue InGaN SQW LEDs, green InGaN SQW LEDs, and red GaAlAs LEDs, many kinds of applications, such as LED full-color displays and LED white lamps for use in place of light bulbs or fluorescent lamps are now possible with the characteristics of high reliability, high durability, and low energy consumption. Also, very recently, III-V nitride-based LDs were fabricated for the first time. These LDs emitted coherent light at 420 nm from an InGaN-based multiple quantum well (MQW) structure under pulsed current injection at room temperature (Nakamura et al., 1996). Here, recent studies on (Al, Ga, In)N compound semiconductors are described.

II. Gallium Nitride Growth

1. UNDOPED GALLIUM NITRIDE

Usually, GaN films are grown on a sapphire substrate with (0001) orientation (c face) at temperatures around 1000°C by the metalorganic chemical vapor deposition (MOCVD) method as one of the growth methods. The lattice constants along the a-axis of the sapphire and GaN are 4.758 and 3.189 Å, respectively. Therefore, the lattice-mismatch between the sapphire and the GaN is very large. The lattice constant along the a-axis of 6H-SiC is 3.08 Å, which is relatively close to that of GaN (Fig. 1). However, the price of an SiC substrate is extraordinarily expensive to use for the practical growth of GaN. Therefore, at present, there are no alternative substrates except for sapphire, considering the price of substrates and the high growth temperature, although the lattice-mismatch is large. The grown GaN layers usually show n-type conduction without intentional doping. The donors are probably native defects or residual impurities such as nitrogen vacancies or residual oxygen.

The surface morphology of the GaN films was markedly improved when an AlN buffer layer was initially deposited on the sapphire (Yoshida *et al.*, 1983). More recently, high-quality GaN films have been obtained using this AlN buffer layer by means of the MOCVD method (Amano *et al.*, 1986; Akasaki *et al.*, 1989), showing that the uniformity, crystalline quality, luminescence, and electrical properties of the GaN films were markedly improved. The carrier concentration and Hall mobility, whose values were $2-5 \times 10^{17}/cm^3$ and 350 to 430 cm^2/V·sec at room temperature, were obtained by the prior deposition of a thin AlN layer as a buffer layer before the growth of the GaN film (Amano and Akasaki, 1991). Those values became about $5 \times 10^{16}/cm^3$ and 500 cm^2/V·sec at 77 K.

In addition, high-quality GaN films were obtained using GaN buffer layers instead of AlN buffer layers on a sapphire substrate (Nakamura, 1991). This involved development of a novel two-flow MOCVD reactor for the GaN growth (Fig. 2) that has two different gas flows. One is the main flow, which carries the reactant gas parallel to the substrate at high velocity

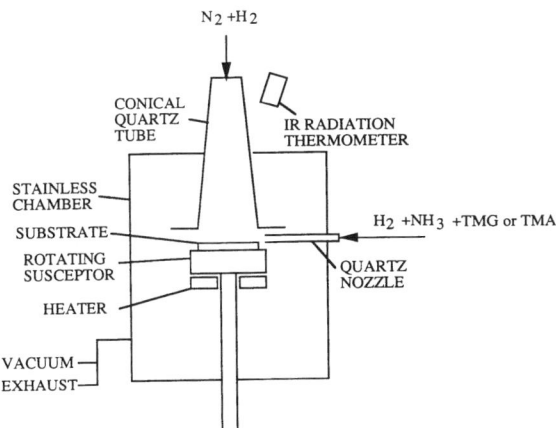

FIG. 2. Schematic diagram of novel two-flow metalorganic chemical vapor-phase deposition reactor for GaN growth. IR, infrared; TMGa, trimethylgallium; TMAl, trimethylaluminum. Reprinted from Nakamura *et al.* (1991c) with the permission of the American Institute of Physics.

through the quartz nozzle. The other flow is the subflow, which transports the inactive gas perpendicular to the substrate for the purpose of changing the direction of the main flow to bring the reactant gas into contact with the substrate (Fig. 3). Sapphire with (0001) orientation (c face) was used as the substrate. Trimethylgallium (TMGa) and NH_3 were used as Ga and N sources, respectively. First, the substrate was heated to 1050°C in a stream of hydrogen. Then, the substrate temperature was lowered to about 550°C to grow the GaN buffer layer. Next, the substrate temperature was elevated to about 1000°C to grow the GaN film. The total thickness of the GaN film was about 4 μm. Hall measurement was performed on GaN films grown with a GaN buffer layer as a function of the thickness of the GaN buffer layer. For the GaN film grown with a 200 Å GaN buffer layer, the carrier concentration and Hall mobility were $4 \times 10^{16}/cm^3$ and 600 $cm^2/V \cdot sec$, respectively, at room temperature. The values became $8 \times 10^{15}/cm^3$ and 1500 $cm^2/V \cdot sec$ at 77 K, respectively (Fig. 4).

The carrier concentration and Hall mobility of GaN films grown with GaN buffer layers are shown as a function of the temperature in Fig. 5 (Nakamura et al., 1992b). The GaN films were grown with an approximately 200-Å thick GaN buffer layer. The total thickness is about 4 μm. The Hall mobility is 700 $cm^2/V \cdot sec$ at room temperature. The crystal quality of this GaN film was characterized by the double-crystal X-ray rocking curve (XRC) method. The FWHM for (0002) diffraction from this GaN film was 4 minutes. Therefore, the value of the FWHM of XRC is not directly related to the Hall mobility because the Hall mobility becomes smaller than 600 cm^2/V and the value of FWHM of XRC becomes smaller than 4 minutes as the buffer layer thickness is decreased below 200 Å. The Hall mobility gradually increases as the temperature decreases from room

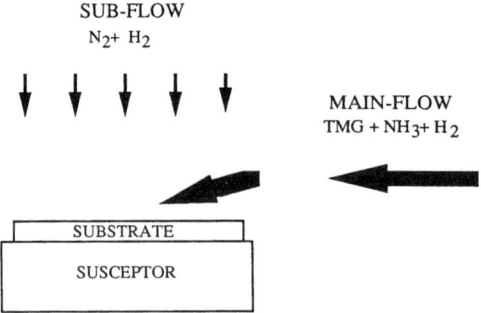

FIG. 3. Schematic of two-flow metalorganic chemical vapor-phase deposition. TMGa, trimethylgallium; NH_3, Ammonia; H_2, hydrogen; N_2, nitrogen. Reprinted from Nakamura et al. (1991c) with the permission of the American Institute of Physics.

8 GROUP III-V NITRIDE-BASED ULTRAVIOLET LEDs AND LASER DIODES 397

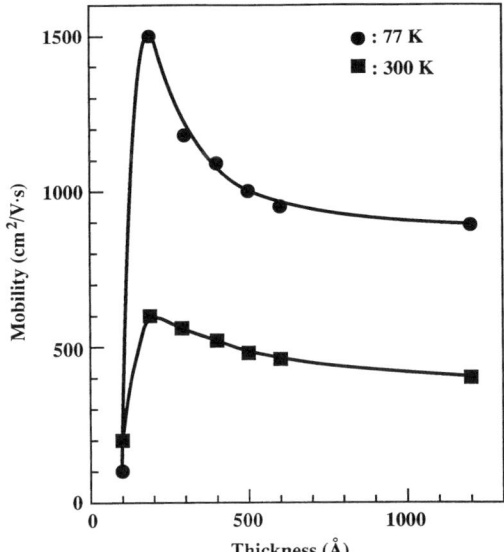

FIG. 4. The Hall mobility measured at 77 and 300 K as a function of the thickness of the GaN buffer layer. Reprinted from Nakamura (1991) with the permission of the Japanese Journal of Applied Physics.

temperature (Fig. 5). The Hall mobility is about 3000 cm²/V·sec at 70 K. According to other previous studies (Amano et al., 1986; Akasaki et al., 1989), the maximum Hall mobility (about 900 cm²/V·sec) was obtained at around 150 K using aluminum nitride (AlN) buffer layers. On the other hand, GaN films grown with GaN buffer layers had a maximum value at around 70 K. Therefore, the contribution of ionized impurity scattering in GaN films grown with GaN buffer layers is much smaller than that in GaN films grown with AlN buffer layers. The Hall mobility varies roughly following $\mu = \mu_0 T^{-1}$ between 70 and 300 K, where μ is the Hall mobility, and μ_0 is a constant practically independent of absolute temperature T. Thus, in this temperature range the Hall mobility is mainly determined by the contribution of polar phonon scattering. Below 70 K, ionized impurity scattering dominates and the Hall mobility decreases. The carrier concentration decreases drastically below 100 K and varies slightly between 100 and 300 K. Therefore, it seems that a different donor level contributes to the generation of the carriers corresponding to the two different temperature ranges. To consider these two different donor levels, the carrier concentration as a function of the reciprocal of the temperature was plotted in Fig. 6. The thermal activation energy of the electron from the donor level to the conduction band can be obtained from the gradient of the linear regions in

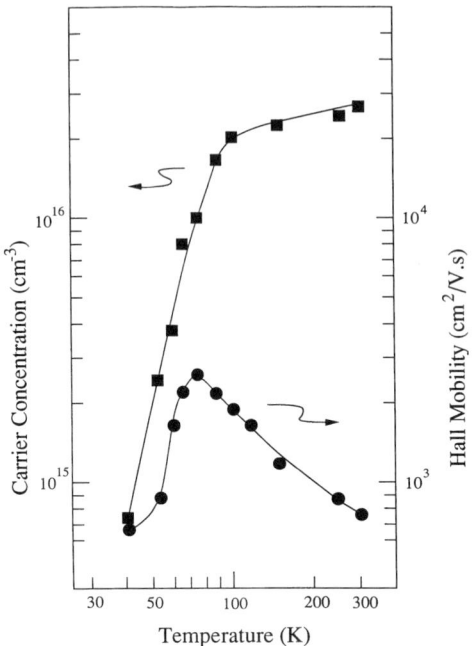

FIG. 5. Carrier concentration and Hall mobility of gallium nitride (GaN) films grown with GaN buffer layers as a function of the temperature. Reprinted from Nakamura (1991) with the permission of the Japanese Journal of Applied Physics.

Fig. 6, assuming that the carrier concentration varies according to the following formula:

$$N = N_0 \exp(-E/2kT)$$

where N is the carrier concentration, N_0 is a constant practically independent of temperature, E is the thermal activation energy, and k is Boltzmann's constant. Thermal activation energies of 34 meV between 100 and 42 K and of 5 meV between 300 and 100 K were obtained.

2. n-TYPE GALLIUM NITRIDE

Figure 7 shows the carrier concentration of Si-doped GaN films as a function of the flow rate of silane (SiH_4). The carrier concentration varies between $1 \times 10^{17}/cm^3$ and $2 \times 10^{19}/cm^3$ (Nakamura et al., 1992c). Good linearity is observed between the carrier concentration and the flow rate of

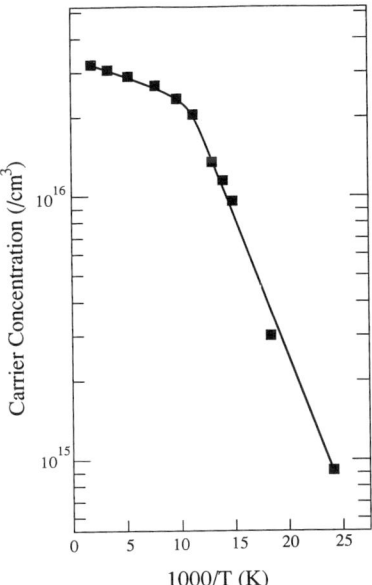

FIG. 6. Carrier concentration as a function of the reciprocal of the temperature. Reprinted from Nakamura *et al.* (1992b) with the permission of the American Institute of Physics.

SiH_4. Therefore, Si is considered to be a good donor impurity for GaN in order to control the carrier concentration. Also, germanium-doped GaN films showed the same carrier dependence on the flow rate of GeH_4.

Photoluminescence (PL) measurements were performed at room temperature. The excitation source was a 10-mW helium–cadmium (He–Cd) laser. Figure 8 shows the PL spectra of Si-doped GaN films, whose carrier concentrations are $4 \times 10^{18}/cm^3$ and $2 \times 10^{19}/cm^3$, respectively. A relatively strong deep-level (DL) emission is observed at around 560 nm. The ultraviolet (uv) emission, which is a band-edge emission of GaN, is observed around 380 nm. The intensity of DL emissions is always stronger than that of band-edge emissions in this range of the SiH_4 flow rate. Ge-doped GaN films showed the same strong DL emissions. Undoped GaN also shows the strong DL emissions and weak band-edge emissions at room temperature. The origin of these strong DL emissions has not been determined yet.

3. p-TYPE GALLIUM NITRIDE

It was impossible to obtain p-type GaN films for a long time. The unavailability of p-type GaN films has prevented the use of III-V nitrides in

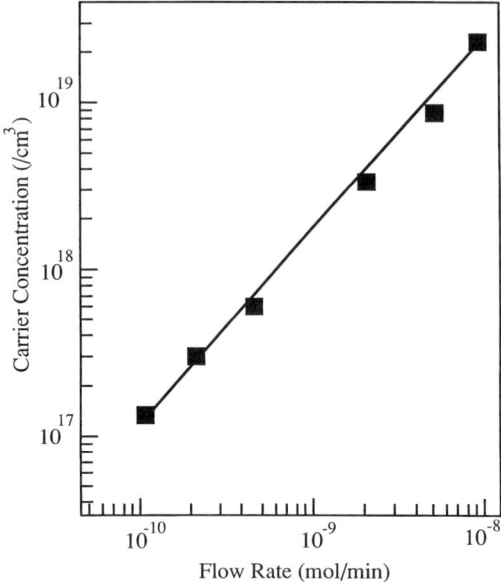

FIG. 7. Carrier concentrations of silicon-doped gallium nitride films as a function of the flow rate of silane (SiH$_4$). Reprinted from Nakamura et al. (1992c) with the permission of the Japanese Journal of Applied Physics.

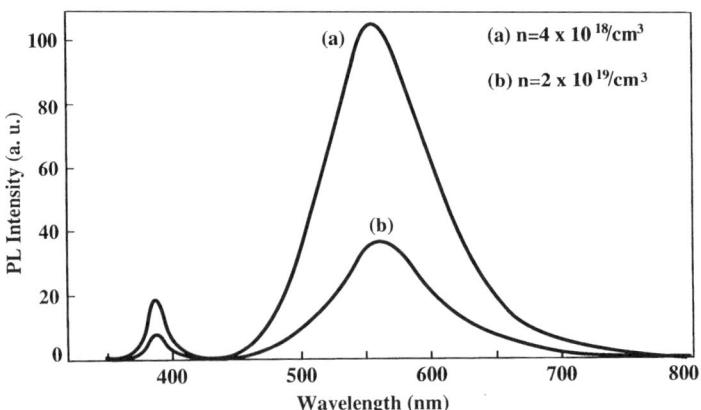

FIG. 8. Photoluminescence (PL) spectra of silicon-doped gallium nitride (GaN) films grown with GaN buffer layers under the same growth conditions except for the flow rate of silane (SiH$_4$). The flow rates for SiH$_4$ were (a) 2 nmol/min and (b) 10 nmol/min. The carrier concentrations were (a) 4×10^{18}/cm^3 and (b) 2×10^{19}/cm^3. Reprinted from Nakamura et al. (1992c) with the permission of the Japanese Journal of Applied Physics.

the realization of light-emitting devices, such as blue LEDs and LDs. In 1989, p-type GaN films were obtained using Mg doping and post–low-energy electron beam irradiation (LEEBI) treatment by means of MOCVD for the first time (Amano *et al.*, 1989). After the growth, LEEBI treatment was performed for Mg-doped GaN films to obtain a low-resistivity p-type GaN film. The hole concentration and lowest resistivity were $10^{17}/cm^3$ and $12\,\Omega\cdot cm$, respectively. These values were still insufficient for fabricating blue LDs and high-power blue LEDs. The effect of the LEEBI treatment was considered to be Mg displacement by the energy of electron beam irradiation. At the first stage of as-grown Mg-doped GaN, the Mg atoms lie in sites different from Ga sites where they act as acceptors. Under the LEEBI treatment, the Mg atoms move to the exact Ga site.

In 1992, low-resistivity Mg-doped p-type GaN films were obtained by N_2-ambient thermal annealing at temperatures above 400°C (Nakamura *et al.*, 1992d). Before thermal annealing, the resistivity of Mg-doped GaN films was approximately $1 \times 10^6\,\Omega\cdot cm$. After thermal annealing at temperatures above 700°C, the resistivity, hole carrier concentration, and hole mobility became $2\,\Omega\cdot cm$, $3 \times 10^{17}/cm^3$, and $10\,cm^2/V\cdot sec$, respectively, as shown in Fig. 9. In PL measurements, the intensity of the 750-nm DL emission sharply decreased on thermal annealing at temperatures above 700°C, as did the change in resistivity, and the 450-nm blue emission showed a maximum intensity at approximately 700°C for thermal annealing, as shown in Fig. 10.

Soon, a hydrogenation process whereby acceptor-H neutral complexes are formed in p-type GaN films was proposed as a compensation mechanism (Nakamura *et al.*, 1992a). Low-resistivity p-type GaN films, which were obtained by N_2-ambient thermal annealing or LEEBI treatment, showed a resistivity as high as $1 \times 10^6\,\Omega\cdot cm$ after NH_3 ambient thermal annealing at

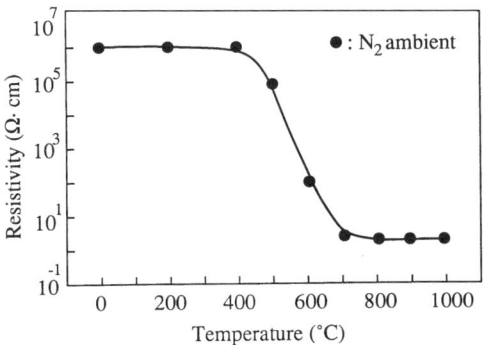

FIG. 9. Resistivity of as-grown magnesium-doped gallium nitride films as a function of annealing temperature. N_2, nitrogen. Reprinted from Nakamura *et al.* (1992d) with the permission of the Japanese Journal of Applied Physics.

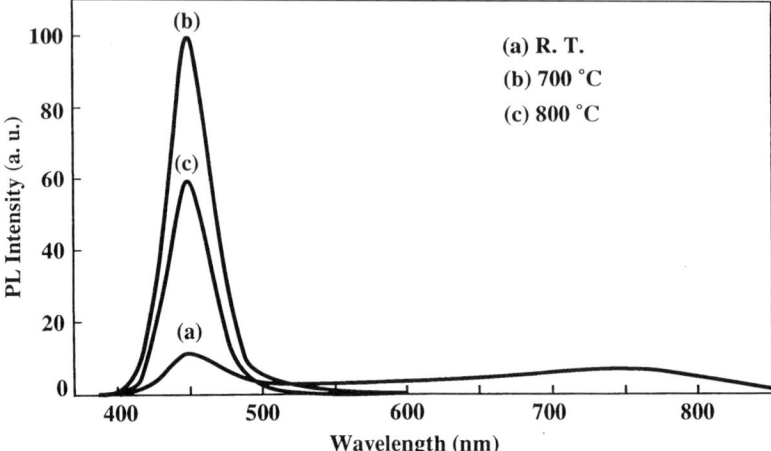

FIG. 10. Photoluminescence (PL) of as-grown magnesium-doped gallium nitride films that were annealed at different temperatures: (a) room temperature, (b) 700°C, and (c) 800°C. R.T., room temperature. Reprinted from Nakamura et al. (1992d) with the permission of the Japanese Journal of Applied Physics.

temperatures above 600°C. In the case of N_2 ambient thermal annealing at temperatures between room temperature and 1000°C, the low-resistivity p-type GaN films showed no change in resistivity, which was almost constant between 2 and 8 $\Omega \cdot$ cm, as shown in Fig. 11.

Figure 12(a) shows the PL spectrum of 800°C N_2 ambient thermal-annealed GaN film, Fig. 12(b) shows the film after NH_3 ambient thermal annealing at 800°C for the sample in Fig. 12(a), and Fig. 12(c) shows the

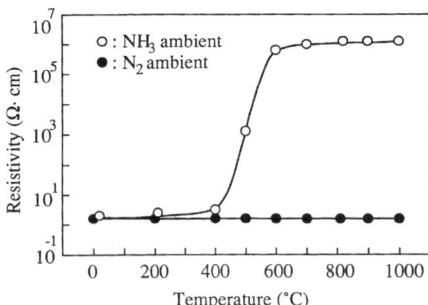

FIG. 11. The resistivity change in N_2 ambient thermal-annealed low-resistivity magnesium-doped gallium nitride films as a function of annealing temperature. The ambient gases, ammonia (NH_3) and nitrogen (N_2), were used for thermal annealing. Reprinted from Nakamura et al. (1992a) with the permission of the Japanese Journal of Applied Physics.

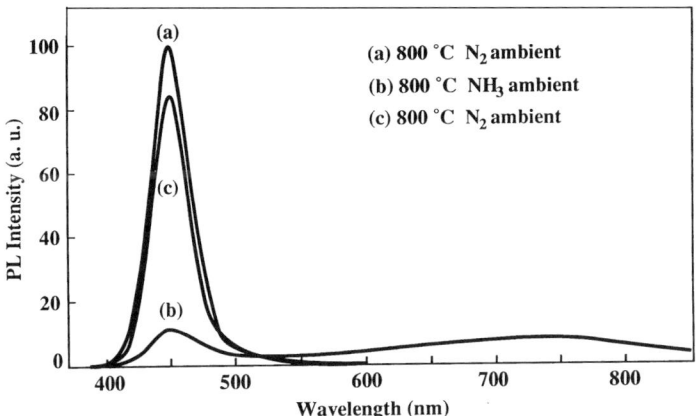

FIG. 12. Photoluminescence (PL) spectra of magnesium-doped gallium nitride (GaN) films that were continuously annealed under different conditions: (a) GaN film after 800°C nitrogen ambient thermal annealing of the Mg-doped GaN film; (b) GaN film after 800°C ammonia ambient thermal annealing of the GaN film in (a); (c) GaN film after 800°C N_2-ambient thermal annealing of the GaN film in (b). Reprinted from Nakamura et al. (1992a) with the permission of the Japanese Journal of Applied Physics.

film after N_2 ambient thermal annealing at 800°C for the sample in Fig. 12(b). Before NH_3 ambient thermal annealing, the intensity of the blue emission is strong, and the broad DL emission is not observed at around 750 nm (Fig. 12(a)). After NH_3 ambient thermal annealing at 800°C for the sample in Fig. 12(a), the intensity of the blue emission becomes weaker, and the DL emission around 750 nm appears (Fig. 12(b)). The PL spectrum recovers after N_2 ambient thermal annealing at 800°C. These changes in the PL spectra were found to be reversible with a change in the annealing ambient gas from NH_3 to N_2, as is the case with the resistivity change.

These results indicate that atomic hydrogen produced by NH_3 dissociation at temperatures above 400°C is related to the acceptor compensation mechanism. A hydrogenation process whereby acceptor-H neutral complexes are formed in p-type GaN films was proposed. The formation of acceptor-H neutral complexes causes acceptor compensation, and DL and weak blue emission in PL. At temperatures above 400°C, the dissociation of NH_3 into hydrogen atoms occurs at the surface of GaN films because dangling bonds exist mainly at the surface, and the atomic hydrogen easily diffuses into the GaN films because the number of hydrogen atoms is too great at the surface and the size of hydrogen atoms is very small. First, the atomic hydrogen, produced by dissociation of NH_3 at temperatures above 400°C, diffuses into p-type GaN films. Second, the formation of acceptor-H neutral complexes, that is, Mg–H complexes in GaN films occurs. As a result, the formation of Mg–H complexes causes hole compensation, and the resistivity of Mg-doped GaN films reaches a maximum of $1 \times 10^6 \, \Omega \cdot cm$ (Fig. 11). For PL measurements, it is assumed that the DL emission is caused by related levels of Mg–H complexes, and the blue emission is caused by Mg-related levels. When these proposals are applied to the results of the PL measurements in Figs. 10 and 12, the results are quite well explained.

The change of PL spectra of Fig. 10 is easily explained using the above-mentioned models. When the N_2 ambient thermal annealing temperature exceeds 400°C, removal of atomic hydrogen from Mg–H complexes begins, the number of blue emission centers that are Mg-related radiative recombination centers begins to increase, and the intensity of the blue emission gradually increases. However, when the temperature reaches approximately 700°C, the effects of N vacancies produced by the dissociation of the GaN films (mainly near the surface) begin to exceed those of the increased number of Mg-related radiative recombination centers. As a result, the intensity of the blue emission shows a maximum at around 700°C in N_2 ambient thermal annealing of as-grown GaN films (see Fig. 10).

In the NH_3 ambient thermal annealing of Figs. 11 and 12, Mg–H complexes are not formed below 400°C, and Mg–H complexes are formed above 400°C in Mg-doped GaN films because the amount of atomic hydrogen diffused into the bulk of the GaN films is not great until the temperature exceeds 400°C. As a result, the resistivity of NH_3 ambient thermal-annealed GaN films above 400°C becomes higher (almost insulating) in Fig. 11, and the intensity of the blue emission of NH_3 ambient thermal-annealed GaN films at 800°C becomes weaker than that of N_2 ambient thermal-annealed GaN films at 800°C because the number of Mg-related levels is reduced by formation of Mg–H complexes under NH_3 ambient thermal annealing (Fig. 12). Under NH_3 ambient thermal annealing at a temperature of 800°C, the DL emission can be observed because Mg–H complexes are formed. Under N_2 ambient thermal annealing at 800°C, the DL emission cannot be observed because atomic hydrogen is removed from the Mg–H complexes and the number of Mg–H complexes is reduced dramatically (Fig. 12).

Usually, NH_3 is used as the N-source for GaN growth in MOCVD. Therefore, an *in situ* hydrogenation process, in which Mg–H complexes are formed during MOCVD growth, naturally occurs, and the resistivity of the as-grown Mg-doped GaN films becomes high (almost insulating). After the growth, N_2-ambient thermal annealing can reactivate the acceptors by removing atomic hydrogen from the acceptor-H neutral complexes in p-type GaN films. As a result, the resistivity of p-type GaN films becomes lower and the intensity of the blue emission becomes stronger. Today, this hydrogenation process is accepted as the acceptor compensation mechanism of p-type III-V nitride by many researchers (Rubin *et al.*, 1994; Brandt *et al.*, 1994; Zavada *et al.*, 1994).

4. GALLIUM NITRIDE p-n JUNCTION LIGHT-EMITTING DIODES

High-quality GaN films were grown using GaN or AlN buffer layers, as mentioned above. Hall mobility values for undoped GaN films grown with GaN buffer layers were 600 cm²/V·sec at room temperature (Nakamura, 1991). The carrier concentration of n-type GaN films was controlled between $1 \times 10^{17}/\text{cm}^3$ and $2 \times 10^{19}/\text{cm}^3$ by Si doping into GaN (Nakamura *et al.*, 1992c). The hole concentrations of p-type Mg-doped GaN films grown with GaN buffer layers were on the order of $1 \times 10^{18}/\text{cm}^3$ (Nakamura *et al.*, 1991a). Due to these results, there is a great interest in fabricating emitting devices using GaN films grown with buffer layers. Using these techniques, GaN p-n junction blue LEDs were fabricated (Amano *et al.*, 1989; Nakamura *et al.*, 1991b). Here, GaN p-n junction blue LEDs are described.

The structure of a GaN p-n junction LED is shown in Fig. 13 (Nakamura *et al.*, 1991b). The carrier concentration of the n-type GaN layer was $5 \times 10^{18}/\text{cm}^3$ and that of the p-type GaN layer about $5 \times 10^{18}/\text{cm}^3$. Silicon was doped into the n-type GaN layer as a donor impurity. Magnesium was doped into the p-type GaN layer as an acceptor impurity. After the growth, thermal annealing was performed to obtain a low resistivity p-type GaN layer.

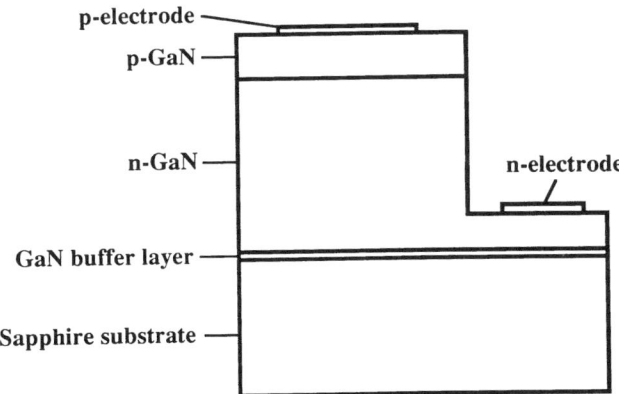

FIG. 13. The structure of the gallium nitride (GaN) p-n junction light-emitting diode. Reprinted from Nakamura *et al.* (1992a) with the permission of the Japanese Journal of Applied Physics.

Electroluminescence (EL) of the LED is shown in Fig. 14. The peak wavelength and the FWHM of the EL are 430 and 55 nm, respectively, at 10 mA. According to previous work (Amano et al., 1989), the EL of GaN LEDs showed two peaks, one at 370 nm (uv EL) and the other 430 nm (blue EL), when the forward current was lower than 30 mA. In the present LEDs, however, there was a strong blue EL and no uv EL when the forward current was lower than 30 mA. There also was a weak DL emission with a peak wavelength of 550 nm. At 50 mA, a weak uv EL peak at 390 nm was observed (Fig. 14). This peak wavelength (390 nm) of the uv EL is longer than that of previously reported LEDs (370 nm).

The forward current density of these LEDs is almost the same as that of previously reported LEDs because the chip size of the present LEDs (0.6 × 0.5 mm) is almost the same as that of LEDs with AlN buffer layers (Amano et al., 1989). This work suggested that the origin of the blue EL to the emission in the p-type GaN layer was electron injection from the n-type GaN layer to the p-type GaN layer where blue emission occurred (Amano et al., 1989). Therefore, it is considered that the number of radiative recombination centers for blue emission in the p-type GaN layer is much larger than that in the previous LEDs because the intensity of the blue EL is much stronger than that of the uv EL in GaN LEDs with GaN buffer

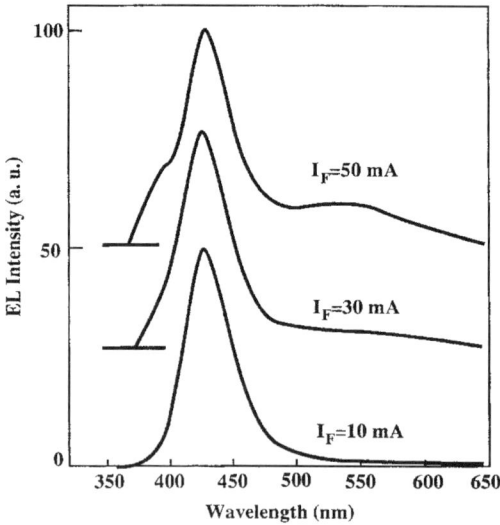

FIG. 14. Emission spectra of the gallium nitride (GaN) p-n junction light-emitting diode (LED) at different forward currents. Hole concentration of the p-type GaN of the LED was $8 \times 10^{18}/cm^2$. EL, electroluminescence; I_F, forward current. Reprinted from Nakamura et al. (1991a) with the permission of the Japanese Journal of Applied Physics.

layers (see Fig. 14), assuming that the intensity of the uv EL is almost the same between the present LEDs and those reported previously. Considering that the hole concentration of the p-type GaN layer with GaN buffer layers (about $5 \times 10^{18}/cm^3$) is much higher than that with AlN buffer layers (the typical value is on the order of $10^{16}/cm^3$), such a proposal is probably correct because the blue emission centers are related to the energy levels introduced by Mg doping in the energy gap of GaN, and the intensity of the blue emission in PL measurement of p-GaN layers becomes strong when the hole concentration is high. Therefore, in the present LEDs, uv EL was not observed at lower than 30 mA, and uv EL intensity was very weak over 30 mA in comparison with previous reports. The peak wavelength of the uv EL of the present LEDs (390 nm) is longer than that of the previous work (370 nm). An undoped GaN layer was used as the n-type GaN layer in the previous work, and the peak wavelength of the uv EL was 370 nm. Considering this result, the longer peak wavelength of the uv EL (390 nm) is possibly caused by Si doping into the n-type GaN layer because the uv EL was caused by hole injection from the p-type GaN layer to the n-type GaN layer, and the uv emission occurred in the n-type GaN layer. A relatively strong uv EL against the blue EL was observed in the present GaN LEDs when the crystal quality of the GaN film was poor and the hole concentration of the p-type GaN layer was as low as $1 \times 10^{17}/cm^3$. This is shown in Fig. 15. The shape of the EL and the hole concentration of this LED were almost the same as those of previously reported LEDs (Amano *et al.*, 1989). Considering that the hole concentration is as low as $1 \times 10^{17}/cm^3$, this weak blue EL is related to the small number of radiative recombination centers in the p-type GaN layer that contribute to the blue EL, and the uv EL becomes dominant. The output power of this LED, which had a low hole concentration ($1 \times 10^{17}/cm^3$), was very low (about one fourth) in comparison with that of LEDs that had a high hole concentration ($5 \times 10^{18}/cm^3$). This LED was easily broken down by a current of 50 mA. On the other hand, LEDs that had a high hole concentration ($5 \times 10^{18}/cm^3$) were not broken down, even at 100 mA. The 550-nm DL emission may be caused by hole injection from the p-type GaN layer to the n-type GaN layer, similarly as for the uv EL, because the intensity of the DL emission becomes strong when the uv EL becomes strong (Figs. 14 and 15) and the PL measurements for the n-type GaN layers show DL emission.

The output power (P) is shown as a function of the forward current in Fig. 16. A commercially available SiC LED whose brightness is 8 mcd with a peak wavelength of 480 nm is also shown for comparison with the GaN LED. The output power of the GaN LED is almost 10 times stronger in the range of forward current between 1 and 4 mA. At 4 mA, the output power

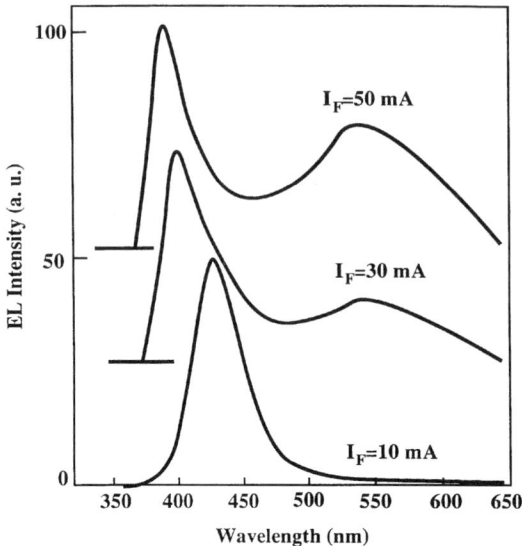

FIG. 15. Emission spectra of the gallium nitride (GaN) p-n junction light-emitting diode (LED) at different forward currents. Hole concentration of the p-type GaN of the LED was $1 \times 10^{17}/cm^2$. EL, electroluminescence; I_F, forward current. Reprinted from Nakamura et al. (1991a) with the permission of the Japanese Journal of Applied Physics.

of the GaN LED is 20 µW and that of the SiC LED is 2 µW. At 20 mA, the output power of the GaN LED is 42 µW and that of the SiC LED 7 µW. Generally in LEDs, the output power is proportional to I_F^m (I_F is the forward current). If the recombination current is dominant, m becomes 2; if the diffusion current is dominant, m becomes 1. In the range of dc current between 0.2 and 0.8 mA (low current range) in GaN LEDs, m is approximately 2.23. Between 1 and 4 mA (intermediate current range), it is 1.15. Over 6 mA (high current range), it is 0.41. Therefore, the recombination current is dominant in the low current range, and the diffusion current becomes dominant in the intermediate current range. The generation of heat possibly caused the low output power in the high current range. In SiC LEDs, m was almost equal to 0.73 between 0.2 and 30 mA. The highest external quantum efficiency of 0.18% was obtained in an intermediate current range for the GaN LED and 0.02% for the SiC LED. Blue EL was dominant at less than 50 mA (Fig. 14). Therefore, these changes in output power are caused by the change in the intensity of the blue EL.

High-power GaN p-n junction blue LEDs were fabricated using GaN films grown with GaN buffer layers. The external quantum efficiency was as high as 0.18%. The output power was almost 10 times higher than that of

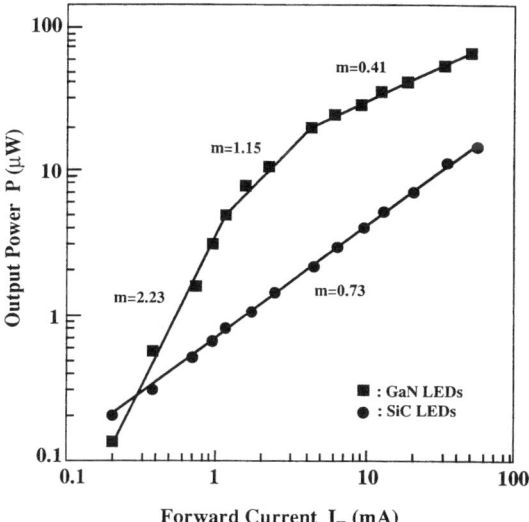

FIG. 16. The output power (P) of the gallium nitride (GaN) p-n junction light-emitting diode (LED) and the conventional 8-mcd silicon carbide (SiC) LED as a function of the forward current I_F. m is an exponent of I_F when it is assumed that P is proportional to I_F^m. Reprinted from Nakamura et al. (1991a) with the permission of the Japanese Journal of Applied Physics.

conventional 8-mcd SiC blue LEDs. The forward voltage was 4 V at 20 mA. The recombination current was dominant below 1 mA. Therefore, further improvement of the crystal quality is required to obtain high-power GaN p-n junction LEDs.

III. InGaN Growth

1. UNDOPED InGaN

Utilizing the (In,Ga,Al)N system, a bandgap energy from 1.95 to 6.2 eV can be obtained (see Fig. 1). For high-performance optical devices, a DH is indispensable. This material enables a DH construction. In this system, the ternary III-V semiconductor compound InGaN is one candidate for the active layer for both blue and green emission because its bandgap varies from 1.95 to 3.4 eV, depending on the In mole fraction. If the InGaN semiconductor compound is used as an active layer in the DH, an InGaN/AlGaN DH can be considered for blue-emitting devices because p-type conduction has been obtained for AlGaN.

Up to now, little research has been performed on InGaN growth (Matsuoka et al., 1989; Nagatomo et al., 1989; Yoshimoto et al., 1991). To prevent InN dissociation during growth, InGaN crystal growth was originally conducted at low temperatures (about 500°C) by means of MOCVD (Matsuoka et al., 1989; Nagatomo et al., 1989). Recently, relatively high-quality InGaN films were obtained on a (0001) sapphire substrate using a high growth temperature (800°C) and a high In mole fraction in the vapor (Yoshimoto et al., 1991). These authors reported that DL emissions were dominant in PL measurements of the InGaN film at room temperature, and that the FWHM of the XRC for (0002) diffraction from the InGaN films was about 30 minutes. High-quality single-crystal InGaN films were grown on GaN films with a high In source flow rate and high growth temperatures between 780 and 830°C for the first time (Nakamura and Mukai, 1992). These authors observed strong and sharp band-edge emission between 400 and 445 nm in room temperature PL.

Figure 17 shows the XRC for (0002) diffraction of InGaN films grown on the GaN films. Curve (a) shows the InGaN films grown at a temperature of 830°C (sample A) and curve (b) at a temperature of 780°C (sample B). Both curves clearly show two peaks. One is the (0002) peak of the X-ray diffraction of GaN, and the other is that of InGaN. The In mole fraction of

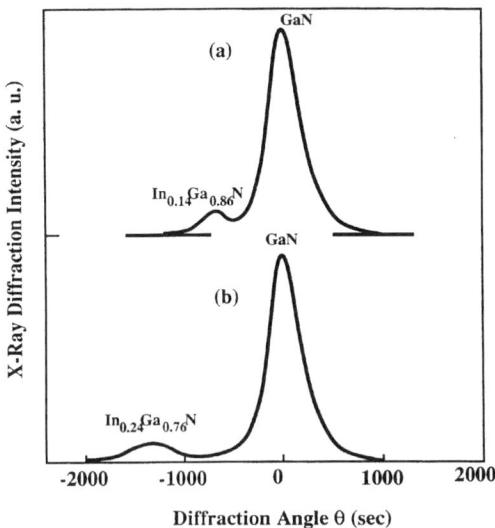

FIG. 17. X-ray rocking curve (XRC) for (0002) diffraction from the InGaN films grown on the gallium nitride (GaN) films under the same conditions except for the InGaN growth temperature. The growth temperatures of InGaN were (a) 830°C and (b) 780°C. Reprinted from Nakamura and Mukai (1992) with the permission of the Japanese Journal of Applied Physics.

the InGaN films was estimated by calculating the difference in peak position between the InGaN and GaN peaks, assuming the (0002) peak of the X-ray diffraction of GaN is constant at $2\theta = 34.53°$ and Vegard's law is valid. These calculated values of the In mole fraction of the InGaN films are 0.14 for sample A and 0.24 for sample B, as shown in Fig. 17. Therefore, the incorporation rate of In into the InGaN film during growth is increased when the growth temperature is decreased. The FWHM of the XRC for the (0002) diffraction from the InGaN film was about 8 min and that from the GaN film was 6 min for sample A. The FWHMs were about 9 and 7 min for sample B. The values of FWHM for the InGaN films are almost the same as those for the GaN films, which are used as substrates.

Figure 18 shows the room temperature PL results with excitation by a 10-mW He–Cd laser. Curve (a) shows sample A and curve (b) sample B. Sample A shows a strong sharp peak at 400 nm and sample B at 438 nm. These emissions are considered to be the band-edge emissions of InGaN films because they have a very narrow FWHM (about 70 and 100 meV for samples A and B, respectively). The broad DL emission considered to originate from defects such as nitrogen vacancies in InGaN films, which were dominant in the results of previous work (Yoshimoto et al., 1991), were rarely observed in this study. These results also indicate that the crystal quality of the InGaN films grown on the GaN films is very good, as is also shown by the XRC measurements.

InGaN films were also grown with In mole fractions as high as 0.33 at temperatures between 720 and 850°C (Nakamura, 1994b). The growth rate

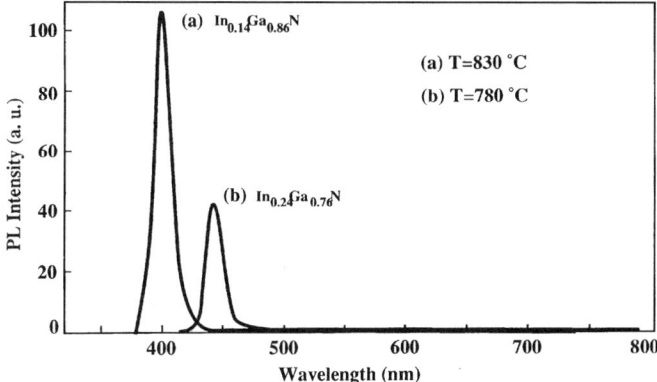

FIG. 18. Room temperature photoluminescent (PL) spectra of the InGaN films grown on gallium nitride films under the same growth conditions except for the InGaN growth temperature. The growth temperatures of InGaN were (a) 830 and (b) 780°C. Reprinted from Nakamura and Mukai (1992) with the permission of the Japanese Journal of Applied Physics.

had to be decreased sharply to obtain high-quality InGaN films when the growth temperature was decreased. Figure 19 shows the bandgap energy of grown InGaN films as a function of the In mole fraction x. The bandgap energy was obtained by room temperature PL measurements, assuming that the narrow sharp emissions in the violet and blue regions are due to band-edge emissions. The In mole fraction of the InGaN films was determined by X-ray diffraction measurements. It has been shown that E_g in $In_xGa_{(1-x)}N$ obeys a parabolic form:

$$E_g(x) = xE_{g,InN} + (1-x)E_{g,GaN} - bx(1-x) \tag{1}$$

where $E_g(x)$ represents the bandgap energy of $In_xGa_{(1-x)}N$, $E_{g,InN}$ and $E_{g,GaN}$ represent the bandgap energies of InN and GaN, respectively, and b is the bowing parameter. In that calculation, $E_{g,InN}$ is 1.95 eV, $E_{g,GaN}$ is 3.40 eV, and b is 1.00 eV (Osamura et al., 1975). These calculated values by Eq. (1) are shown by the solid curve in Fig. 19, which fits the experimental data quite well between In mole fractions of 0.07 and 0.33.

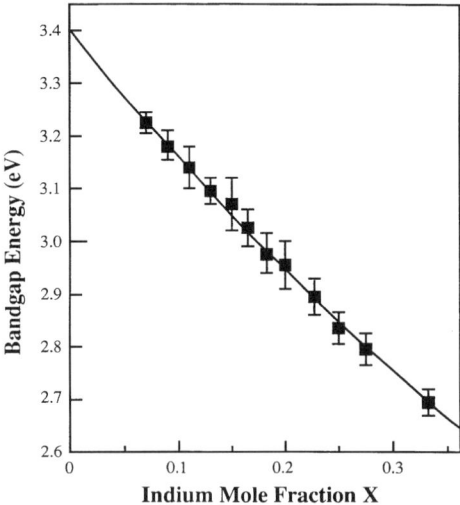

FIG. 19. Bandgap energy of $In_xGa_{1-x}N$ films as a function of the indium (In) mole fraction x. The In mole fraction x was determined by measurements of the X-ray diffraction peaks. Solid curve represents values obtained by Eq. (1) as discussed in the text, assuming that the bandgap energy for gallium nitride and indium nitride is 3.40 and 1.95 eV, respectively. Reprinted from Nakamura (1995a) with the permission of the American Institute of Physics.

2. IMPURITY DOPING OF InGaN

High-quality Si-doped InGaN films were grown on GaN films (Nakamura *et al.*, 1993a). Figure 20 shows the typical results of PL measurements of an Si-doped InGaN film grown at a temperature of 830°C and a SiH_4 flow rate of 1.5 nmol/min. The PL measurements were performed at room temperature. Very strong and sharp violet emission at 400 nm was observed, whereas DL emission was not. This violet emission is considered to be a band-edge emission of Si-doped InGaN because the value of the FWHM is very small (about 140 meV). The PL measurements of undoped InGaN films grown under the same growth conditions as this study without SiH_4 gas flow have been reported (Nakamura and Mukai, 1992). Comparing the Si-doped InGaN with undoped InGaN, the peak position of the band-edge emission at an In mole fraction of 0.14 is not changed by Si doping; however, the intensity becomes much stronger than for undoped InGaN films. This is shown in Fig. 21.

Figure 21 shows the relative band-edge emission intensity of the PL of InGaN films as a function of the SiH_4 flow rate. These InGaN films were grown at 830°C. The peak wavelength of the band-edge emission of these InGaN films was 400 nm and the In mole fraction determined by X-ray diffraction measurements was 0.14. At an SiH_4 flow rate of 0.22 nmol/min, the intensity of the band-edge emission became 20 times stronger than that for undoped InGaN films. At an SiH_4 flow rate of 1.50 nmol/min, the

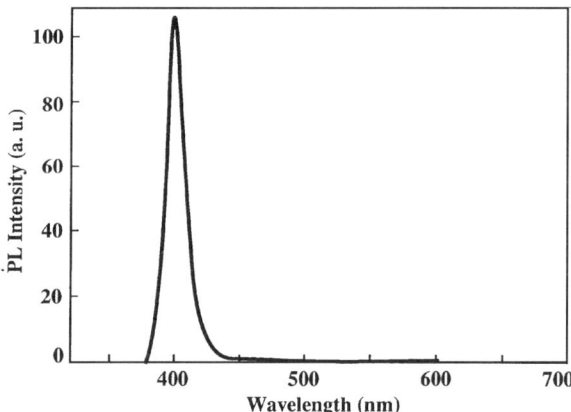

FIG. 20. Room temperature photoluminescent (PL) spectrum of the silicon-doped InGaN film grown at the silane (SiH_4) flow rate of 1.5 nmol/min and at the growth temperature of 830°C. Reprinted from Nakamura *et al.* (1993a) with the permission of the Japanese Journal of Applied Physics.

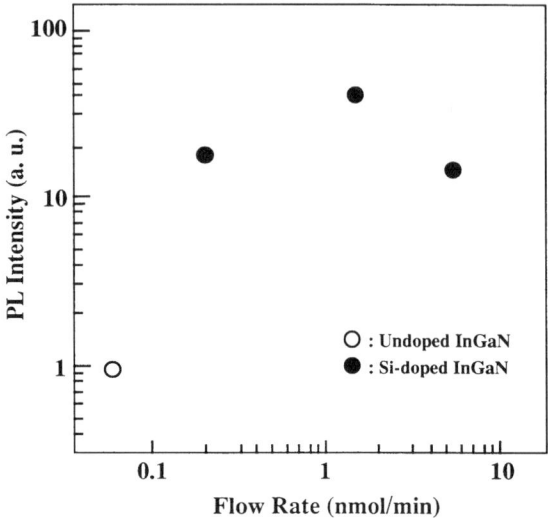

FIG. 21. The relative photoluminescent (PL) intensity of the band-edge emissions of silicon-doped InGaN films as a function of the silane (SiH_4) flow rate. The growth temperatures of the Si-doped InGaN films were 830°C. Reprinted from Nakamura et al. (1993a) with the permission of the Japanese Journal of Applied Physics.

band-edge intensity became 36 times stronger. However, at an SiH_4 flow rate of 4.46 nmol/min, the intensity became weaker: The optimum SiH_4 flow rate for the active layer of a DH structure is around 1.50 nmol/min. Silicon doping of the InGaN films may, perhaps, form shallow donor levels, as does Si doping of GaN films. The carrier concentrations of the InGaN increased from 10^{17} to 10^{19}/cm^3 as a result of Si doping with an SiH_4 flow rate of 1.50 nmol/min. This may explain why the PL intensity of the Si-doped InGaN films is stronger.

Next, Si and Zn were codoped into the InGaN films to obtain longer-wavelength blue emission centers, giving a luminous intensity high enough to be detected by the human eye (Nakamura, 1994c, 1995a). The luminous intensity of the violet emission originating from $In_xGa_{(1-x)}N$ ($x < 0.2$) band-to-band (BB) recombination is insufficient for practical use in visible LEDs. Figure 22 shows the typical room temperature PL spectrum of an Si- and Zn-codoped InGaN film grown on GaN, using SiH_4 and DEZ as the dopants. Broad strong emission is observed at around 460 nm, which is considered to originate from impurity-assisted recombination. In the shorter-wavelength region, a weak peak can be observed at around 385 nm, which is considered to be the band-edge emission. The In mole fraction determined from XRC measurements for this sample is 0.06.

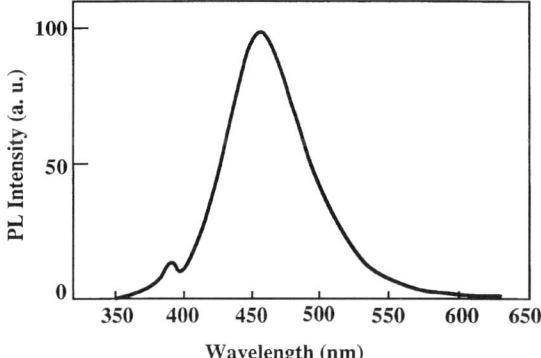

FIG. 22. Typical room temperature photoluminescent (PL) spectrum of silicon- and zinc-codoped $In_{0.06}Ga_{0.94}N$ film grown on GaN film. Reprinted from Nakamura (1995a) with the permission of the American Institute of Physics.

Figure 23 shows the Zn-related emission energy as a function of the In mole fraction x. Curve (a) shows the bandgap energy calculated using Eq. (1). Curves (b) and (c) show the energy levels, which are 0.4 and 0.5 eV below the bandgap energy, respectively. From Fig. 23, we observe that 2.50 to 2.83 eV (500 to 438 nm) is the Zn-related emission energy range in Zn-doped InGaN for compositions between $x = 0.17$ and 0.07. The maxi-

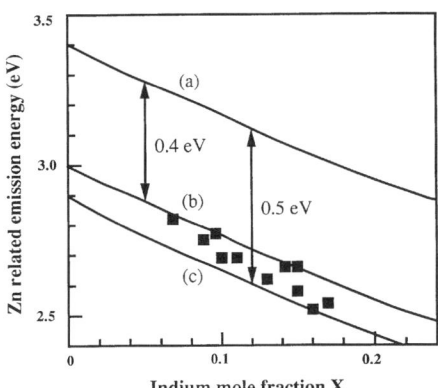

FIG. 23. Zinc-related emission energy as a function of the indium mole fraction x of $In_xGa_{1-x}N$. Curve (a) shows the bandgap energy of $In_xGa_{1-x}N$ that was calculated using Eq. (1). Curves (b) and (c) show the energy levels that are 0.4 and 0.5 eV below the bandgap energy of $In_xGa_{1-x}N$, respectively. Reprinted from Nakamura (1994c) with the kind permission of Elsevier Science – NL.

mum In mole fraction that presently can be obtained for Si- and Zn-co-doped InGaN films is 0.3. The Zn-related emission energy is always between 0.4 to 0.5 eV lower than the band-edge emission energy in this Zn doping range. Zinc doping of GaN has been performed by many researchers to obtain blue LEDs (Pankove and Hutchby, 1976; Bergman et al., 1987). The peak energy shifts from the blue at low Zn concentrations to the red at high Zn concentrations. Therefore, higher Zn doping of GaN produces deeper Zn-related levels. Room temperature Hall effect and PL measurements were performed on InGaN films grown on GaN films for various flow rates of SiH_4 and DEZ. All of the InGaN films grown under these conditions were n-type. Figure 24 shows the PL intensity of the impurity-assisted emission as a function of the carrier concentration. The PL intensity reaches a maximum at around 1×10^{19} cm^{-3}. For carrier concentrations below 1×10^{18} cm^{-3} and over 1×10^{20} cm^{-3}, the PL intensity gradually decreases. The optimum is approximately 1×10^{19} cm^{-3} for the active layer of InGaN/AlGaN LEDs.

IV. Light-Emitting Diodes

1. InGaN/Gallium Nitride Double Heterostructure Light-Emitting Diodes

InGaN/GaN DH LEDs were fabricated for the first time in 1993 (Nakamura et al., 1993b) using the two-flow MOCVD method. Sapphire with the (0001) orientation (c face) was used as the substrate. Trimethylgallium (TMGa), trimethylindium (TMIn), SiH_4, Cp_2Mg, and NH_3 were used as the source materials. First, the substrate was heated to 1050°C in a stream of hydrogen. Then the substrate temperature was lowered to 510°C to grow the GaN buffer layer, approximately 250-Å thick. Next, the substrate temperature was elevated to 1020°C to grow the GaN film. During the deposition, the flow rates of NH_3, TMGa, and SiH_4 (10 ppm SiH_4 in H_2) in the main flow were 4.0 liter/min, 30 μmol/min, and 4 nmol/min, respectively. The flow rates of H_2 and N_2 in the subflow were both maintained at 10 liter/min. The Si-doped GaN films were grown for 60 minutes, giving a thickness of approximately 4 μm. After GaN growth, the temperature was decreased to 800°C, and the Si-doped InGaN film was grown for 7 minutes. The flow rates of NH_3, TMIn, TMGa, and SiH_4 in the main flow were maintained at 4.0 liter/min, 24 μm/min, 2 μmol/min, and 1 nmol/min, respectively. The thickness of the Si-doped InGaN layer was approximately 200 Å. The temperature was then increased to 1020°C for 15 minutes to grow the Mg-doped p-type GaN film. The Cp_2Mg gas flow rate was

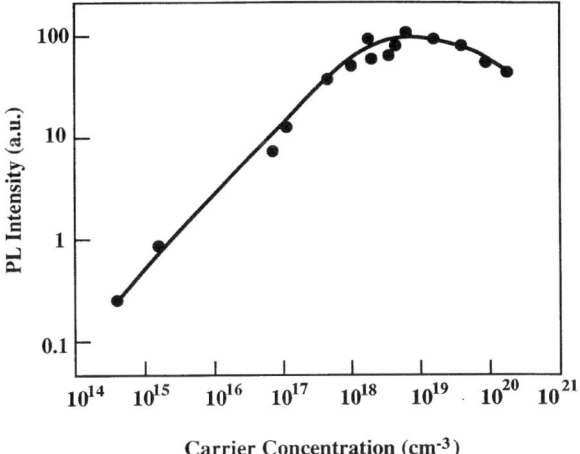

FIG. 24. Photoluminescent (PL) intensity of InGaN films codoped with silicon and zinc as a function of the electron carrier concentration. Reprinted from Nakamura et al. (1994b) with the permission of the American Institute of Physics.

3.6 μmol/min. The total thickness was about 4.8 μm. During GaN growth, H_2 was used as a carrier gas of the main flow, and during InGaN growth, N_2 at a flow rate of 2 liter/min was used. Fabrication of LED chips was accomplished as follows. The surface of the p-type GaN layer was partially etched until the n-type layer was exposed (Fig. 25). Next, an Au contact was evaporated onto the p-type GaN layer and an Al contact onto the n-type GaN layer. The wafer was cut into a rectangular shape (0.6 × 0.5 mm).

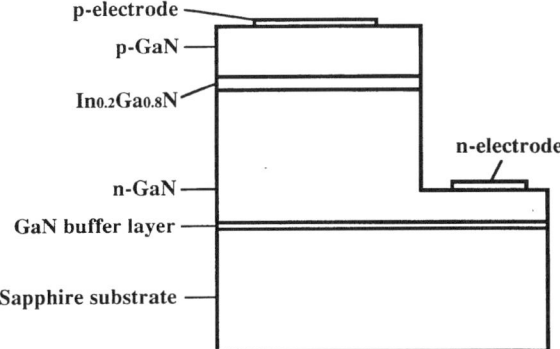

FIG. 25. The structure of the p-GaN/n-InGaN/n-GaN double heterostructure blue light-emitting diode. GaN, gallium nitride. Reprinted from Nakamura et al. (1993b) with the permission of the Japanese Journal of Applied Physics.

These chips were set on the lead frame, and were then molded. The characteristics of the LEDs were measured under dc-biased conditions at room temperature.

Figure 26 shows the ELs of the InGaN/GaN DH LEDs at forward currents of 5, 10, and 20 mA. The peak wavelength of 440 nm and the FWHM of the peak emission of 20 nm were almost constant under these dc-biased conditions. When the growth temperature of the Si-doped InGaN film was 800°C, the In mole fraction became 0.2 and the peak PL wavelength of the band-edge emission became 425 nm (Nakamura et al., 1993a). The FWHM of the band-edge of the EL emission was about 20 nm. The values of FWHM for the EL and PL spectra are almost the same. The peak wavelength of the EL (440 nm) is slightly different from that of the PL (425 nm). Thus, the blue emission can be assumed to result from recombination between the electrons injected into the conduction band and holes injected into the valence band of the InGaN active layer. Under reverse bias conditions, blue EL was not observed.

Homostructure GaN p-n junction blue LEDs have been reported with a peak wavelength and FWHM of 430 nm and 380 meV, respectively (Nakamura et al., 1991b). The peak wavelength of the InGaN/GaN DH LEDs is 10 nm longer, and the value of the FWHM is about half that of the homostructure GaN LEDs.

The output power is shown as a function of the forward current in Fig. 27. It increases almost linearly up to 20 mA. The output power of the InGaN/GaN DH LEDs is 70 μW at 10 mA and 125 μW at 20 mA. The external quantum efficiency is 0.22% at 20 mA. The homostructure GaN p-n

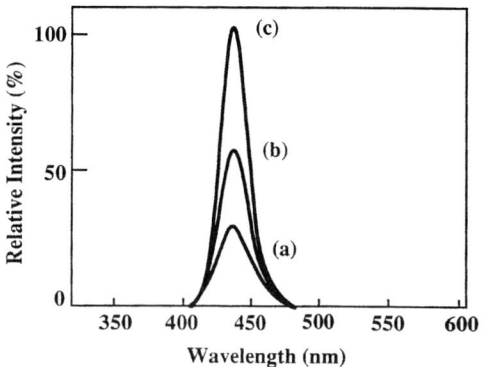

FIG. 26. Electroluminescence spectra of the p-GaN/n-InGaN/n-GaN double heterostructure blue light-emitting diode. Forward currents are (a) 5 mA, (b) 10 mA, and (c) 20 mA. GaN, gallium nitride. Reprinted from Nakamura et al. (1993b) with the permission of the Japanese Journal of Applied Physics.

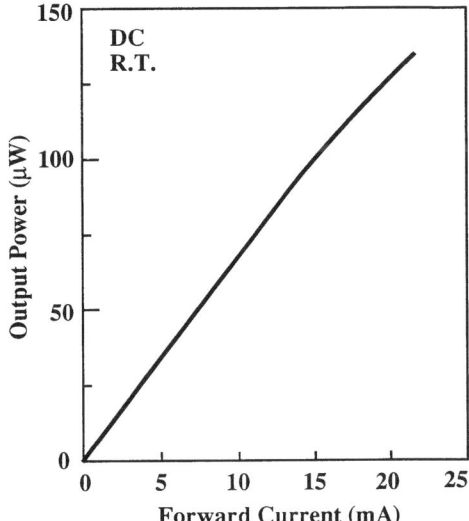

FIG. 27. The output power of the p-GaN/n-InGaN/n-GaN double heterostructure blue light-emitting diode as a function of the forward current. GaN, gallium nitride; R.T., room temperature. Reprinted from Nakamura et al. (1993b) with the permission of the Japanese Journal of Applied Physics.

junction blue LEDs had an output power of 42 μW at 20 mA (Nakamura et al., 1991b). Hence, the output power of the InGaN/GaN DH LEDs is about three times larger than that of the homostructure GaN LEDs.

2. InGaN/AlGaN Double Heterostructure Light-Emitting Diodes

Figure 28 shows the structure of the InGaN/AlGaN DH LEDs (Nakamura, 1994a, 1995a). The active layer is InGaN codoped with Si and Zn to enhance the blue emission. The blue emission intensity is at maximum at an electron carrier concentration of 1×10^{19} cm^{-3}. The need for codoping suggests that the high efficiency of this LED is the result of impurity-assisted—e.g., free-carrier-acceptor (FA)—recombination. A p-type GaN layer was used as the contact layer for the p-type electrode in order to improve the ohmic contact. After the growth, N_2 ambient thermal annealing at 700°C was performed to obtain highly p-type GaN and AlGaN layers. Fabrication of LED chips was accomplished as follows. The surface of the p-type GaN layer was partially etched until the n-type GaN layer was exposed. Next, an Ni–Au contact was evaporated onto the p-type GaN layer and a Ti–Al contact onto the n-type GaN layer. The wafer was cut

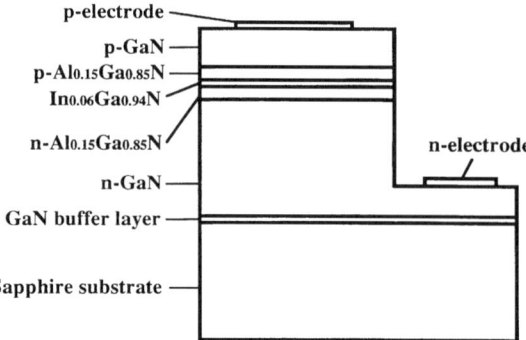

FIG. 28. The structure of the InGaN/AlGaN double heterostructure light-emitting diodes. Reprinted from Nakamura (1994c) with the kind permission of Elsevier Science – NL.

into a rectangular shape (350 × 350 μm). These chips were set on the lead frame, and were then molded. The characteristics of LEDs were measured under dc-biased conditions at room temperature.

Figure 29 shows the electroluminescence (EL) spectra of the InGaN/AlGaN DH blue LEDs at forward currents of 0.1, 1, and 20 mA. The carrier concentration of the InGaN active layer in this LED was 1×10^{19} cm^{-3}.

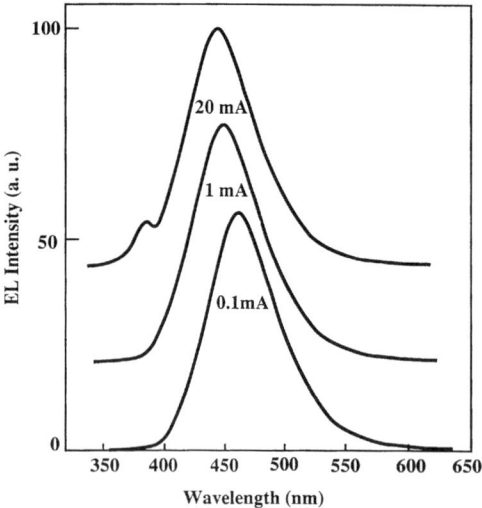

FIG. 29. Electroluminescent (EL) spectra of the InGaN/AlGaN double heterostructure blue light-emitting diodes under different forward currents. Reprinted from Nakamura (1994c) with the kind permission of Elsevier Science – NL.

Typical peak wavelengths and values of FWHM of the EL at 20 mA were 450 and 70 nm, respectively. The peak wavelength shifts to shorter wavelengths with increasing forward current. The peak wavelength is 460 nm at 0.1 mA, 449 nm at 1 mA, and 447 nm at 20 mA. At 20 mA, a narrower, higher-energy peak emerges at around 385 nm, as shown in Fig. 29. This peak, due to BB recombination in the InGaN active layer, becomes resolved at injection levels at which the impurity-related recombination is saturated. The output power of these blue LEDs is 1.5 mW at 10 mA, 3 mW at 20 mA, and 4.8 mW at 40 mA. The external quantum efficiency is 5.4% at 20 mA. The typical on-axis luminous intensity of InGaN/AlGaN LEDs with a 15-degree conical viewing angle is 2.5 cd at 20 mA. The forward voltage is 3.6 V at 20 mA. High-brightness blue LEDs with a luminous intensity over 1 cd will pave the way for realization of full-color LED displays, especially for outdoor use.

Blue-green LEDs were fabricated for application to traffic lights by increasing the In mole fraction of the InGaN active layer from 0.06 to 0.19 (Nakamura *et al.*, 1994b). Figure 30 shows the EL spectra of these InGaN/AlGaN DH LEDs at forward currents of 0.5, 1, and 20 mA. Typical peak wavelengths and values of FWHM of the EL at 20 mA were 500 and 80 nm. The peak wavelength shifts from 537 nm at 0.5 mA, to 525 nm at 1 mA, and to 500 nm at 20 mA. The output power of these blue-green LEDs

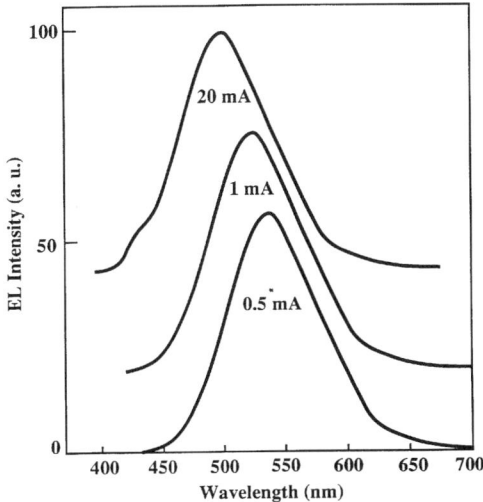

FIG. 30. Electroluminescent (EL) spectra of the InGaN/AlGaN double heterostructure blue-green light-emitting diodes under different forward currents. Reprinted from Nakamura *et al.* (1994b) with the permission of the American Institute of Physics.

is 1.0 mW at 20 mA, where the external quantum efficiency is 2.1%. A typical on-axis luminous intensity with a 15-degree conical viewing angle is 2 cd at 20 mA. This luminous intensity is sufficiently bright for outdoor application, such as traffic lights and displays. The forward voltage was 3.5 V at 20 mA.

Figure 31 shows a chromaticity diagram where blue and blue-green InGaN/AlGaN LEDs are shown (Nakamura, 1995a). Commercially available green GaP LEDs and red GaAlAs LEDs are also shown. From Fig. 31, only the blue-green InGaN/AlGaN LEDs are within the applicable regions for roadway and railway signals. Therefore, these blue-green LEDs can be used for those applications from the viewpoint of color. Traffic lights may prove to be a great application for the blue-green LEDs. In Japan, total power consumption by traffic lights reaches the gigawatt range. These InGaN/AlGaN blue-green LED traffic lights, with an electrical power consumption of only 12% that of present incandescent bulb traffic lights, promise to save vast amounts of energy. With its extremely long lifetime of several tens of thousands of hours, the replacement of burned-out traffic

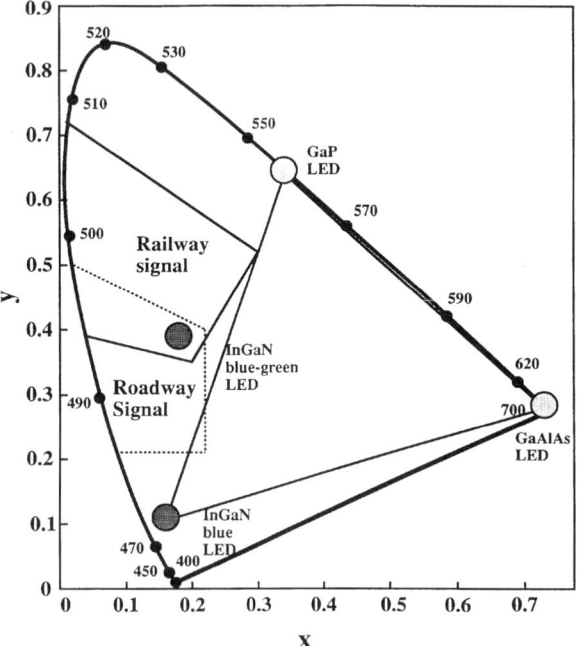

FIG. 31. Chromaticity diagram where blue InGaN/AlGaN light-emitting diodes (LEDs), blue-green InGaN/AlGaN LEDs, green gallium phosphide (GaP) LEDs and red GaAlAs LEDs are shown. Reprinted from Nakamura (1995a) with the permission of the American Institute of Physics.

light bulbs will be dramatically reduced. Using these high-brightness blue-green LEDs, safe energy-efficient roadway and railway signals can be achieved. Figure 32 (see color plate section) shows the first actual LED traffic light installed in Japan in 1994 using InGaN/AlGaN blue-green, AlInGaP yellow, and GaAlAs red LEDs.

Presently, using high-brightness InGaN/AlGaN blue, GaP green, and GaAlAs red LEDs, full-color displays—especially for outdoor use—can be fabricated. Figure 33 (see color plate section) shows the actual LED full-color display (10 × 10 m) installed for the first time in Japan in 1994. The color range of light emitted by a full-color LED lamp in the chromaticity diagram is shown as the region inside a triangle, which is drawn by connecting the positions of the three primary color LED lamps. The color range is not very wide because the color of the green GaP LEDs is yellowish green (555 nm), not pure green. This means that this LED full-color display cannot express an actual nature-color, especially in the green region. For pure green LEDs, emission wavelengths between 510 and 530 nm are required.

Figure 34 shows the EL spectrum of an InGaN/AlGaN DH violet LED at forward currents of 1 and 20 mA (Nakamura, 1995a). These violet LEDs were grown under the same conditions as were the blue and blue-green

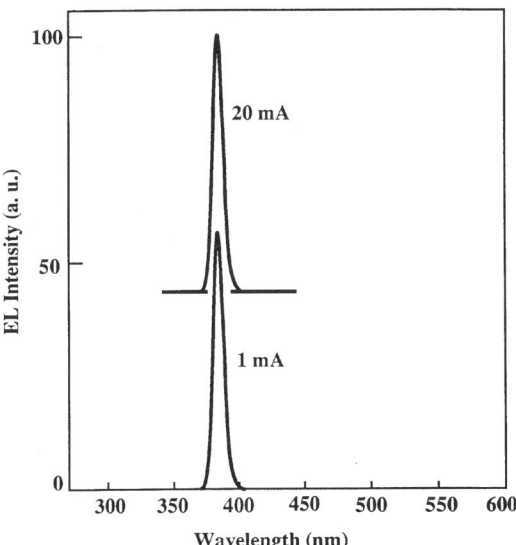

FIG. 34. Electroluminescent (EL) spectra of the InGaN/AlGaN double heterostructure violet light-emitting diodes (LEDs) under different forward currents. Reprinted from Nakamura (1995b) with the permission of the IEEE.

LEDs, except for the InGaN active layer. During InGaN growth, only Si was added without Zn. The typical output power was 1.5 mW, and the external quantum efficiency was as high as 2.3% at a forward current of 20 mA at room temperature. The peak wavelength and the FWHM of the EL were 385 and 10 nm, respectively. This violet LED is expected to be useful for the realization of violet LDs in the near future because the emission is very sharp and strong.

Figure 35 shows the external quantum efficiency as a function of the peak wavelength for various commercially available LEDs. Judging from Fig. 35, there are no LED materials except for InGaN that have efficiencies over 1% below the peak wavelength of 550 nm. Therefore, InGaN is the most promising material for LEDs and LDs with peak wavelengths between 550 and 360 nm. For II-VI materials, a ZnSe/ZnTeSe DH green LED has been reported (Eason *et al.*, 1995). The output power, external quantum efficiency, and peak wavelength of these II-VI LEDs are 1.3 mW, 5.3%, and 512 nm, respectively, at a forward current of 10 mA. A ZnSe/ZnCdSe DH blue LED was reported with an output power, an external quantum efficiency and a peak wavelength of 0.3 mW, 1.3%, and 489 nm, respectively, at a forward current of 10 mA. However, the lifetime of these II-VI LEDs is only a few hundred hours under room temperature operation. Because of this poor reliability, II-VI LEDs and LDs have never been commercialized. Therefore, II-VI LEDs are not shown in Fig. 35.

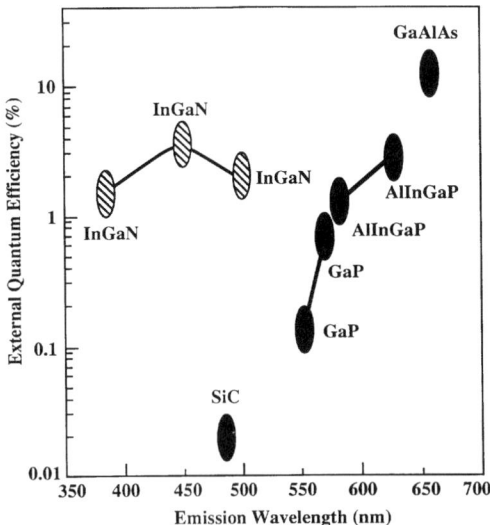

FIG. 35. The external quantum efficiencies as a function of the peak wavelength of various commercially available LEDs. GaP, gallium phosphide; SiC, silicon carbide. Reprinted from Nakamura (1995b) with the permission of the IEEE.

3. InGaN Quantum Well Structures

Two kinds of $In_{0.22}Ga_{0.78}N/In_{0.06}Ga_{0.94}N$ multiple quantum well (MQW) structures were grown on GaN films (Nakamura et al., 1993c). One was MQW-100, in which the thickness of the barrier layer L_B and of the well layer L_W each was 100 Å ($L_B = L_W = 100$ Å) and the number of periods was 10. The other was MQW-30, in which the thickness of the barrier and of the well layer each was 30 Å ($L_B = L_W = 30$ Å) and the number of periods was 20. Figure 36 shows the XRC results for (0002) diffraction from these $In_xGa_{1-x}N/In_yGa_{1-y}N$ MQW structures. Curve (a) shows MQW-100 and curve (b) MQW-30. Both curves clearly show three peaks, which are the (0002) peak of the X-ray diffraction of GaN, zeroth-order peak marked "0," and satellite peak marked "−1" associated with the MQW structures. The FWHMs of the zeroth-order peak and the GaN underlayer peak were 7.1 and 5.4 min for MQW-100, and 6.3 and 4.3 min for MQW-30.

The $In_xGa_{1-x}N/In_yGa_{1-y}N$ MQW period $(L_B + L_W)$ can be accurately determined using the equation

$$(2 \sin \Theta_n - 2 \sin \Theta_{SL}) = \pm n\lambda/(L_B + L_W),$$

where λ is the X-ray wavelength, n is the order of the satellite peaks, Θ_n is

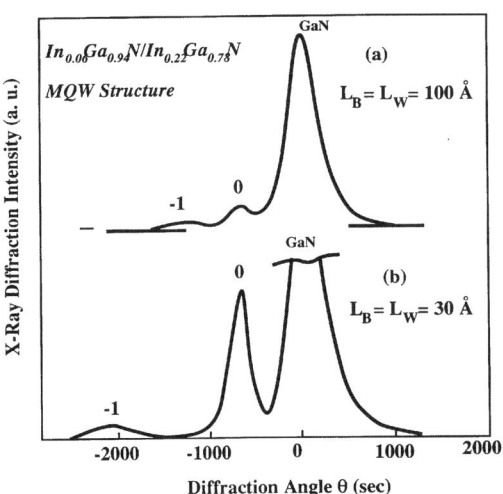

FIG. 36. X-ray rocking curve (XRC) for (0002) diffraction from the $In_xGa_{1-x}N/In_yGa_{1-y}N$ multiple quantum well (MQW) structures grown on the gallium nitride (GaN) films under the same growth conditions except for the period $(L_B + L_W)$. The periods were (a) 200 Å and (b) 60 Å. Reprinted from Nakamura et al. (1993c) with the permission of the American Institute of Physics.

their diffraction angle, and Θ_{SL} is the Bragg angle of the zeroth-order peak. Using this equation, the thicknesses of the periods ($L_B + L_W$) were estimated to be 194 and 64 Å for MQW-100 and MQW-30, respectively. These values are almost equal to the values determined from the GaN growth rate.

Figure 37 shows the results of room temperature PL measurements of the $In_{0.22}Ga_{0.78}N/In_{0.06}Ga_{0.94}N$ MQW structures, curve (a) for MQW-100 and curve (b) for MQW-30. MQW-100 shows a strong sharp peak at 420 nm (2.952 eV) and MQW-30 exhibits a peak at 412 nm (3.010 eV). No DL emission peak is shown on either curve. The intensities of these peak emissions were about twice as strong as the band-edge emission of bulk InGaN, and the FWHMs were 26 and 22 nm for samples A and B. These emissions are considered to be radiative transitions between quantum energy levels in the MQW structures.

In summary, high-quality $In_{0.22}Ga_{0.78}N/In_{0.06}Ga_{0.94}N$ MQW structures were grown on GaN films with periods of 60 and 200 Å. The XRC measurements showed satellite peaks, which indicated the existence of the $In_{0.22}Ga_{0.78}N/In_{0.06}Ga_{0.94}N$ MQW structure. The quantum effects were observed through room temperature PL mesurements. These high-quality QW structures can be used for the active layers of blue LEDs and LDs.

4. InGaN Light-Emitting Diodes with Quantum Well Structures

High-brightness blue and blue-green InGaN/AlGaN DH LEDs with a luminous intensity of 2 cd have been fabricated and are now commercially available, as mentioned earlier (Nakamura, 1994a, 1995a). In order to obtain blue and blue-green emission centers in these InGaN/AlGaN DH LEDs, Zn doping into the InGaN active layer was performed. Although these LEDs produced high-power light output in the blue and blue-green regions with a broad emission spectrum (FWHM = 70 nm), green or yellow LEDs with peak wavelengths longer than 500 nm have not been fabricated. The longest peak wavelength of the EL of these LEDs achieved thus far is 500 nm (blue-green), because the crystal quality of the InGaN active layer of DH LEDs becomes poor when the In mole fraction is increased to obtain a green band-edge emission.

On the other hand, in conventional green GaP LEDs the external quantum efficiency is only 0.1% due to the indirect transition bandgap material, and the peak wavelength is 555 nm (yellowish green) (Craford, 1992). As another material for green emission devices, AlInGaP has been used. Presently, green AlInGaP LEDs have an emission wavelength of 564 nm (yellowish green) and an external quantum efficiency of 0.6% (Kish *et al.*, 1994). When the emission wavelength is reduced to the green region, the

8 GROUP III-V NITRIDE-BASED ULTRAVIOLET LEDS AND LASER DIODES   427

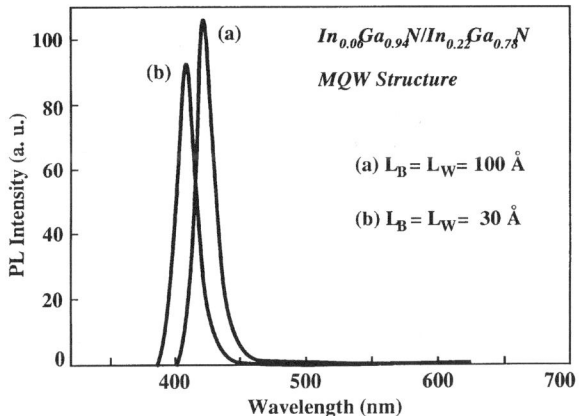

FIG. 37. Room temperature photoluminescent (PL) spectra of the $In_xGa_{1-x}N/In_yGa_{1-y}N$ multiple quantum well (MQW) structures grown on the gallium nitride films under the same growth conditions except for the period ($L_B + L_W$). The periods were (a) 200 Å and (b) 60 Å. Reprinted from Nakamura et al. (1993c) with the permission of the American Institute of Physics.

external quantum efficiency decreases sharply because the band structure of AlInGaP approaches the indirect region. Therefore, high-brightness pure green LEDs, having a high efficiency above 1% and a peak wavelength between 510 and 530 nm with a narrow FWHM, have not been commercialized yet.

Among II-VI materials, ZnSSe- and ZnCdSe-based materials have been intensively studied for use in green light-emitting devices, and much progress has been made recently, as discussed earlier. However, the lifetime of II-VI-based devices is still short, which prevents commercialization of II-VI-based devices at present.

Violet InGaN/AlGaN DH LEDs with a narrow spectrum (FWHM = 10 nm) at a peak wavelength of 400 nm originating from the BB emission of InGaN were discussed previously (Nakamura, 1994b). However, the output power and the external quantum efficiency of these violet LEDs were only 1 mW and 1.6%, respectively, probably due to the formation of misfit dislocation in the thick InGaN active layer (about 1000 Å) caused by the stress introduced into the InGaN active layer due to lattice mismatch, and the difference in thermal expansion coefficients between the InGaN active layer and AlGaN cladding layers. When the InGaN active layer becomes thin, the elastic strain is not relieved by the formation of misfit dislocations and the crystal quality of the InGaN active layer improves. High-quality InGaN MQW structures with 30-Å well and 30-Å barrier layers have been described earlier (Nakamura et al., 1993c). Here, we describe QWS LEDs having a thin InGaN active layer (about 20 Å) in order to obtain high-power emission in the region from blue to yellow with a narrow emission spectrum (Nakamura et al., 1995a).

The green LED device structures (Fig. 38) consist of a 300-Å GaN buffer layer grown at a low temperature (550°C), a 4-μm-thick layer of n-type GaN:Si, a 1000-Å-thick layer of n-type $Al_{0.1}Ga_{0.9}N$:Si, a 500-Å-thick layer of n-type $In_{0.05}Ga_{0.95}N$:Si, a 20-Å-thick active layer of undoped $In_{0.43}Ga_{0.57}N$, a 1000-Å-thick layer of p-type $Al_{0.1}Ga_{0.9}N$:Mg, and a 0.5-μm-thick layer of p-type GaN:Mg. The active region forms an SQW structure consisting of a 20-Å $In_{0.43}Ga_{0.57}N$ well layer sandwiched between a 500-Å n-type $In_{0.05}Ga_{0.95}N$ and a 1000-Å p-type $Al_{0.1}Ga_{0.9}N$ barrier layer. The In mole fraction of the InGaN active layer was varied between 0.2 and 0.7 in order to change the peak wavelength of the InGaN SQW LEDs from blue to yellow.

Figure 39 shows the typical EL of the blue, green, and yellow SQW LEDs with different In mole fractions of the InGaN well layer at a forward current of 20 mA. The longest emission wavelength is 590 nm (yellow). The peak wavelength and the FWHM of the typical blue SQW LEDs are 450 and 20 nm, respectively; of green, 525 nm and 45 nm, respectively; and of yellow, 590 and 90 nm, respectively. When the peak wavelength becomes longer, the value of the FWHM of the EL spectrum increases, probably due to strain between the well and barrier layers of the SQW, which is caused by the mismatch of the lattice and the thermal expansion coefficients between the well and barrier layers.

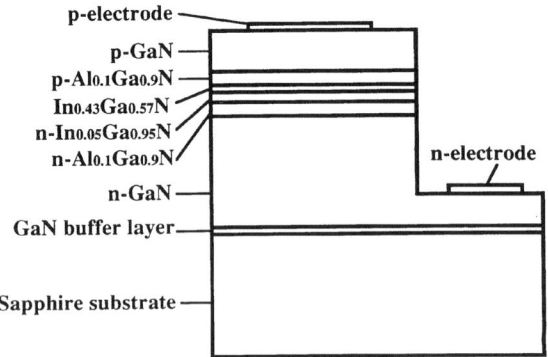

FIG. 38. The structure of green single quantum well light-emitting diode. GaN, gallium nitride; AlGaN, aluminium gallium nitride; InGaN, indium gallium nitride. Reprinted from Nakamura et al. (1995a) with the permission of the Japanese Journal of Applied Physics.

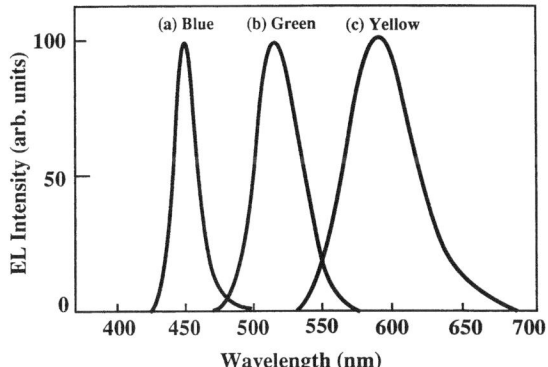

FIG. 39. Electroluminescence (EL) of (a) blue, (b) green, and (c) yellow single quantum well light-emitting diodes at a forward current of 20 mA. Reprinted from Nakamura *et al.* (1995a) with the permission of the Japanese Journal of Applied Physics.

In the green SQW, the In mole fraction of the InGaN active layer is 0.43, corresponding to the band-edge emission wavelength of $In_{0.43}Ga_{0.57}N$ of 490 nm under stress-free conditions (Nakamura, 1994b). On the other hand, the emission wavelength of the green SQW LED is 525 nm. The energy difference between the peak wavelength of the EL and the stress-free bandgap energy is approximately 170 meV. In order to explain this bandgap narrowing of InGaN in the SQW, quantum size effects and exciton effects of the active layer and the mismatch of the lattice and thermal expansion coefficients between well and barrier layers must be considered. Among these effects, the exciton effects and the tensile stress caused by the difference in thermal expansion coefficients between well and barrier layers may be primarily responsible for the bandgap narrowing of the InGaN in the SQW structure.

The output power of the SQW LEDs is shown as a function of the forward current in Fig. 40. The output power of the blue SQW LEDs increases slightly sublinearly up to 40 mA as a function of the forward current. Above 60 mA, the output power almost saturates, probably due to the generation of heat. At 20 mA, the output power and the external quantum efficiencies of the blue SQW LED are 4 mW and 7.3%, respectively, which are much higher than those of the InGaN/AlGaN DH LEDs (1.5 mW and 2.7%). Those of the green SQW LEDs are 1 mW and 2.1%, respectively, and those of the yellow are 0.5 mW and 1.2%, respectively. The output power of the green and yellow SQW LEDs is relatively small in comparison with that of the blue ones, probably due to the poor crystal quality of the InGaN well layer, which has a large lattice mismatch and a large difference in thermal expansion coefficients between well and barrier

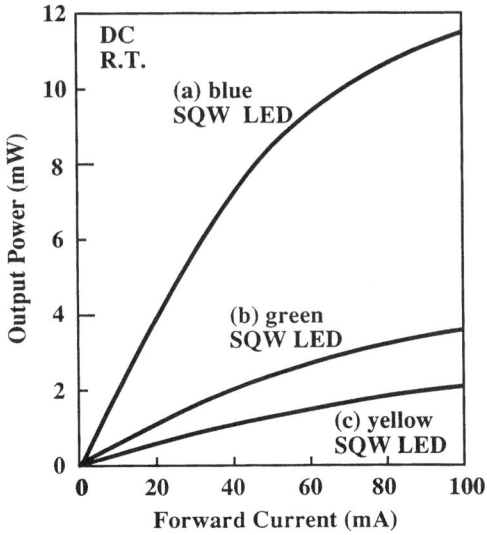

FIG. 40. The output power of (a) blue, (b) green, and (c) yellow single quantum well (SQW) light-emitting diodes (LEDs) as a function of the forward current. R.T., room temperature. Reprinted from Nakamura et al. (1995a) with the permission of the Japanese Journal of Applied Physics.

layers. A typical on-axis luminous intensity of the green SQW LED with a 10-degree cone viewing angle is 4 cd at 20 mA.

The output power decreases when the peak wavelength becomes longer, probably due to the large strain between well and barrier layers. The output powers of the green and yellow LEDs are 1 and 0.5 mW (at 525 and 590 nm), respectively. The conventional green GaP LED with a peak wavelength of 555 nm has an output power of 0.04 mW (Craford, 1992). Also, the output power of the green AlInGaP LEDs with a peak wavelength of 564 nm is 0.3 mW (Kish et al., 1994). Therefore, the performance of the green InGaN SQW LEDs is much higher than that of conventional yellowish green LEDs. Also, the luminous intensity of the InGaN green SQW LEDs (4 cd) is about 40 times higher than that of conventional green GaP LEDs (0.1 cd), and the color of the InGaN SQW LEDs is greener than for conventional GaP and AlInGaP LEDs. A typical example of the I-V characteristics of the green SQW LEDs shows that the forward voltage is 3.6 V at 20 mA.

5. HIGH-BRIGHTNESS GREEN AND BLUE LIGHT-EMITTING DIODES

In order to obtain a longer peak emission wavelength in the green and yellow, InGaN SQW LEDs with the structure of p-AlGaN/InGaN/n-

InGaN/n-AlGaN have been developed, as mentioned earlier (Nakamura et al., 1995a). However, the external quantum efficiency of green SQW LEDs (2.1%) is not very high in comparison with that of the II-VI-based green LEDs discussed in Part IV, Section 2. However, the short lifetime of the II-VI-based devices prevents commercialization at present. Considering that the emission of InGaN SQW LEDs originates from BB emission of InGaN, the FWHM of the EL of those SQW LEDs (45 nm) is wide, probably due to the strain between the well and barrier layers of the SQW structure, which is caused by the mismatch of the lattice and the difference in the thermal expansion coefficients between the well and barrier layers. Here, blue and green SQW LEDs with the structure of p-AlGaN/InGaN/n-GaN are described in order to improve the output power and spectrum width of green InGaN SQW LEDs (Nakamura et al., 1995c).

The green LED device structures (Fig. 41) consist of a 300-Å GaN buffer layer grown at a low temperature (550°C), a 4-μm-thick layer of n-type GaN:Si, a 30-Å-thick active layer of undoped $In_{0.45}Ga_{0.55}N$, a 1000-Å-thick layer of p-type $Al_{0.2}Ga_{0.8}N$:Mg, and a 0.5-μm-thick layer of p-type GaN:Mg. The active region is an SQW structure consisting of a 30-Å-thick $In_{0.45}Ga_{0.55}N$ well layer sandwiched between a 4-μm-thick n-type GaN barrier layer and a 1000-Å-thick p-type $Al_{0.2}Ga_{0.8}N$ barrier layer. Unlike the previous p-AlGaN/InGaN/n-InGaN/n-AlGaN SQW LEDs (Nakamura et al., 1995a), n-InGaN and n-AlGaN barrier layers were replaced by the n-GaN barrier layer in the present SQW structure. Fabri-

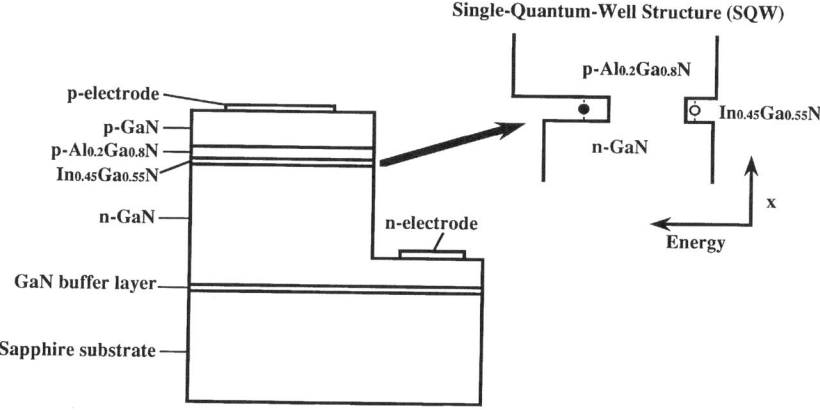

FIG. 41. The structure of InGaN green single quantum well (SQW) light-emitting diodes (LEDs). GaN, gallium nitride; AlGaN, aluminium gallium nitride. Reprinted from Nakamura et al. (1995c) with the permission of the Japanese Journal of Applied Physics.

cation of LED chips was accomplished as follows. The surface of the p-type GaN layer was partially etched until the n-type GaN layer was exposed. Next, an Ni–Au contact was evaporated onto the p-type GaN layer and a Ti–Al contact onto the n-type GaN layer. The wafer was cut into rectangular chips (350 × 350 μm). These chips were set on a lead frame, and were then molded. The characteristics of LEDs were measured under dc-biased conditions at room temperature.

Figure 42 shows the typical EL of the blue and green SQW LEDs with different In mole fractions of the InGaN well layer at a forward current of 20 mA. The peak wavelength and the FWHM of the typical blue SQW LEDs are 450 and 20 nm, respectively, and those of the green SQW LEDs are 520 and 30 nm, respectively. When the peak wavelength becomes longer, the FWHM of the EL spectra increases, probably due to the strain between the well and barrier layers of the SQW, which is caused by mismatch of the lattice and the thermal expansion coefficients between the well and barrier layers.

In the green SQW LED, the In mole fraction of the InGaN active layer is 0.45, corresponding to a band-edge emission wavelength of 495 nm under stress-free conditions. On the other hand, the peak wavelength of the green SQW LEDs is 520 nm. The energy difference between the peak wavelength of the EL and the stress-free bandgap energy is approximately 120 meV. In order to explain this bandgap narrowing of InGaN in the SQW, the quantum size and exciton effects of the active layer and the mismatch of the lattice and the thermal expansion coefficients between well and barrier layers must be considered. Among these effects, the exciton effects and the tensile stress caused by the difference in thermal expansion coefficients

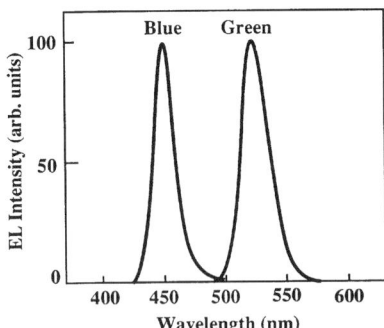

FIG. 42. Electroluminescence (EL) of (a) blue and (b) green single quantum well light-emitting diodes at a forward current of 20 mA. Reprinted from Nakamura et al. (1995c) with the permission of the Japanese Journal of Applied Physics.

between well and barrier layers may be primarily responsible for the bandgap narrowing of InGaN in the SQW structure.

The output power of the SQW LEDs is shown as a function of the forward current in Fig. 43. The output power of the blue and green SQW LEDs increases sublinearly up to 40 mA as a function of the forward current. Above 60 mA, the output power almost saturates, probably due to the generation of heat. At 20 mA, the output power and the external quantum efficiency of the blue SQW LEDs are 5 mW and 9.1%, respectively, which are much higher than those of InGaN/AlGaN DH LEDs (1.5 mW and 2.7%) (Nakamura, 1994a; Nakamura et al., 1994b). Those of the green SQW LEDs are 3 mW and 6.3%, respectively. The output power of the green LEDs is relatively low in comparison with that of the blue SQW LEDs, probably due to the poor crystal quality of the InGaN well layer, which has a large lattice mismatch and a large difference in thermal expansion coefficients between the well and barrier layers. A typical on-axis luminous intensity of the green SQW LEDs with a 10-degree cone viewing angle is 12 cd at 20 mA. These values of output power, external quantum efficiency, and luminous intensity of green SQW LEDs are the highest ever reported for green LEDs.

The output power and external quantum efficiency are much higher than those of previous green InGaN SQW LEDs with the structure of p-AlGaN/

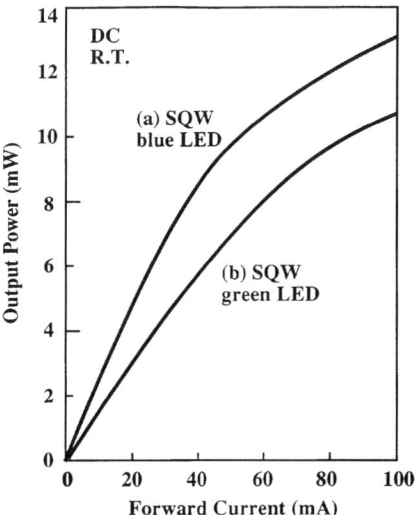

FIG. 43. The output power of (a) blue and (b) green single quantum well light-emitting diodes (LEDs) as a function of the forward current. R.T., room temperature. Reprinted from Nakamura et al. (1995c) with the permission of the Japanese Journal of Applied Physics.

InGaN/n-InGaN/n-AlGaN. The spectrum width of the present green SQW LEDs (30 nm) is narrower than that of the green SQW LEDs discussed previously (45 nm). The n-type barrier layer was composed of two thin layers, namely, a 500-Å-thick layer of $In_{0.05}Ga_{0.95}N$ and a 1000-Å-thick layer of n-type $Al_{0.1}Ga_{0.9}N$. Therefore, the crystal quality of the n-type barrier layer was poor, probably due to the formation of misfit dislocations in the 500-Å-thick layer of $In_{0.05}Ga_{0.95}N$ caused by the stress due to the mismatch of the lattice and the difference in the thermal expansion coefficients between the 500-Å-thick layer of $In_{0.05}Ga_{0.95}N$ and the 1000-Å-thick layer of $Al_{0.1}Ga_{0.9}N$, in comparison with the present n-type barrier layer, which is composed of one thick n-type GaN layer.

Figure 44 is a chromaticity diagram in which the blue and green InGaN SQW LEDs are shown. Commercially available green GaP LEDs, green AlInGaP LEDs, and red GaAlAs LEDs are also shown. The color range of light emitted by a full-color LED lamp in the chromaticity diagram is shown as the region inside each triangle, which is drawn by connecting the

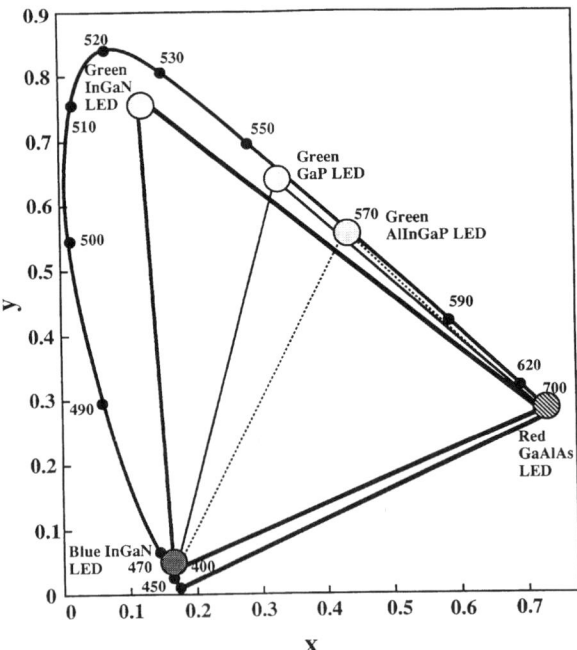

FIG. 44. Chromaticity diagram in which blue InGaN single quantum well (SQW) light-emitting diode (LED), green InGaN SQW LED, green gallium phosphide (GaP) LED, green AlInGaP LED, and red GaAlAs LED are shown. Reprinted from Nakamura et al. (1995c) with the permission of the Japanese Journal of Applied Physics.

positions of three primary color LED lamps. Three color ranges (triangles) are shown for differences only in the green LED (green InGaN, green GaP, and green AlInGaP LEDs). In Fig. 44, the color range of lamps composed of a blue InGaN SQW LED, a green InGaN SQW LED, and a red GaAlAs LED is the widest. This means that the InGaN blue and green SQW LEDs show much better color and color purity in comparison with other blue and green LEDs.

Table I summarizes the comparison of commercially available green and blue LEDs in terms of luminous intensity, output power, and external quantum efficiency. It shows that the peak wavelength of green and blue InGaN SQW LEDs is much shorter than those of conventional green GaP and blue SiC LEDs. The output powers and external quantum efficiencies of III-V nitride LEDs also are much higher than those of conventional green and blue LEDs. Judging from Table I, InGaN SQW LEDs have the highest performance in terms of luminous intensity, output power, and external quantum efficiency compared with green and blue LEDs fabricated using other materials. Concerning II-VI materials, a ZnTeSe DH green LED has been reported (Eason et al., 1995). The output power, external quantum efficiency, and peak wavelength of those II-VI LEDs are 1.3 mW, 5.3%, and 512 nm, respectively, at a forward current of 10 mA. The ZnCdSe DH blue LEDs are reported with output power, external quantum efficiency, and peak wavelength of 0.3 mW, 1.3%, and 489 nm, respectively, at a forward current of 10 mA. The lifetime of these II-VI-based LEDs is still short, which prevents their commercialization at present.

TABLE I

COMPARISON OF VARIOUS GREEN AND BLUE LEDs AT A FORWARD CURRENT OF 20 mA EXCEPT II-VI BASED LEDs WHICH MEASURED AT A FORWARD CURRENT OF 10 mA

Color	Material	Peak wavelength (nm)	Luminous intensity (mcd)	Output power (μW)	External quantum efficiency (%)
Green	AlInGaP	570	1000	400	1.0
	GaP	555	100	40	0.1
	ZnTeSe*	512	4000	1300	5.3
	InGaN	520	12000	3000	6.3
Blue	SiC	470	20	20	0.04
	ZnCdSe*	489	700	327	1.3
	InGaN	450	2500	5000	9.1

Under a forward current of 20 mA.
*Under a forward current of 10 mA.

Superbright green InGaN SQW LEDs with a luminous intensity of 12 cd have been fabricated. The output power, external quantum efficiency, peak wavelength, and spectral width of these green SQW LEDs were 3 mW, 6.3%, 520 nm, and 30 nm, respectively, at a forward current of 20 mA. By combining high-power and high-brightness blue InGaN SQW, green InGaN SQW, and red GaAlAs LEDs, many kinds of applications, such as LED full-color displays and LED white lamps for use in place of light bulbs or fluorescent lamps, are now possible with characteristics of high reliability, high durability, and low energy consumption.

V. InGaN Multiple Quantum Well Structure Laser Diodes

High-brightness blue and green LEDs have been fabricated using III-V nitride materials and are now commercially available, as mentioned previously. At present, the main focus of III-V nitride research is the realization of a current-injected laser diode, which is expected to be the shortest-wavelength semiconductor laser diode ever demonstrated. Optically pumped stimulated emission from GaN was first observed over 20 years ago (Dingle et al., 1971). Since then, many researchers have been working toward the optimization of the design for vertical cavity surface-emitting lasers and conventional separate confinement heterostructure edge-emitting lasers on various substrates (Khan et al., 1994; Amano et al., 1994; Zubrilov et al., 1995). However, stimulated emission has been observed only by optical pumping, and not by current injection. Here, LDs fabricated using wide bandgap III-V nitride materials are described (Nakamura et al., 1996a). These LDs emitted coherent light at 420 nm from an InGaN-based MQW structure under pulsed current injection at room temperature.

The InGaN MQW LD device (Fig. 45) consisted of a 300-Å GaN buffer layer grown at a low temperature of 550°C, a 3-μm-thick layer of n-type GaN:Si, a 0.1-μm-thick layer of n-type $In_{0.1}Ga_{0.9}N$:Si, a 0.4-μm-thick layer of n-type $Al_{0.15}Ga_{0.85}N$:Si, a 0.1-μm-thick layer of n-type GaN:Si, 7 periods of an $In_{0.2}Ga_{0.8}N$–$In_{0.05}Ga_{0.95}N$ MQW structure consisting of 25-Å-thick $In_{0.2}Ga_{0.8}N$ well layers and 50-Å-thick $In_{0.05}Ga_{0.95}N$ barrier layers, a 200-Å-thick layer of p-type $Al_{0.2}Ga_{0.8}N$:Mg, a 0.1-μm-thick layer of p-type GaN:Mg, a 0.4-μm-thick layer of p-type $Al_{0.15}Ga_{0.85}N$:Mg, and a 0.5-μm-thick layer of p-type GaN:Mg.

The 0.1-μm-thick layer of n-type $In_{0.1}Ga_{0.9}N$ served as a buffer layer for the thick AlGaN film growth to prevent cracking of the film. The 200-Å-thick layer of p-type $Al_{0.2}Ga_{0.8}N$ was used to prevent dissociation of

InGaN violet MQW LDs

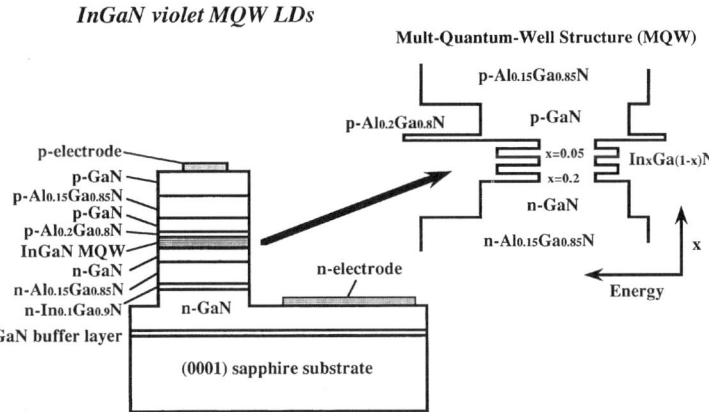

FIG. 45. The structure of the InGaN violet multiple quantum well structure laser diode. GaN, gallium nitride; AlGaN, aluminium gallium nitride.

InGaN layers during the growth of the p-type layers. The 0.1-μm-thick n-type and p-type GaN layers were light-guiding layers. The 0.4-μm-thick n-type and p-type $Al_{0.15}Ga_{0.85}N$ layers were cladding layers for confinement of light emitted from the active region of the InGaN MQW structure.

It is difficult to cleave the GaN crystal grown on the c-face sapphire substrate. Therefore, reactive ion etching (RIE) was employed to form mirror cavity facets. The surface of the p-type GaN layer was partially etched with Cl_2 plasma until the n-type GaN layer was exposed to make a stripe LD, the area of which was 20 × 700 μm. High-reflection facet coatings (70%) were used to reduce the threshold current. An Ni–Au contact was evaporated onto the p-type GaN layer with a 10-μm stripe width, and a Ti–Al contact onto the n-type GaN layer. The electrical characteristics of LDs were measured under pulsed current-biased conditions (pulse width, 2 μsec; pulse period, 2 msec) at room temperature. The output power from one facet was measured by an Si photodetector.

Figure 46 shows the voltage-current (V-I) characteristics and the light-output power per coated facet of one of these devices as a function of the pulsed forward current (L-I). The stimulated emission was not observed up to the threshold current of 610 mA, which corresponded to a threshold current density of 8.7 kA/cm², which was calculated by assuming that the current flow was confined only below the electrode. The differential quantum efficiency of 17% per facet and pulsed output power of 57 mW per facet were obtained at a current of 700 mA. The operating voltage of this device at the threshold current was 21 V.

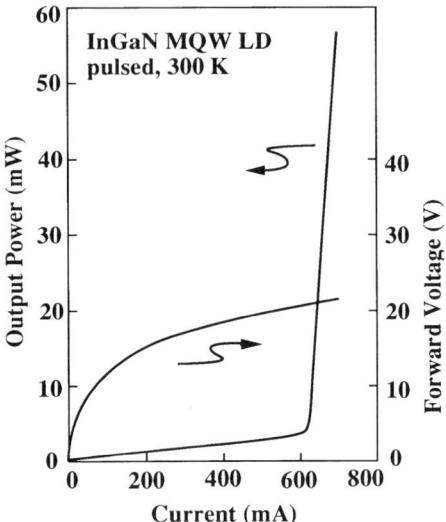

FIG. 46. The L-I and V-I characteristics of the InGaN multiple quantum well (MQW) laser diode (LD). Reprinted from Nakamura *et al.* (1996c) with the permission of the American Institute of Physics.

Figure 47 shows optical spectra of one of the InGaN MQW LDs. These spectra were measured using an HR-640 Monochromator (JOBIN YVON) (Hewlett-Packard Company, San Jose, CA), which had a resolution of 0.016 nm. At injection currents below the threshold, spontaneous emission that had a FWHM of 22 nm and a peak wavelength of 417 nm was observed, as shown in Fig. 47(a). Above the threshold current, strong stimulated emissions were observed. At a current above 610 mA, many longitudinal mode emissions with a narrow spectrum width were observed, as shown in Figs. 47(b) and 47(c).

Figure 48 shows the polarized light-output intensity as a function of the current for the sample with a laser threshold current of 610 mA. The transverse electric (TE) polarized light-output intensity increased to a much larger value than did the transverse magnetic (TM) polarized light-output intensity, above the threshold current. This demonstrates that the emission is strongly TE polarized, and indicates the laser operation at a current above 610 mA.

Typical far-field radiation patterns of the InGaN MQW laser structure in the planes parallel and perpendicular to the junction are shown in Fig. 49. The beam full-width at half-power (FWHP) level for the parallel and perpendicular far-field patterns are 5 and 17 degrees, respectively.

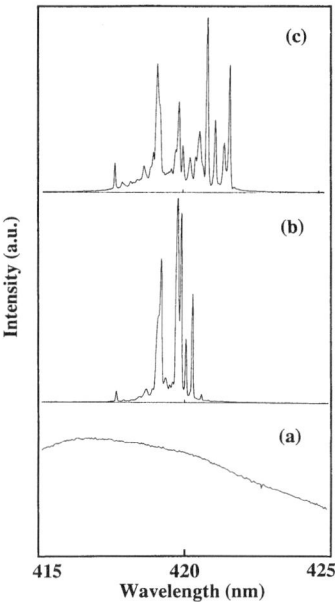

FIG. 47. The optical spectra for the InGaN multiple quantum well laser diode, at a current of (a) 580 mA, (b) 630 mA, and (c) 660 mA. Intensity scales for these three spectra are in arbitrary units, and are different. Reprinted from Nakamura et al. (1996c) with the permission of the American Institute of Physics.

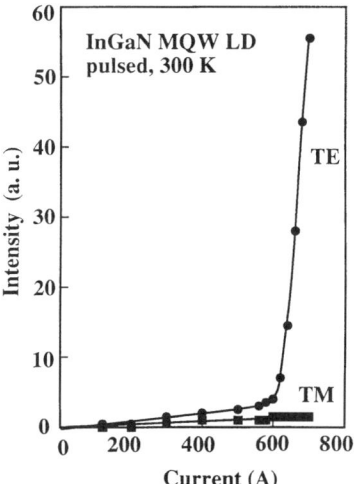

FIG. 48. Polarized output intensity against current. MQW, multiple quantum well; LD, laser diode; InGaN, indium gallium nitride. Reprinted from Nakamura et al. (1996c) with the permission of the American Institute of Physics.

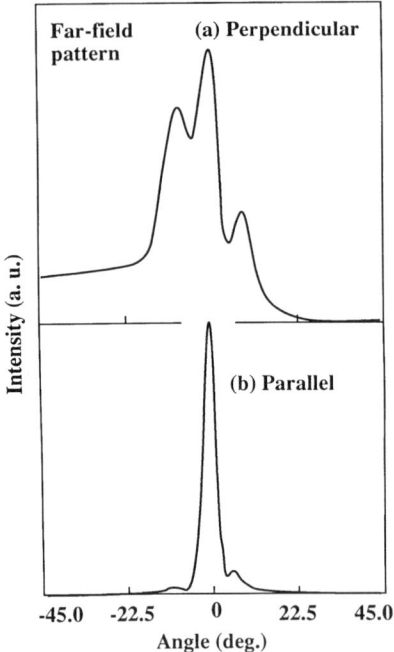

FIG. 49. (a) Perpendicular far-field radiation pattern of the InGaN multiple quantum well (MQW) laser diode. (b) Parallel far-field radiation pattern of the InGaN MQW laser diode. Reprinted from Nakamura et al. (1996c) with the permission of the American Institute of Physics.

Figure 50 (see color plate section) shows the streak line of the laser emission from one of the LD chips, which was operated under pulsed current injection at room temperature.

A violet LD was fabricated from III-V nitride materials for the first time. It has been confirmed that the device operates with lasing action from an L-I curve, a narrowing spectrum, and TE polarization results. The stimulated emission was observed at a wavelength of around 420 nm at room temperature, under pulsed current injection, from an AlGaN/GaN/InGaN MQW, separate-confinement heterostructure.

VI. Summary

Highly efficient InGaN/AlGaN DH blue LEDs with an external quantum efficiency of 5.4% were fabricated by codoping Zn and Si into the InGaN active layer. The output power was as high as 3 mW at a forward current of 20 mA. The peak wavelength and the FWHM of the EL of the blue LEDs were 450 nm and 70 nm, respectively. Blue-green LEDs with a brightness of 2 cd were fabricated by increasing the In mole fraction of the InGaN active layer.

High-brightness blue LEDs with a luminous intensity over 1 cd will pave the way toward realization of full-color LED displays, especially for outdoor use. In Japan, total power consumption by traffic lights reaches the gigawatt range. The InGaN/AlGaN blue-green LED traffic lights, with an electrical power consumption 12% that of present incandescent bulb traffic lights, hold promise for vast energy savings. With its extremely long lifetime of several tens of thousands of hours, the replacement of burned-out traffic light bulbs will be dramatically reduced. Using these high-brightness blue-green LEDs, safe energy-efficient roadway and railway signals can be achieved in the near future.

Superbright InGaN green SQW LEDs have been fabricated. The luminous intensity was 12 cd and the external quantum efficiency was as high as 6.3% at a forward current of 20 mA at room temperature. The peak wavelength and the FWHM of the green LEDs were 520 and 30 nm, respectively, and those of the blue LEDs were 450 and 20 nm, respectively. The color of the green InGaN SQW LEDs was greener than for conventional GaP and AlInGaP LEDs. Fabrication of practical visible LEDs in the range from blue to yellow is possible at present using III-V nitride materials. The luminous intensity of those green InGaN SQW LEDs (12 cd) was about 100 times higher than that of conventional green GaP LEDs (0.1 cd). By combining a high-power high-brightness blue InGaN SQW LED, a green InGaN SQW LED, and a red GaAlAs LED, many kinds of applications, such as LED full-color displays and LED white lamps for use in place of light bulbs or fluorescent lamps, are now possible with characteristics of high reliability, high durability, and low energy consumption.

Also, very recently, III-V nitride-based LDs were fabricated for the first time. These LDs emitted coherent light at 420 nm from an InGaN-based MQW structure under pulsed current injection at room temperature (Nakamura et al., 1996a). These results indicate a possibility that the short-wavelength LDs from green to uv will be realized in the near future using III-V nitride material.

References

Akasaki, I., Amano, H., Koide, Y., Hiramatsu, K., and Sawaki, N. (1989). *J. Cryst. Growth* **98**, 209.
Amano, H., and Akasaki, I. (1991). *Oyo Buturi* **60**, 163.

Amano, H., Sawaki, N., Akasaki, I., and Toyoda, Y. (1986). *Appl. Phys. Lett.* **48**, 353.
Amano, H., Kito, M., Hiramatsu, K., and Akasaki, I. (1989). *Jpn. J. Appl. Phys.* **28**, L2112.
Amano, H., Tanaka, T., Kunii, Y., Kato, K., Kim, S. T., and Akasaki, I. (1994). *Appl. Phys. Lett.* **64**, 1377.
Bergman, P., Ying, G., Monemar, B., and Holz, P. O. (1987). *J. Appl. Phys.* **61**, 4589.
Brandt, M. S., Johnson, N. M., Molnar, R. J., Singh, R., and Moustakas, T. D. (1994). *Appl. Phys. Lett.* **64**, 2264.
Craford, M. G. (1992). *Circuits & Devices*, September, p. 24.
Dingle, D., Shaklee, K. L., Leheny, R. F., and Zetterström, R. B. (1971). *Appl. Phys. Lett.* **19**, 5.
Eason, D. E., Yu, Z., Hughes, W. C., Roland, W. H., Boney, C., Cook, J. W., Jr., Schetzina, J. F., Cantwell, G., and Harasch, W. C. (1995). *Appl. Phys. Lett.* **66**, 115.
Edmond, J., Kong, H., and Dmitrieve, V. (1994). *Conf. Ser.—Inst. Phys.* **137**, 515.
Khan, M. A., Krishnankutty, S., Skogman, R. A., Kuznia, J. N., and Olson, D. T. (1994). *Appl. Phys. Lett.* **65**, 520.
Kish, F. A., Steranka, F. M., DeFevere, D. C., Vanderwater, D. A., Park, K. G., Kuo, C. P., Osentowski, T. D., Peanasky, M. J., Yu, J. G., Fletcher, R. M., Steigerwald, D. A., Craford, M. G., and Robbins, V. M. (1994). *Appl. Phys. Lett.* **64**, 2839.
Koga, K., and Yamaguchi, T. (1991). *Prog. Cryst. Growth Charact.* **23**, 127.
Matsuoka, T., Tanaka, H., Sasaki, T., and Katsui, A. (1989). *Conf. Ser.—Inst. Phys.* **106**, 141.
Morkoç, H., Strite, S., Gao, G. B., Lin, M. E., Sverdlov, B., and Burns, M. (1994). *J. Appl. Phys.* **76**, 1363.
Nagatomo, T., Kuboyama, T., Minamino, H., and Omoto, O. (1989). *Jpn. J. Appl. Phys.* **28**, L1334.
Nakamura, S. (1991). *Jpn. J. Appl. Phys.* **30**, L1705.
Nakamura, S. (1994a). *Nikkei Electron. Asia* **6**, 65.
Nakamura, S. (1994b). *Microelectron. J.* **25**, 651.
Nakamura, S. (1994c). *J. Cryst. Growth* **145**, 911.
Nakamura, S. (1995a). *J. Vac. Sci. Technol., A* [2] **13**, 705.
Nakamura, S. (1995b). *Circuits and Devices*, May, p. 19.
Nakamura, S., and Mukai, T. (1992). *Jpn. J. Appl. Phys.* **31**, L1457.
Nakamura, S., Senoh, M., and Mukai, T. (1991a). *Jpn. J. Appl. Phys.* **30**, L1708.
Nakamura, S., Mukai, T., and Senoh, M. (1991b). *Jpn. J. Appl. Phys.* **30**, L1998.
Nakamura, S., Harada, Y., and Senoh, M. (1991c). *Appl. Phys. Lett.* **58**, 2021.
Nakamura, S., Iwasa, N., Senoh, M., and Mukai, T. (1992a). *Jpn. J. Appl. Phys.* **31**, 1258.
Nakamura, S., Mukai, T., and Senoh, M. (1992b). *J. Appl. Phys.* **71**, 5543.
Nakamura, S., Mukai, T., and Senoh, M. (1992c). *Jpn. J. Appl. Phys.* **31**, 2883.
Nakamura, S., Mukai, T., Senoh, M., and Iwasa, N. (1992d). *Jpn. J. Appl. Phys.* **31**, L139.
Nakamura, S., Mukai, T., and Senoh, M. (1993a). *Jpn. J. Appl. Phys.* **32**, L16.
Nakamura, S., Senoh, M., and Mukai, T. (1993b). *Jpn. J. Appl. Phys.* **32**, L8.
Nakamura, S., Mukai, T., Senoh, M., Nagahama, S., and Iwasa, N. (1993c). *J. Appl. Phys.* **74**, 3911.
Nakamura, S., Mukai, T., and Senoh, M. (1994a). *Appl. Phys. Lett.* **64**, 1687.
Nakamura, S., Mukai, T., and Senoh, M. (1994b). *J. Appl. Phys.* **76**, 8189.
Nakamura, S., Senoh, M., Iwasa, N., and Nagahama, S. (1995a). *Jpn. J. Appl. Phys.* **34**, L797.
Nakamura, S., Senoh, M., Iwasa, N., and Nagahama, S. (1995b). *Appl. Phys. Lett.* **67**, 1868.
Nakamura, S., Senoh, M., Iwasa, N., Nagahama, S., Yamada, T., and Mukai, T. (1995c). *Jpn. J. Appl. Phys.* **34**, L1332.
Nakamura, S., Senoh, M., Nagahama, S., Iwasa, N., Yamada, T., Matsushita, T., Kiyoku, H., and Sugimoto, Y. (1996a). *Jpn. J. Appl. Phys.* **35**, L74.
Nakamura, S., Senoh, M., Nagahama, S., Iwasa, N., Yamada, T., Matsushita, T., Kiyoku, H.,

and Sugimoto, Y. (1996b). *Appl. Phys. Lett.* **68**, 2105.
Nakamura, S., Senoh, M., Nagahama, S., Iwasa, N., Yamada, T., Matsushita, T., Kiyoku, H., and Sugimoto, Y. (1996c). *Appl. Phys. Lett.* **68**, 3269.
Okuyama, H., and Ishibashi, A. (1994). *Microelectron. J.* **25**, 643.
Osamura, K., Naka, S., and Murakami, Y. (1975). *J. Appl. Phys.* **46**, 3432.
Pankove, J. I., and Hutchby, J. A. (1976). *J. Appl. Phys.* **47**, 5387.
Pankove, J. I., Miller, E. A., and Berkeyheiser, J. E. (1971). *RCA Rev.* **32**, 283.
Rubin, M., Newman, N., Chan, J. S., Fu, T. C., and Ross, J. T. (1994). *Appl. Phys. Lett.* **64**, 64.
Strite, S., and Morkoç, H. (1992). *J. Vac. Sci. Technol., B* [2] **10**, 1237.
Sugawara, H., Itaya, K., and Hatakoshi, G. (1994). *Jpn. J. Appl. Phys.* **33**, 5784.
Xie, W., Grillo, D. C., Gunshor, R. L., Kobayashi, M., Jeon, H., Ding, J., Nurmikko, A. V., Hua, G. C., and Otsuka, N. (1992). *Appl. Phys. Lett.* **60**, 1999.
Yoshida, S., Misawa, S., and Gonda, S. (1983). *Appl. Phys. Lett.* **42**, 427.
Yoshimoto, N., Matsuoka, T., Sasaki, T., and Katsui, A. (1991). *Appl. Phys. Lett.* **59**, 2251.
Zavada, J. M., Wilson, R. G., Abernathy, C. R., and Pearton, S. J. (1994). *Appl. Phys. Lett.* **64**, 2724.
Zubrilov, A. S., Nikolaev, V. I., Tsvetkov, D. V., Dmitriev, V. A., Irvine, K. G., Edmond, J. A., and Carter, C. H. (1995). *Appl. Phys. Lett.* **67**, 533.

Index

A

Active layer
  AlGaAs red light-emitting diodes, 72–73
  AlGaInP light-emitting diodes, 153–70
    band structure, 154–59
      bandgap energies, 151, 154–57
      band offsets, 158–59
      carrier effective masses, 157
    double heterostructure devices, 159–69
      carrier confinement, 166–69
      carrier injection, 161–63
      carrier recombination, 163–65
      quantum well devices, 169–70
AIN. *See* Aluminum nitride
AIP. *See* Aluminum phosphide
Aircraft, light-emitting diode, high brightness in, 337
Al. *See* Aluminum
AlAs. *See* Aluminum arsenide
AlGaAs, 9, 33
  light-emitting diode, 59
AlGaAs red light-emitting diodes, 65–96
  clear epoxy, elasticity, 91
  crystal growth, 74–78
    metalorganic chemical vapor deposition, 74–75
    molecularbeam epitaxy, 74–75
    slow-cooling liquid-phase epitaxy, 75–77
    temperature-difference method, 77–79
  design, 68–73
    active layer composition, 72–73
    heterostructure design, 66, 69–71
    materials parameters, 68–69
    optical extraction, 71–73
  fabrication, 79–81
    contact formation, 79–80
    dicing, 80–81
    etching, selective, 81
  linear expansion, clear epoxy, 91
  other types, compared, 67, 81–83
  reliability, 89–95
    high-temperature, height–humidity operation, 94–95
    low-temperature operation, 89–94
  temperature-dependent properties, 83–89
    current voltage, curves, 83–84
    efficiency, 85–86
    electroluminescent spectra, 84–5
    forward voltage, 83–84
    minority-carrier lifetime, 86–89
  van der Pauw measurements, liquid-phase epitaxy-grown AlGaAs single layers, 69
AlGaInN, 18, 34–40
AlGaInP light-emitting diodes, 149–226
  active layer design, 153–70
    band structure, 154–59
      bandgap energies, 151, 154–57
      band offsets, 158–59
      carrier effective masses, 157
    double heterostructure devices, 159–69
      carrier confinement, 166–69
      carrier injection, 161–63
      carrier recombination, 163–65
      quantum well devices, 169–70
  current-blocking structures, 178–80
  current spreading, 170–78
    indium tin oxide, 176–78
    OHMIC contact modifications, 171
    P-type substrates, growth on, 172
    window layer, 172–76, 174–77
      material parameters, 174

current–voltage characteristics, 213–15
electro-luminescence spectra, 215–16
hydride vapor-phase epitaxy, 150
light extraction, 180–206
  bonding
    mechanism, 198
    technique, 198
  Bragg reflector, distributed
    design, 187–92
    light-emitting diodes, 187–95
  electrical conductivity, interfacial, 198
  epoxy bonding, 198
  metal–eutectic bonding, 198
  optical transparency, interfacial, 198
  oxide bonding, 198
  performance, 185, 188, 191–95
  semiconductor–semiconductor direct
    bonding, 198
  strength, of bond, 198
  substrate absorption, 183–87
  thermal cycling, 198
  upper window design, 181–83
  van der Waals bonding, 198
  wafer-bonded, 196–204
    techniques compared, 198
    transparent substrate, 186, 195–206
luminous efficiency, 211–12
performance characteristics, 208–19
quantum efficiency, 186, 208–11
reliability, 217–19
wafer fabrication techniques, 171, 206–8
AlGaN/GaInN DH, nitride-based
    superbright light-emitting diodes,
    380–81
  asymmetric, 372, 380–82
AlInGaN, 59
AlInGaP, 59
Alloys
  composition, diode material, 12
  optimization of, high-brightness light-
    emitting diodes, 3, 15–17
AlN. See Aluminum nitride
Aluminum
distribution coefficient, vs. temperature,
    calculated for AlInP, 20
light-emitting diode, 55
Aluminum arsenide, AlGaAs red light-
    emitting diodes, 66
Aluminum gallium arsenide. See AlGaAs
Aluminum nitride, 4, 5, 36

OMVPE of gallium nitride, 357
Aluminum phosphide, 19
  OMVPE growth, AlGaInP, 104
Ammonia, light-emitting diode, 21
Antimonide
  gallium, 2
  indium, 2
Appliances, light-emitting diode, high
    brightness in, 337
Applications, high-brightness light-emitting
    diodes, 227–356
Arcade games, light-emitting diode, high
    brightness in, 337
Arsenide
  gallium, 2
  indium, 2
Arsine, light-emitting diodes, 7
AsH_3. See Arsine
Audio equipment, light-emitting diode, high
    brightness in, 337
Automotive interior lighting, 277–96
  instrument cluster warning light, 280,
    284–89
    cavity design, 289–94, 321
    legend optimization, 280, 294–96
Automotive luminous intensity, signal
    lighting, 252–53
Automotive signal lighting, 251–77
  automotive luminous intensity, signal
    lighting, 252–53
  electrical design, signal lighting, 268–73
  luminous flux requirements, signal
    lighting, 252, 259–60, 259–63, 269
  optics design, signal lighting, 263–68
  thermal design, signal lighting, 273–77

B

Backlighting, liquid crystal display, large-
    area displays, 337, 340–41
Bandgap, direct, light-emitting diode, 12
Bandgap energy
  high bandgap materials, 36
  vs. lattice constant
    high-bandgap IIIV compounds, 3
    ternary alloys, 3
Bandgap materials, high, properties of, 36
Band structure, active layer design,
    AlGaInP light-emitting diodes,
    154–59

bandgap energies, 151, 154–57
band offsets, 158–59
carrier effective masses, 157
Bank terminal, light-emitting diode, high brightness in, 337
Binary compounds, high-brightness light-emitting diodes, 9, 17–19
Blue-green-yellow light-emitting diodes, ultraviolet, group III-V nitride-based, 391–443
Blue light-emitting diodes, high-brightness, ultraviolet, 430–36
Bonding, AlGaInP light-emitting diodes, light extraction
  mechanism, 198
  technique, 198
Bragg reflector, distributed, AlGaInP light-emitting diodes, light extraction
  design, 187–92
  light-emitting diodes, 187–95
Business commercial market, light-emitting diode, high brightness in, 337

C

Calculators, light-emitting diode, high brightness in, 337
Camcorder viewfinder, light-emitting diode, high brightness in, 337
Cash registers, light-emitting diode, high brightness in, 337
CBE. See Chemical beam epitaxy
Cellular telephone, light-emitting diode, high brightness in, 337
CH_3. See Methyl
Character height, large-area displays, 328–30
Characteristics, high-brightness light emitting diodes, 59
Chemical beam epitaxy, light-emitting diodes, 8
Chemical vapor deposition, metalorganic, AlGaAs red light-emitting
diodes, 74–75
Clear epoxy, elasticity, AlGaAs red light-emitting diodes, 91
Color
  control, OMVPE growth, AlGaInP, 141–43
  light-emitting diodes, 59

measurement, 247–51
principles of, 230–51
Color mixing, large-area displays, 325–29
Color requirements, signal lighting, 252–53
Computer market, light-emitting diode, high-brightness in, 337
Conductivity control, OMVPE of gallium nitride, 374–79
Consumer market, light-emitting diode, high-brightness in, 337
Contact formation, AlGaAs red light-emitting diodes, 79–80
Controlled doping, light-emitting diode, 12
Copiers, light-emitting diode, high brightness in, 337
Crystal growth, AlGaAs red light-emitting diodes, 74–78
  metalorganic chemical vapor deposition, 74–75
  molecular-beam epitaxy, 74–75
  slow-cooling liquid-phase epitaxy, 75–77
  temperature-difference method, 77–79
Current-blocking structures, AlGaInP light-emitting diodes, 178–80
Current spreading, AlGaInP light-emitting diodes, 170–78
  indium tin oxide, 176–78
  OHMIC contact modifications, 171
  ptype substrates, growth on, 172
  window layer, 172–77
    material parameters, 174
Current–voltage characteristics, AlGaInP light-emitting diodes, 213–15

D

Data collection terminals, light-emitting diode, high brightness in, 337
Desktop computers, light-emitting diode, high-brightness in, 337
Device issues, high-brightness in light-emitting diodes, 47–63
  future potential, 58–63
  internal quantum efficiency, 54–56
  light extraction, 56–57
  luminous efficacy, 53–54
  other light sources, compared, 54, 61–63
  overview, 47–50
  performance, 58–61
  technology status, 58–63

448   INDEX

DH. *See* Double heterostructure
Dicing, AlGaAs red light-emitting diodes, 80–81
Dimethyldrazine, light-emitting diode, 39
Direct bandgap, light-emitting diode, 12
Displays, large-area, 319–36
  alternative technologies, 320–24
  direct-view liquid crystal display backlighting, 341, 345–47
  light-emitting diode technology, 342–44
  liquid crystal display, operation, 338–40
  market characteristics, 319–20
Distributed Bragg reflector, AlGaInP light-emitting diodes, light extraction, 187–95
  design, 187–92
Doping, 24–28
  background, control of, 24–26
  high-brightness light-emitting diodes, 24–28
  intentional, 26–28
Doping range, diode material, 12
Double heterostructure, light-emitting diode, 38, 55
  high brightness, 159–69
    carrier confinement, 166–69
    carrier injection, 161–63
    carrier recombination, 163–65
    InGaN/AlGaN, ultraviolet, 419–24
    InGaN/gallium nitride, ultraviolet, 416–19
Drive circuit, large-area displays, 330–36

E

Edgelighting, liquid crystal display, large-area displays, 347–51
Effluent abatement system, OMVPE growth, AlGaInP, 138
EL. *See* Electroluminescence
Elasticity, clear epoxy, AlGaAs red light-emitting diodes, 91
Electrical design, signal lights, 268–73, 307–19
Electroluminescence spectra
  AlGaAs light-emitting diodes, 84, 85
  AlGaInP light-emitting diodes, 215–16
Electronic games, light-emitting diode, high-brightness in, 337

Etching, AlGaAs red light-emitting diodes, 81
Evolution of performance, visible light-emitting diode, 48
Exhaust system, OMVPE growth, AlGaInP, 136–37
Expansion, linear, clear epoxy, AlGaAs red light-emitting diodes, 91
Expense, of diode material, External quantum efficiency, light-emitting diode, 59
Extraction, of light, high-brightness in light-emitting diodes, 56–57

F

Fax machines, light-emitting diode, high-brightness in, 337
Fiber-optics, large area displays, light-emitting diode, 351–56
Forward voltage, AlGaAs red light-emitting diodes, 83–84

G

Ga/Sb. *See* Gallium antimonide
GaAs. *See* Gallium arsenide
GaInP
  double-crystal X-ray diffractometry, OMVPE growth, AlGaInP, 140
  photoluminescent peak wavelength contour map, OMVPE growth, AlGaInP, 139
Gallium antimonide, light-emitting diodes, 2
Gallium arsenide, 2, 48
  AlGaAs red light-emitting diodes, 66
  OMVPE growth, AlGaInP, 104
Gallium nitride, 4, 36
  growth, ultraviolet diode, 394–409
  p–n homojunction-type, nitridebased superbright light-emitting diodes, 379–80
  p–n junction light-emitting diodes, ultraviolet, 405–9
Gallium phosphide, 2, 47
  OMVPE growth, AlGaInP, 104
GaN. *See* Gallium nitride
Gas source molecularbeam epitaxy, light-emitting diodes, 7

Germanium, light-emitting diodes, 2
Green, light-emitting diodes, high-brightness, ultraviolet, 430–36
Growth rate, temperature dependence, light-emitting diode, 10
Growth reactions, kinetics, 28–29

H

Heterostructure
  design, AlGaAs red light-emitting diodes, 66, 69–71
  diode material, 12
High-bandgap III-V compounds, lattice constant, vs. bandgap energy, 3
High-bandgap materials, properties of, 36
High eye response, diode material, 12
High-temperature, height–humidity operation, AlGaAs red light-emitting diodes, 94, 94–95
High-volume manufacturing, OMVPE growth, AlGaInP, 136–43
HVPE. *See* Hydride vapor-phase epitaxy
Hydride vapor-phase epitaxy
  AlGaInP light-emitting diodes, 150
  OMVPE of gallium nitride, 359, 360–61
Hydrogen
  compensation by, diode material, 12
  passivation, acceptors, OMVPE growth, AlGaInP, 122–27

I

III-V semiconductors
  energy gap, as function of lattice constant, 67
  high-brightness light-emitting diodes, 5–11
    liquid-phase epitaxy, 6–7
    molecular-beam epitaxy, 7–8
    OMVPE, 8–11
Impurity doping, InGaN, ultraviolet diode, 413–16, 417
InAs. *See* Indium arsenide
Incandescent signal lights, light-emitting diode, compared, 298–302
Indium antimonide, 2
Indium arsenide, 2
Indium nitride, 4, 36
  diode material, 18
  OMVPE of gallium nitride, 357
Indium phosphide, 2
  OMVPE growth, AlGaInP, 104
Indium tin oxide, AlGaInP light-emitting diodes, current spreading, 176–78
Industrial market, light-emitting diode, high-brightness in, 337
Infrared, light-emitting diode, 49
InGaN
  growth, ultraviolet diode, 409–16
  impurity doping, ultraviolet diode, 413–17
  light-emitting diodes with quantum well structures, ultraviolet, 426–30
  multiple quantum well structure light-emitting diodes,
  ultraviolet, 436–40
  quantum well structures, ultraviolet, 425–27
  undoped, ultraviolet diode, 392, 409–12
InGaN/AlGaN double heterostructure light-emitting diodes,
ultraviolet, 419–24
InGaN/gallium nitride, double heterostructure light-emitting
diodes, ultraviolet, 416–19
InN. *See* Indium nitride
InSb. *See* Indium antimonide
Instrumentation, light-emitting diode, high-brightness in, 337
Instrument cluster, auto, light-emitting diode, high-brightness in, 337
Instrument cluster warning light, automotive, 280, 284–89
  cavity design, 289–94, 321
  legend optimization, 280, 294–96
Intentional doping, high-brightness light-emitting diodes, 26–28
Interior lighting, automotive, 277–96
  truck market, characteristics of, 277–79
Internal quantum efficiency, high-brightness in light-emitting diodes, 54–56
Inverse temperature, vs. vapor pressure, 18
IR. *See* Infrared

K

Kinetics, high-brightness light-emitting diodes, 17–33

L

Large-area displays, 319–36
  alternative technologies, 320–23
  character height, 329–30
  color mixing, 326–29
  direct-view liquid crystal display backlighting, 341, 345–47
  drive circuit, 330–36
  edgelighting liquid crystal display backlighting, 347–51
  fiber-optics light-emitting diode backlighting, 351–56
  light-emitting diode technology, 342–45
  liquid crystal display backlighting
    alternative technologies, 340–42
    market characteristics, 337–38
    operation, 338–40
  market characteristics, 319–20
Laser diodes, 32
  OMVPE of gallium nitride, 385–87
Lattice constant
  vs. bandgap energy
    high-bandgap IIIV compounds, 3
    ternary alloys, 3
  contours, OMVPE growth, AlGaInP, 104
Lattice matching, diode material, 12
Lattice mismatch, 29–31
  high-brightness light-emitting diodes, 29–31
Lattice parameters, high-bandgap materials, 36
Layer thickness, diode material, 12
PLCD. See Liquid crystal display
LD. See Laser diode
Light extraction
  AlGaInP light-emitting diodes, 180–206
    bonding
      mechanism, 198
      technique, 198
    Bragg reflector, distributed
    design, 187–92
    light-emitting diodes, 187–95
    electrical conductivity, interfacial, 198
    epoxy bonding, 198
    metal–eutectic bonding, 198
    optical transparency, interfacial, 198
    oxide bonding, 198
    performance, 185, 188, 191–95
    semiconductor–semiconductor direct bonding, 198
    strength, of bond, 198
    substrate absorption, 183–87
    thermal cycling, 198
    upper window design, 181–83
    van der Waals bonding, 198
    waferbonded, 196–204
    techniques, compared, 198
    transparent substrate, 186, 195–206
  high-brightness in light-emitting diodes, 56–57
Linear expansion, clear epoxy, AlGaAs red light-emitting diodes, 91
Line defects, light-emitting diode, 12
Liquid crystal display, 229
  backlighting, large area displays, 337–38
Liquid-phase epitaxy
  AlGaAs red light-emitting diodes, 67
  III-V semiconductors, high-brightness light-emitting diodes, 6–7
  light-emitting diodes, 6
Long minority carrier lifetime, light-emitting diode, 12
Low-dimensional structures
  high-brightness light-emitting diodes, 32–33
  light-emitting diode, 32–33
Low-temperature operation, AlGaAs red light-emitting diodes, 89–94
LPE. See Liquid-phase epitaxy
Luminous efficacy, light-emitting diode, 53–54, 59
  AlGaInP, 211–12
  vs. wavelength, 54
Luminous flux requirements, signal lighting, 252, 259–60, 259–63, 269, 302–7
Luminous intensity, luminous flux, conversion, 233–47

M

Marine instrumentation, light-emitting diode, high-brightness in, 337
Materials, high-brightness light-emitting diodes, 1–45
  AlGaAs, 9, 33
  AlGaInN, 18, 34–40
  AlGaInP, 34
  binary compounds, 9, 17–19

doping, 24–28
　background, control, 24–26
　intentional, 26–28
III-V semiconductors, 5–11
　liquid-phase epitaxy, 6–7
　molecular-beam epitaxy, 7–8
　OMVPE, 8–11
kinetics, 17–33
lattice mismatch, 29–31
low-dimensional structures, 32–33
materials systems, 33–40
pyrolysis, kinetics, 28–29
selection, 12–17
　alloys, optimization of, 3, 15–17
surface recombination, 31–32
ternary alloys, 18, 19–24
thermodynamics, 17–33
MBE. *See* Molecularbeam epitaxy
Medical equipment, light-emitting diode, high-brightness in, 337
MeNH. *See* Dimethyldrazine
Metalorganic chemical vapor deposition
　AlGaAs red light-emitting diodes, 74–75
　light-emitting diodes, 6
Methyl radicals, light-emitting diode, 11
Minority-carrier lifetime
　AlGaAs red light-emitting diodes, 86–89
　light-emitting diode, 12
Mismatch, lattice, 29–31
　high-brightness light-emitting diodes, 29–31
MOCVD. *See* Metalorganic chemical vapor deposition
Molecularbeam epitaxy, 6
　gallium nitride, OMVPE, 361–62
　gas source, light-emitting diodes, 7
　III-V semiconductors, high-brightness light-emitting diodes, 7–8
MOVPE. *See* Metalorganic vapor-phase epitaxy
MQW. *See* Multiple quantum well
Multiple quantum well
　InGaN, ultraviolet, 436–40
　light-emitting diode, 56

N

Native defects, compensation by, diode material, 12

Nitride-based superbright light-emitting diodes, OMVPE of gallium nitride, 379–84
　electrode, 383–84
　layered structure, 379–82
　AlGaN/GaInN DH, 380–81
　asymmetric, 372, 380–82
　gallium nitride p–n homojunctiontype, 379–80
　light-emitting diodes, thin active layers, 382–83
Nonradiative recombination center, diode material, 12, 13
Notebook computers, light-emitting diode, high-brightness in, 337
NRC. *See* Nonradiative recombination center
N-type gallium nitride, ultraviolet, 398–400

O

OMVPE, gallium nitride, 357–59, 362
　AlGaInP, 97–148
　binary compounds, lattice parameters, 104
　color control, 141–43
　constant energy bandgap, 104
　direct–indirect crossover, 117–20
　dopant-incorporation behavior, 110, 112–17
　double-crystal X-ray diffraction rocking curve, 109
　effluent abatement system, 138
　exhaust system, 138
　exhaust treatment, reactor, 136–38
　GaInP
　　doublecrystal X-ray diffractometry, 140
　　photoluminescent peak wavelength contour map, 139
　gas delivery system, 101
　growth conditions, 102–7
　high-volume manufacturing, 136–43
　hole concentration, 115
　hydrogen passivation, acceptors, 122–27
　lattice constant contours, 104

magnesium
  doping, temperature dependence, 116
  vapor pressure, 116
materials, 99–102
optoelectronic devices, growth of, 107–12
ordering, 119–22
oxygen incorporation, 127–36
quality, 140–41
schematic diagram, structure, AlGaInP light-emitting diode, 108
silicon doping, 114
uniformity, 137–40
yield loss categories, 143
zinc, vapor pressure, 116
aluminum nitride, 357
conductivity control, 374–79
efficiency, 384–85
high-brightness blue light-emitting diodes, 357–90
historical overview, 359–79
hydride vapor-phase epitaxy, 359, 360–61
indium nitride, 357
laser diodes, 385–87
lifetime, 384–85
molecular-beam epitaxy, 361–62
nitride-based superbright light-emitting diodes, 379–84
  electrode, 383–84
  layered structure, 379–82
    AlGaN/GaInN DH, 380–81
    asymmetric, 372, 380–82
    gallium nitride p–n homojunctiontype, 379–80
    light-emitting diodes, thin active layers, 382–83
  nitrides, properties of, 358, 362–76
  wavelength, 384–85
Optical extraction, AlGaAs red light-emitting diodes, 71–73
Optical transparency, AlGaInP light-emitting diodes, light
extraction, interfacial, 198
Optics design, signal lighting, 263–68
Organometallic vapor-phase epitaxy. *See* OMVPE
Oxygen incorporation, OMVPE growth, AlGaInP, 127–36

P

Packaged light-emitting diode chip, schematic illustration, 51
Peak wavelength, light-emitting diodes, 59
Performance, evolution of, visible light-emitting diode, 48
Performance characteristics, AlGaInP light-emitting diodes, 208–19
Personal organizer, light-emitting diode, high-brightness in, 337
PH_3. *See* Phosphine
Phosphine, light-emitting diodes, 7
Photon extraction, light-emitting diode, 12
Printer, light-emitting diode, high-brightness in, 337
Process controller, light-emitting diode, high-brightness in, 337
Projection televisions, light-emitting diode, high-brightness in, 337
Ptype gallium nitride, ultraviolet, 399–405
Pyrolysis
  high-brightness light-emitting diodes, 28–29
  kinetics, 28–29

Q

Quantum efficiency, external, light-emitting diode, 59
Quantum well device
  AlGaInP light-emitting diodes, active layer design, 169–70
  InGaN light-emitting diodes with, ultraviolet, 426–30

R

Radiative recombination efficiency, light-emitting diode, 12
Radiometry, 230–33
Reactor integrity, diode material, 12
Recombination, surface, 12, 31–32
  high-brightness light-emitting diodes, 31–32
Residual doping, light-emitting diode, 12

S

Self-compensation, diode material, 12
Semiconductor device market, by device type, 49

Si. *See* Silicon
Signal lighting, automotive, 251–77, 296–319
　electrical design, signal lighting, 268–73
　luminous flux requirements, signal lighting, 252, 259–63, 269
　luminous intensity, signal lighting, 252, 253
　market, 251–52, 296–97
　optics design, signal lighting, 246, 263–68
　signal head, 298
　thermal design, signal lighting, 273–77
SiH_4. *See* Silane
Silane, light-emitting diode, 27
Silicon, light-emitting diodes, 2
Single heterostructure, light-emitting diode, 55
Single quantum well, light-emitting diode, 56
SnP. *See* Indium phosphide
Source purity, diode material, 12
Source quality, OMVPE growth, AlGaInP, 140–41
SQW. *See* Single quantum well
Substrate absorption, AlGaInP light-emitting diodes, light extraction, 183–87
Surface recombination, light-emitting diode, 31–32

T

TBA. *See* Tertiarybutylarsine
TDMAP. *See* Trisdimethylaminophosphine
TEA1. *See* Triethylaluminum
TEGa. *See* Triethylgallium
Television
　portable, light-emitting diode, high-brightness in, 337
　projection, light-emitting diode, high-brightness in, 337
Temperature
　dependence
　　AlGaAs red light-emitting diodes, 83–89
　　current voltage, curves, 83–84
　　growth rate, light-emitting diode, 10
　　inverse, vs. vapor pressure, 18
Ternary alloys
　high-brightness light-emitting diodes, 18–24
　　lattice constant, vs. bandgap energy, 3
Tertiarybutylarsine, light-emitting diode, 11
Test equipment, light-emitting diode, high-brightness in, 337
Thermal cycling, AlGaInP light-emitting diodes, light extraction, 198
Thermal design, signal lighting, 273–77
Thermal expansion coefficients, high-bandgap materials, 36
Thermodynamics, high-brightness light-emitting diodes, 17–33
Ticketing machine, light-emitting diode, high-brightness in, 337
TMAA. *See* Trimethylaminealane
TMGa. *See* Trimethylgallium
Traffic signal lights, 296–319
　electrical design, 307–19
　incandescent, light-emitting diode, compared, 298–302
　luminous flux requirements, 302–7
　market, 296–97
　signal head, 298
Train, light-emitting diode, high-brightness in, 337
Transportation market, light-emitting diode, high-brightness in, 337
Triethylaluminum, light-emitting diode, 39
Triethylgallium, light-emitting diodes, 9
Trimethylaminealane, light-emitting diode, 24, 39
Trimethylgallium, light-emitting diode, 10
Trisdimethylaminophosphine, light-emitting diode, 24

U

Ultrahigh-vacuum, light-emitting diodes, 7
Ultraviolet blue–green–yellow light-emitting diodes, group IIIV-nitride-based, 391–443
Undoped gallium nitride, ultraviolet, 392, 394–98
Uniformity, OMVPE growth, AlGaInP, 137–40
UVH. *See* Ultra high-vacuum

V

van der Pauw measurements, liquid-phase epitaxy grown AlGaAs,

single layers, 69
van der Waals bonding, AlGaInP light-emitting diodes, light extraction, 198
Vapor deposition, chemical, metalorganic, AlGaAs red light-emitting diodes, 74–75
Vapor-phase epitaxial, light-emitting diodes, 6, 47
Vapor pressure, vs. inverse temperature, 18
VHz. *See* Ammonia
Video equipment, light-emitting diode, high-brightness in, 337
Viewfinder, camcorder, light-emitting diode, high-brightness in, 337
VPE. *See* Vapor-phase epitaxial

W

Warning light, instrument cluster, automotive, 280, 284–89
 cavity design, 289–94, 321
 legend optimization, 280, 294–96
Watches, light-emitting diode, high-brightness in, 337

Wavelength
 peak, light-emitting diodes, 59
 vs. luminous efficacy, light-emitting diode, 54
Window heterostructure, diode material, 12
Window layer, AlGaInP light-emitting diodes, current spreading, 172–77
 material parameters, 174
Word processor, portable, light-emitting diode, high-brightness in, 337

Y

Yield loss categories, OMVPE growth, AlGaInP, 143

Z

Zinc oxygen, light-emitting diodes, 2
ZnCdSe, light-emitting diode, 59
Zinc selenide (ZnSe), light-emitting diode, 52, 61
Zn–O. *See* Zinc oxygen
ZnTeSe, light-emitting diode, 59

Contents of Volumes in This Series

Volume 1 Physics of III–V Compounds

C. Hilsum, Some Key Features of III–V Compounds
Franco Bassani, Methods of Band Calculations Applicable to III–V Compounds
E. O. Kane, The k-p Method
V. L. Bonch-Bruevich, Effect of Heavy Doping on the Semiconductor Band Structure
Donald Long, Energy Band Structures of Mixed Crystals of III–V Compounds
Laura M. Roth and Petros N. Argyres, Magnetic Quantum Effects
S. M. Puri and T. H. Geballe, Thermomagnetic Effects in the Quantum Region
W. M. Becker, Band Characteristics near Principal Minima from Magnetoresistance
E. H. Putley, Freeze-Out Effects, Hot Electron Effects, and Submillimeter Photoconductivity in InSb
H. Weiss, Magnetoresistance
Betsy Ancker-Johnson, Plasma in Semiconductors and Semimetals

Volume 2 Physics of III–V Compounds

M. G. Holland, Thermal Conductivity
S. I. Novkova, Thermal Expansion
U. Piesbergen, Heat Capacity and Debye Temperatures
G. Giesecke, Lattice Constants
J. R. Drabble, Elastic Properties
A. U. Mac Rae and G. W. Gobeli, Low Energy Electron Diffraction Studies
Robert Lee Mieher, Nuclear Magnetic Resonance
Bernard Goldstein, Electron Paramagnetic Resonance
T. S. Moss, Photoconduction in III–V Compounds
E. Antoncik ad J. Tauc, Quantum Efficiency of the Internal Photoelectric Effect in InSb
G. W. Gobeli and I. G. Allen, Photoelectric Threshold and Work Function
P. S. Pershan, Nonlinear Optics in III–V Compounds
M. Gershenzon, Radiative Recombination in the III–V Compounds
Frank Stern, Stimulated Emission in Semiconductors

Volume 3 Optical of Properties III–V Compounds

Marvin Hass, Lattice Reflection
William G. Spitzer, Multiphonon Lattice Absorption
D. L. Stierwalt and R. F. Potter, Emittance Studies
H. R. Philipp and H. Ehrenveich, Ultraviolet Optical Properties
Manuel Cardona, Optical Absorption above the Fundamental Edge
Earnest J. Johnson, Absorption near the Fundamental Edge
John O. Dimmock, Introduction to the Theory of Exciton States in Semiconductors
B. Lax and J. G. Mavroides, Interband Magnetooptical Effects
H. Y. Fan, Effects of Free Carries on Optical Properties
Edward D. Palik and George B. Wright, Free-Carrier Magnetooptical Effects
Richard H. Bube, Photoelectronic Analysis
B. O. Seraphin and H. E. Bennett, Optical Constants

Volume 4 Physics of III–V Compounds

N. A. Goryunova, A. S. Borschevskii, and D. N. Tretiakov, Hardness
N. N. Sirota, Heats of Formation and Temperatures and Heats of Fusion of Compounds $A^{III}B^{V}$
Don L. Kendall, Diffusion
A. G. Chynoweth, Charge Multiplication Phenomena
Robert W. Keyes, The Effects of Hydrostatic Pressure on the Properties of III–V Semiconductors
L. W. Aukerman, Radiation Effects
N. A. Goryunova, F. P. Kesamanly, and D. N. Nasledov, Phenomena in Solid Solutions
R. T. Bate, Electrical Properties of Nonuniform Crystals

Volume 5 Infrared Detectors

Henry Levinstein, Characterization of Infrared Detectors
Paul W. Kruse, Indium Antimonide Photoconductive and Photoelectromagnetic Detectors
M. B. Prince, Narrowband Self-Filtering Detectors
Ivars Melngalis and T. C. Harman, Single-Crystal Lead-Tin Chalcogenides
Donald Long and Joseph L. Schmidt, Mercury-Cadmium Telluride and Closely Related Alloys
E. H. Putley, The Pyroelectric Detector
Norman B. Stevens, Radiation Thermopiles
R. J. Keyes and T. M. Quist, Low Level Coherent and Incoherent Detection in the Infrared
M. C. Teich, Coherent Detection in the Infrared
F. R. Arams, E. W. Sard, B. J. Peyton, and F. P. Pace, Infrared Heterodyne Detection with Gigahertz IF Response
H. S. Sommers, Jr., Macrowave-Based Photoconductive Detector
Robert Sehr and Rainer Zuleeg, Imaging and Display

Volume 6 Injection Phenomena

Murray A. Lampert and Ronald B. Schilling, Current Injection in Solids: The Regional Approximation Method
Richard Williams, Injection by Internal Photoemission
Allen M. Barnett, Current Filament Formation

R. Baron and J. W. Mayer, Double Injection in Semiconductors
W. Ruppel, The Photoconductor-Metal Contact

Volume 7   Application and Devices
Part A

John A. Copeland and Stephen Knight, Applications Utilizing Bulk Negative Resistance
F. A. Padovani, The Voltage-Current Characteristics of Metal-Semiconductor Contacts
P. L. Hower, W. W. Hooper, B. R. Cairns, R. D. Fairman, and D. A. Tremere, The GaAs Field-Effect Transistor
Marvin H. White, MOS Transistors
G. R. Antell, Gallium Arsenide Transistors
T. L. Tansley, Heterojunction Properties

Part B

T. Misawa, IMPATT Diodes
H. C. Okean, Tunnel Diodes
Robert B. Campbell and Hung-Chi Chang, Silicon Junction Carbide Devices
R. E. Enstrom, H. Kressel, and L. Krassner, High-Temperature Power Rectifiers of $GaAs_{1-x}P_x$

Volume 8   Transport and Optical Phenomena

Richard J. Stirn, Band Structure and Galvanomagnetic Effects in III–V Compounds with Indirect Band Gaps
Roland W. Ure, Jr., Thermoelectric Effects in III–V Compounds
Herbert Piller, Faraday Rotation
H. Barry Bebb and E. W. Williams, Photoluminescence I: Theory
E. W. Williams and H. Barry Bebb, Photoluminescence II: Gallium Arsenide

Volume 9   Modulation Techniques

B. O. Seraphin, Electroreflectance
R. L. Aggarwal, Modulated Interband Magnetooptics
Daniel F. Blossey and Paul Handler, Electroabsorption
Bruno Batz, Thermal and Wavelength Modulation Spectroscopy
Ivar Balslev, Piezopptical Effects
D. E. Aspnes and N. Bottka, Electric-Field Effects on the Dielectric Function of Semiconductors and Insulators

Volume 10   Transport Phenomena

R. L. Rhode, Low-Field Electron Transport
J. D. Wiley, Mobility of Holes in III–V Compounds
C. M. Wolfe and G. E. Stillman, Apparent Mobility Enhancement in Inhomogeneous Crystals
Robert L. Petersen, The Magnetophonon Effect

Volume 11 Solar Cells

Harold J. Hovel, Introduction; Carrier Collection, Spectral Response, and Photocurrent; Solar Cell Electrical Characteristics; Efficiency; Thickness; Other Solar Cell Devices; Radiation Effects; Temperature and Intensity; Solar Cell Technology

Volume 12 Infrared Detectors (II)

W. L. Eiseman, J. D. Merriam, and R. F. Potter, Operational Characteristics of Infrared Photodetectors
Peter R. Bratt, Impurity Germanium and Silicon Infrared Detectors
E. H. Putley, InSb Submillimeter Photoconductive Detectors
G. E. Stillman, C. M. Wolfe, and J. O. Dimmock, Far-Infrared Photoconductivity in High Purity GaAs
G. E. Stillman and C. M. Wolfe, Avalanche Photodiodes
P. L. Richards, The Josephson Junction as a Detector of Microwave and Far-Infrared Radiation
E. H. Putley, The Pyroelectric Detector—An Update

Volume 13 Cadmium Telluride

Kenneth Zanio, Materials Preparations; Physics; Defects; Applications

Volume 14 Lasers, Junctions, Transport

N. Holonyak, Jr. and M. H. Lee, Photopumped III–V Semiconductor Lasers
Henry Kressel and Jerome K. Butler, Heterojunction Laser Diodes
A Van der Ziel, Space-Charge-Limited Solid-State Diodes
Peter J. Price, Monte Carlo Calculation of Electron Transport in Solids

Volume 15 Contacts, Junctions, Emitters

B. L. Sharma, Ohmic Contacts to III–V Compounds Semiconductors
Allen Nussbaum, The Theory of Semiconducting Junctions
John S. Escher, NEA Semiconductor Photoemitters

Volume 16 Defects, (HgCd)Se, (HgCd)Te

Henry Kressel, The Effect of Crystal Defects on Optoelectronic Devices
C. R. Whitsett, J. G. Broerman, and C. J. Summers, Crystal Growth and Properties of $Hg_{1-x}Cd_xSe$ alloys
M. H. Weiler, Magnetooptical Properties of $Hg_{1-x}Cd_xTe$ Alloys
Paul W. Kruse and John G. Ready, Nonlinear Optical Effects in $Hg_{1-x}Cd_xTe$

Volume 17 CW Processing of Silicon and Other Semiconductors

James F. Gibbons, Beam Processing of Silicon
Arto Lietoila, Richard B. Gold, James F. Gibbons, and Lee A. Christel, Temperature Distribu-

tions and Solid Phase Reaction Rates Produced by Scanning CW Beams
Arto Leitoila and James F. Gibbons, Applications of CW Beam Processing to Ion Implanted Crystalline Silicon
N. M. Johnson, Electronic Defects in CW Transient Thermal Processed Silicon
K. F. Lee, T. J. Stultz, and James F. Gibbons, Beam Recrystallized Polycrystalline Silicon: Properties, Applications, and Techniques
T. Shibata, A. Wakita, T. W. Sigmon, and James F. Gibbons, Metal-Silicon Reactions and Silicide
Yves I. Nissim and James F. Gibbons, CW Beam Processing of Gallium Arsenide

Volume 18 Mercury Cadmium Telluride

Paul W. Kruse, The Emergence of $(Hg_{1-x}Cd_x)Te$ as a Modern Infrared Sensitive Material
H. E. Hirsch, S. C. Liang, and A. G. White, Preparation of High-Purity Cadmium, Mercury, and Tellurium
W. F. H. Micklethwaite, The Crystal Growth of Cadmium Mercury Telluride
Paul E. Petersen, Auger Recombination in Mercury Cadmium Telluride
R. M. Broudy and V. J. Mazurczyck, (HgCd)Te Photoconductive Detectors
M. B. Reine, A. K. Soad, and T. J. Tredwell, Photovoltaic Infrared Detectors
M. A. Kinch, Metal-Insulator-Semiconductor Infrared Detectors

Volume 19 Deep Levels, GaAs, Alloys, Photochemistry

G. F. Neumark and K. Kosai, Deep Levels in Wide Band-Gap III–V Semiconductors
David C. Look, The Electrical and Photoelectronic Properties of Semi-Insulating GaAs
R. F. Brebrick, Ching-Hua Su, and Pok-Kai Liao, Associated Solution Model for Ga-In-Sb and Hg-Cd-Te
Yu. Ya. Gurevich and Yu. V. Pleskon, Photoelectrochemistry of Semiconductors

Volume 20 Semi-Insulating GaAs

R. N. Thomas, H. M. Hobgood, G. W. Eldridge, D. L. Barrett, T. T. Braggins, L. B. Ta, and S. K. Wang, High-Purity LEC Growth and Direct Implantation of GaAs for Monolithic Microwave Circuits
C. A. Stolte, Ion Implantation and Materials for GaAs Integrated Circuits
C. G. Kirkpatrick, R. T. Chen, D. E. Holmes, P. M. Asbeck, K. R. Elliott, R. D. Fairman, and J. R. Oliver, LEC GaAs for Integrated Circuit Applications
J. S. Blakemore and S. Rahimi, Models for Mid-Gap Centers in Gallium Arsenide

Volume 21 Hydrogenated Amorphous Silicon
Part A

Jacques I. Pankove, Introduction
Masataka Hirose, Glow Discharge; Chemical Vapor Deposition
Yoshiyuki Uchida, di Glow Discharge
T. D. Moustakas, Sputtering
Isao Yamada, Ionized-Cluster Beam Deposition
Bruce A. Scott, Homogeneous Chemical Vapor Deposition

Frank J. Kampas, Chemical Reactions in Plasma Deposition
Paul A. Longeway, Plasma Kinetics
Herbert A. Weakliem, Diagnostics of Silane Glow Discharges Using Probes and Mass Spectroscopy
Lester Gluttman, Relation between the Atomic and the Electronic Structures
A. Chenevas-Paule, Experiment Determination of Structure
S. Minomura, Pressure Effects on the Local Atomic Structure
David Adler, Defects and Density of Localized States

Part B

Jacques I. Pankove, Introduction
G. D. Cody, The Optical Absorption Edge of a-Si:H
Nabil M. Amer and Warren B. Jackson, Optical Properties of Defect States in a-Si:H
P. J. Zanzucchi, The Vibrational Spectra of a-Si:H
Yoshihiro Hamakawa, Electroreflectance and Electroabsorption
Jeffrey S. Lannin, Raman Scattering of Amorphous Si, Ge, and Their Alloys
R. A. Street, Luminescence in a-Si:H
Richard S. Crandall, Photoconductivity
J. Tauc, Time-Resolved Spectroscopy of Electronic Relaxation Processes
P. E. Vanier, IR-Induced Quenching and Enhancement of Photoconductivity and Photoluminescence
H. Schade, Irradiation-Induced Metastable Effects
L. Ley, Photoelectron Emission Studies

Part C

Jacques I. Pankove, Introduction
J. David Cohen, Density of States from Junction Measurements in Hydrogenated Amorphous Silicon
P. C. Taylor, Magnetic Resonance Measurements in a-Si:H
K. Morigaki, Optically Detected Magnetic Resonance
J. Dresner, Carrier Mobility in a-Si:H
T. Tiedje, Information about band-Tail States from Time-of-Flight Experiments
Arnold R. Moore, Diffusion Length in Undoped a-Si:H
W. Beyer and J. Overhof, Doping Effects in a-Si:H
H. Fritzche, Electronic Properties of Surfaces in a-Si:H
C. R. Wronski, The Staebler-Wronski Effect
R. J. Nemanich, Schottky Barriers on a-Si:H
B. Abeles and T. Tiedje, Amorphous Semiconductor Superlattices

Part D

Jacques I. Pankove, Introduction
D. E. Carlson, Solar Cells
G. A. Swartz, Closed-Form Solution of I–V Characteristic for a-Si:H Solar Cells
Isamu Shimizu, Electrophotography
Sachio Ishioka, Image Pickup Tubes

P. G. LeComber and W. E. Spear, The Development of the a-Si:H Field-Effect Transistor and Its Possible Applications
D. G. Ast, a-Si:H FET-Addressed LCD Panel
S. Kaneko, Solid-State Image Sensor
Masakiyo Matsumura, Charge-Coupled Devices
M. A. Bosch, Optical Recording
A. D'Amico and G. Fortunato, Ambient Sensors
Hiroshi Kukimoto, Amorphous Light-Emitting Devices
Robert J. Phelan, Jr., Fast Detectors and Modulators
Jacques I. Pankove, Hybrid Structures
P. G. LeComber, A. E. Owen, W. E. Spear, J. Hajto, and W. K. Choi, Electronic Switching in Amorphous Silicon Junction Devices

Volume 22 Lightwave Communications Technology
Part A

Kazuo Nakajima, The Liquid-Phase Epitaxial Growth of IngaAsp
W. T. Tsang, Molecular Beam Epitaxy for III–V Compound Semiconductors
G. B. Stringfellow, Organometallic Vapor-Phase Epitaxial Growth of III–V Semiconductors
G. Beuchet, Halide and Chloride Transport Vapor-Phase Deposition of InGaAsP and GaAs
Manijeh Razeghi, Low-Pressure Metallo-Organic Chemical Vapor Deposition of $Ga_xIn_{1-x}AsP_{1-y}$ Alloys
P. M. Petroff, Defects in III–V Compound Semiconductors

Part B

J. P. van der Ziel, Mode Locking of Semiconductor Lasers
Kam Y. Lau and Ammon Yariv, High-Frequency Current Modulation of Semiconductor Injection Lasers
Charles H. Henry, Special Properties of Semiconductor Lasers
Yasuharu Suematsu, Katsumi Kishino, Shigehisa Arai, and Fumio Koyama. Dynamic Single-Mode Semiconductor Lasers with a Distributed Reflector
W. T. Tsang, The Cleaved-Coupled-Cavity (C^3) Laser

Part C

R. J. Nelson and N. K. Dutta, Review of InGaAsP InP Laser Structures and Comparison of Their Performance
N. Chinone and M. Nakamura, Mode-Stabilized Semiconductor Lasers for 0.7–0.8- and 1.1–1.6-μm Regions
Yoshiji Horikoshi, Semiconductor Lasers with Wavelengths Exceeding 2 μm
B. A. Dean and M. Dixon, The Functional Reliability of Semiconductor Lasers as Optical Transmitters
R. H. Saul, T. P. Lee, and C. A. Burus, Light-Emitting Device Design
C. L. Zipfel, Light-Emitting Diode-Reliability
Tien Pei Lee and Tingye Li, LED-Based Multimode Lightwave Systems
Kinichiro Ogawa, Semiconductor Noise-Mode Partition Noise

Part D

Federico Capasso, The Physics of Avalanche Photodiodes
T. P. Pearsall and M. A. Pollack, Compound Semiconductor Photodiodes
Takao Kaneda, Silicon and Germanium Avalanche Photodiodes
S. R. Forrest, Sensitivity of Avalanche Photodetector Receivers for High-Bit-Rate Long-Wavelength Optical Communication Systems
J. C. Campbell, Phototransistors for Lightwave Communications

Part E

Shyh Wang, Principles and Characteristics of Integrable Active and Passive Optical Devices
Shlomo Margalit and Amnon Yariv, Integrated Electronic and Photonic Devices
Takaoki Mukai, Yoshihisa Yamamoto, and Tatsuya Kimura, Optical Amplification by Semiconductor Lasers

Volume 23 Pulsed Laser Processing of Semiconductors

R. F. Wood, C. W. White, and R. T. Young, Laser Processing of Semiconductors: An Overview
C. W. White, Segregation, Solute Trapping, and Supersaturated Alloys
G. E. Jellison, Jr., Optical and Electrical Properties of Pulsed Laser-Annealed Silicon
R. F. Wood and G. E. Jellison, Jr., Melting Model of Pulsed Laser Processing
R. F. Wood and F. W. Young, Jr., Nonequilibrium Solidification Following Pulsed Laser Melting
D. H. Lowndes and G. E. Jellison, Jr., Time-Resolved Measurement During Pulsed Laser Irradiation of Silicon
D. M. Zebner, Surface Studies of Pulsed Laser Irradiated Semiconductors
D. H. Lowndes, Pulsed Beam Processing of Gallium Arsenide
R. B. James, Pulsed CO_2 Laser Annealing of Semiconductors
R. T. Young and R. F. Wood, Applications of Pulsed Laser Processing

Volume 24 Applications of Multiquantum Wells, Selective Doping, and Superlattices

C. Weisbuch, Fundamental Properties of III–V Semiconductor Two-Dimensional Quantized Structures: The Basis for Optical and Electronic Device Applications
H. Morkoc and H. Unlu, Factors Affecting the Performance of (Al,Ga)As/GaAs and (Al,Ga)As/InGaAs Modulation-Doped Field-Effect Transistors: Microwave and Digital Applications
N. T. Linh, Two-Dimensional Electron Gas FETs: Microwave Applications
M. Abe et al., Ultra-High-Speed HEMT Integrated Circuits
D. S. Chemla, D. A. B. Miller, and P. W. Smith, Nonlinear Optical Properties of Multiple Quantum Well Structures for Optical Signal Processing
F. Capasso, Graded-Gap and Superlattice Devices by Band-Gap Engineering
W. T. Tsang, Quantum Confinement Heterostructure Semiconductor Lasers
G. C. Osbourn et al., Principles and Applications of Semiconductor Strained-Layer Superlattices

Volume 25 Diluted Magnetic Semiconductors

W. Giriat and J. K. Furdyna, Crystal Structure, Composition, and Materials Preparation of Diluted Magnetic Semiconductors

W. M. Becker, Band Structure and Optical Properties of Wide-Gap $A^{II}_{1-x}Mn_xB^{IV}$ Alloys at Zero Magnetic Field

Saul Oseroff and Pieter H. Keesom, Magnetic Properties: Macroscopic Studies

Giebultowicz and T. M. Holden, Neutron Scattering Studies of the Magnetic Structure and Dynamics of Diluted Magnetic Semiconductors

J. Kossut, Band Structure and Quantum Transport Phenomena in Narrow-Gap Diluted Magnetic Semiconductors

C. Riquaux, Magnetooptical Properties of Large-Gap Diluted Magnetic Semiconductors

J. A. Gaj, Magnetooptical Properties of Large-Gap Diluted Magnetic Semiconductors

J. Mycielski, Shallow Acceptors in Diluted Magnetic Semiconductors: Splitting, Boil-off, Giant Negative Magnetoresistance

A. K. Ramadas and R. Rodriquez, Raman Scattering in Diluted Magnetic Semiconductors

P. A. Wolff, Theory of Bound Magnetic Polarons in Semimagnetic Semiconductors

Volume 26 III–V Compound Semiconductors and Semiconductor Properties of Superionic Materials

Zou Yuanxi, III–V Compounds

H. V. Winston, A. T. Hunter, H. Kimura, and R. E. Lee, InAs-Alloyed GaAs Substrates for Direct Implantation

P. K. Bhattachary and S. Dhar, Deep Levels in III–V Compound Semiconductors Grown by MBE

Yu. Yu. Gurevich and A. K. Ivanov-Shits, Semiconductor Properties of Supersonic Materials

Volume 27 High Conducting Quasi-One-Dimensional Organic Crystals

E. M. Conwell, Introduction to Highly Conducting Quasi-One-Dimensional Organic Crystals

I. A. Howard, A Reference Guide to the Conducting Quasi-One-Dimensional Organic Molecular Crystals

J. P. Pouquet, Structural Instabilities

E. M. Conwell, Transport Properties

C. S. Jacobsen, Optical Properties

J. C. Scott, Magnetic Properties

L. Zuppiroli, Irradiation Effects: Perfect Crystals and Real Crystals

Volume 28 Measurement of High-Speed Signals in Solid State Devices

J. Frey and D. Ioannou, Materials and Devices for High-Speed and Optoelectronic Applications

H. Schumacher and E. Strid, Electronic Wafer Probing Techniques

D. H. Auston, Picosecond Photoconductivity: High-Speed Measurements of Devices and Materials

J. A. Valdmanis, Electro-Optic Measurement Techniques for Picosecond Materials, Devices, and Integrated Circuits.

J. M. Wiesenfeld and R. K. Jain, Direct Optical Probing of Integrated Circuits and High-Speed Devices

G. Plows, Electron-Beam Probing

A. M. Weiner and R. B. Marcus, Photoemissive Probing

Volume 29 Very High Speed Integrated Circuits: Gallium Arsenide LSI

M. Kuzuhara and T. Nazaki, Active Layer Formation by Ion Implantation
H. Hasimoto, Focused Ion Beam Implantation Technology
T. Nozaki and A. Higashisaka, Device Fabrication Process Technology
M. Ino and T. Takada, GaAs LSI Circuit Design
M. Hirayama, M. Ohmori, and K. Yamasaki, GaAs LSI Fabrication and Performance

Volume 30 Very High Speed Integrated Circuits: Heterostructure

H. Watanabe, T. Mizutani, and A. Usui, Fundamentals of Epitaxial Growth and Atomic Layer Epitaxy
S. Hiyamizu, Characteristics of Two-Dimensional Electron Gas in III–V Compound Heterostructures Grown by MBE
T. Nakanisi, Metalorganic Vapor Phase Epitaxy for High-Quality Active Layers
T. Nimura, High Electron Mobility Transistor and LSI Applications
T. Sugeta and T. Ishibashi, Hetero-Bipolar Transistor and LSI Application
H. Matsueda, T. Tanaka, and M. Nakamura, Optoelectronic Integrated Circuits

Volume 31 Indium Phosphide: Crystal Growth and Characterization

J. P. Farges, Growth of Discoloration-free InP
M. J. McCollum and G. E. Stillman, High Purity InP Grown by Hydride Vapor Phase Epitaxy
T. Inada and T. Fukuda, Direct Synthesis and Growth of Indium Phosphide by the Liquid Phosphorous Encapsulated Czochralski Method
O. Oda, K. Katagiri, K. Shinohara, S. Katsura, Y. Takahashi, K. Kainosho, K. Kohiro, and R. Hirano, InP Crystal Growth, Substrate Preparation and Evaluation
K. Tada, M. Tatsumi, M. Morioka, T. Araki, and T. Kawase, InP Substrates: Production and Quality Control
M. Razeghi, LP-MOCVD Growth, Characterization, and Application of InP Material
T. A. Kennedy and P. J. Lin-Chung, Stoichiometric Defects in InP

Volme 32 Strained-Layer Superlattices: Physics

T. P. Pearsall, Strained-Layer Superlattices
Fred H. Pollack, Effects of Homogeneous Strain on the Electronic and Vibrational Levels in Semiconductors
J. Y. Marzin, J. M. Gerárd, P. Voisin, and J. A. Brum, Optical Studies of Strained III–V Heterolayers
R. People and S. A. Jackson, Structurally Induced States from Strain and Confinement
M. Jaros, Microscopic Phenomena in Ordered Suprlattices

Volume 33 Strained-Layer Superlattices: Materials Science and Technology

R. Hull and J. C. Bean, Principles and Concepts of Strained-Layer Epitaxy
William J. Schaff, Paul J. Tasker, Marc C. Foisy, and Lester F. Eastman, Device Applications of Strained-Layer Epitaxy

S. T. Picraux, B. L. Doyle, and J. Y. Tsao, Structure and Characterization of Strained-Layer Superlattices
E. Kasper and F. Schaffer, Group IV Compounds
Dale L. Martin, Molecular Beam Epitaxy of IV–VI Compounds Heterojunction
Robert L. Gunshor, Leslie A. Kolodziejski, Arto V. Nurmikko, and Nobuo Otsuka, Molecular Beam Epitaxy of II–VI Semiconductor Microstructures

Volume 34   **Hydrogen in Semiconductors**

J. I. Pankove and N. M. Johnson, Introduction to Hydrogen in Semiconductors
C. H. Seager, Hydrogenation Methods
J. I. Pankove, Hydrogenation of Defects in Crystalline Silicon
J. W. Corbett, P. Deák, U. V. Desnica, and S. J. Pearton, Hydrogen Passivation of Damage Centers in Semiconductors
S. J. Pearton, Neutralization of Deep Levels in Silicon
J. I. Pankove, Neutralization of Shallow Acceptors in Silicon
N. M. Johnson, Neutralization of Donor Dopants and Formation of Hydrogen-Induced Defects in n-Type Silicon
M. Stavola and S. J. Pearton, Vibrational Spectroscopy of Hydrogen-Related Defects in Silicon
A. D. Marwick, Hydrogen in Semiconductors: Ion Beam Techniques
C. Herring and N. M. Johnson, Hydrogen Migration and Solubility in Silicon
E. E. Haller, Hydrogen-Related Phenomena in Crystalline Germanium
J. Kakalios, Hydrogen Diffusion in Amorphous Silicon
J. Chevalier, B. Clerjaud, and B. Pajot, Neutralization of Defects and Dopants in III–V Semiconductors
G. G. DeLeo and W. B. Fowler, Computational Studies of Hydrogen-Containing Complexes in Semiconductors
R. F. Kiefl and T. L. Estle, Muonium in Semiconductors
C. G. Van de Walle, Theory of Isolated Interstitial Hydrogen and Muonium in Crystalline Semiconductors

Volume 35   **Nanostructured Systems**

Mark Reed, Introduction
H. van Houten, C. W. J. Beenakker, and B. J. van Wees, Quantum Point Contacts
G. Timp, When Does a Wire Become an Electron Waveguide?
M. Büttiker, The Quantum Hall Effects in Open Conductors
W. Hansen, J. P. Kotthaus, and U. Merkt, Electrons in Laterally Periodic Nanostructures

Volume 36   **The Spectroscopy of Semiconductors**

D. Heiman, Spectroscopy of Semiconductors at Low Temperatures and High Magnetic Fields
Arto V. Nurmikko, Transient Spectroscopy by Ultrashort Laser Pulse Techniques
A. K. Ramdas and S. Rodriguez, Piezospectroscopy of Semiconductors
Orest J. Glembocki and Benjamin V. Shanabrook, Photoreflectance Spectroscopy of Microstructures
David G. Seiler, Christopher L. Littler, and Margaret H. Wiler, One- and Two-Photon Magneto-Optical Spectroscopy of InSb and $Hg_{1-x}Cd_xTe$

Volume 37 The Mechanical Properties of Semiconductors

A.-B. Chen, Arden Sher and W. T. Yost, Elastic Constants and Related Properties of Semiconductor Compounds and Their Alloys
David R. Clarke, Fracture of Silicon and Other Semiconductors
Hans Siethoff, The Plasticity of Elemental and Compound Semiconductors
Sivaraman Guruswamy, Katherine T. Faber and John P. Hirth, Mechanical Behavior of Compound Semiconductors
Subhanh Mahajan, Deformation Behavior of Compound Semiconductors
John P. Hirth, Injection of Dislocations into Strained Multilayer Structures
Don Kendall, Charles B. Fleddermann, and Kevin J. Malloy, Critical Technologies for the Micromachining of Silicon
Ikuo Matsuba and Kinji Mokuya, Processing and Semiconductor Thermoelastic Behavior

Volume 38 Imperfections in III/V Materials

Udo Scherz and Matthias Scheffler, Density-Functional Theory of sp-Bonded Defects in III/V Semiconductors
Maria Kaminska and Eicke R. Weber, El2 Defect in GaAs
David C. Look, Defects Relevant for Compensation in Semi-Insulating GaAs
R. C. Newman, Local Vibrational Mode Spectroscopy of Defects in III/V Compounds
Andrzej M. Hennel, Transition Metals in III/V Compounds
Kevin J. Malloy and Ken Khachaturyan, DX and Related Defects in Semiconductors
V. Swaminathan and Andrew S. Jordan, Dislocations in III/V Compounds
Krzysztof W. Nauka, Deep Level Defects in the Epitaxial III/V Materials

Volume 39 Minority Carriers in III–V Semiconductors: Physics and Applications

Niloy K. Dutta, Radiative Transitions in GaAs and Other III–V Compounds
Richard K. Ahrenkiel, Minority-Carrier Lifetime in III–V Semiconductors
Tomofumi Furuta, High Field Minority Electron Transport in p-GaAs
Mark S. Lundstrom, Minority-Carrier Transport in III–V Semiconductors
Richard A. Abram, Effects of Heavy Doping and High Excitation on the Band Structure of GaAs
David Yevick and Witold Bardyszewski, An Introduction to Non-Equilibrium Many-Body Analyses of Optical Processes in III–V Semiconductors

Volume 40 Epitaxial Microstructures

E. F. Schubert, Delta-Doping of Semiconductors: Electronic, Optical, and Structural Properties of Materials and Devices
A. Gossard, M. Sundaram, and P. Hopkins, Wide Graded Potential Wells
P. Petroff, Direct Growth of Nanometer-Size Quantum Wire Superlattices
E. Kapon, Lateral Patterning of Quantum Well Heterostructures by Growth of Nonplanar Substrates
H. Temkin, D. Gershoni, and M. Panish, Optical Properties of Gal-$_x$In$_x$As/InP Quantum Wells

Volume 41   High Speed Heterostructure Devices

F. Capasso, F. Beltram, S. Sen, A. Pahlevi, and A. Y. Cho, Quantum Electron Devices: Physics and Applications
P. Solomon, D. J. Frank, S. L. Wright, and F. Canora, GaAs-Gate Semiconductor–Insulator–Semiconductor FET
M. H. Hashemi and U. K. Mishra, Unipolar InP-Based Transistors
R. Kiehl, Complementary Heterostructure FET Integrated Circuits
T. Ishibashi, GaAs-Based and InP-Based Heterostructure Bipolar Transistors
H. C. Liu and T. C. L. G. Sollner, High-Frequency-Tunneling Devices
H. Ohnishi, T. More, M. Takatsu, K. Imamura, and N. Yokoyama, Resonant-Tunneling Hot-Electron Transistors and Circuits

Volume 42   Oxygen in Silicon

F. Shimura, Introduction to Oxygen in Silicon
W. Lin, The Incorporation of Oxygen into Silicon Crystals
T. J. Schaffner and D. K. Schroder, Characterization Techniques for Oxygen in Silicon
W. M. Bullis, Oxygen Concentration Measurement
S. M. Hu, Intrinsic Point Defects in Silicon
B. Pajot, Some Atomic Configurations of Oxygen
J. Michel and L. C. Kimerling, Electical Properties of Oxygen in Silicon
R. C. Newman and R. Jones, Diffusion of Oxygen in Silicon
T. Y. Tan and W. J. Taylor, Mechanisms of Oxygen Precipitation: Some Quantitative Aspects
M. Schrems, Simulation of Oxygen Precipitation
K. Simino and I. Yonenaga, Oxygen Effect on Mechanical Properties
W. Bergholz, Grown-in and Process-Induced Effects
F. Shimura, Intrinsic/Internal Gettering
H. Tsuya, Oxygen Effect on Electronic Device Performance

Volume 43   Semiconductors for Room Temperature Nuclear Detector Applications

R. B. James and T. E. Schlesinger, Introduction and Overview
L. S. Darken and C. E. Cox, High-Purity Germanium Detectors
A. Burger, D. Nason, L. Van den Berg, and M. Schieber, Growth of Mercuric Iodide
X. J. Bao, T. E. Schlesinger, and R. B. James, Electrical Properties of Mercuric Iodide
X. J. Bao, R. B. James, and T. E. Schlesinger, Optical Properties of Red Mercuric Iodide
M. Hage-Ali and P. Siffert, Growth Methods of CdTe Nuclear Detector Materials
M. Hage-Ali and P Siffert, Characterization of CdTe Nuclear Detector Materials
M. Hage-Ali and P. Siffert, CdTe Nuclear Detectors and Applications
R. B. James, T. E. Schlesinger, J. Lund, and M. Schieber, $Cd_{1-x}Zn_xTe$ Spectrometers for Gamma and X-Ray Applications
D. S. McGregor, J. E. Kammeraad, Gallium Arsenide Radiation Detectors and Spectrometers
J. C. Lund, F. Olschner, and A. Burger, Lead Iodide
M. R. Squillante, and K. S. Shah, Other Materials: Status and Prospects
V. M. Gerrish, Characterization and Quantification of Detector Performance
J. S. Iwanczyk and B. E. Patt, Electronics for X-ray and Gamma Ray Spectrometers
M. Schieber, R. B. James, and T. E. Schlesinger, Summary and Remaining Issues for Room Temperature Radiation Spectrometers

Volume 44 II–IV Blue/Green Light Emitters: Device Physics and Epitaxial Growth

J. Han and R. L. Gunshor, MBE Growth and Electrical Properties of Wide Bandgap ZnSe-based II–VI Semiconductors

Shizuo Fujita and Shigeo Fujita, Growth and Characterization of ZnSe-based II–VI Semiconductors by MOVPE

Easen Ho and Leslie A. Kolodziejski, Gaseous Source UHV Epitaxy Technologies for Wide Bandgap II–VI Semiconductors

Chris G. Van de Walle, Doping of Wide-Band-Gap II–VI Compounds — Theory

Roberto Cingolani, Optical Properties of Excitons in ZnSe-Based Quantum Well Heterostructures

A. Ishibashi and A. V. Nurmikko, II–VI Diode Lasers: A Current View of Device Performance and Issues

Supratik Guha and John Petruzello, Defects and Degradation in Wide-Gap II–VI-based Structures and Light Emitting Devices

Volume 45 Effect of Disorder and Defects in Ion-Implanted Semiconductors: Electrical and Physiochemical Characterization

Heiner Ryssel, Ion Implantation into Semiconductors: Historical Perspectives

You-Nian Wang and Teng-Cai Ma, Electronic Stopping Power for Energetic Ions in Solids

Sachiko T. Nakagawa, Solid Effect on the Electronic Stopping of Crystalline Target and Application to Range Estimation

G. Müller, S. Kalbitzer and G. N. Greaves, Ion Beams in Amorphous Semiconductor Research

Jumana Boussey-Said, Sheet and Spreading Resistance Analysis of Ion Implanted and Annealed Semiconductors

M. L. Polignano and G. Queirolo, Studies of the Stripping Hall Effect in Ion-Implanted Silicon

J. Stoemenos, Transmission Electron Microscopy Analyses

Roberta Nipoti and Marco Servidori, Rutherford Backscattering Studies of Ion Implanted Semiconductors

P. Zaumseil, X-ray Diffraction Techniques

Volume 46 Effect of Disorder and Defects in Ion-Implanted Semiconductors: Optical and Photothermal Characterization

M. Fried, T. Lohner and J. Gyulai, Ellipsometric Analysis

Antonios Seas and Constantinos Christofides, Transmission and Reflection Spectroscopy on Ion Implanted Semiconductors

Andreas Othonos and Constantinos Christofides, Photoluminescence and Raman Scattering of Ion Implanted Semiconductors. Influence of Annealing

Constantinos Christofides, Photomodulated Thermoreflectance Investigation of Implanted Wafers. Annealing Kinetics of Defects

U. Zammit, Photothermal Deflection Spectroscopy Characterization of Ion-Implanted and Annealed Silicon Films

Andreas Mandelis, Arief Budiman and Miguel Vargas, Photothermal Deep-Level Transient Spectroscopy of Impurities and Defects in Semiconductors

R. Kalish and S. Charbonneau, Ion Implantation into Quantum-Well Structures

Alexandre M. Myasnikov and Nikolay N. Gerasimenko, Ion Implantation and Thermal Annealing of III-V Compound Semiconducting Systems: Some Problems of III-V Narrow Gap Semiconductors

Volume 47 Uncooled Infrared Imaging Arrays and Systems

R. G. Buser and M. P. Tompsett, Historical Overview
P. W. Kruse, Principles of Uncooled Infrared Focal Plane Arrays
R. A. Wood, Monolithic Silicon Microbolometer Arrays
C. M. Hanson, Hybrid Pyroelectric-Ferroelectric Bolometer Arrays
D. L. Polla and J. R. Choi, Monolithic Pyroelectric Bolometer Arrays
N. Teranishi, Thermoelectric Uncooled Infrared Focal Plane Arrays
M. F. Tompsett, Pyroelectric Vidicon
T. W. Kenny, Tunneling Infrared Sensors
J. R. Vig, R. L. Filler and Y. Kim, Application of Quartz Microresonators to Uncooled Infrared Arrays
P. W. Kruse, Application of Uncooled Monolithic Thermoelectric Linear Arrays to Imaging Radiometers

Volume 48 High Brightness Light Emitting Diodes

G. B. Stringfellow, Materials Issues in High-Brightness Light-Emitting Diodes
M. G. Craford, Overview of Device issues in High-Brightness Light-Emitting Diodes
F. M. Steranka, AlGaAs Red Light Emitting Diodes
C. H. Chen, S. A. Stockman, M. J. Peanasky, and C. P. Kuo, OMVPE Growth of AlGaInP for High Efficiency Visible Light-Emitting Diodes
F. A. Kish and R. M. Fletcher, AlGaInP Light-Emitting Diodes
M. W. Hodapp, Applications for High Brightness Light-Emitting Diodes
I. Akasaki and H. Amano, Organometallic Vapor Epitaxy of GaN for High Brightness Blue Light Emitting Diodes
S. Nakamura, Group III-V Nitride Based Ultraviolet-Blue-Green-Yellow Light-Emitting Diodes and Laser Diodes

ISBN 0-12-752156-9